Understanding Power Electronics via Simulations

RIVER PUBLISHERS SERIES IN POWER

Series Editors:

MASSIMO MITOLO
Irvine Valley College, USA

VIJAY K. SOOD
Ontario Tech University, Canada

The "River Publishers Series in Power" is a series of comprehensive academic and professional books focussing on the theory and applications behind power generation and distribution. The series features content on energy engineering, systems and development of electrical power, looking specifically at current technology and applications.

The series serves to be a reference for academics, researchers, managers, engineers, and other professionals in related matters with power generation and distribution.

Topics covered in the series include, but are not limited to:

- Power generation;
- Energy services;
- Electrical power systems;
- Photovoltaics;
- Power distribution systems;
- Energy distribution engineering;
- Smart grid;
- Transmission line development.

For a list of other books in this series, visit www.riverpublishers.com

Understanding Power Electronics via Simulations

S. Raghuram Naidu

Professor of Electrical Engineering (Retired)
Universidade Federal de Campina Grande
Campina Grande, PB, Brazil

River Publishers

Routledge
Taylor & Francis Group
NEW YORK AND LONDON

Published 2025 by River Publishers
River Publishers
Broagervej 10, 9260 Gistrup, Denmark
www.riverpublishers.com

Distributed exclusively by Routledge
605 Third Avenue, New York, NY 10017, USA
4 Park Square, Milton Park, Abingdon, Oxon OX14 4RN

Understanding Power Electronics via Simulations / by S. Raghuram Naidu.

Routledge is an imprint of the Taylor & Francis Group, an informa business

ISBN 978-87-7004-677-0 (hardback)
ISBN 978-87-4380-074-3 (paperback)
ISBN 978-87-7004-679-4 (online)
ISBN 978-87-7004-678-7 (master ebook)

Contents

Preface xiii

Acknowledgement xv

List of Figures xvii

List of Tables xxvii

List of Abbreviations xxix

1 Power Semiconductor Devices 1

1.1 Power Diodes 1

1.1.1 Switching characteristics of the power diode 3

1.1.2 Schottky diode 5

1.1.3 Controlled axial lifetime (CAL) diodes 6

1.2 Thyristors 6

1.2.1 Basic structure and operating principle 6

1.2.2 I–V characteristic of the thyristor 7

1.2.3 Thyristor's turn–off 9

1.2.4 Simulation model 10

1.3 Gate Turn-off Thyristors 11

1.3.1 Basic structure 11

1.3.2 Turn-off of the GTO thyristor 12

1.4 Metal-oxide Semiconductor Field Effect Transistor 14

1.4.1 Structure of the N-channel, enhancement type MOSFET 14

1.4.2 The field effect 15

1.4.3 The output characteristics 16

1.4.4 Circuit model 18

1.4.5 Turn-on transients 19

1.4.6 Turn-off transients 21

1.5 Insulated Gate Bipolar Transistor 22
1.5.1 Basic structure 22
1.5.2 Device operation 23
1.5.3 I–V characteristics 24
1.5.4 Switching transients 25
1.6 Wide Bandgap (WBG) Semiconductor Devices 26
1.6.1 Gallium nitride devices 27
1.6.1.1 High electron mobility transistors (HEMT) 27
1.6.1.2 Cascode-mode device 28
1.6.1.3 Enhancement mode HEMT 29
1.6.1.4 Homo-epitaxial GaN devices 29
1.6.2 Silicon carbide devices 30
1.6.2.1 SiC rectifiers 30
1.6.2.2 SiC MOSFETs 31

2 Magnetic Devices **35**
2.1 Maxwell's Equations 35
2.1.1 Gauss's law for electric fields 36
2.1.2 Gauss's law for magnetic fields 36
2.1.3 Faraday's law 36
2.1.4 Ampere's law 36
2.2 Magnetic Circuits 38
2.3 Magnetic Materials 41
2.4 Magnetization Characteristics of Ferromagnetic Materials . . 42
2.5 Losses in Magnetic Devices 45
2.5.1 The skin effect in windings 46
2.6 The Air Gap and Its Effects 47
2.7 Inductor and Transformer Design 49
2.7.1 The area product for inductors 50
2.7.2 The inductor's core geometry factor K_g 51
2.7.3 The transformer's area product 53
2.7.4 Core geometry factor K_g for transformers 55
2.8 Design Examples 56
2.8.1 Isolation transformer 56
2.8.2 Filter inductor for a buck converter operating in CCM 60
2.8.3 Three-winding transformer for a forward converter . 64
2.8.4 Flyback transformer 67

3 Half-wave Converters **73**
3.1 Half-wave Converter with Resistive Load 73
3.2 Half-wave Converter with Resistive-inductive Load 74
3.2.1 Example 3.1 75
3.3 Half-wave Converter with a Generalized Load 77
3.3.1 Example 3.2 78
3.4 Half-wave Converter with a Freewheeling Diode 79
3.4.1 Example 3.3 80
3.5 Commutation 81
3.6 Simulation of Half-wave Converters 83
3.6.1 Half-wave converter with RL load 85
3.6.2 Exercise 89
3.6.3 Simulation of the half-wave converter with a freewheeling diode 90
3.6.4 Half-wave converter with capacitor filter 93
3.6.5 Example 3.3 96
3.7 Computational Programs 98
3.7.1 Program 3.01 98
3.7.2 Program 3.02 99
3.7.3 Program 3.03 100
3.7.4 Program 3.04 100
3.7.5 Program 3.06 101
3.7.6 Program 3.13 103
3.7.7 Program 3.15 105
3.7.8 Program 3.19 107
3.7.9 Program 3.23 109
3.7.10 Program 3.24 111

4 Full-wave AC–DC Converters **113**
4.1 Simplified Full-wave Converter 113
4.2 Simplified Full-wave Converter with Non-ideal Voltage Source 117
4.3 Simulation of Single-phase, Full-wave Converters 119
4.3.1 Non-ideal alternating voltage source and highly capacitive load 119
4.3.2 The general, single-phase, full-wave AC/DC converter 123
4.4 Three-phase, Uncontrolled AC/DC Converters 126
4.4.1 The simplified three-phase, AC/DC converter 127

4.4.2 Simplified AC/DC converter, non-ideal three-phase AC source . . . 131
4.4.3 Simulation of the three-phase AC/DC converter . . . 134
4.5 Solved Problems . . . 138
4.6 Computational Programs . . . 139
4.6.1 Program 4.05 . . . 139
4.6.2 Program 4.09 . . . 142
4.6.3 Program 4.13 . . . 145
4.6.4 Program 4.19 . . . 148
4.6.5 Program 4.24 . . . 152

5 Controlled Thyristor-based AC–DC Converters 157
5.1 The Basic Thyristor Circuit . . . 157
5.1.1 Simulation of the basic circuit . . . 158
5.2 Exercise 5.1 . . . 161
5.3 Single-phase, Full-wave, Controlled AC/DC Converter . . . 162
5.3.1 Effect of the source inductance of a non-ideal AC source . . . 164
5.3.2 Inverter mode of operation for $\alpha > \pi/2$ . . . 167
5.4 Three-phase, Controlled, AC/DC Converters . . . 168
5.4.1 AC side phase currents . . . 170
5.4.2 Effect of the inductance of the AC voltage source . . . 172
5.4.3 AC side currents . . . 176
5.5 Simulation of the three-phase, Controlled, AC/DC Converter . . . 177
5.5.1 Feedback control of the three-phase, AC/DC converter . . . 179
5.6 Computational Programs . . . 184
5.6.1 Program 5.04 . . . 184
5.6.2 Program 5.06 . . . 187
5.6.3 Program 5.08 . . . 190
5.6.4 Program 5.10 . . . 193
5.6.5 Program 5.13 . . . 196
5.6.6 Program 5.16 . . . 201
5.6.7 Program 5.19 . . . 206
5.6.8 Program 5.21 . . . 211
5.6.9 Program 5.25 . . . 216
5.6.10 Program 5.27 . . . 221

6 Non-isolated DC–DC Converters **229**
6.1 The Non-isolated Buck Converter 229
6.1.1 Steady-state analysis of the basic buck converter . . 230
6.1.2 Inclusion of losses 233
6.1.3 Selection of the power stage's components 234
6.1.4 Simulation of the buck converter 236
6.1.5 Buck converter model in CCM 239
6.1.6 Control to output transfer function (CCM) 243
6.1.7 Feedback control of the buck converter in CCM . . . 243
6.1.8 Placement of the poles and zeros of the compensating network 246
6.1.8.1 Type II compensator's pole-zero placement 247
6.1.8.2 Pole-zero placement for the Type IIIA compensator 248
6.1.8.3 Pole-zero placement for the Type IIIB compensator 248
6.1.9 Type II compensator design example 248
6.1.10 The buck converter operating in DCM 251
6.1.10.1 Mathematical model for the buck converter operating in DCM 253
6.1.10.2 Linearization of the averaged state equations 255
6.1.10.3 Transfer function of the lossless buck converter operating in DCM 256
6.2 The Non-isolated Boost Converter 257
6.2.1 Steady-state analysis 258
6.2.2 Inclusion of losses 260
6.2.3 Selection of power stage components 261
6.2.3.1 Inductors 261
6.2.3.2 Output capacitor 262
6.2.3.3 Power switch 262
6.2.3.4 Diode 263
6.2.4 Simulation of the boost converter 263
6.2.5 Boost converter model in CCM 265
6.2.6 Feedback control of the boost converter in CCM . . 269
6.2.6.1 Voltage mode control 269
6.2.6.2 Current mode control 273
6.2.6.3 Modeling the current mode controlled, boost converter 274

6.2.6.4 Compensator design in current mode control 278
6.2.7 Boost converter in DCM 279
6.2.8 Small-signal analysis in DCM 281
6.3 The Non-isolated Buck–Boost Converter 283
6.3.1 Steady-state analysis 283
6.3.2 Inclusion of losses 286
6.3.3 Selection of components 288
6.3.3.1 Filter capacitor 288
6.3.3.2 Inductor 289
6.3.3.3 Power switch 289
6.3.3.4 Diode 289
6.3.4 Simulation of the buck–boost converter 290
6.3.5 Buck–boost converter model in CCM 292
6.3.6 Feedback control of the buck–boost converter in CCM 294
6.3.6.1 Voltage mode control 294
6.3.7 Current mode control 296
6.3.8 Buck–boost converter in DCM 299
6.3.9 Small-signal analysis of the buck–boost converter operating in DCM 300
6.4 Computational Programs 302
6.4.1 Program 6.07 302
6.4.2 Program 6.10 306
6.4.3 Program 6.12 308
6.4.4 Program 6.20 312
6.4.5 Program 6.22 315
6.4.6 Program 6.23 317
6.4.7 Program 6.36 322
6.4.8 Program 6.38 325
6.4.9 Program 6.39 327
6.4.10 Program 6.40 331
6.4.11 Program 6.41 333

7 Isolated DC–DC Converters **339**
7.1 Single-ended Forward Converter 340
7.1.1 Simulation of the forward converter 343
7.1.2 Effects of leakage inductances 346
7.1.2.1 Leakage inductance between primary and secondary 346

7.1.2.2 Leakage inductance between the primary and the tertiary 349
7.1.3 Forward converter using resonant reset 351
7.2 Isolated Single-ended Hybrid-bridge Converter 355
7.3 The Double-ended, Full-bridge Forward Converter 357
7.3.1 Simulation of the double-ended, full-bridge, forward converter 360
7.4 The Push–Pull Converter 362
7.4.1 Low magnetizing current, high output load current . 363
7.4.2 High magnetizing current, low output load current . 365
7.5 The Flyback Converter 367
7.5.1 Simulation of the flyback converter 368
7.6 Computational Programs 372
7.6.1 Program 7.07 372
7.6.2 Program 7.11 376
7.6.3 Program 7.13 381
7.6.4 Program 7.14 385
7.6.5 Program 7.19 390
7.6.6 Program 7.21 394
7.6.7 Program 7.27 398
7.6.8 Program 7.29 404
7.6.9 Program 7.30 409
7.6.10 Program 7.35 414

8 Resonant DC–DC Converters **421**
8.1 Basic Resonant Tank Networks 421
8.2 LLC Resonant Half-bridge Converter 425
8.2.1 Linear model of the LLC converter 428
8.2.2 Simulation of the LLC converter 432
8.3 The LCC Converter 436
8.4 Soft-switched, Quasi-resonant DC–DC Converters 438
8.4.1 The quasi-resonant buck converter 438
8.4.1.1 Simulation of the quasi-resonant, ZVS buck converter 441
8.5 Constant Switching Frequency, Multi-resonant, ZVS Buck Converter 443
8.6 Computational Programs 445
8.6.1 Program 8.04 445
8.6.2 Program 8.13 447

8.6.3 Program 8.21 452
8.6.4 Program 8.24 457

9 Inverters **465**
9.1 Single-phase Inverters 465
9.1.1 Square-wave inverter 465
9.1.2 Amplitude control and harmonic elimination 467
9.2 Pulse-width Modulation 470
9.2.1 Bipolar switching 470
9.2.1.1 Simulation of the full-bridge, single-phase inverter with bipolar switching 473
9.2.2 Full-bridge, single-phase inverter with unipolar switching 474
9.3 Three-phase Inverters 476
9.3.1 Three-phase, square-wave inverter 479
9.3.2 Three-phase, triangle-comparison PWM 480
9.3.3 Space-vector PWM 482
9.3.4 Hybrid PWM 488
9.4 Computational Programs 492
9.4.1 Program 9.02 492
9.4.2 Program 9.04 493
9.4.3 Program 9.05 498
9.4.4 Program 9.07 501
9.4.5 Program 9.08 505
9.4.6 Program 9.13 510
9.4.7 Program 9.14 516
9.4.8 Program 9.16 521
9.4.9 Program 9.17 527

Index **535**

About the Author **541**

Preface

This book is a senior undergraduate text on power electronics, with its emphasis on simulations. There are several approaches for simulating power electronic circuits. One approach is to use a general purpose circuit simulation program that will accept the circuit parameters as input data and will provide the response waveforms. Another approach is to write programs in a high-level language and plot the response waveforms. The latter approach has been adopted in this book.

The programs provided at the end of each chapter have been written in the Python programming language. It is a simple language and the programs are interpreted. The programs have been created and tested using the Anaconda distribution. The IDE and interpreter versions used in the simulations are Spyder 4.2.5 and Python 3.8.8. The programs are simple and each program has been divided into three functions: (1) the device characteristics, (2) the solver function, and (3) the main input–output function. It should be emphasized that this book does not teach programming in Python. The student should consult the free on-line resources besides an introductory book that is available in the market.

Power electronics is an interdisciplinary area of study and the undergraduate will encounter this course after exposure to circuit analysis, semiconductor devices, signal processing, electromagnetism, and elementary control theory. Semiconductor devices and electromagnetic devices have been briefly reviewed in Chapter 1 and Chapter 2, respectively. However, control theory and Fourier analysis are treated as pre-requisites.

This book follows the path laid out in some of the classic power electronics books. Chapter 3 and Chapter 4 cover diode-based AC/DC converters. Thyristor-based AC/DC converters are discussed in Chapter 5. Basic DC/DC switch-mode converters are dealt with in some detail in Chapter 6 and their isolated versions are covered in Chapter 7. Resonant DC/DC converters are discussed in Chapter 8. Inverters, i.e., DC/AC converters, form the subject matter of Chapter 9. Though lists of problems have not been provided, the text is seeded with many circuits that may be simulated.

Acknowledgement

Having been trained as a high-voltage engineer, the author began teaching power electronics rather late in his academic career. His colleague Prof. Antonio Marcus N. Lima led him very gently into power electronics, for which he is much obliged. The author was constantly encouraged to write this book by his other colleague Prof. Antonio Faustino C. Neto. Many colleagues and students aided the author in his journey through power electronics. The author remembers the many discussions he has had with Prof. Darlan A. Fernandes. Prof. Marcelo Alencar Sampaio is thanked profusely for encouraging the author to finish the manuscript.

The editors at River Publishers have done a great job converting the author's manuscript into the elegant book in the reader's hands. For this, the author is much obliged. The author would also like to thank Mr. J. Nakajima for all the assistance extended to him. The author has been blessed with a wonderful family, i.e., wife Sasikala, son Prasad, and daughter Anupama. Without their encouragement and support, it would not have been possible to write this book.

List of Figures

Figure 1.1	Structure of the power diode: (a) Epitaxial diode; (b) Diffused diode.	2
Figure 1.2	*I–V* characteristic of a power diode.	2
Figure 1.3	Approximation of the diode's characteristic.	3
Figure 1.4	Switching transients of the power diode: (a) electrical circuit; (b) turn-on transients; (c) turn-off transients.	4
Figure 1.5	Structure of the Schottky diode.	6
Figure 1.6	Structure of the thyristor.	7
Figure 1.7	*I–V* characteristic of the thyristor.	8
Figure 1.8	Thyristor's turn-off transients: (a) current transient; (b) voltage transient.	9
Figure 1.9	Thyristor's simulation model.	10
Figure 1.10	Structure of the GTO thyristor.	12
Figure 1.11	GTO thyristor's turn-off transients.	13
Figure 1.12	The N-channel MOSFET; (a) structure; (b) circuit symbol; (c) simulation model.	15
Figure 1.13	Formation of the n-channels and the accumulation layer.	16
Figure 1.14	*I–V* characteristics of the MOSFET: (a) output characteristic; (b) transfer characteristic.	17
Figure 1.15	Circuit model of the MOSFET: (a) OFF–state model; (b) ON-state model; (c) approximation of nonlinear C_{GD}.	18
Figure 1.16	Turn-on process of the MOSFET: (a) electrical circuit; (b) turn-on trajectory in the $i_{\mathrm{DS}}, v_{\mathrm{DS}}$ plane .	20
Figure 1.17	MOSFET's turn-on transients.	21
Figure 1.18	Structure of the N-channel, PT-IGBT.	23
Figure 1.19	*I–V* characteristics of the IGBT: (a) output characteristics; (b) transfer characteristic; (c) circuit symbol.	24
Figure 1.20	Switching waveforms of the IGBT.	25

Figure 1.21 Structure of GaN transistor: (a) depletion mode; (b) symbol for GaN transistor; (c) cascode connection. 28
Figure 1.22 Structure of the enhancement mode GaN HEMT. . 29
Figure 1.23 Structure of the SiC diodes: (a) SB diode; (b) PiN diode; (c) JBS diode. 31
Figure 1.24 Structure of SiC MOSFETs: (a) planar type; (b) trench type. 32
Figure 2.1 Electromagnetic fields. 37
Figure 2.2 Basic magnetic circuit. 38
Figure 2.3 Basic magnetic circuit: Ampere's law. 39
Figure 2.4 Magnetic hysteresis. 43
Figure 2.5 Magnetic hysteresis: iron losses. 44
Figure 2.6 Hysteresis loop for gapped and ungapped cores. . . 49
Figure 2.7 Fringing of magnetic flux at an airgap. 50
Figure 2.8 Filter inductor in a converter circuit. 53
Figure 2.9 The 19EI-1 core for the 500 VA, 50 Hz isolation transformer. 57
Figure 3.1 Half-wave converter, resistive load: circuit diagram. 73
Figure 3.2 Half-wave converter, resistive load: response waveforms. 74
Figure 3.3 Half-wave converter, RL load: circuit diagram. . . 75
Figure 3.4 Half-wave converter, generalized load: circuit diagram. 78
Figure 3.5 Half-wave converter with free wheeling diode: circuit diagram. 79
Figure 3.6 Half-wave converter, generalized load: response waveforms. 81
Figure 3.7 Half-wave converter, free-wheeling diode, non-ideal source: circuit diagram. 82
Figure 3.8 Half-wave converter of Figure (3.7): response waveforms. 83
Figure 3.9 Transient equivalents for inductors and capacitors. . 85
Figure 3.10 Half-wave converter, RL load: circuit diagram. . . 86
Figure 3.11 Half-wave converter, RL load: transient equivalent circuit. 86
Figure 3.12 Half-wave converter, RL load: small–signal equivalent. 88
Figure 3.13 Half-wave converter, RL load: response waveforms. 88

Figure 3.14 Half-wave converter, generalized load: circuit diagram. 89
Figure 3.15 Half-wave converter with generalized load: response waveforms. 89
Figure 3.16 Half-wave converter with free wheeling diode: circuit diagram. 90
Figure 3.17 Half-wave converter with free wheeling diode: transient equivalent circuit. 91
Figure 3.18 Half-wave converter with free wheeling diode: small-signal equivalent. 92
Figure 3.19 Half-wave converter with free wheeling diode: response waveforms. 92
Figure 3.20 Half-wave converter with capacitor filter: circuit diagram. 93
Figure 3.21 Half-wave converter with capacitor filter: transient equivalent circuit. 94
Figure 3.22 Half-wave converter with capacitor filter: small-signal equivalent. 95
Figure 3.23 Half-wave converter with capacitor filter: response waveforms. 96
Figure 4.1 Single-phase, full-wave, AC/DC converter: circuit diagram. 114
Figure 4.2 Simplified, full-wave AC/DC converter: circuit diagram 115
Figure 4.3 Simplified full-wave AC/DC converter: response waveforms. 116
Figure 4.4 Simplified single-phase, full-wave converter: non-ideal voltage source 118
Figure 4.5 Simplified full-wave converter, non-ideal voltage source: response waveforms. 119
Figure 4.6 Full-wave converter, non-ideal voltage source, highly capacitive load: circuit diagram. 120
Figure 4.7 Full-wave converter, non-ideal voltage source, highly capacitive load: transient equivalent circuit . 121
Figure 4.8 Full-wave converter, non-ideal voltage source, highly capacitive load: small-signal equivalent. . . 122
Figure 4.9 Full-wave converter, non-ideal voltage source, highly capacitive load: response waveforms. 123

Figure 4.10 Single-phase, full-wave AC/DC converter: circuit diagram . . . 124
Figure 4.11 Single-phase, full-wave AC/DC converter: equivalent circuit. . . . 124
Figure 4.12 Single-phase, full-wave AC/DC converter: small-signal equivalent. . . . 126
Figure 4.13 Single-phase, full-wave converter, response waveforms . . . 127
Figure 4.14 Three-phase AC/DC converter . . . 128
Figure 4.15 Simplified three-phase AC/DC converter. . . . 128
Figure 4.16 Simplified three-phase AC/DC converter: response waveforms. . . . 129
Figure 4.17 Phase-a: voltage and current waveforms . . . 131
Figure 4.18 Three-phase AC/DC converter, non-ideal three-phase AC source . . . 132
Figure 4.19 Three-phase AC/DC converter: current commutation 132
Figure 4.20 Current transfer from diode D5 to diode D1. . . . 133
Figure 4.21 Typical three-phase AC/DC converter . . . 135
Figure 4.22 Three-phase AC/DC converter: transient equivalent circuit. . . . 136
Figure 4.23 Three-phase AC/DC converter: small-signal equivalent . . . 137
Figure 4.24 Three-phase AC/DC converter: response waveforms 138
Figure 5.1 Thyristor-based, half-wave AC/DC converter . . . 158
Figure 5.2 Thyristor-based half-wave converter, transient equivalent circuit . . . 159
Figure 5.3 Thyristor-based, half-wave converter, small-signal equivalent circuit . . . 160
Figure 5.4 Thyristor-based, half-wave converter: response waveforms . . . 160
Figure 5.5 Converter with inductive load and a back emf . . . 161
Figure 5.6 Converter with inductive load and back emf: response waveforms . . . 161
Figure 5.7 Simplified, full-wave, controlled AC/DC converter . 162
Figure 5.8 Simplified, full-wave, controlled AC/DC converter: response waveforms . . . 163
Figure 5.9 Single-phase, full-wave, controlled AC/DC converter: commutation process . . . 165

Figure 5.10 Controlled AC/DC converter, non-ideal AC source: response waveforms 166
Figure 5.11 Controlled AC/DC converter, $\alpha > \pi/2$: response waveforms 167
Figure 5.12 Three-phase, idealized, controlled AC/DC converter 168
Figure 5.13 Three-phase, idealized, controlled AC/DC converter, response waveforms 169
Figure 5.14 Input current, output voltage as functions of delay angle 170
Figure 5.15 Three-phase, controlled AC/DC converter: phase-a current waveform 171
Figure 5.16 Phase-a current waveform as a function of delay angle 171
Figure 5.17 Three-phase, controlled AC/DC converter, non-ideal AC voltage source 173
Figure 5.18 Three-phase, AC/DC converter, commutation of current, phase-c to phase-a. 174
Figure 5.19 Three-phase, AC/DC converter: response waveforms. 175
Figure 5.20 Three-phase, controlled AC/DC converter, non-ideal AC voltage source: AC side currents 176
Figure 5.21 Three-phase, controlled AC/DC converter: inverter mode of operation 177
Figure 5.22 A practical three-phase, controlled AC/DC converter 178
Figure 5.23 Three-phase, controlled AC/DC converter: transient equivalent circuit 179
Figure 5.24 Three-phase, controlled AC/DC converter: small-signal equivalent circuit 180
Figure 5.25 Three-phase, controlled AC/DC converter: response waveforms 181
Figure 5.26 Three-phase, controlled AC/DC converter: feedback control system. 182
Figure 5.27 Three-phase, controlled AC/DC converter with feedback control: average load current. 183
Figure 6.1 Power stage of the basic buck converter. 230
Figure 6.2 Buck converter: current and voltage waveforms. . . 231
Figure 6.3 Inclusion of buck converter losses. (a) ON state (b) OFF state. 233
Figure 6.4 Buck converter: circuit diagram. 237
Figure 6.5 Buck converter: transient equivalent circuit. 237
Figure 6.6 Buck converter: small-signal equivalent circuit. . . 239

Figure 6.7	Buck converter: response waveforms.	239
Figure 6.8	Buck converter: power stage and feedback control system.	244
Figure 6.9	Buck converter's feedback control system.	246
Figure 6.10	Buck converter with voltage mode control: Bode plots.	249
Figure 6.11	Transient equivalent circuit of the error amplifier.	250
Figure 6.12	Buck converter, voltage mode control: transient response.	251
Figure 6.13	Buck converter in DCM: response waveforms.	252
Figure 6.14	Basic boost converter: power stage.	258
Figure 6.15	Boost converter: current and voltage waveforms.	259
Figure 6.16	Boost converter: inclusion of losses: (a) ON state; (b) OFF state.	260
Figure 6.17	Boost converter for simulation.	264
Figure 6.18	Boost converter: transient equivalent circuit.	265
Figure 6.19	Boost converter: small-signal equivalent circuit.	266
Figure 6.20	Boost converter: response waveforms.	266
Figure 6.21	Boost converter with voltage mode control.	270
Figure 6.22	Boost converter, voltage mode control: Bode plots.	273
Figure 6.23	Boost converter, voltage mode control: transient response.	274
Figure 6.24	Boost converter: peak current mode control.	275
Figure 6.25	Boost converter, current mode control: ON state current waveforms.	276
Figure 6.26	Boost converter, current mode control: Bode plots.	279
Figure 6.27	Boost converter, current mode control: transient response.	280
Figure 6.28	Absence of compensating ramp: second harmonic generation.	280
Figure 6.29	The boost converter in DCM.	281
Figure 6.30	Buck–boost converter's power stage.	284
Figure 6.31	Buck–boost converter's current and voltage waveforms.	284
Figure 6.32	Buck–boost converter with losses: (a) ON state; (b) OFF state.	287
Figure 6.33	Buck–boost converter for simulation.	290
Figure 6.34	Buck–boost converter: transient equivalent circuit.	290
Figure 6.35	Buck–boost converter: small-signal equivalent circuit.	291

Figure 6.36 Buck–boost converter: response waveforms. 291
Figure 6.37 Buck–boost converter: voltage mode control. . . . 294
Figure 6.38 Buck–boost converter, voltage mode control: Bode plots. 295
Figure 6.39 Buck–boost converter, voltage mode control: transient response. 296
Figure 6.40 Buck–boost converter, current mode control: Bode plots. 298
Figure 6.41 Buck–boost converter, current mode control: transient response. 299
Figure 6.42 The buck–boost converter in DCM. 300
Figure 7.1 Equivalent circuit of isolation transformer. 340
Figure 7.2 Single-ended forward converter, with reset winding. 341
Figure 7.3 Single-ended forward converter waveforms. 341
Figure 7.4 Forward converter for simulation, with reset winding. 344
Figure 7.5 Transient equivalent circuit of the forward converter. 344
Figure 7.6 Small signal equivalent of the forward converter. . . 345
Figure 7.7 Forward converter with reset winding: response waveforms. 346
Figure 7.8 Commutation of the inductor current. 347
Figure 7.9 Equivalent circuit of forward converter: leakage inductance L_K. 348
Figure 7.10 Forward converter with leakage inductance L_K: small–signal equivalent circuit. 348
Figure 7.11 Response waveforms of the forward converter including the effects of L_K. 349
Figure 7.12 Forward converter with leakage inductance L_{PT}. . 350
Figure 7.13 Forward converter with leakage inductance L_{PT}, without snubber. 350
Figure 7.14 Forward converter with leakage inductance L_{PT}, with snubber. 351
Figure 7.15 Forward converter with resonant reset. 352
Figure 7.16 Resonant circuit during the OFF state. 353
Figure 7.17 Forward converter, resonant reset, transient equivalent circuit. 354
Figure 7.18 Forward converter with resonant reset, small–signal equivalent. 354
Figure 7.19 Forward converter with resonant reset: response waveforms. 355

Figure 7.20 Isolated hybrid-bridge converter. 356
Figure 7.21 Isolated hybrid-bridge converter: response waveforms. 356
Figure 7.22 Full-bridge forward converter: circuit diagram. . . . 357
Figure 7.23 Full-bridge forward converter: sequence A: response waveforms. 359
Figure 7.24 Full-bridge forward converter: sequence B, response waveforms. 359
Figure 7.25 Full-bridge forward converter: transient equivalent circuit. 361
Figure 7.26 Full-bridge forward converter: small–signal equivalent circuit. 361
Figure 7.27 Full-bridge forward converter: response waveforms. 362
Figure 7.28 The push–pull converter. 363
Figure 7.29 Push-pull converter, high L_M: response waveforms. 364
Figure 7.30 Push-pull converter, low L_M: response waveforms. 366
Figure 7.31 The flyback converter. 367
Figure 7.32 The flyback converter with RCD snubber. 368
Figure 7.33 Flyback converter: transient equivalent circuit. . . . 370
Figure 7.34 Flyback converter: small–signal equivalent circuit. . 371
Figure 7.35 Flyback converter: response waveforms. 371
Figure 8.1 Structure of resonant converters. 422
Figure 8.2 Two-element resonant tank circuits. 422
Figure 8.3 Three-element resonant tank circuits. 423
Figure 8.4 Series resonant converter: gain curves 424
Figure 8.5 Parallel resonant converter: gain curves 425
Figure 8.6 The LLC resonant half-bridge converter 426
Figure 8.7 LLC converter: response waveforms. 427
Figure 8.8 LLC converter: operating modes 1b–4. 429
Figure 8.9 LLC converter: operating modes 5b–8. 430
Figure 8.10 LLC converter: linear model. 431
Figure 8.11 LLC converter: gain curves. 433
Figure 8.12 LLC converter: equivalent circuits. 434
Figure 8.13 Response waveforms of the LLC converter. 435
Figure 8.14 LLC converter: output voltage. 435
Figure 8.15 A typical LCC converter. 436
Figure 8.16 LCC converter: fundamental harmonic equivalent. . 437
Figure 8.17 LCC converter: gain characteristics. 437
Figure 8.18 LCC converter: response waveforms. 438

Figure 8.19 Quasi-resonant, ZVS, DC–DC converters. 439
Figure 8.20 Quasi-resonant buck converter: equivalent circuits. 442
Figure 8.21 Quasi-resonant buck converter: response waveforms 442
Figure 8.22 Quasi-resonant buck converter: variation of V_0 and d_r 443
Figure 8.23 Constant frequency, multi-resonant, ZVS buck converter. 444
Figure 8.24 Multi-resonant converter: response waveforms . . . 445
Figure 9.1 Single-phase, full-bridge, square-wave inverter. . . 466
Figure 9.2 Square–wave converter waveforms. 467
Figure 9.3 Square wave with voltage cancellation of α radians. 468
Figure 9.4 Harmonic elimination by voltage cancellation; elimination of (a) $3\mathrm{rd}$ and $5\mathrm{th}$ harmonics; (b) $3\mathrm{rd}$ to $11\mathrm{th}$ harmonics. 469
Figure 9.5 Bipolar switching, PWM waveforms. 471
Figure 9.6 Full-bridge, single–phase inverter: Transient and small-signal equivalent circuits. 474
Figure 9.7 Full-bridge, single-phase inverter with bipolar switching: response waveforms. 475
Figure 9.8 Full-bridge, single-phase inverter, unipolar switching: waveforms. 476
Figure 9.9 Three-phase inverter with generalized load. 477
Figure 9.10 Switching scheme for the square-wave generator. . 478
Figure 9.11 Three-phase inverter: transient equivalent circuit. . 478
Figure 9.12 Three-phase inverter: small-signal equivalent circuit. 479
Figure 9.13 Three-phase, square-wave inverter: response waveforms. 480
Figure 9.14 Switching interval for $SW1$. 481
Figure 9.15 Three-phase inverter: triangle-comparison switching scheme. 483
Figure 9.16 Space vector representation in the α, β plane. . . . 486
Figure 9.17 Three-phase inverter, space-vector PWM: response waveforms. 488
Figure 9.18 Hybrid PWM scheme. (a) Load currents; (b) reference and modified reference control signals. 491

List of Tables

Table 1.1 Semiconductor parameters. 26
Table 5.1 Conduction pattern for the three-phase controlled converter. 169
Table 7.1 Switching sequences. 357
Table 7.2 Switching sequence. 363
Table 9.1 Switching scheme of the square-wave inverter. 465
Table 9.2 Normalized amplitudes V_f/E_S as a function of m_a. . 472
Table 9.3 Normalized amplitudes V_{LL}/E_S as a function of m_a. 483
Table 9.4 State vectors in the α, β domain. 485
Table 9.5 Switching sequences in sectors 1, 3, and 5. 487
Table 9.6 Switching sequences in sectors 2, 4, and 6. 487

List of Abbreviations

2DEG	Two-dimensional electron gas
BJT	Bipolar junction transistor
CCM	Continuous conduction mode
CMC	Current mode control
DCM	Discontinuous conduction mode
ESR	Equivalent series resistance
GTO	Gate turn-off
HEMT	High electron mobility transistors
IGBT	Insulated gate bipolar transistor
JBS	Junction barrier Schottky
LHS	Left-hand side
MOSFET	Metal oxide semiconductor field effect transistor
MPP	Molybdenum permalloy
NPT	Non-punch-through
PCC	Point of common coupling
PT	Punch-through
PWM	Pulse-width modulation
RHS	Right-hand side
RLS	Recursive least square
RMS	Root mean square
SB	Schottky barrier
SPA	Solid polymer aluminum
SVPWM	Space-vector PWM
THD	Total harmonic distortion
WBG	Wide bandgap
ZCS	Zero-current switched
ZVS	Zero-voltage switched

1

Power Semiconductor Devices

A review of power semiconductor devices used in power electronic circuits is provided in this chapter. This is an area that has been intensively studied and researched over the past several decades and several books have been written [1–3]. The coverage in this review is brief but, hopefully, sufficient for the purposes of this volume. Bipolar junction transistors (BJTs) have been excluded in this review since they have been largely substituted by metal oxide semiconductor field effect transistors (MOSFETs) and insulated gate bipolar transistors (IGBTs).

1.1 Power Diodes

The typical structure of a power diode is shown in Figure 1.1. The lightly doped n^- layer increases the blocking capacity of the diode. There are two types of diodes: (a) epitaxial diodes and (b) diffused diodes. The fabrication of the epitaxial diode starts with a highly doped n^+ substrate. The lightly doped n^- layer is epitaxially deposited on the substrate and then the p^+ layer is diffused. Epitaxial diodes are mainly applied for blocking between 100 to 600 V. Diodes for blocking higher voltages are fabricated by the diffusion process. The process starts with a lightly doped n^- wafer on which the p^+ and n^+ layers are grown by diffusion.

Figure 1.2 shows the *I–V* characteristic of a power diode. The manufacturer's data sheet will provide data shown in the figure, such as:

- v_{RM}–the maximum reverse voltage that can be blocked;
- v_B–the reverse breakdown voltage;
- v_T–the threshold voltage, above which the diode becomes conductive;
- v_F, i_F–the forward voltage drop and the corresponding current.

A reasonable approximation of the diode's characteristic that is used in the simulations is shown in Figure 1.3. Observe that the polarities of the

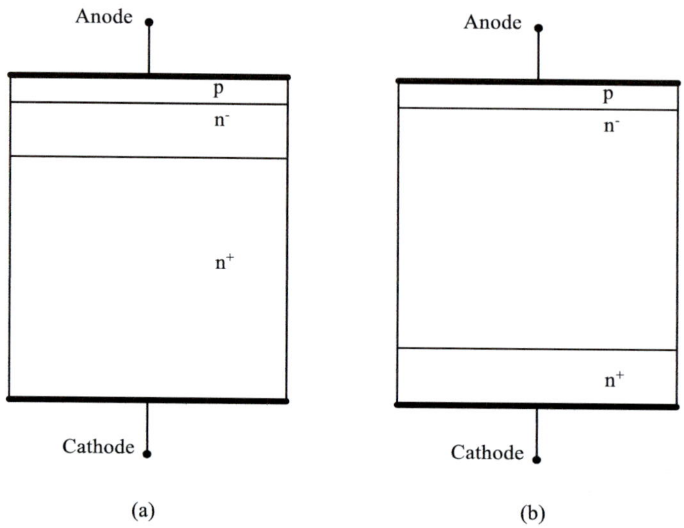

Figure 1.1 Structure of the power diode: (a) Epitaxial diode; (b) Diffused diode.

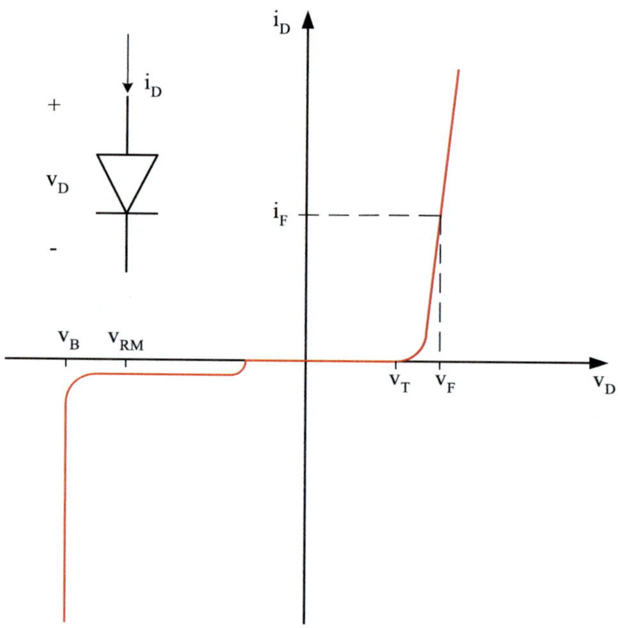

Figure 1.2 *I–V* characteristic of a power diode.

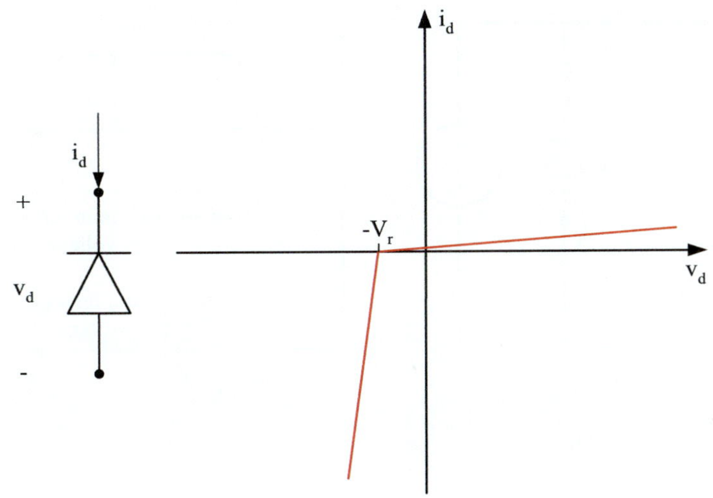

Figure 1.3 Approximation of the diode's characteristic.

diode's voltage and current have been inverted in the representation. The functional representation is

$$\begin{aligned} \text{for } v_d > -V_r, \; g_d &= g_{\text{low}} \\ \text{for } v_d \leq -V_r, \; g_d &= g_{\text{high}}. \end{aligned}$$

Values for $(-V_r, g_{\text{low}}, g_{\text{high}})$ may be obtained from the manufacturer's data sheet. The values used in the simulations are (0.3 V/0.5 V, 10^{-7} S, 40 S/100 S). For a given v_d, the diode's current i_d and its conductance g_d are determined by calling the function

$$i_d, \, g_d = fdiode(v_d).$$

1.1.1 Switching characteristics of the power diode

Consider the circuit shown in Figure 1.4 (a). Let the switch S be closed for $t < 0$ so that the diode is in the blocked state with reverse bias V_d applied to its terminals. Space charge layers are formed on either sides of the junction and a depletion capacitance is formed. Let the switch S be opened at $t = 0$ and the diode be forward biased.

To turn on the diode, charges are first delivered to the junction to neutralize the space charges. The voltage across the diode shows an initial overshoot

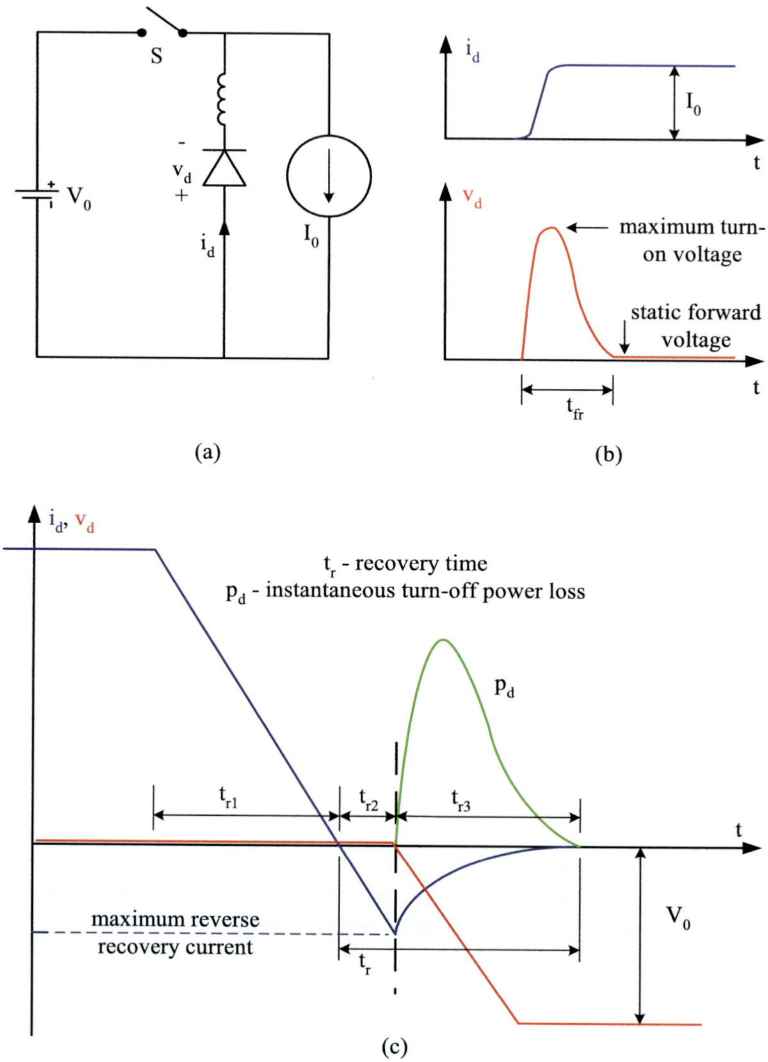

Figure 1.4 Switching transients of the power diode: (a) electrical circuit; (b) turn-on transients; (c) turn-off transients.

that is due to the ohmic resistance of the drift region and the inductance of the bonding wires attached to the silicon wafer. When the removal of the space charge is complete, excess carriers are injected into the drift region from both ends. Holes are injected from the p^+n^- junction and electrons from the n^+n^- junction. The diode voltage drops to the steady-state value v_F and the diode current becomes equal to the forward current i_F. Figure 1.4 (b) shows the typical forward recovery characteristics.

The turn-off transient starts when the switch S that was open, is closed. Commutation of the diode current to the switch is spread over the intervals t_{r1}, t_{r2}, and t_{r3}. During the interval t_{r1}, excess stored charge in the drift region is removed by recombination and by the reverse current that begins to flow through the diode. The total diode current begins to fall, but the diode voltage remains essentially at the ON-state value. The interval t_{r2} begins when the diode current falls to zero and subsequently becomes negative. The carriers at the ends of the drift region are swept away and the reverse recovery current reaches a maximum value at the end of the interval t_{r2}. The diode's voltage continues to remain at the ON-state value. Very few carriers remain in the drift region at this time and the junctions become reverse biased. At the beginning of the interval t_{r3}, the diode current ceases to grow in the negative direction and falls to zero at the end of the interval. The diode's voltage becomes more and more negative and finally equals the reverse bias applied by the external circuit. Figure 1.4 (c) shows the approximate reverse recovery transients of the power diode.

The integral of the (negative) reverse recovery current is the reverse recovery charge Q_{rr} and its duration is the interval t_r. A good approximation for the turn-off energy loss is $\mathcal{E}_r = Q_{rr} V_0$, where V_0 is the voltage blocked by the diode after turn-off. The turn-off power loss is $P_d = Q_{rr} V_0 f_S$, where f_S is the switching frequency.

1.1.2 Schottky diode

It is a unipolar device in which a metallurgical junction is formed between a metal layer and a semiconductor (usually of the n type). Schottky diodes have an exponential *I–V* characteristic similar to the pn junction diode. However, current is conducted only by electrons. There are no minority-carrier, charge-storage effects. Thus, the Schottky diode has a very short turn-off time and it could be used at high switching frequencies. Silicon Schottky diodes have high saturation leakage currents, low ON-state voltage drop, and limited reverse voltage capability. Wide band gap Schottky diodes (based on SiC or GaN) can block 1000 V or more. However, the ON-state voltage drop of SiC-based Schottky diode could exceed 2 V.

Figure 1.5 shows the structure of the Schottky diode. The Schottky junction requires that the junction be formed between a metallic contact material and a lightly doped n^- layer. Contact between a metal and a highly doped n^+ layer will result in an ohmic contact.

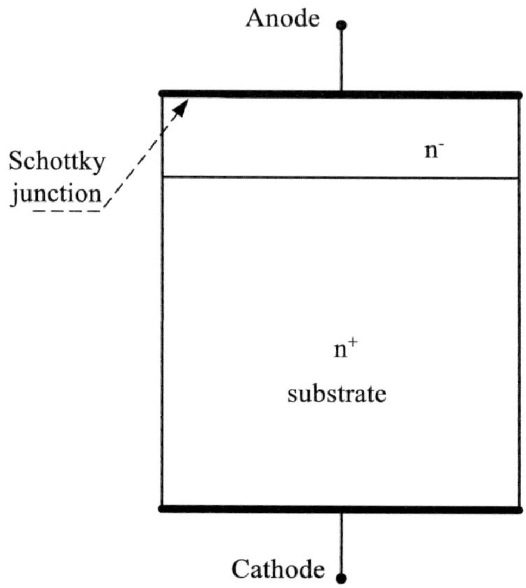

Figure 1.5 Structure of the Schottky diode.

1.1.3 Controlled axial lifetime (CAL) diodes

These are a family of soft and fast recovery diodes used as anti-parallel diodes with IGBTs. A freewheeling diode for an IGBT should have a soft recovery characteristic during turn-off so that high over-voltages caused by stray inductances are avoided. The soft recovery and low reverse recovery charge are obtained by locally modifying the carrier lifetime along the current path between the anode and the cathode. CAL diodes are available from 600 to 1700 V and current ratings from 6 to 200 A.

1.2 Thyristors

The thyristor is a semiconductor switching device that can be used at low switching frequencies, very high blocking voltages, and very high currents. The high power handling capabilities of the thyristor are particularly suitable for high–voltage DC power transmission.

1.2.1 Basic structure and operating principle

Figure 1.6 shows the typical structure of the thyristor. It is a three-terminal, four–layer device. The terminals are the anode (A), the cathode (K), and the

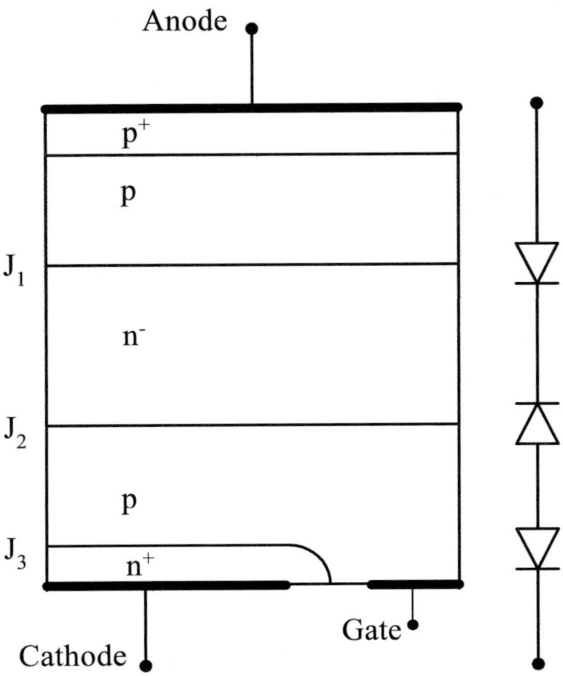

Figure 1.6 Structure of the thyristor.

gate (G). The four layers, formed by the p^+p, n^-, p, and n^+ layers, result in three pn junctions. These are shown in Figure 1.6 as J_1, J_2, and J_3.

The thyristor is in the forward blocking state when the anode to cathode voltage is positive. The junctions J_1 and J_3 are forward biased in this state and the junction J_2 is reverse biased. A space charge region builds across J_2 and it penetrates deeply into the weakly doped n_- layer. In the reverse blocking state, the anode to cathode voltage is negative, J_1 and J_3 reverse biased, and J_2 forward biased. However, the applied voltage appears mainly across the J_1 junction when the thyristor is reverse-blocked because the high doping on either sides of the J_3 junction causes its avalanche breakdown to be low.

1.2.2 I–V characteristic of the thyristor

Figure 1.7 is the *I–V* characteristic of the thyristor, and it shows the thyristor's current i_{AK} as a function of its anode to cathode voltage v_{AK}. The characteristic consists of three segments corresponding to the following modes of operation:

- the forward blocking mode;
- the reverse blocking mode;
- the conduction mode.

The forward blocking mode is the segment for which $0 < v_{AK} < v_{FB}$ and the thyristor's current i_{AK} is very small. The voltage v_{FB} is the thyristor's maximum forward blocking capability. For $v_{AK} > v_{FB}$, the thyristor will self-trigger and jump to the conduction mode. It is usual to avoid spontaneous trigger by limiting v_{AK} to some value less than v_{FB}. The thyristor is usually triggered by injecting a pulse of current into the gate. The gating pulse is maintained for sufficient time (i.e., about 10 μs) during which the anode current should become greater than a minimum value I_L known as the latching current. Only then will the thyristor switch safely into the conduction mode and the gating pulse may be removed. The triggering pulse may only be applied when the thyristor is in the forward blocking state.

The gate loses control when the thyristor enters the conduction mode. The thyristor will return to the blocking state only when the anode current is reduced to a value lower than I_H, which is known as the holding current. (It

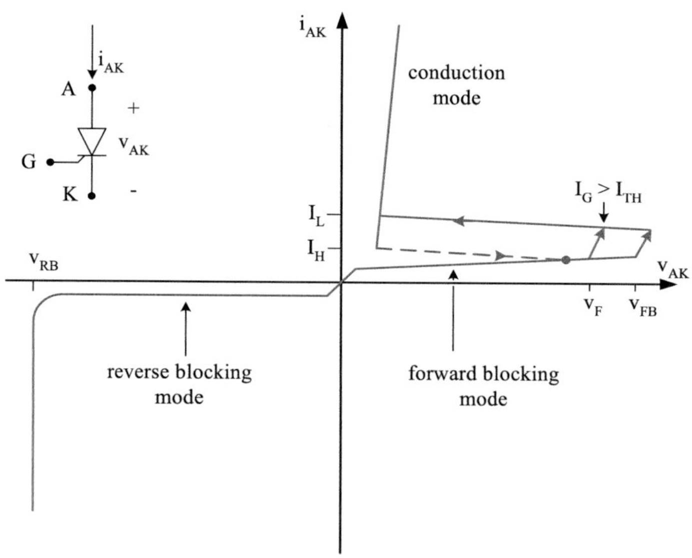

Figure 1.7 *I–V* characteristic of the thyristor.

is usual for I_H to be about half of I_L.) Observe that the thyristor's current is controlled by the external circuit when it is in the conduction mode. Thus, it is the external circuit that can switch the thyristor from the conduction state to the forward blocking state.

1.2.3 Thyristor's turn–off

The gate loses control of the thyristor when the device enters the conduction mode. Subsequently, the thyristor's current is controlled by the external circuit. To turn off the thyristor, the external circuit needs to bring the thyristor current to a value below the holding current. This can be achieved by applying a negative anode to cathode voltage v_{AK} that will reduce the thyristor current and remove the charge carriers flooding the n^- region. To effectively turn

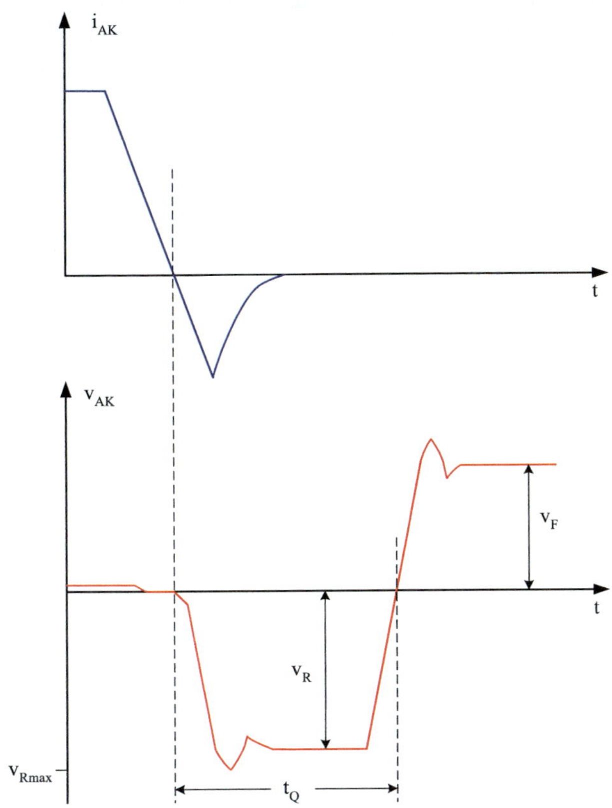

Figure 1.8 Thyristor's turn-off transients: (a) current transient; (b) voltage transient.

off the thyristor, its current should become negative to remove the stored charge almost completely. The thyristor enters the reverse blocking mode on completion of the removal of the stored charge.

A minimum recovery time t_Q is necessary for the charge removal during which the application of a positive anode to cathode voltage is not permitted. The thyristor will not be able to support the rated forward voltage or the rated dv_{AK}/dt until the stored charge is completely removed from the n^- layer. Figure 1.8 shows the turn-off transient and the recovery time. Note that the anode current's zero crossing and the slope di_{AK}/dt are determined by the external circuit.

1.2.4 Simulation model

Figure 1.9 shows the simplified simulation model of the thyristor. Given the thyristor's cathode to anode voltage v_{AK} and the parameter nsw, the model returns the cathode to anode current i_{AK} and its conductance g_{Th}. Observe that the polarities of the current and voltage have been reversed. The function

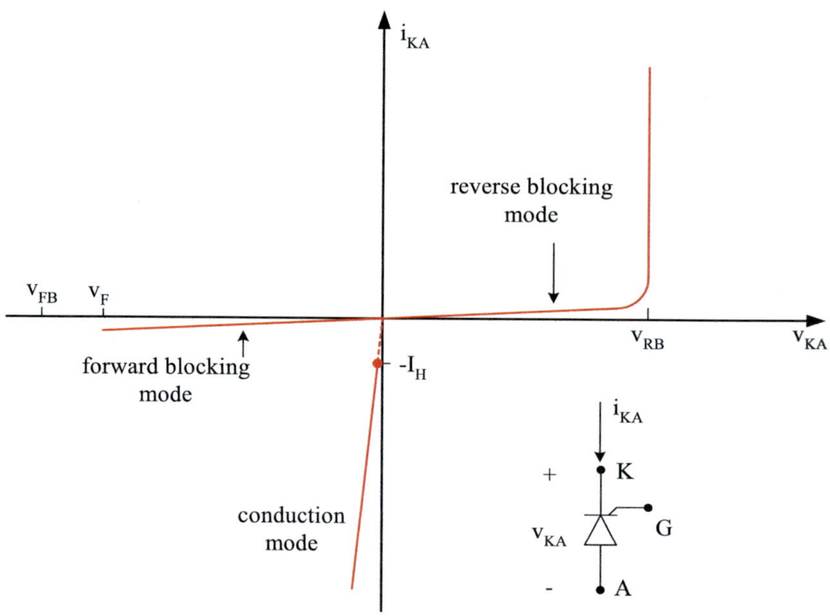

Figure 1.9 Thyristor's simulation model.

to be called is

$$i_{KA}, g_{\text{Th}} = fthy(v_{KA}, nsw).$$

The parameter nsw is set to 1 at the instants of triggering. It is set to -1 when the thyristor enters the forward or reverse blocking mode. For $nsw = 1$

$$\begin{aligned} g_{\text{Th}} &= g_{\text{high}} \ \ if \ \ v_{KA} < 0 \\ &= g_{\text{low}} \ \ if \ \ v_{KA} \geq 0. \end{aligned}$$

For $nsw = -1$, $g_{\text{Th}} = g_{\text{low}}$. Typical values are $g_{\text{high}} = 100$ S and $g_{\text{low}} = 0.1e - 07$ S. A given circuit with the thyristor is solved and the thyristor's current is checked at each time-step. If $i_{KA} > -i_H$, nsw is set to -1 for the following time-steps until the next instant of triggering (refer to Chapter 5).

1.3 Gate Turn-off Thyristors

The thyristor can block high voltages in the OFF–state and conduct high currents with low voltage drop in the ON–state. It is turned on by injecting a current into the gate terminal. However, the thyristor cannot be turned off by applying a control signal at the gate. In fact, the gate loses control as soon as the thyristor is turned on. Structural modifications of the thyristor are necessary to include gate turn-off capability. The resulting device is the gate turn-off (GTO) thyristor.

1.3.1 Basic structure

Figure 1.10 shows the structure of the GTO thyristor. There are significant differences between the structures of the conventional thyristor and the GTO thyristor. The emitter and gate layers are interspersed in the GTO so that the distance between the two regions is much reduced. The n^+ emitter regions appear as islands. Another major difference may be observed in the p^+ layer at the collector terminal where n^+ regions penetrate the p^+ collector layer and make contact with the n^- layer. This is the collector short (also referred to as the anode short), which accelerates the turn–off of the thyristor. The GTO thyristor with collector short has no reverse blocking capability. In most applications, an anti-parallel diode is connected to the GTO thyristor. The GTO thyristor's turn-on process is essentially the same as that for the conventional thyristor. It is usual to connect a snubber circuit to prevent dangerous over–voltages developing during turn-off.

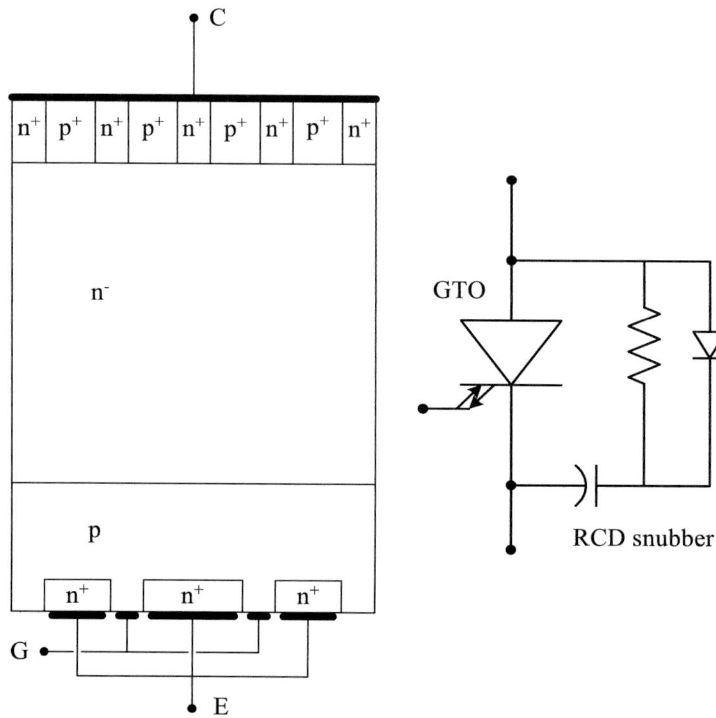

Figure 1.10 Structure of the GTO thyristor.

1.3.2 Turn-off of the GTO thyristor

The GTO thyristor is turned off by applying a gate voltage $v_{GG} = -15$ V. A large current of the order of 20%–30% of the collector current being turned off flows out of the gate. However, the large current is required only for a short interval.

The negative voltage applied between the gate and the emitter will remove the charges stored in the n^+ islands and the adjacent p layers. The collector current begins to fall when a sufficient amount of charge has been removed. The steep fall of the collector current ends when excess carriers at the gate-emitter junction have been swept out and the junction recovers its reverse blocking capacity. Some excess stored charge still remain in the n^- and p layers, due to which a small tail current will flow between the negatively biased gate and the collector. During the tail current flow, the collector-emitter voltage v_{CE} rises to a peak due to the stray inductance in the power circuit.

The over-voltage in v_{CE} may be reduced by an RCD snubber connected in parallel with the GTO thyristor. A major part of the turn-off losses occur during the tail-time interval. The approximate turn-off transients are shown in Figure 1.11.

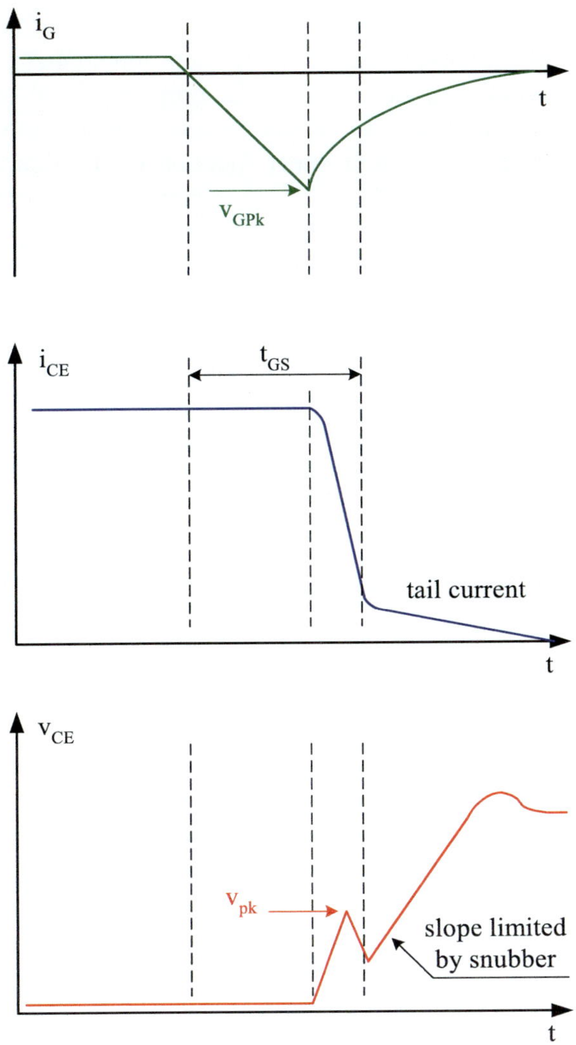

Figure 1.11 GTO thyristor's turn-off transients.

1.4 Metal-oxide Semiconductor Field Effect Transistor

Metal-oxide semiconductor field effect transistors (MOSFETs) [4, 5] capable of carrying high currents in the ON–state and blocking voltages in the OFF–state have been available for several decades and widely used in power electronic applications. In applications requiring high switching frequencies, MOSFETs have been replacing bipolar junction transistors as the switching devices.

The MOSFET is a three–terminal device, the terminals being the drain, the source, and the gate. The gate controls the flow of current between the drain and the source, which are the output terminals. In the ON–state, the drain–source voltage drop $v_{\rm DS}$ is a linear function of the drain to source current $i_{\rm DS}$. That is, $v_{\rm DS} = R_{\rm DS,on} \cdot i_{\rm DS}$, where $R_{\rm DS,on}$ is a constant resistance for a given temperature and gate to source voltage $v_{\rm GS}$. Moreover, $R_{\rm DS,on}$ has a positive temperature coefficient, which enables the parallel operation of MOSFETs in high current applications. The gate terminal of the MOSFET is electrically isolated so that the MOSFET is a voltage-controlled switching device. It requires a simple and inexpensive drive circuit. Being a unipolar device, the MOSFET may be switched at high frequencies though the conduction loss during the ON–state could be significant. The MOSFET can also withstand high voltages and currents simultaneously for short durations.

1.4.1 Structure of the N-channel, enhancement type MOSFET

Figure 1.12 shows the basic structure of an isolated cell of the vertical power MOSFET. It consists of the n^+, p, n^-, and n^+ layers. The n^+ layers are the source and the drain regions. The n^- layer above the drain is the drain drift region. The n^+ drain layer is the substrate on which the n^- layer is grown epitaxially. The p layer is the body where the n–channels are formed. The body layer is diffused into the wafer from the source side, followed by the n^+ source diffusion. Observe that a parasitic npn bipolar junction transistor is formed by the n^+ source, the p body, and the n^- drift layer. It is essential that this parasitic transistor should never be turned on. For this reason, the p body is connected to the source metallization so that the emitter and the base of the parasitic BJT are shorted. A parasitic diode, referred to as the body diode, is the result of the body-source short and it appears connected anti-parallel between the drain and the source of the MOSFET. The gate electrode is isolated from the body by a layer of silicon dioxide, which is a good insulator. Many cells, such as shown in Figure 1.12, are

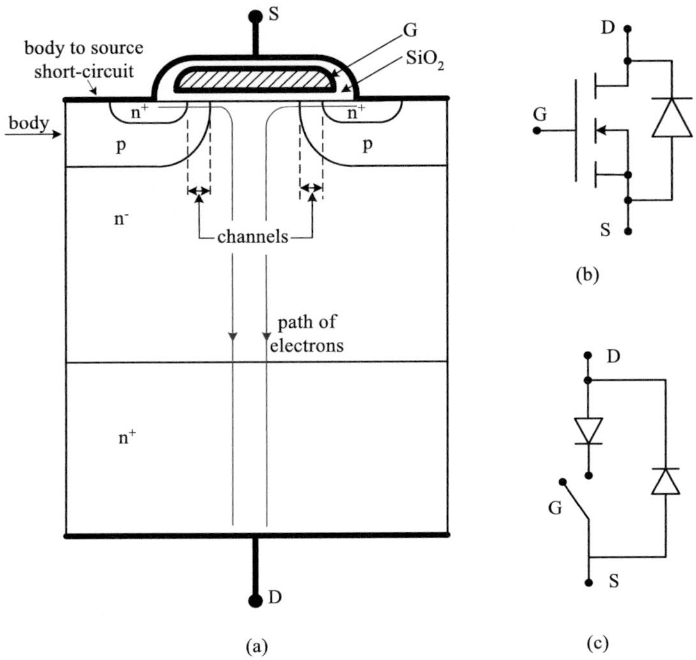

Figure 1.12 The N-channel MOSFET; (a) structure; (b) circuit symbol; (c) simulation model.

connected in parallel in a complete MOSFET. Figure 1.12 (b) shows the circuit symbol of the N-channel MOSFET. The simulation model is shown in Figure 1.12 (c).

1.4.2 The field effect

Consider a small positive voltage v_{GS} applied between the gate and the source of an N-channel MOSFET. The electric field repels the majority-carrier holes from the interface region and exposes the negatively charged acceptors. A depletion region forms at the interface between the gate oxide and the p body. Observe that there is an overlap of the gate metallization across the n^- drift region. Electrons are attracted toward the n^--gate oxide interface forming a region of enhanced conductivity known as the (a)ccumulation layer shown in Figure 1.13.

Further increase in v_{GS} causes the electric field at the oxide–$p$body interface to intensify and generate electron–hole pairs by thermal ionization.

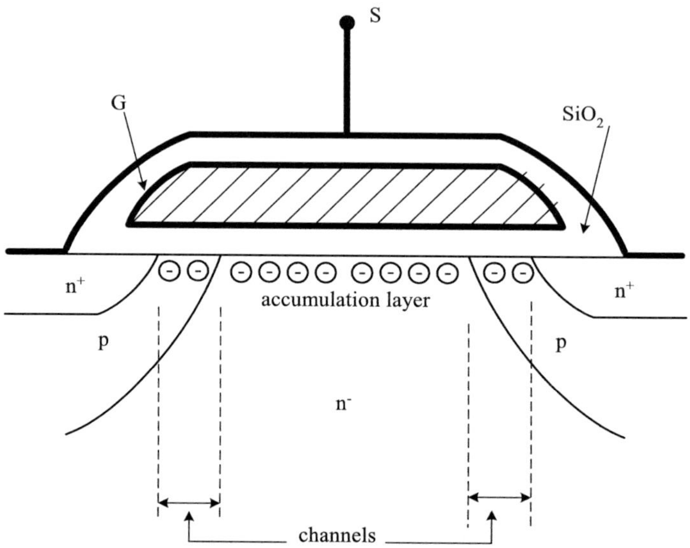

Figure 1.13 Formation of the n-channels and the accumulation layer.

The holes are pushed into the semiconductor ahead of the depletion region, where they are neutralized by electrons attracted from the n^+ source. The free electrons are attracted toward the gate oxide–$p$body interface. Eventually, as the gate–source voltage v_{GS} is increased, the layer of free electrons at the interface becomes highly conducting and it is termed the N-channel. The N-channel is a conductive path that permits the flow of current between the drain and source regions.

1.4.3 The output characteristics

Figure 1.14 shows the output characteristics of an N-channel MOSFET, i.e., the drain current i_{DS} as a function of the drain–source voltage v_{DS} with the gate–source voltage v_{GS} as the parameter. The MOSFET is in the OFF–state when $v_{\mathrm{GS}} < v_{\mathrm{Th}}$, where v_{Th} is the threshold voltage of a few volts. The device is an open circuit in the OFF–state and it should withstand the voltage that appears across the drain and source terminals. Obviously, the drain–source voltage v_{DS} in the OFF–state should be reasonably lower than the breakdown voltage v_{BDS}.

On applying $v_{\mathrm{GS}} \geq v_{\mathrm{Th}}$, the MOSFET is driven either into the ohmic region or into the active region. The MOSFET is in the ON–state when it is

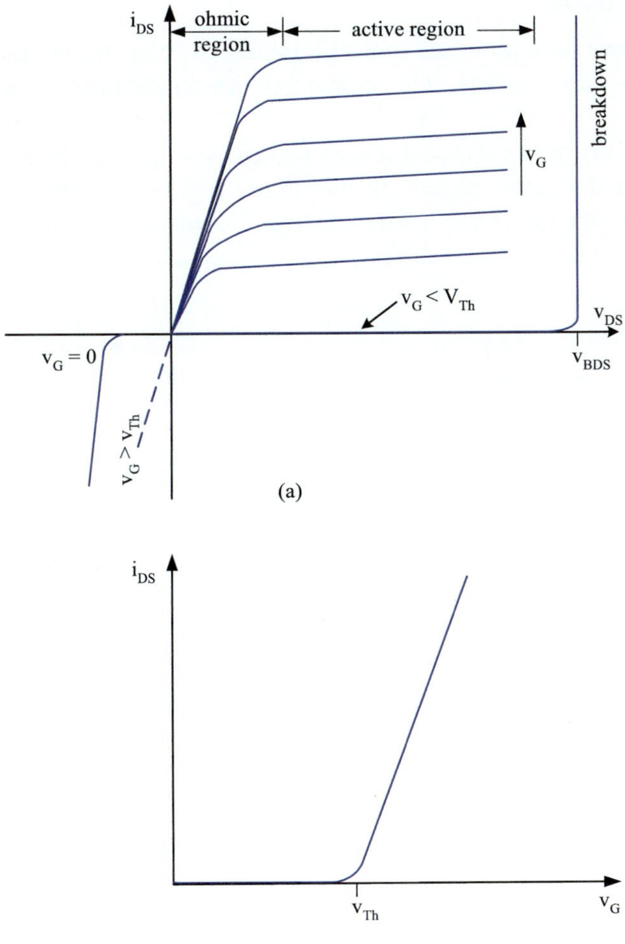

Figure 1.14 *I–V* characteristics of the MOSFET: (a) output characteristic; (b) transfer characteristic.

in the ohmic region where $(v_{\mathrm{GS}} - v_{\mathrm{Th}}) > v_{\mathrm{DS}} > 0$. (In the active region, $(v_{\mathrm{GS}} - v_{\mathrm{Th}}) < v_{\mathrm{DS}}$.) In power electronic applications, the MOSFET is either in the OFF–state or in the ON–state. It is never made to function in the active region. However, during a transition from the OFF–state to the ON–state (or vice versa), the MOSFET traverses the $i_D - v_{\mathrm{DS}}$ characteristics through the active region to the ohmic (or cut-off) region.

1.4.4 Circuit model

Many models have been proposed in the literature. Of these, the simple model shown in Figure 1.15 is sufficient to explain the turn-on and turn-off switching waveforms.

Figure 1.15 (a) shows a model that is valid when the MOSFET is in the OFF–state or in the active region. In the ohmic region, the inversion layer is represented by the ON–state resistance R_{DS-on} between the drain and the

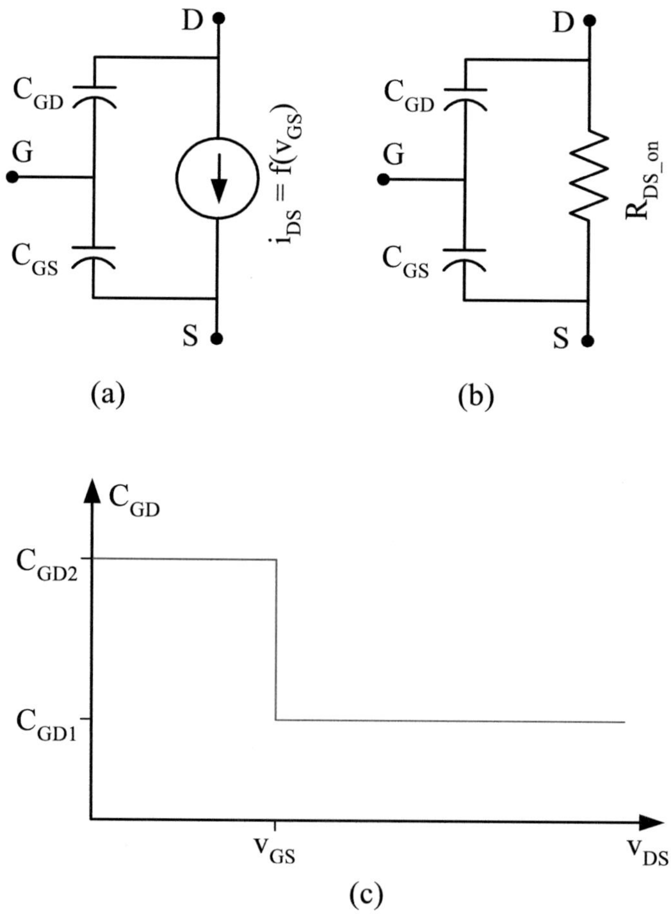

Figure 1.15 Circuit model of the MOSFET: (a) OFF–state model; (b) ON–state model; (c) approximation of nonlinear C_{GD}.

source and the relevant model is shown in Figure 1.15 (b). The MOSFET enters the ohmic region when $v_{\mathrm{DS}} < (v_{\mathrm{GS}} - v_{\mathrm{Th}})$.

The gate–source capacitance C_{GS} is the series combination of the capacitance of the gate oxide layer and the capacitance of the depletion layer at the gate oxide–n^- interface. It is assumed to be constant. The gate–drain capacitance is the combination of the gate oxide layer and the $p-n^-$ junction capacitance. Due to the variation in $v_{\mathrm{GD}}(\simeq v_{\mathrm{DS}})$, the capacitance C_{GD} varies significantly. For approximate calculation of the switching transients, C_{GD} may be approximated by two capacitances C_{GD1} and C_{GD2} (as shown in Figure 1.15 (c)):

$$
\begin{aligned}
\text{for } v_{\mathrm{GS}} \leq v_{\mathrm{DS}},\ C_{\mathrm{GD}} &= C_{GD1} \\
\text{for } v_{\mathrm{GS}} > v_{\mathrm{DS}},\ C_{\mathrm{GD}} &= C_{GD2}.
\end{aligned}
$$

There is a third capacitance C_{DS} that appears between the drain and the source. This is the capacitance of the body diode. The switching transients are not affected by C_{DS}, for which reason it is not included in the circuit models.

The capacitances C_{GS}, C_{GD}, and C_{DS} are usually not given in the MOSFET data sheets. However, the data sheets provide values of the input capacitance C_{ISS}, the output capacitance C_{OSS}, and the reverse transfer capacitance C_{RSS}. The capacitances in the circuit model are obtained as

$$
\begin{aligned}
C_{\mathrm{GD}} &= C_{\mathrm{RSS}} \\
C_{\mathrm{GS}} &= C_{\mathrm{ISS}} - C_{\mathrm{RSS}} \\
C_{\mathrm{DS}} &= C_{\mathrm{OSS}} - C_{\mathrm{RSS}}.
\end{aligned}
$$

1.4.5 Turn-on transients

Consider the MOSFET switch embedded in the step-down DC–DC converter shown in Figure 1.16 (a). The turn-on transient process may be divided into four intervals:

- The capacitance C_{GS} is charged from 0 V to v_{Th} during the turn-on delay. The major part of the gate current charges C_{GS}. A small current also flows into C_{GD}. The operating point is at A in the $i_{\mathrm{DS}} - v_{\mathrm{DS}}$ graph (Figure 1.16 (b)).
- When v_{GS} increases above v_{Th}, the device moves into the linear region where $i_{\mathrm{DS}} = g_m(v_{\mathrm{GS}} - v_{\mathrm{Th}})$ where g_m is the slope of the transfer characteristic. The drain current increases, but the drain–source voltage

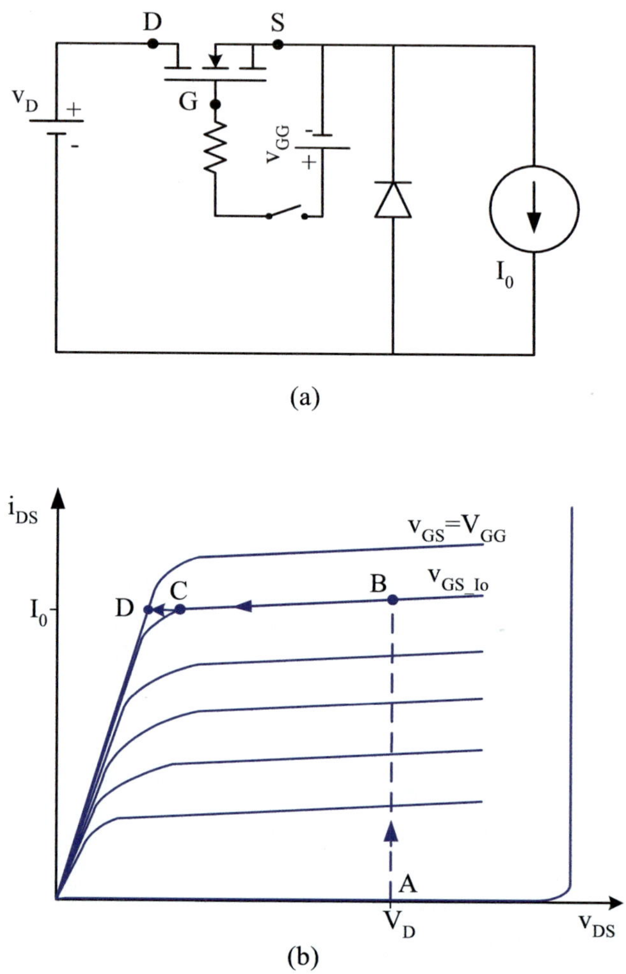

Figure 1.16 Turn-on process of the MOSFET: (a) electrical circuit; (b) turn-on trajectory in the i_{DS}, v_{DS} plane

remains at v_D. This interval continues until the entire load current i_0 is transferred from the diode to the MOSFET. The gate–source voltage rises to v_{GS-i_0} corresponding to the current i_0 in the transfer characteristic (i.e., point B in Figure 1.16 (b)).

- The MOSFET carries the entire load current and the gate–source voltage is v_{GS-i_0}. The drain–source voltage falls and moves to C in Figure 1.16 (b)).

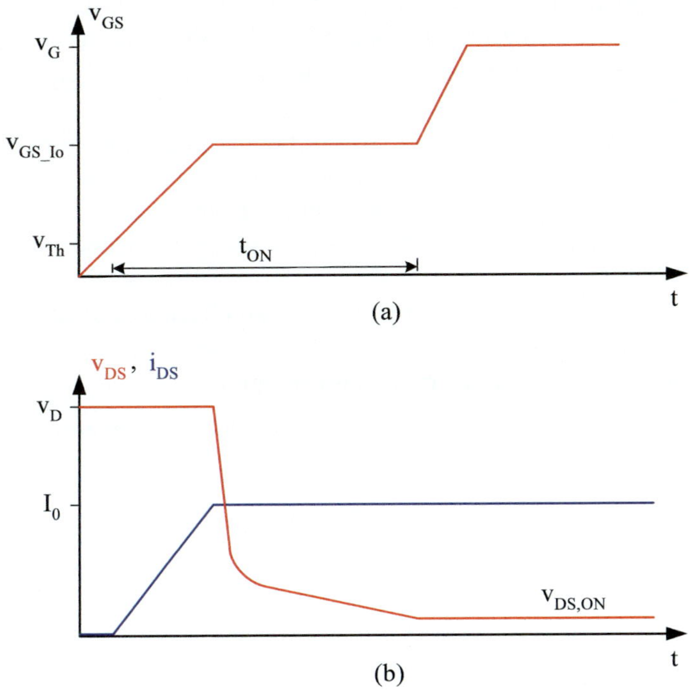

Figure 1.17 MOSFET's turn-on transients.

- In the last step, the conducting channel is fully enhanced by the higher gate drive voltage v_{GG}, which determines the ON–resistance $R_{\mathrm{DS-on}}$ of the MOSFET during the ON–state. The operating point moves to D in Figure 1.16 (b).

Figure 1.17 shows the approximate turn-on transients.

1.4.6 Turn-off transients

The turn-off process involves the same intervals as the turn-on process, but in the reverse order. The gate drive voltage is a negative step from v_{GG} to 0 V. The turn-off transient intervals are:

- The gate–source voltage falls from v_{GG} to v_{GS-i_0} due to the discharge of the input capacitance C_{ISS}. The gate current is a negative current. The drain current continues to flow unchanged and the drain–source voltage v_{DS} remains low.

- In the second interval, $v_{\rm DS}$ increases from $R_{\rm DS-on} \cdot i_0$ to v_D and it remains clamped at the output voltage. During this interval, the gate–source voltage $v_{\rm GS}$ is constant and the drain current is still equal to the load current.
- The diode turns on and the load current begins to get transferred from the switch to the diode. The gate–source voltage falls from v_{GS-i_0} to $v_{\rm Th}$. The decreasing $v_{\rm GS}$ causes the drain current to fall nearly to zero by the end of the interval. The drain–source voltage remains constant at v_D.
- In the last interval, the gate–source voltage falls from $v_{\rm Th}$ to zero, which implies that the input capacitance is fully discharged. The drain current and the drain–source voltage remain unchanged.

Plotting the approximate turn-off waveforms is left as an exercise.

1.5 Insulated Gate Bipolar Transistor

The insulated gate bipolar transistor (IGBT) [6] is (as the name implies) a three–terminal, bipolar device like the BJT. Bipolar devices have lower conduction losses in the ON–state. Furthermore, the IGBT is driven by an insulated gate and it can be switched faster than the BJT. Thus, the IGBT combines the desirable qualities of the BJT and the MOSFET.

1.5.1 Basic structure

There are two types of IGBTs and these are the punch-through (PT) and the non-punch-through (NPT) structures. The N-channel, PT-IGBT is shown in Figure 1.18. The three terminals of the IGBT are the collector, the emitter, and the gate. (The collector and the emitter may also be referred to in the literature, as the drain and the source.)

The structure of the PT-IGBT consists of the p^+ injection layer, the n^+ buffer layer, the n^- drift layer, and p body region with n^+ diffused islands. The NPT-IGBT is manufactured without the n^+ buffer layer. The insulated gate is placed above the body layer and it usually extends over the body surface between the n^+ islands. The structure of the PT-IGBT results in three junctions. These are the junction J_1 between the injection layer and the buffer layer, the junction J_2 between the p body layer and the n^- drift layer, and the junction J_3 between the n^+diffused islands and the body layer. (The J_1 junction is between the injection layer and the drift layer in the case of the NPT-IGBT.) The emitter metallization is spread over the surfaces of the n^+

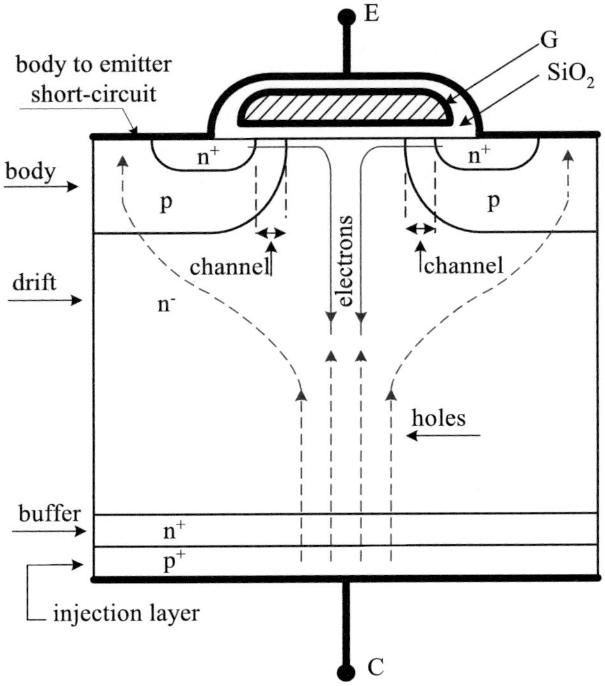

Figure 1.18 Structure of the N-channel, PT-IGBT.

islands and the p body. The NPT-IGBT has reverse blocking capability. The PT-IGBT does not have reverse blocking capability, but it has lower ON–state losses in the 1000–1200 V rating.

The vertical IGBT structure leads to a parasitic thyristor formed by the n^+, p, n^-n^+, and p^+ layers. The parasitic thyristor could get triggered when the collector current becomes very high. The spontaneous triggering of the parasitic thyristor is known as latch up and it usually ends in the destruction of by the IGBT. The manufacturer takes measures to ensure that the parasitic thyristor will never turn on. One such measure is the extension of the emitter metallization over the body region.

1.5.2 Device operation

The IGBT is in the forward blocking state if $v_{\mathrm{CE}} > 0$ and $v_{GE} < v_{\mathrm{Th}}$, where v_{Th} is the threshold voltage. A reverse voltage $v_{\mathrm{CE}} < 0$ cannot be blocked by the PT-IGBT. For $v_{GE} > v_{\mathrm{Th}}$ and with nominal voltage applied between the collector and the emitter, an inversion layer is formed in the body region below the gate. An electron current flows through the inversion layer from the n^+ emitter region to the n^- drift region. The electron current causes holes to be injected from the p^+ collector contact layer into the n^- drift region. A portion of the injected holes flows by drift and diffusion into the p body region where they attract electrons from the emitter metallization and are recombined. Another portion of the injected holes recombine with the electrons that flow through the inversion layer into the n^- drift region.

1.5.3 *I–V* characteristics

Figure 1.19 shows the *I–V* characteristics of the N-channel IGBT with the gate-emitter voltage as the controlling parameter. The transfer characteristic

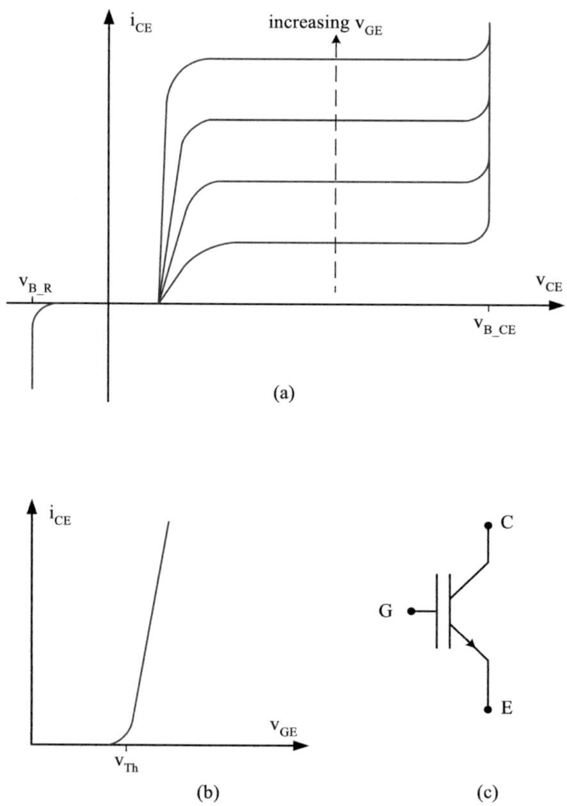

Figure 1.19 *I–V* characteristics of the IGBT: (a) output characteristics; (b) transfer characteristic; (c) circuit symbol.

and the circuit symbol for the IGBT are also shown in the figure. The NPT-IGBT can be designed to block a reverse voltage as large as the nominal forward blocking voltage.

1.5.4 Switching transients

The turn-on current and voltage waveforms of an IGBT switch embedded in a step-down converter are shown in Figure 1.20 (a). The IGBT behaves as a MOSFET during much of the turn-on period. The collector current rises linearly after the time delay until it becomes equal to the inductive load current and it remains clamped at this value. Soon after the moment when $i_{\mathrm{CE}} = I_0$, the voltage v_{CE} begins to fall. In the interval t_1, the IGBT is traversing the active state and moves toward the ohmic state. The interval t_2 is due to (a) the increase in the gate-collector voltage and (b) the *pnp* transistor traversing the characteristic from the active state toward its saturation state.

The gate-emitter voltage is reduced from a positive value to zero or to a negative value in order to turn off the IGBT. The turn-off transients (shown in Figure 1.20 (b)) are similar to the turn-off transients of the MOSFET. The

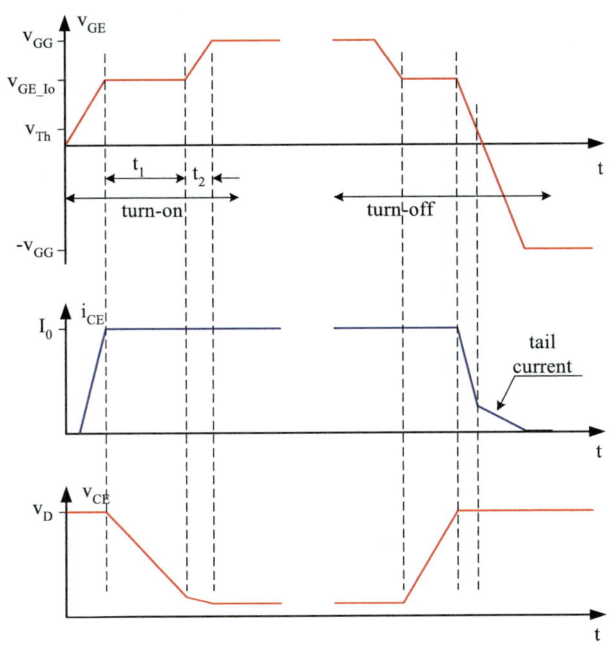

Figure 1.20 Switching waveforms of the IGBT.

main difference is in the collector tail current due to the stored charge in the n^- drift region. High collector-emitter voltage accompanies the tail current and the losses are substantial. Either the magnitude of the tail current should be reduced (as in the NPT-IGBT by special design methods) or its duration should be reduced (by providing the n^+ buffer layer in the PT-IGBT).

1.6 Wide Bandgap (WBG) Semiconductor Devices

The semiconductor devices described so far are silicon-based devices. Silicon is a semiconductor material that has an energy bandgap of 1.12 eV. (The bandgap is defined as the energy difference between the minimum energy level of the conduction band and the maximum energy level of the valence band.) The narrow bandgap of silicon leads to low electric breakdown field of about 300 kV/cm. The maximum operating voltage of silicon-based devices is therefore limited by its electric breakdown strength. On the other hand, WBG semiconductor materials are characterized by high electric breakdown fields and high operating voltages [7, 8].

WBG semiconductor-based devices possess many other advantages. Since their electric breakdown fields are higher, the drift region of these devices is thinner and it is doped to a higher concentration than in Si-based devices. Thus, the ON-resistance of WBG semiconductor-based devices is lower and their operating temperature is higher. The forward and reverse characteristics of WBG semiconductor devices do not vary significantly with temperature. Their reverse recovery currents, and consequently, the switching losses are low. Low switching losses permits the devices to operate at high switching frequencies. In spite of the many advantages that WBG semiconductor devices may possess, their widespread deployment is yet to happen. Their defect-free wafers are difficult to manufacture, and, therefore, they are expensive. Moreover, the packaging of these devices along with other components that can operate at high temperatures is still under development.

The WBG semiconductors that have been much researched are GaN and SiC. Some important parameters of GaN and SiC are given in Table 1.1,

Table 1.1 Semiconductor parameters.

Parameter	Si	4H SiC	GaN
Bandgap E_g (eV)	1.12	3.26	3.45
Electrical breakdown E_c (kV/cm)	250	2200	3300
Electron mobility μ_n cm^2/V s	1400	900	1000
Thermal conductivity K (W/cm $^\circ$K)	1.5	3.8	1.3

along with the corresponding values of Si for comparison. Gallium nitride is a direct bandgap semiconductor, i.e., it is a semiconductor in which the maximum energy level of the valence band is aligned with the minimum energy level of the conduction band with respect to electron momentum. Direct recombination of an electron and hole can take place in GaN with the release of energy equal to the energy difference of the recombining particles. Direct bandgap semiconductors are preferred for making optical devices. Silicon carbide is an indirect bandgap semiconductor. Momentum is first conserved with the release of energy when particles collide in SiC. Only when the two momenta are aligned can recombination occur with further release of energy. SiC consists of equal parts of Si and C atoms and it exists in many crystal structures known as polytypes. Of these, the 4H SiC polytype has high electron mobility, and hence, is the preferred semiconductor for power electronic devices.

1.6.1 Gallium nitride devices

The breakdown electric field of GaN is very high and that leads to compact switches operating at high voltages. Electron mobility in GaN is also high so that the ON-state and switching losses are low. However, the thermal conductivity of GaN is lower than that of Si nor is there a native insulating oxide that can be used to provide the insulated gate structure in GaN devices. GaN Schottky power diodes developed over the past decade are mostly lateral devices with the GaN rectifier implemented on Si or Sapphire substrates. More recently, GaN rectifiers on GaN substrates have been reported [9, 10].

1.6.1.1 High electron mobility transistors (HEMT)

The structure of the GaN transistor is shown in Figure 1.21 (a). It consists of an Si substrate on which a number of buffer layers are deposited to provide strain relief between the GaN layer and the substrate. Above the GaN layer, a layer of AlGaN is deposited forming a hetero-junction. Since AlGaN and GaN have different lattice constants, a tensile strain arises at the interface that causes the formation of a layer of high mobility electrons known as the two-dimensional electron gas (2DEG). Note that the 2DEG carriers do not result from doping. The 2DEG forms a channel between the source and drain terminals even when no external voltage is applied. On applying a non-zero drain–source voltage, a current flows between the terminals. A device that relies on the hetero-junction structure is referred to as a high electron mobility

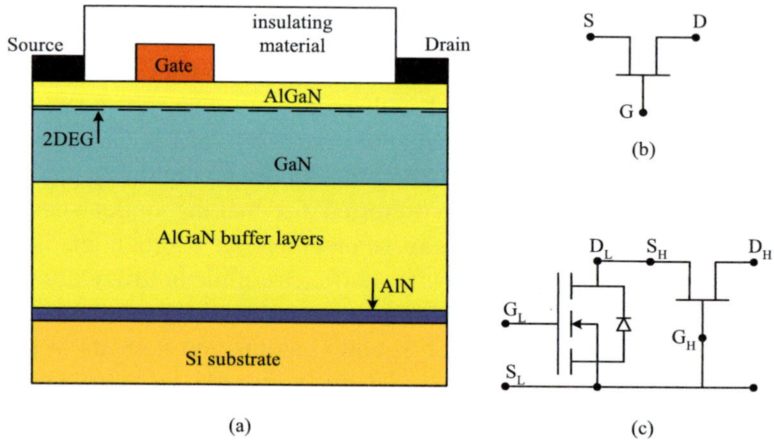

Figure 1.21 Structure of GaN transistor: (a) depletion mode; (b) symbol for GaN transistor; (c) cascode connection.

transistor (HEMT). The HEMT is normally-on, referred to as a depletion-mode device. It can only be turned off by applying a negative gate–source voltage. Figure 1.21 (b) shows the symbol for the HEMT.

1.6.1.2 Cascode-mode device

Normally-on devices are seldom used in power electronic applications. A normally-off device is obtained by connecting a low–voltage Si MOSFET in series with the high–voltage GaN HEMT, as shown in Figure 1.21 (c). The drain and the source of the Si MOSFET are connected to the source and gate, respectively, of the GaN transistor. When the Si MOSFET is turned on, its channel is established and its parasitic drain–source capacitance begins to discharge. Thus, the drain–source voltage of the Si MOSFET decreases and the gate–source voltage of the GaN transistor increases from a negative value toward 0 V. When the parasitic capacitor is completely discharged, the GaN transistor is turned on.

A positive gate–source voltage applied to the Si MOSFET turns on the GaN transistor. The cascode connection requires only a regular Si MOSFET driver since the Si MOSFET is the component that controls the GaN transistor. However, the losses due to the Si MOSFET's ON-resistance and due to the reverse recovery of Si MOSFET's body diode all add to the losses of the cascode connection. For this reason, the cascode connection may be adopted for the GaN transistor only in the 600–650 V range.

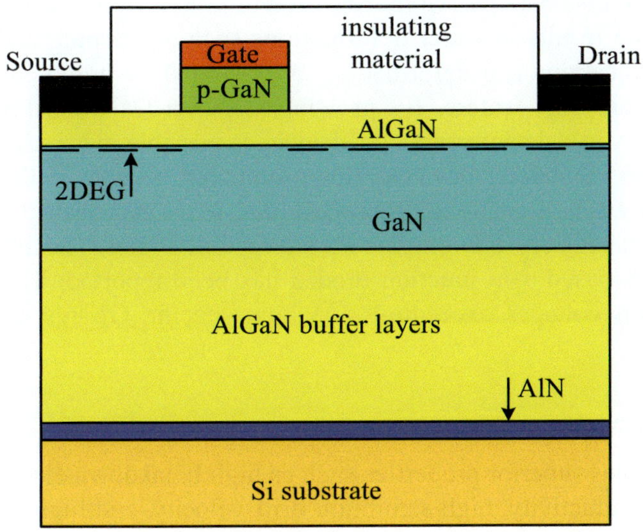

Figure 1.22 Structure of the enhancement mode GaN HEMT.

1.6.1.3 Enhancement mode HEMT

Figure 1.22 shows the basic structure of the enhancement mode GaN HEMT. The gate of the depletion mode HEMT has been modified to shift the threshold voltage to a positive value and thus convert the device to a normally-off or enhancement mode HEMT. The commonly used modification is to introduce an additional layer of p-doped GaN between the gate and the AlGaN barrier layer. This modification shifts the threshold voltage and results in the enhancement mode HEMT being a normally-off device.

The main disadvantage of the structure is that the threshold voltage is about 1.1 V, which is low. Since the margin between 0 V and v_{Th} is low, the device is susceptible to noise. Besides, the gate–source voltage required to fully turn on the device is about 5–6 V and the maximum allowed gate–source voltage is 7 V. Therefore, the margin between full turn-on and maximum allowed gate–source voltages is very low.

1.6.1.4 Homo-epitaxial GaN devices

The GaN devices described above have lateral structures with GaN layer grown epitaxially on Si, Sapphire, or SiC substrates. The mismatched lattice constants of GaN and the substrate material lead to defects in the crystalline structure known as dislocations. The electrical properties of GaN are degraded by dislocations. Moreover, growing thick layers of GaN on a foreign substrate results in wafer deformations such as warping and cracking. Wafer dislocations and deformations are avoided by homo-epitaxially growing GaN on GaN wafers. By growing GaN on GaN substrates, the manufacture of vertical semiconductor devices is possible, whereas GaN on Si/SiC is limited to lateral devices. Homo-epitaxial growth of GaN on a flat GaN substrate is a challenging process that requires polished surfaces cleaved along specific crystallographic axes at specific angles. The fabrication of vertical GaN-based, p-n junction diodes has been reported. The diodes have breakdown voltages exceeding $4\,k$V and specific ON-resistances of 2.8 mΩ.cm^2.

1.6.2 Silicon carbide devices

Silicon carbide has superior properties such as high breakdown electric field, high thermal conductivity, high saturation drift velocity, and high operating frequency of SiC-based devices [11, 12]. Insulated gate structures in SiC devices are possible, thanks to the existence of silicon dioxide. The processing of SiC-based devices is very much compatible with that of Si-based devices.

1.6.2.1 SiC rectifiers

There are three types of SiC power rectifiers: (i) Schottky barrier (SB) diodes; (ii) PiN diodes; (iii) junction barrier Schottky (JBS) diodes. The SB diode may be used at high switching frequencies, but its voltage blocking capability is low. The leakage current in the non-conducting mode is high. The PiN diode can operate at high voltages and its leakage current is low. However, there is a reverse recovery current whenever the PiN diode is turned off. The switching losses are therefore significant.

Figure 1.23 shows the structure of the JBS diode. It consists of Schottky regions interspersed with p^+ doped regions at the anode end. The JBS diode behaves like a Schottky diode when it is in the ON–state. The ON-state voltage drop is determined by the resistance of the drift regions, the metal-SiC barrier height, and the relative areas of the Schottky and the p^+ doped regions. At high current densities, some minority carrier injection takes place at the p^+n^- junctions resulting in higher surge current capability. The Schottky contacts to the drift regions are able to extract the minority carriers when the diode recovers from the high current transients. The anode metal-SiC barrier height should be low enough to result in a low ON-state voltage

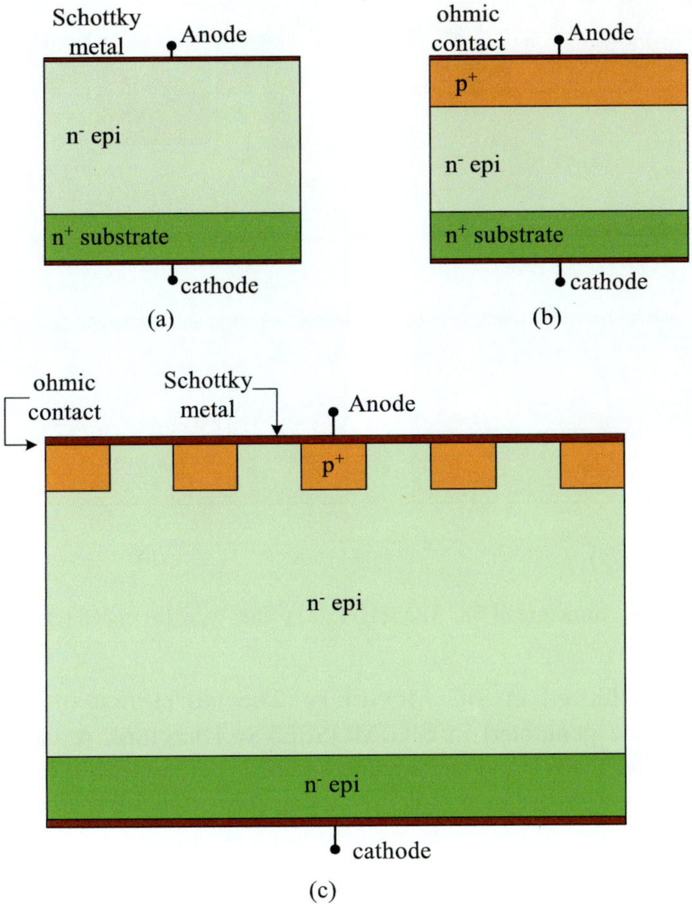

Figure 1.23 Structure of the SiC diodes: (a) SB diode; (b) PiN diode; (c) JBS diode.

drop but high enough to block the reverse voltage during the OFF-state. On increasing the reverse bias, the depletion regions around the p^+ doped regions expand and pinch the leakage current. Additionally, the p^+ regions reduce the electric field at the anode metal-SiC junction. Thus, the OFF-state switching characteristics of the JBS diode resemble those of the PiN diode.

1.6.2.2 SiC MOSFETs

The structure of the SiC MOSFET is similar to the Si MOSFET with the insulated gate. High breakdown voltage and low ON-resistance can be

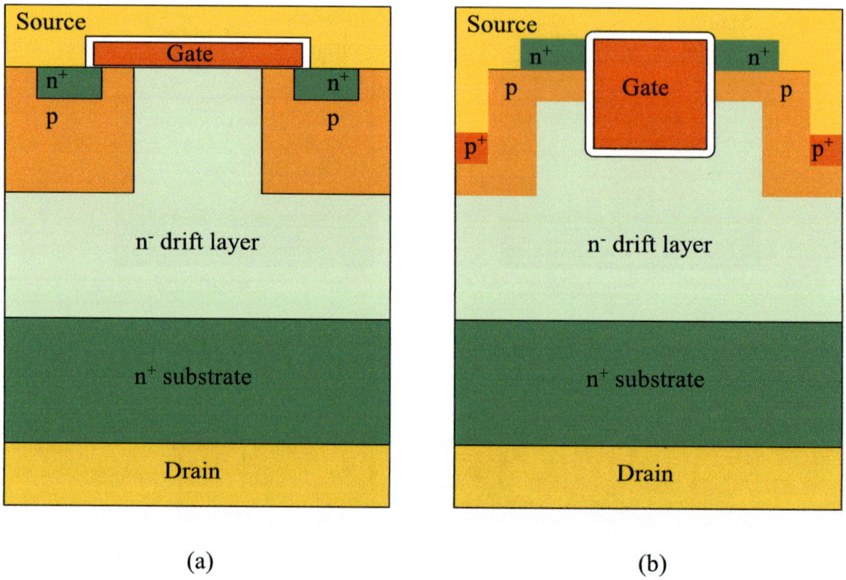

Figure 1.24 Structure of SiC MOSFETs: (a) planar type; (b) trench type.

simultaneously achieved in SiC MOSFETs. The tail current observed in IGBTs will not be generated in SiC MOSFETS. Therefore, replacing the IGBT with SiC MOSFET can significantly reduce the switching losses and also reduce the size of the heat sink. By using high switching frequencies, the size of the passive components are also reduced. Figure 1.24 shows the structure of two commercially offered SiC MOSFETs. The planar type shown in Figure 1.24 (a) has an ON-resistance of 125 $m\Omega$ at 125 °C. The trench type shown in Figure 1.24 (b) has an ON-resistance of about half this value (62 $m\Omega$).

References

[1] J. Lutz, H. Schlangenotto, U. Schevermann, R. De Doncker, *Semiconductor Power Devices: Physics, Characteristics, Reliability,* Springer-Verlag, 2011

[2] B. J. Baliga, *Power Semiconductor Devices,* PWS Publishing Company, Boston, 1996

[3] N. Mohan, T. M. Undeland, W. P. Robbins, *Power Electronics: Converters, Applications and Design,* 2nd Edition, John Wiley and Sons, 1995

[4] V. Barkhordarian, *Power MOSFET basics,* International Rectifier, El Segundo, CA

[5] Fairchild Semiconductor Corp., *MOSFET basics,* Application Note AN9010, 2000

[6] International Rectifier, *IGBT Characteristics,* Application Note AN990, August 2012

[7] J. Millán, P. Godignon, X. Perpiñà, A. Pérez-Tomás, J. Rebollo, *A survey of wide bandgap power semiconductor devices,* IEEE Trans. on Power Electronics, vol.29, no.5, May 2014

[8] B. Ozpineci, L. M. Tolbert, *Comparison of wide bandgap semiconductors for power electronic applications,* Oak Ridge National Laboratory, ORNL/TM-2003/257, December 2003

[9] M. Comola, *E-mode GaN technology: tips for best driving,* STMicroelectronics, Application Note AN5583, Feb.2021

[10] Kedar Patel, *Unlocking the potential of GaN with Vertical GaN ,* NexGen Power Systems Inc., white paper

[11] F. Dalquist, *Junction barrier Schottky rectifiers in Silicon Carbide,* Dissertation, KTH, Royal Institute of Technology, Dept. of Microelectronics & Information, Stockholm, 2002

[12] C. Langpoklakpam, A.-C. Liu, K.-H. Chu, L.-H. Hsu, W.-C. Lee, S.-C. Chen, C.-W. Sun, *Review of Silicon Carbide processing for power MOSFET,* Crystals 2022, 12, 245

2

Magnetic Devices

Inductors are to be found in most power electronic circuits, and transformers are obviously required in the isolated converter circuits (of Chapter 7). The inductor consists of a ferromagnetic core that forms the magnetic circuit. A coil of insulated wire is wound on the core. The inductor stores energy during an interval of the switching period and transfers the energy to the load or to the source during another interval. In the steady-state, the inductor is a short circuit to any direct voltage that appears at its terminals. An inductor, in association with a capacitor, can perform the functions of a filter or a resonant circuit. By itself, an inductor can limit the rate of rise of current or limit transient currents.

The transformer has a construction similar to that of the inductor. However, it has multiple windings on the core. It provides galvanic isolation and transformation of voltages, currents and impedances, but it does not usually store energy. Power received on the primary side is transferred to a load on the secondary side. Some energy does get stored in the magnetic field that is set up in the core of the transformer.

The analysis of electromagnetic devices is the search for solutions of Maxwell's equations, subject to specified boundary conditions. This is a well-known task, though it usually requires a complex numerical algorithm. Design of inductors and transformers is a far greater challenge due to the number of parameters involved, the many design objectives that are possible, and the large volume of data required from the manufacturers of magnetic cores. A brief review of magnetics, magnetic core characteristics, and windings is presented in this chapter. The review assumes prior exposure to electromagnetism.

2.1 Maxwell's Equations

The equations establish the relations between the vector fields $\vec{E}$ (electric field, Volts/m), $\vec{B}$ (magnetic flux density, Tesla or webers/m^2), $\vec{H}$

(the magnetic field intensity, Ampere-Turns/m), and $\vec{J}$ (the current density, Amps/m^2). The integral form of the equations are given below [1, 2].

2.1.1 Gauss's law for electric fields

$$\oint_S \epsilon\vec{E}\cdot\hat{n}\,da = \int_V \rho_v\,dV, \tag{2.1}$$

where $\hat{n}$ is the normal to the surface element da as shown in Figure 2.1 (a). The constant ϵ is the permittivity of the medium in which the volume is immersed. The volumetric charge density is ρ_v in coulomb/m^3. The equation states that the net flux of $\epsilon\vec{E}$ emerging from the closed surface S is equal to the charge contained within the volume V.

2.1.2 Gauss's law for magnetic fields

$$\oint_S \vec{B}\cdot\hat{n}\,da = 0. \tag{2.2}$$

The equation states that the net flux of $\vec{B}$ emerging from any closed surface is zero. Thus, magnetic flux lines are always closed Figure 2.1(b).

2.1.3 Faraday's law

$$\oint_C \vec{E}\cdot\vec{dl} = -\frac{\partial}{\partial t}\int_S \vec{B}\cdot\hat{n}\,da. \tag{2.3}$$

Consider an open surface S bounded by the closed curve C Figure 2.1(c). The law states that the line integral of the electric field $\vec{E}$ over the closed curve C is equal to the negative rate of change of the magnetic flux that crosses the surface S.

2.1.4 Ampere's law

$$\oint_C \vec{H}\cdot\vec{dl} = \int_S (\vec{J} + \frac{\partial}{\partial t}\epsilon\vec{E})\cdot\hat{n}\,da. \tag{2.4}$$

Consider the open surface S bounded by the closed curve C Figure 2.1(d). The law states that the line integral of the magnetic field intensity $\vec{H}$ over the curve C is equal to the total current crossing the surface S.

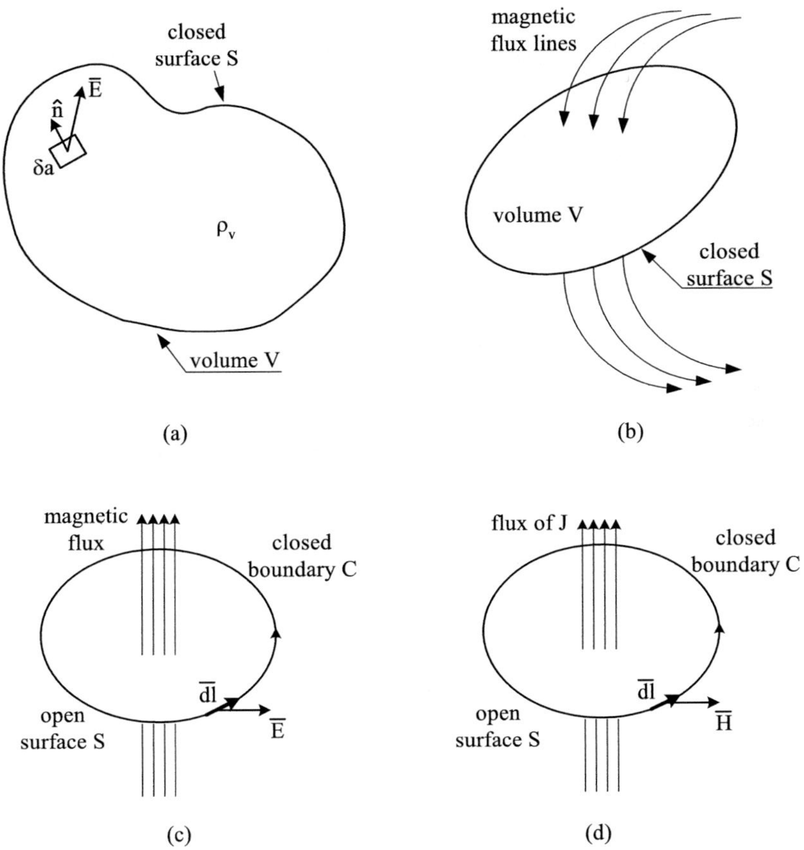

Figure 2.1 Electromagnetic fields.

Taking into account that the switching frequencies in power electronic applications do not exceed a few MHz and that current densities of the order of 300 Amps/cm^2 are normal (i.e. $\vec{J} \gg \epsilon\vec{E}$), Maxwell's equations may be approximated as

$$\oint_S \vec{B}\cdot\hat{n}\,da = 0$$

$$\oint_C \vec{E}\cdot\vec{dl} = -\frac{\partial}{\partial t}\int_S \vec{B}\cdot\hat{n}\,da$$

$$\oint_C \vec{H}\cdot\vec{dl} = \int_S \vec{J}\cdot\hat{n}\,da.$$

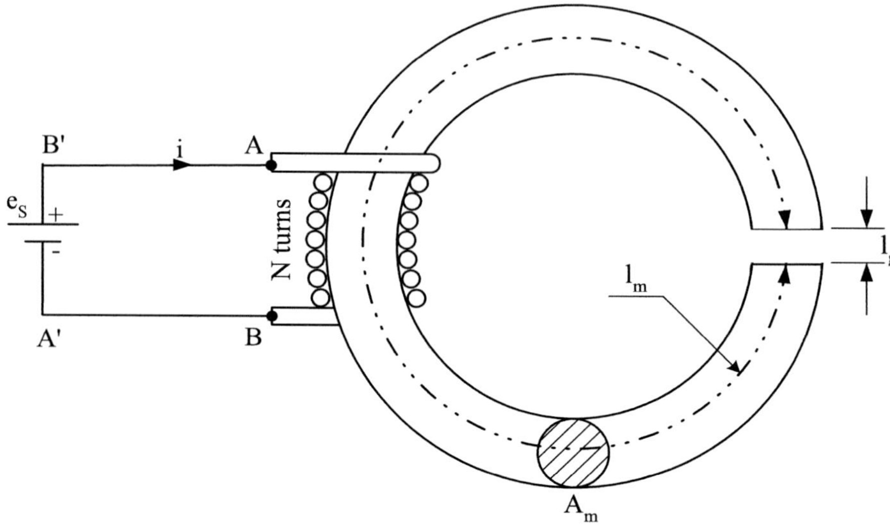

Figure 2.2 Basic magnetic circuit.

2.2 Magnetic Circuits

Consider the ferromagnetic core shown in Figure 2.2 [1–3]. It is in the form of a toroid with circular cross section. Let the cross-sectional area be A_m and the mean circumference be l_m. Also, let a thin slice of the toroid be removed to form an air gap of length l_g. A winding of N turns of insulated copper wire is wound on the toroid and it is excited by a voltage source of V_i volts. A current of i amperes flows in the winding and a magnetic flux density B Tesla is established in the core. Assume that the voltage drop in the winding is negligible, that there is no flux leakage, and that there is no fringing of flux when crossing the air gap.

Applying Faraday's law,

$$\int_B^A \vec{E} \cdot \vec{dl} + \int_{A'}^{B'} \vec{E} \cdot \vec{dl} = -\frac{\partial}{\partial t} \int_S \vec{B} \cdot \hat{n} \, da. \tag{2.5}$$

Since the voltage drop in the winding is negligible, the first line integral results in zero volts. The second line integral is equal to $-e_S$ volts. The surface integral on the RHS is evaluated over a screw-like surface whose boundary is the N turns of wire. By integrating $\vec{B}$ over this surface, we get

$$-e_S = -\frac{d}{dt}(N \phi), \tag{2.6}$$

where ϕ is the magnetic flux in webers. Putting $\lambda = N\phi$, where λ is the flux linkage, we get

$$e_S = \frac{d}{dt}\lambda. \tag{2.7}$$

Applying Ampere's law to the magnetic circuit Figure 2.3,

$$\oint_{C_t} \vec{H} \cdot d\vec{l} = \int_{S_t} \vec{J} \cdot \hat{n}\, da, \tag{2.8}$$

where $\vec{H}$ is the magnetic field intensity and $\vec{J}$ is the current density in the wire. It is convenient to choose the magnetic flux line at the center of the core. Assuming that $\vec{H}$ is uniform in the toroid and the air gap, the line integral gives

$$\oint_{C_t} \vec{H} \cdot d\vec{l} = H_m l_m + H_g l_g. \tag{2.9}$$

The surface integral gives

$$\int_{S_t} \vec{J} \cdot \hat{n}\, da = N\, i, \tag{2.10}$$

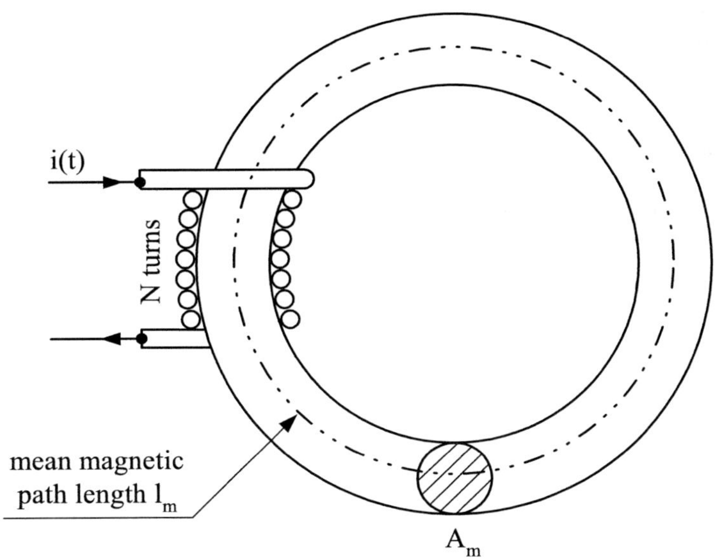

Figure 2.3 Basic magnetic circuit: Ampere's law.

where $N\,i$ is the magneto-motive force in Amp-Turns. Putting $H_m = B_m/\mu_m = \phi/(\mu_m A_m)$ and $H_g = B_g/\mu_g = \phi/(\mu_g A_g)$, we get

$$\phi \cdot \left\{ \frac{l_m}{\mu_m A_m} + \frac{l_g}{\mu_g A_g} \right\} = N\,i, \tag{2.11}$$

where μ_m, μ_g are the permeabilities of the magnetic material and of air. The quantity $l/(\mu\,A)$ is the reluctance $\mathcal{R}$ of the material. Then,

$$\phi\{\mathcal{R}_m + \mathcal{R}_g\} = N\,i. \tag{2.12}$$

That is,

$$\phi = N\,i/\Sigma\mathcal{R}. \tag{2.13}$$

From the equation $e_S = \frac{d}{dt}(N\,\phi)$, we get

$$\begin{aligned} e_S &= N \cdot \frac{d}{dt}\frac{Ni}{\mathcal{R}} \\ &= \frac{N^2}{\mathcal{R}}\frac{di}{dt}. \end{aligned} \tag{2.14}$$

The parameter $\frac{N^2}{\mathcal{R}}$ is defined as the inductance L of the winding.

In the SI system of units, the unit of magnetic flux is the weber. The magnetic flux density is in webers/m^2, which is the Tesla. The magnetic field intensity is in Amp-Turns/m. Inductances are measured in henries, the unit henry being equal to 1 volt sec/Amp. Permeabilities are in henries/m. The permeability of free space is $\mu_0 = 4\pi \cdot 10^{-7}$ henries/m. The permeability of a material is expressed as $\mu = \mu_r \cdot \mu_0$, where μ_r, a dimensionless number, is the relative permeability. Based on their relative permeabilities, materials are classified as paramagnetic ($\mu_r > 1$), diamagnetic ($\mu_r < 1$) or ferromagnetic ($\mu_r \gg 1$). The permeability of air μ_g is approximately equal to μ_0.

Ferromagnetic materials become paramagnetic when heated above a temperature known as the Curie temperature. The Curie temperature of iron is 770 °C, that of nickel is 358 °C, and that of cobalt is 1130 °C. The internal structure of ferromagnetic materials has fully magnetized domains. In a de-magnetized sample, the domains are small and randomly oriented so that the residual magnetic flux density is zero. In fact, a ferromagnetic material may be de-magnetized by raising its temperature above the Curie temperature. Subsequently, on cooling to the ambient temperature, the material is de-magnetized.

2.3 Magnetic Materials

Materials appropriate for the cores of inductors and transformers are the ferromagnetic materials such as silicon steels, permalloy (Ni-Fe), permendur (Co-Fe), amorphous metallic alloys, and ferrites [4, 5]. Cold-rolled, grain-oriented silicon steels (3% Si, 97% Fe) have high electrical resistivity and very low losses. They are used in power frequency devices.

Tape-wound permalloy cores have a magnetization characteristic that is almost a square loop with relative permeability approaching 60,000 and saturation flux density of 0.9 Tesla. These cores are used in the frequency range of 50–400 Hz. Gapped tape-wound cores are used in filter inductors when the ripple current and the associated flux density swing is low enough for the core losses to be acceptable.

Powdered molybdenum permalloy (MPP) cores (82% Ni, 16% Fe, 2% Mo) store considerable energy and are used in inductor and flyback transformer applications. The energy is stored in the binder that holds the metallic particles together. These cores have relative permeabilities in the 15–200 range. Their B-H characteristics show significant rounding as the magnetic field is increased. Therefore, the incremental permeability decreases as the DC operating point is raised.

Ferrite cores are extensively used in power electronic applications. Ferrite are inexpensive and the losses are low. However, they are mechanically fragile and are likely to break easily. Ferrites are ceramics made by sintering a mixture of iron oxide and the oxides or carbonates of Mn, Zn, or Ni. MnZn ferrites are used in applications involving switching frequencies of up to 2 MHz. NiZn ferrites are used in applications with switching frequencies from 1 MHz to several hundred MHz. Ferrite cores are gapped for inductor and flyback transformer applications. The relative permeability of ferrites is in the 1500–3000 range and the saturation flux density is about 0.3 Tesla. MnZn ferrites have higher permeability and NiZn ferrites have higher resistivity.

Metallic glasses are produced by rapidly cooling the molten metal alloy. The atomic structure of metallic glasses is amorphous, i.e., they do not have long–range structural order. The electrical resistivity of metallic glasses is high. The high saturation flux density, high permeability, and low core loss of metallic glasses are advantages in high–frequency applications such as high–frequency inductors with DC bias. One of the metallic glasses is an iron-based alloy having a saturation flux density of 1.2 Tesla, and it is well-suited for high–frequency transformer applications.

2.4 Magnetization Characteristics of Ferromagnetic Materials

Consider the magnetic core shown in Figure 2.3. It has a winding of N turns. Let the core be completely de-magnetized before being excited by the current i. The magnetic flux in the core due to the excitation is

$$\phi = Ni \cdot \frac{(\mu A_m)}{l_m}. \tag{2.15}$$

Assuming that the flux is uniformly distributed over the cross section of the core and that there is negligible flux leakage or fringing, the flux density is

$$B = \phi / A_m. \tag{2.16}$$

The magnetic field intensity is

$$H = \frac{Ni}{l_m}. \tag{2.17}$$

The magnetization characteristic is the flux density B plotted as a function of H [1, 3]. It is obtained by exciting the winding with direct current and the magnetic field intensity should be changed very slowly so that eddy currents are not generated in the core.

The magnetization curve obtained by DC excitation of a de-magnetized core is the initial magnetization curve (shown in Figure 2.4). In this curve, the flux density increases slowly from the origin, then increases very rapidly, and, finally, almost stops increasing when the core becomes saturated. The flux density at saturation is B_S corresponding to the magnetic field intensity H_S. The initial slope of the curve is the initial permeability. When the core is saturated, its permeability is approximately equal to that of air. That is, $\mu_S \simeq \mu_g \simeq \mu_0$.

After the core is taken into saturation, let the magnetic field intensity be decreased. Then, the magnetization curve follows the upper limiting curve of the major hysteresis loop. The core remains magnetized with a residual flux density B_R when H is reduced to zero. The residual flux density is also known as the remanent flux density. On further increasing H, but with reversed polarity, the core flux density is reduced to zero by applying a field intensity $H = -H_C$, which is the coercive force. On continuing the reverse excitation, the core is saturated in the reverse direction. That is, the flux density is $-B_S$ when the magnetic field intensity is $-H_S$. Now let the magnetic field intensity H be made less negative. Then, the magnetization

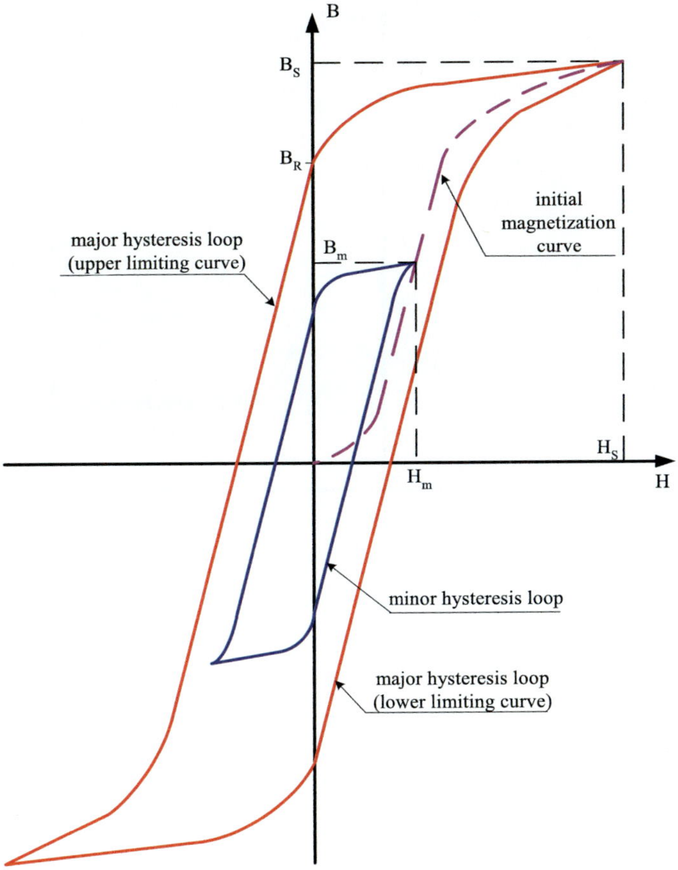

Figure 2.4 Magnetic hysteresis.

curve follows the lower limiting curve of the major hysteresis loop. This curve passes through $(0, -B_R)$ and eventually reaches positive saturation at (H_S, B_S).

Suppose that the magnetic core is magnetized along the lower limiting curve of the major hysteresis loop from $-B_R$ to B_S. The energy per unit volume of the core supplied by the exciting source is

$$\mathcal{E}_1 = \int_{-B_R}^{B_S} H dB, \tag{2.18}$$

which is the area $(PQRUTP)$ in Figure 2.5. After reaching saturation, let the core be de-magnetized from B_S to B_R along the upper limiting curve.

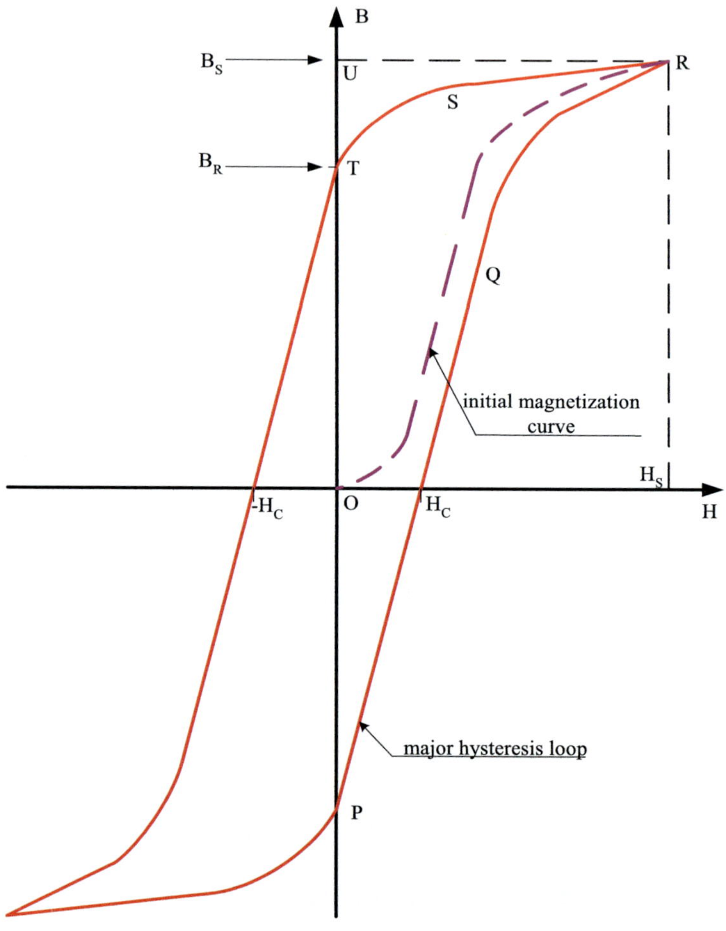

Figure 2.5 Magnetic hysteresis: iron losses.

Then the energy per unit volume, i.e., area $(RSTU)$ in Figure 2.5, given by

$$\mathcal{E}_2 = \int_{B_S}^{B_R} HdB, \tag{2.19}$$

is returned to the source. Extending the same reasoning to the second and third quadrants, we conclude that the area enclosed by the major hysteresis loop is the energy loss per unit volume per cycle and it corresponds to the flux density B_S. A minor hysteresis loop is generated when the magnetic field intensity is cycled between H_m and $-H_m$, where $H_m < \ H_S$ (Figure 2.4). The minor hysteresis loop has the extreme points (H_m, B_m) and $(-H_m, -B_m)$. The area enclosed by the minor loop is the energy loss per unit volume per cycle corresponding to the flux density B_m.

Magnetic devices operate at a flux density that is well below the saturation flux density B_S. Manufacturers specify B_S (and the corresponding value of H_S). It is about 1.7Tesla for Si-steels, about 0.7Tesla for permalloy, and about 0.3Tesla for ferrites. When saturated, the permeability of the core approaches that of air and a significant portion of the magnetic flux leaks through the air. For this reason, there is a limiting value of Ampere-turns that can be applied to an inductor's winding. The limiting value is given by

$$Ni_{\max} = \frac{B_S}{\mu_0} \cdot l_m. \tag{2.20}$$

The maximum energy that can be stored in the magnetic core is

$$\begin{aligned} \mathcal{E} &= \frac{1}{2} L i_{\max}^2 \\ &= \frac{1}{2} \frac{(Ni_{\max})^2}{\mathcal{R}} \\ &= \frac{1}{2} B_S^2 \frac{A_m l_m}{\mu_m}. \end{aligned}$$

2.5 Losses in Magnetic Devices

Consider a ferromagnetic core and its associated winding. If the exciting current is alternating, then the magnetic flux and its density B are alternating quantities. The operating point in the $B - H$ plane will traverse a hysteresis loop once every $1/f_S$ seconds, where f_S is the frequency of the exciting source. Hysteresis loss per unit volume is thus the energy per unit volume required for one complete traversal of the loop multiplied by the frequency f_S. That is, the hysteresis loss is given by [3, 5]

$$P_H = 2\, f_S V_m \left(\int_{-B_r}^{B_m} H\, dB - \int_{B_r}^{B_m} H\, dB \right), \tag{2.21}$$

where V_m is the volume of the core, B_m is the peak flux density at the loop's extremity, and B_r is the remanent flux density at the crossing of the loop and the B axis.

Another consequence of the alternating exciting current are the eddy currents that are induced in the core. The magnitude of the eddy currents will depend on the frequency of the exciting current, the magnitude of the flux density, and the resistivity of the core material. The power dissipated by the eddy currents is a loss. The eddy current losses may be reduced by using magnetic cores built with thin laminations and by using magnetic materials having high resistivity [2, 3, 5]. The total core loss is the sum of the hysteresis loss and the eddy current loss. At low frequencies, core loss is mainly due to hysteresis. Eddy current loss overtakes the hysteresis loss at high frequencies, i.e., $200-300\,\mathrm{kHz}$ for ferrites and about $300\,\mathrm{Hz}$ for metal alloys.

The inductor current in power electronic circuits is usually an alternating component i_a, which is half the peak-to-peak current ripple, riding over a DC bias current I_0. The DC current I_0 does not generate losses in the core. The core loss is due to the alternating current i_a. The flux densities established in the core by I_0 and i_a are

$$B_0 = \frac{\mu_0 N I_0}{(l_g + l_m/\mu_r)} \tag{2.22}$$

$$B_a = \frac{\mu_0 N i_a}{(l_g + l_m/\mu_r)}, \tag{2.23}$$

where N is the number of turns in the inductor winding. The peak flux density in the core is $B_m = B_0 + B_a$. The core loss data is usually supplied by the manufacturer in the form of a log-log plot showing the loss in Watts/kg as a function of B_a, with frequency as a parameter. Some manufacturers of magnetic cores provide the loss data in the form of an equation:

$$\text{Loss in Watts/kg} = k \cdot f_S^m \cdot B_a^n. \tag{2.24}$$

The parameters $(k,\ m,\ n)$ are provided by the manufacturer in the form of tables.

2.5.1 The skin effect in windings

Consider an alternating current flowing in a conducting wire. At low frequencies, the current distributes itself uniformly over the cross section from the conductor's surface to its center. The current tends to concentrate nearer to the surface as the frequency increases. This is the skin effect [4, 5]. The penetration depth δ in cms, is the distance from the conductor surface to the radius in the interior at which the current density falls to 36.8% of the current density at the surface.

The penetration depth is given by

$$\delta = \sqrt{\frac{\rho}{\mu_0 \mu_r \pi f}}, \tag{2.25}$$

where ρ is the conductor resistivity in $\Omega \cdot m$, $\mu_0 = 4\pi * 10^{-7}\text{H/m}$ is the permeability of free space, μ_r is the relative permeability of copper, and f is the frequency of the current in Hz. Since the resistivity is temperature dependent, so too is the penetration depth. For copper wires,

$$\begin{aligned} \delta &= 6.608/\sqrt{f}\text{ cm at } 25\ ^\circ\text{C} \\ &= 7.6328/\sqrt{f}\text{ cm at } 100\ ^\circ\text{C}. \end{aligned}$$

In a converter operating in the continuous conduction mode, the filter inductor's current is a direct current I_0 with an alternating ripple ΔI. If $\Delta I \ll I_0$, the skin effect may not be significant. When the converter operates in the discontinuous mode, then the alternating ripple becomes significant and the skin effect should be taken into account in the design of the inductor's winding. The skin effect should also be taken into account in the design of the windings of high-frequency transformers.

Suppose δ is the penetration depth in a copper wire at a frequency of f Hz. Then, the limiting area of a wire that can be used at this frequency is $A_s = \pi\delta^2$. Let $A_{w(b)}$ be the required bare-wire area for a given application. The area $A_{w(b)}$ can be obtained by using several parallel strands of the wire whose cross-sectional area is A_s. The number of strands in parallel is given by $n_{str} = A_{w(b)}/A_s$. Twisting the several strands will not ensure the correct distribution of the current in the parallel strands. Each of the parallel strands should have the same induced voltage integrated along its length. Equal distribution of the total current in the parallel strands is achieved in the Litz wire that consists of a large number of fine wires (such as the AWG40) correctly woven together.

2.6 The Air Gap and Its Effects

It may be necessary in some situations to provide an air gap in the magnetic circuit [5, 6]. The air gap prevents saturation of the core and ensures that the magnetic circuit is approximately linear. This is particularly true for ferrites that have a low saturation flux density and high permeability. A gap of about 1 mm may be cut in a ferromagnetic core and material such as paper, mylar, or glass introduced into the gap. Another manufacturing technique is to physically cut a gap-less toroid into two halves, lap and acid-etch the cut surface to remove the burrs, polish the cut surfaces, and band the two halves together. In this manner, an air gap of a few μm is inserted into the magnetic circuit. The air gap may also be distributed as in powder cores. MPP core is an example. The alloy in the form of hot-rolled cast ingots is ground to a fine powder and the particles are insulated and finely screened. The losses in MPP cores are very low.

Consider the magnetic circuit shown in Figure 2.3. Neglecting leakage and fringing the reluctances involved in this circuit are $\mathcal{R}_m = l_m/(\mu_0\mu_r A_m)$ and $\mathcal{R}_g = l_g/(\mu_0 A_m)$. The total reluctance of the circuit is

$$\begin{aligned}\mathcal{R}_t &= l_m/(\mu_0\mu_r A_m) + l_g/(\mu_0 A_m) \\ &= \frac{1}{\mu_0 A_m}\left(\frac{l_m}{\mu_r} + l_g\right). \end{aligned} \tag{2.26}$$

The contribution of the air gap to the total reluctance is much higher than that of the ferromagnetic core. As an example, suppose $l_m = 10$ cms, $\mu_r = 2500$ (typical of ferrites), and $l_g = 1$ mm. Then, $l_m/\mu_r = 0.04$ mm, whereas $l_g = 1$ mm.

Defining the effective relative permeability μ_e such that $\mathcal{R}_t = \frac{l_m + l_g}{\mu_0\mu_e A_m}$, we get

$$\frac{l_m + l_g}{\mu_0\mu_e A_m} = \frac{1}{\mu_0 A_m}\left(\frac{l_m}{\mu_r} + l_g\right). \tag{2.27}$$

That is,

$$\mu_e = \mu_r \frac{1 + l_g/l_m}{1 + \mu_r \cdot l_g/l_m}. \tag{2.28}$$

Putting $l_g/l_m = 0.01$ and $\mu_r = 2500$, we get $\mu_e = 97.115$. The air gap has a de-magnetizing effect that results in inclining the hysteresis loop as shown in Figure 2.6.

Another important effect of inserting an air gap in the magnetic core is the fringing effect. Fringing flux appears whenever an air gap is introduced in the magnetic path (as shown in Figure 2.7). It leaves the magnetic core slightly above the air gap and re-enters the core after bypassing the air gap. The effect is apparently to decrease the total reluctance of the magnetic path. The amount of fringing depends on the length of the gap and the shape of the cut surfaces. Fringing may be taken into account by means of a fringing factor given by [4, 7].

$$F = \left(1 + \frac{l_g}{\sqrt{A_m}} \ln\frac{2G}{l_g}\right), \tag{2.29}$$

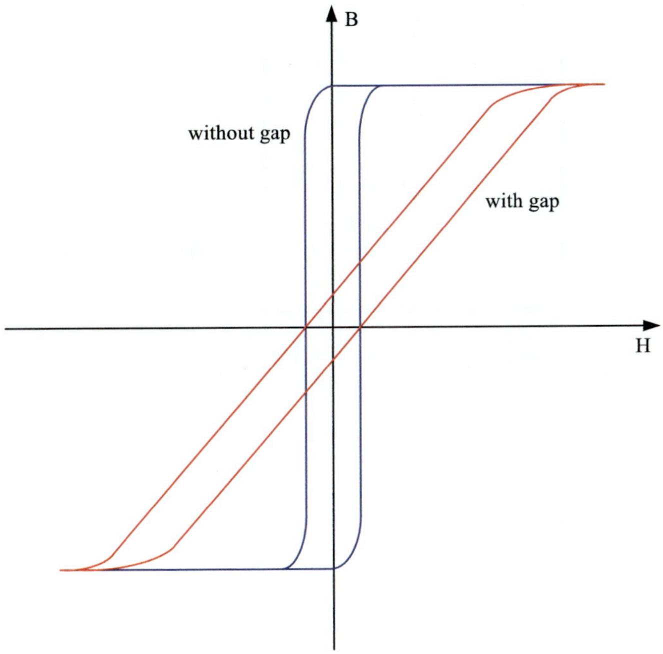

Figure 2.6 Hysteresis loop for gapped and ungapped cores.

where G is the length of the winding. To take fringing into account, substitute $(l_g + l_m/\mu_r)$ by $\frac{1}{F} \cdot (l_g + l_m/\mu_r)$. Thus, when fringing is included, the inductance of a ferromagnetic core inductor becomes

$$L = \frac{F \mu_0 N^2 A_m}{(l_g + l_m/\mu_r)}. \tag{2.30}$$

Care must be taken to prevent the fringing flux from linking with conducting material such as the inductor's winding. Any such linkage will induce eddy currents in the conductor and contribute to some additional eddy current loss.

2.7 Inductor and Transformer Design

Design of inductors and transformers may be based on the area product or the core geometry factor. These important parameters are derived below for inductors and transformers [4, 7].

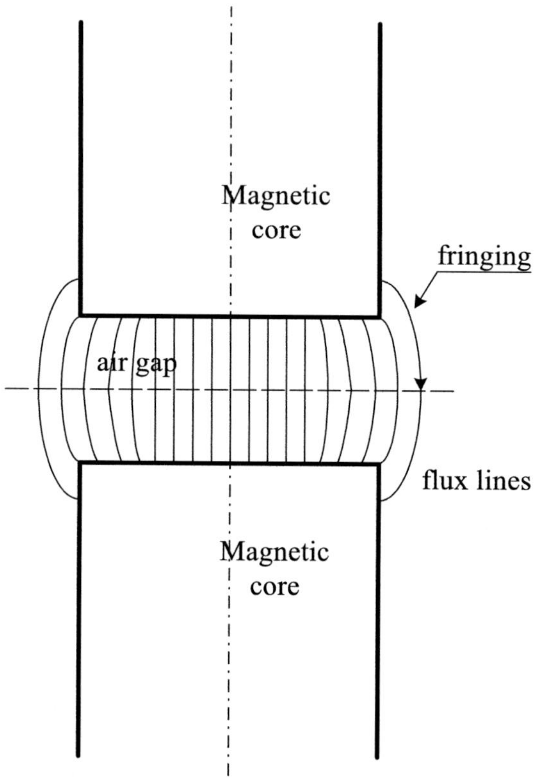

Figure 2.7 Fringing of magnetic flux at an airgap.

2.7.1 The area product for inductors

Consider the ferromagnetic core shown in Figure 2.2. The reluctance of the magnetic circuit is given in eqn (2.26), i.e.,

$$\mathcal{R} = \frac{1}{\mu_0 A_m}\left(\frac{l_m}{\mu_r} + l_g\right). \tag{2.31}$$

The inductance is given by

$$L = N^2 \cdot \frac{\mu_0 A_m}{(l_g + l_m/\mu_r)}. \tag{2.32}$$

The maximum flux density B_m (which is well below the saturation flux density) is given by

$$B_m = \frac{N\,\mu_0 I_m}{(l_g + l_m/\mu_r)}. \tag{2.33}$$

Then,

$$I_m = \frac{B_m \left(l_g + l_m/\mu_r\right)}{N\,\mu_0}. \tag{2.34}$$

The maximum energy stored in the core is

$$\begin{aligned} \mathcal{E} &= \frac{1}{2} L\, I_m^2 \\ &= \frac{1}{2} B_m^2 A_m \left(\frac{l_g + l_m/\mu_r}{\mu_0}\right). \end{aligned} \tag{2.35}$$

Now, $N I_m = K_u W_a J$, where K_u is the window utilization factor, W_a is the core's window area, and J is the current density. Therefore,

$$\begin{aligned} \frac{K_u W_a J}{N} = I_m &= \frac{B_m}{\mu_0 N}\left(l_g + l_m/\mu_r\right) \\ \text{or,}\ \frac{K_u W_a J}{B_m} &= \frac{1}{\mu_0}\left(l_g + l_m/\mu_r\right). \end{aligned} \tag{2.36}$$

Using eqn (2.35), we get

$$\frac{K_u W_a J}{B_m} = \frac{2\mathcal{E}}{B_m^2 A_m}. \tag{2.37}$$

The area product A_P is

$$A_P = A_m W_a = \frac{2\mathcal{E}}{K_u B_m J}. \tag{2.38}$$

The area product obtained from eqn (2.38) has been obtained solely from the design specifications such as B_m, J, L, I_m, etc. On the other hand, manufacturers provide the area product of their magnetic cores in their data sheets. Consequently, a core is chosen that has an area product greater than that obtained in eqn (2.38).

2.7.2 The inductor's core geometry factor K_g

Inductors may also be designed for a specified regulation. Consider the filter inductor in a converter circuit shown in Figure 2.8. The inductor current is the sum of a direct component I_0 and an alternating component i_a. If the output voltage is V_0, then the power delivered to the load is $P_0 = V_0 I_0$. The regulation α expressed as a fraction is given by

$$\begin{aligned}\alpha &= \frac{R_L I_0}{V_0} = \frac{R_L I_0^2}{V_0 I_0} \\ &= \frac{P_{cu}}{P_0}.\end{aligned} \tag{2.39}$$

From eqn (2.32) and an equation similar to eqn (2.33), we get

$$\frac{L}{B_0} = \frac{N A_m}{I_0}, \tag{2.40}$$

where B_0 is the flux density set up by the current I_0. From eqn (2.40), we get

$$N = \frac{L I_0}{B_0 A_m}. \tag{2.41}$$

The resistance of the winding is

$$R = \frac{N\rho(MLT)}{A_{cu}}, \tag{2.42}$$

where the electrical conductivity of copper $\rho = 1.724 * 10^{-8}\Omega \cdot \text{m}$, MLT is the mean length of a turn, and A_{cu} is the cross-sectional area of the bare wire. It is given by $A_{cu} = K_u W_a / N$. Substituting eqn (2.41), and (2.42) into eqn (2.39), we get

$$\alpha = \frac{I_0}{V_0} \cdot \frac{\rho(MLT)}{K_u W_a} \cdot \left(\frac{L I_0}{B_0 A_m}\right)^2. \tag{2.43}$$

The energy stored in the magnetic field by I_0 is $\mathcal{E} = \frac{1}{2} L I_0^2$. Substituting in eqn (2.43), we get

$$\begin{aligned}\mathcal{E}^2 &= \frac{P_0 B_0^2}{4\rho} \cdot \frac{K_u W_a A_m^2}{MLT} \cdot \alpha \\ &= K_e \cdot K_g \cdot \alpha.\end{aligned} \tag{2.44}$$

In eqn (2.44),

$$\text{electrical constraints constant } K_e = \frac{P_0 B_0^2}{4\rho}$$

$$\text{core geometry factor } K_g = \frac{K_u W_a A_m^2}{MLT} \text{ in m}^5.$$

Finally,

$$K_g = \frac{\rho\,(2\mathcal{E})^2}{P_0\,B_0^2\,\alpha}. \tag{2.45}$$

Manufacturers provide the core geometry factor for the magnetic cores in their data sheets. Choose a core that has its core geometry factor greater than that obtained from eqn (2.45).

2.7.3 The transformer's area product

A transformer has several windings. All windings need to be accommodated in a fraction of the core's window. The space in the window allocated to each winding is proportional to the power handled by the winding. Applying Faraday's law to the primary side of the transformer, we get

$$v_p = K_v N_p f_S A_m B_m, \tag{2.46}$$

where f_S is the switching frequency, N_p is the number of primary turns, A_m is the cross-sectional area of the ferromagnetic core, and B_m is the maximum flux density in the core. The constant K_v depends on the inductor's voltage waveform. It may be sinusoidal or periodic pulse-width modulated, as in the case of the filter inductor in the circuit of Figure 2.8. The constant K_v is:

$$\begin{aligned} K_v &= \sqrt{2}\cdot\pi \text{ for sinusoidal voltage waveform,} \\ &= \frac{2}{d} \text{ for pulse-width, modulated waveform, } d \text{ being the duty ratio.} \end{aligned}$$

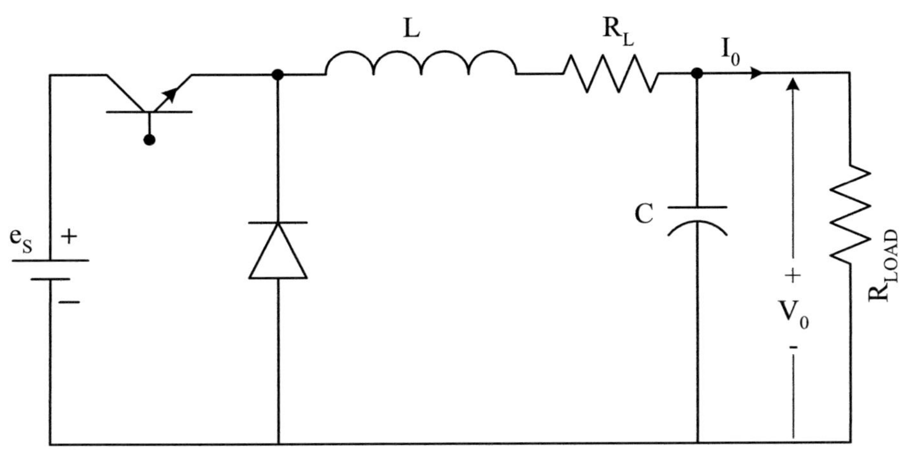

Figure 2.8 Filter inductor in a converter circuit.

When the duty ratio is 0.5, i.e., for a square wave, $K_v = 4$. Note that v_p is the RMS value of the primary voltage.

From eqn (2.46), we get

$$N_p = \frac{v_p}{K_v f_S A_m B_m}. \tag{2.47}$$

The window utilization factor K_u is usually set between 0.25 and 0.4. That leaves sufficient space for the insulation and bobbins. Thus,

$$K_u W_a = N_p A_{w(p)} + N_s A_{w(s)}, \tag{2.48}$$

where $A_{w(p)}$, $A_{w(s)}$ are the bare wire cross-sectional areas of the primary and secondary windings. It is usual to design the transformer so that the current densities in the two windings are equal. Consequently,

$$\begin{aligned} K_u W_a &= 2N_p A_{w(p)} \\ &= 2N_p i_p / J. \end{aligned} \tag{2.49}$$

Using eqn (2.46), we obtain

$$K_u W_a = \frac{2\, v_p i_p}{K_v f_S A_m B_m J}. \tag{2.50}$$

The area product for the transformer $A_P = W_a A_m$ is obtained from eqn (2.50) as

$$\begin{aligned} A_P &= \frac{2\, v_p i_p}{K_v K_u f_S B_m J} \\ &= \frac{P_{\text{tot}}}{K_v K_u f_S B_m J}. \end{aligned} \tag{2.51}$$

In eqn (2.51), the total power has been taken as $P_{\text{tot}} = 2P_{\text{in}}$. Actually, P_{tot} is the sum of the input power P_{in} and the output power P_o. The output power is a design specification. It is usual to assume that the input power is $P_{\text{in}} = P_o/\eta$, where η is the power transfer efficiency. Then,

$$P_{\text{tot}} = P_o \left(\frac{1}{\eta} + 1 \right). \tag{2.52}$$

The peak flux density B_m should be well below the saturation flux density. Manufacturers provide the saturation flux density of their magnetic cores. Based on this information, a reasonable value for B_m may be specified.

Another variable that is required is the current density J. A reasonable value is J $\leqslant$ 200 Amps/cm^2. The RHS of eqn (2.51) may now be evaluated and this value is the minimum area product of the core that is required for the application. From the manufacturer's data sheets, choose a magnetic core that has an area product greater than the value obtained by eqn (2.51).

2.7.4 Core geometry factor K_g for transformers

Consider a two-winding transformer with N_p primary turns and N_s secondary turns. Let a voltage v_p be applied to the primary winding and v_s be the induced secondary voltage. The regulation α is given by

$$\begin{aligned}\alpha &= \frac{\Delta v_p}{v_p} + \frac{\Delta v_s}{v_s} \\ &= \frac{R_p i_p}{v_p} + \frac{R_s i_s}{v_s}.\end{aligned}$$

Assuming that $\frac{R_p i_p}{v_p} = \frac{R_s i_s}{v_s}$, we obtain

$$\begin{aligned}\alpha &= \frac{2\,R_p i_p}{v_p} \\ &= \frac{2\,R_p i_p}{v_p} \cdot \frac{v_p}{v_p} \\ &= \frac{2\,R_p\,P_{\text{in}}}{v_p^2}.\end{aligned} \tag{2.53}$$

In eqn (2.53), R_p is the resistance of the primary winding and P_{in} is the volt-amps delivered to the primary.

As shown previously,

$$\begin{aligned}R_p &= \frac{N_p^2\,(MLT)}{(K_u W_a)/2} \cdot \rho \\ v_p &= K_v N_p A_m f_S B_m \qquad \text{from Faraday's law.}\end{aligned}$$

Substituting for R_p and v_p in eqn (2.53) and re-arranging, we get

$$\begin{aligned}P_{\text{in}} &= \frac{(K_v f_S N_p A_m B_m)^2}{2 \cdot \frac{N_p^2\,(MLT)}{(K_u W_a)/2} \cdot \rho} \\ &= \frac{(K_v f_s B_m)^2}{4\,\rho} \cdot \frac{K_u W_a A_m^2}{MLT} \cdot \alpha.\end{aligned} \tag{2.54}$$

$$\text{Putting } K_e = \frac{(K_v f_s B_m)^2}{4\,\rho}$$
$$K_g = \frac{K_u W_a A_m^2}{MLT},$$

we obtain from eqn (2.54),

$$K_g = \frac{P_{\text{in}}}{K_e\,\alpha}. \tag{2.55}$$

The parameter K_e is determined by the specified electromagnetic operating conditions. The core geometry factor K_g is determined by the core dimensions and it has been tabulated by the manufacturers. Choose a core that has its K_g factor greater than that obtained by eqn (2.55).

2.8 Design Examples

Refer to [7] for magnetic core data. The design examples follow the procedures outlined in [4].

2.8.1 Isolation transformer

The specifications for the isolation transformer are:

$$\begin{aligned}
\text{Transformer ratings} &: \text{500 VA, 220 V, 50 Hz} \\
\text{Frequency} &= 50 \\
\text{Turns ratio} &= 1.0 \\
\text{Stacking factor} &= 0.9 \\
\text{Efficiency of power transfer} &= 0.95 \\
\text{Window utilization factor } K_v &\simeq 0.4 \\
\text{Material} &: 3\%\ \text{Si},97\%\ \text{Fe steel, 1 mil lamination} \\
B_{\max} &= 1.1\ \text{Tesla} \\
J_{\max} &= 200\ \text{Amps/cm}^2
\end{aligned}$$

$$\begin{aligned}
A_P &= \frac{P_{\text{tot}}}{K_v K_u f B_{\max} J_{\max}} \\
&= \frac{500 * (1 + 0.95^{-1})}{4.44 * 0.4 * 50 * 1.1 * 200 * 10^{-4}} \\
&= 525\ \text{cm}^4
\end{aligned}$$

$$
\begin{aligned}
\text{Choose the 19EI1 core. Area product} &= 669 \text{ cm}^4. \\
\text{The core's } Wt_{fe} \text{ is reasonably low.} & \\
\text{Principal dimensions are shown in Figure 2.9.} & \\
\text{The other parameters of the core are:} & \\
\text{core cross section } A_m &= 19.758 \text{ cm}^2 \\
\text{window area } W_a &= 33.871 \text{ cm}^2 \\
\text{mean path length } MPL &= 32.6 \text{ cm} \\
\text{area product } A_P &= 669 \text{ cm}^4 \\
\text{core geometry factor } K_g &= 162.5 \text{ cm}^5 \\
\text{core weight } Wt_{fe} &= 4.889 \text{ kg} \\
\text{surface area } At &= 1066.9 \text{ cm}^2 \\
\text{Using the stacking factor (SF)} = 0.9, &
\end{aligned}
$$

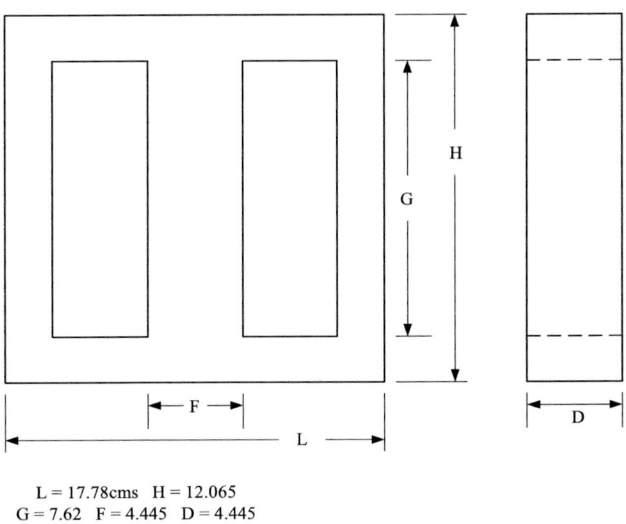

Figure 2.9 The 19EI-1 core for the 500 VA, 50 Hz isolation transformer.

$$
\begin{aligned}
\text{Effective area product } A_P &= 0.9 * 669 \\
&= 602 \text{ cm}^4 \\
\text{Effective iron area } A_m &= 0.9 * 19.758 \\
&= 17.78 \text{ cm}^2
\end{aligned}
$$

$$\begin{aligned}
\text{No. of primary turns } N_p &= \frac{v_p}{K_v * f * B_m * A_m} \\
&= \frac{220}{4.44 * 50 * 1.1 * 17.78 * 10^{-4}} \\
&= 507 \\
\text{Primary current } I_p &= \frac{500}{0.95 * 220} = 2.39\ Amps \\
\text{Current density } J &= \frac{P_{\text{tot}}}{K_v K_u f B_m A_P} \\
&= \frac{500 * (1 + 0.95^{-1})}{4.44 * 0.4 * 50 * 1.1 * 602 * 10^{-8}} \\
&= 175\ \text{Amps/cm}^2 \\
\text{Bare wire area for primary winding} &= \frac{2.39}{175} \\
&= 0.013657\ \text{cm}^2
\end{aligned}$$

Choose $AWG16$ for the primary and secondary windings.

$$\begin{aligned}
A_{w(b)} &= 0.01307\ \text{cm}^2 \\
\text{Specific resistance} &= 131.8\ \mu\Omega/\text{cm} \\
\text{Diameter with insulation} &= 0.137\ cm \\
\text{Primary winding resistance} &= MLT * N_p * \mu\Omega/\text{cm} * 10^{-6} \\
&= 32.6 * 507 * 131.8 * 10^{-6} \\
&= 2.178\ \Omega \\
\text{Primary copper loss } P_{cu,p} &= 2.39^2 * 2.178 \\
&= 12.44\ \text{Watts} \\
\text{Secondary copper loss } P_{cu,s} &= \text{ Primary copper loss} \\
&= 12.44\ \text{Watts} \\
\text{Effective core weight} &= SF * Wt_{fe} \\
&= 0.9 * 4.889 \\
&= 4.4\ \text{kg} \\
\text{Iron loss for 3\% Si, 1 mil lamination} &= 5.93 * 10^{-2} * f^{0.993} * B_m^{1.74} \\
&= 5.93 * 10^{-2} * 50^{0.993} * 1.1^{1.74} \\
&= 3.405\ \text{Watts/kg} \\
\text{Iron loss} &= 3.405 * 4.4
\end{aligned}$$

$$
\begin{aligned}
&= 14.96 \text{ Watts} \\
\text{Total loss} &= 2 * 12.44 + 14.96 \\
&= 39.84 \text{ Watts} \\
\text{Power dissipated per unit surface area} &= \frac{39.84}{1067} \\
&= 0.0373 \text{ Watts/cm}^2 \\
\text{Estimate of temperature rise} &= 450 * (0.0373)^{0.826} \\
&= 29.75\,^{\circ}\text{C} \\
\text{Transformer regulation} &= \frac{100 * 2 * 12.44}{500 + 2 * 12.44} \\
&= 4.74\% \\
\text{Window utilization factor} &= \frac{2 * 507 * 0.01307}{33.87} \\
&= 0.39.
\end{aligned}
$$

The first iteration of the design has been concluded. The final design can be arrived at only many iterations. In the next iteration, the designer may try to reduce the regulation of 4.74% by substituting the 19EI-1 core by the 175EI-2 core. The design calculations for the second iteration are given below.

The dimensions and other parameters of the 175EI-2 core are

$$
\begin{aligned}
D = 8.89 \text{ cms},\ H = 11.113 \text{ cms}, W &= 13.335 \text{ cms}, \\
L = 13.335 \text{ cms},\ G = 6.67 \text{ cms},\ \text{MPL} &= 26.7 \text{ cms}, \\
\text{MLT} = 34.9 \text{ cms},\ A_m = 39.516\ cm^2,\ W_a &= 14.819,\ \text{cm}^2, \\
A_P = 585.57 \text{ cm}^4,\ K_g = 265.4767 \text{ cm}^5, At &= 869.4 \text{ cm}^2, \\
Wt_{fe} &= 7.812 \text{ kg} \\
\text{Assumed value of stacking factor SF} &= 0.9 \\
\text{Effective area product } A_P &= 527 \text{ cm}^4 \\
\text{Effective iron area } A_m &= 35.564 \text{ cm}^2 \\
\text{Effective core weight } Wt_{fe} &= 7.0308 \text{ kg} \\
\text{Number of primary turns } N_p &= 253 \\
\text{Primary current } I_p &= 2.3923 \text{ Amps} \\
\text{Current density } J &= 199 \text{ Amps/cm}^2 \\
\text{Bare wire area for primary winding} &= 12.01 * 10^{-3} \text{ cm}^2
\end{aligned}
$$

Choose $AWG16$ for the primary and secondary windings:

$$\begin{aligned}
\text{Effective iron area } A_m &= 35.564 \text{ cm}^2 \\
\text{Effective core weight } Wt_{fe} &= 7.0308 \text{ kg} \\
\text{Number of primary turns } N_p &= 253 \\
\text{Primary current } I_p &= 2.3923 \text{ Amps} \\
\text{Current density } J &= 199 \text{ Amps/cm}^2 \\
\text{Bare wire area for primary winding} &= 12.01 * 10^{-3} \text{ cm}^2
\end{aligned}$$

Choose $AWG16$ for the primary and secondary windings:

$$\begin{aligned}
A_{w(b)} &= 0.01307 \text{ cm}^2 \\
\text{Specific resistance} &= 131.8\ \mu\Omega/\text{cm} \\
\text{Diameter with insulation} &= 0.137 \text{ cm} \\
\text{Primary winding resistance} &= 1.1638\ \Omega \\
\text{Primary copper loss } P_{cu,p} &= 6.66 \text{ Watts} \\
\text{Secondary copper loss } P_{cu,s} &= 6.66 \text{ Watts} \\
\text{Iron loss for 3\% Si, 1 mil lamination} &= 5.97 * 10^{-3} * f^{1.26} * B_m^{1.73} \\
&= 0.169 \text{ Watts/kg} \\
\text{Iron loss} &= 1.188 \text{ Watts} \\
\text{Total loss} &= 14.508 \text{ Watts} \\
\text{Power dissipated per unit surface area} &= 0.0167 \text{ Watts/}cm^2 \\
\text{Estimate of temperature rise} &= 15.3\ ^\circ\text{C} \\
\text{Transformer regulation} &= 2.59\% \\
\text{Window utilization factor} &= 0.446.
\end{aligned}$$

The window utilization factor is higher than the specified value of 0.4, though it may be acceptable.

2.8.2 Filter inductor for a buck converter operating in CCM

The design specifications are:

$$\begin{aligned}
\text{Input voltage} &= 24 - 40 \text{ V} \\
\text{Nominal input voltage} &= 30 \text{ V} \\
\text{Output voltage} &= 5 \text{ V}
\end{aligned}$$

$$\begin{aligned}
\text{Output current} &= 5\text{ A}\\
\text{Peak-to-peak inductor current ripple } \Delta I &= 1\ \text{Amp}\\
\text{Peak inductor current } I_{pk} &= 5.5\text{ Amps}\\
\text{Switching frequency } f_S &= 200,000\text{ Hz}\\
\text{Nominal duty ratio } D &= 0.1667\\
\text{Inductance} &= 22\mu\text{ H}\\
\text{Regulation } \alpha &= 0.01.
\end{aligned}$$

The design procedure:

$$\begin{aligned}
\text{Maximum stored energy } \mathcal{E} &= \frac{1}{2}LI_{pk}^2\\
&= 0.5 * 22 * 10^{-6} * 5.5^2\\
&= 0.33275 * 10^{-3}\text{Joules}\\
\text{Output power } P_0 &= 25\text{ Watts}\\
\text{Maximum flux density } B_m \text{ for}&\\
\text{ferrite core} &= 0.22\text{ Tesla}\\
\text{Flux density corresponding to DC}&\\
\text{offset current} &= 0.2\text{ Tesla}\\
\text{Window utilization factor } K_u &= 0.29\\
\text{Core geometry factor } K_g &= \frac{\rho(2\mathcal{E})^2}{P_0 B_0^2 \alpha}\ m^5\\
&= \frac{1.724 * 10^{-8} * (0.6655 * 10^{-3})^2}{25 * 0.2^2 * 0.01}\\
&= 0.007635\text{ cm}^5\\
\text{Choose Magnetics EE-42510 core: } K_g &= 0.0097\text{ cm}^5\\
A_P &= 0.312\text{ cm}^4\\
H = 1.905\text{ cm},\ \ W = 1.935\text{ cm},\ \ L &= 2.515\text{ cm}\\
G = 1.25\text{ cm},\ \ MPL = 4.9\text{ cm},\ \ MLT &= 4.9\text{ cm}\\
A_m = 0.384\text{ cm}^2,\ \ W_a = 0.813\text{ cm}^2,\ \ At &= 23.5\text{ cm}^2\\
Wt_{fe} = 9.5\text{ grams},\ \ \mu_r &= 2300\\
A_L &= 1325\text{ mH}/1000\text{ turns}
\end{aligned}$$

$$\text{Current density } J = \frac{2 * \mathcal{E}}{K_u * B_m * A_P} = \frac{0.6655 * 10^{-3}}{0.29 * 0.22 * 0.312 * 10^{-8}} = 334.3 \text{ Amps/cm}^2$$

$$\text{RMS value of the inductor current} = I_S * \sqrt{1 + \frac{(\Delta I/I_S)^2}{12}} = 5 * 1.001665 = 5.008 \text{ Amps}$$

$$\text{Required bare-wire area of conductor} = \frac{5.008}{334.3} = 15 * 10^{-3} \text{ cm}^2$$

Choose AWG 15 wire for the inductor winding:

$$A_{w(b)} = 16.51 * 10^{-3} \text{ cm}^2$$

$$\text{Wire resistance} = 104.3\ \mu\Omega/\text{cm}$$

$$\text{Diameter of insulated wire} = 0.153\,\text{cm}$$

$$\text{Effective window area } K_u * W_a = 0.29 * 0.813 = 0.2358 \text{ cm}^2$$

$$\text{Number of turns, } N = \frac{0.2358}{0.0165} = 14$$

$$\text{Required length of the air gap: } \left(l_g + \frac{\text{MPL}}{\mu_r}\right) = N^2 * \frac{\mu_0 * A_m}{L}$$

Or

$$\left(l_g + \frac{0.049}{2300}\right) = \frac{4\pi * 10^{-7} * 0.384 * 10^{-4} * 14^2}{22 * 10^{-6}}$$

$$\text{That is, } l_g = 0.04086 \text{ cm} = 16 \text{ mils}$$

8 mil-thick fish paper to be spread across the E core.

$$\text{Fringing flux factoF} = 1 + \frac{l_g}{\sqrt{A_m}} * \ln\frac{2G}{l_g} = 1 + \frac{0.04086}{0.6197} * \ln\frac{2 * 1.25}{0.04086}$$

$$
\begin{aligned}
&= 1.2713\\
\text{Number of turns including fringing: } N &= \sqrt{\frac{(l_g + MPL/\mu_r)L}{\mu_0 F * A_m}}\\
&= 12\\
\text{Winding resistance } R &= \text{MLT} * N * \frac{\mu\Omega}{\text{cm}} * 10^{-6}\\
&= 4.9 * 12 * 104.3 * 10^{-6}\\
&= 0.006133\\
\text{Copper loss} &= 0.006133 * 5.008^2\\
&= 0.1538 \text{ Watts}\\
\text{Regulation } \alpha &= \frac{0.1538}{25.1538}\\
&= 0.61\%\\
\text{Peak flux density } B_m &= \frac{\mu_0 N * i_{pk}}{(l_g + \text{MPL}/\mu_r)}\\
&= \frac{4\pi * 10^{-7} * 12 * 5.5}{0.0004299}\\
&= 0.193 \text{ Tesla}\\
\text{AC flux density, } B_{\text{AC}} &= \frac{4\pi * 10^{-7} * 12 * 0.5}{0.0004299}\\
&= 0.01754 \text{ Tesla}\\
\text{Iron loss in mWatts/gram} &= K_{fe} * f_S^{1.979} * B_{\text{AC}}^{2.628}\\
&= 8.097 * 10^{-7} * 200,000^{1.979}\\
&\quad *0.01754^{2.628}\\
&= 0.6086 \text{ mWatts/gram}\\
\text{Iron loss, } P_{fe} &= 9.5 * 0.6086 * 10^{-3}\\
&= 0.00578 \text{ Watts}\\
\text{Total loss} &= 0.1638 + 0.00578\\
&= 0.1596 \text{ Watts}\\
\text{Efficiency} &= \frac{25}{25.1596}\\
&= 99.37\%\\
\text{Power loss per unit surface area, } \psi &= \frac{0.596}{23.5}\\
&= 0.00679 \text{ Watts/cm}^2
\end{aligned}
$$

$$\text{Temperature rise} = 450 * (0.00679)^{0.826} = 7.28\ ^\circ\text{C}.$$

2.8.3 Three-winding transformer for a forward converter

The forward converter, which is the isolated form of the buck converter, is described in Chapter 7. In one version of the forward converter, a Three-winding transformer is used to reset the core flux density to zero at the start of each switching period. The transformer is provided with a primary winding, a secondary winding, and a reset winding. The design specifications are:

$$\begin{aligned}
\text{Input voltage, } V_i &= 165\ \text{V}\\
\text{Output voltage, } V_0 &= 3.3\ \text{V}\\
\text{Output current, } I_0 &= 11\ \text{A}\\
\text{Turns ratio, } N_p : N_s &= 15\\
\text{Turns ratio, } N_p : N_r &= 1\\
\text{Switching frequency, } f_S &= 50,000\ \text{Hz}\\
\text{Peak core flux density, } B_m &= 0.1\text{Tesla}\\
\text{Window utilization factor, } K_u &= 0.32\\
\text{Nominal duty ratio, } d &= 0.3\\
\text{Efficiency, } \eta &= 0.8\\
\text{Magnetizing inductance, } L_m &= 750\ \mu\text{H}\\
\text{Output filter inductance, } L_S &= 8.5\ \mu\text{H}\\
\text{Current density, } J &\simeq 250\ \text{A/cm}^2.
\end{aligned}$$

The voltages and currents of the primary, secondary, and reset windings are given in the following table:

Variable	$0 - d * T_S$	$d * T_S - 2d * T_S$	$2d * T_S - T_S$
V_P	165 V	−165 V	0
V_S	11 V	−11 V	0 V
V_R	165 V	−165 V	0 V
I_P	$I_P(0) = 0.5522$ A $I_P(d * t_S) = 2.2345$ A	0	0
I_S	$I_S(0) = 8.2824$ A $I_S(d * t_S) = 13.7176$ A	0	0
I_R	0	$I_R(d * t_S) = 1.32$ A $I_R(2d * t_S) = 0$ A	0

$$
\begin{aligned}
V_{P,\text{rms}} &= 165 * \sqrt{d} \\
&= 90.37 \text{ V} \\
V_{S,\text{rms}} &= 11 * \sqrt{d} \\
&= 6.0249 \text{ V} \\
V_{R,\text{rms}} &= 90.37 \text{ V} \\
I_{P0} &= 0.5 * (I_p(0) + I_P(d * T_S)) \\
&= 1.3934 \text{ A} \\
\text{Similarly, } I_{S0} &= 11.0 \text{ A} \\
I_{R0} &= 0.66 \text{ A} \\
\Delta I_S &= I_S(0) - I_S(d * T_S) \\
&= 5.4352 \text{ A} \\
\Delta I_R &= 1.32 \text{ A} \\
I_{S,\text{rms}} &= I_{S0} * \sqrt{d} * \sqrt{1 + \frac{(\Delta I_S / I_{S0})^2}{12}} \\
&= 6.0858 \text{ A} \\
I_{R,\text{rms}} &= 0.4174 \text{ A} \\
P_S &= 6.0249V * 6.0858A \\
&= 36.666 \text{ Watts} \\
P_R &= 37.724 \text{ Watts} \\
P_P &= 36.666 + 37.724 \\
&= 74.39 \text{ Watts} \\
P_{\text{tot}} &= 74.39 * (1 + 0.8^{-1}) \\
&= 167.35 \text{ Watts} \\
I_{P,\text{rms}} &= 74.39/90.37 \\
&= 0.823 \text{ A} \\
\text{Core geometry factor } K_g &= \frac{4\rho * V_{P0} I_{P0}}{\left(\frac{1}{d} * f_s * B_m\right)^2 * \alpha} \\
&= \frac{4 * 1.724 * 10^{-8} * 165 * 1.3934}{\left(\frac{1}{0.3} * 50000 * 0.1\right)^2 * 0.01} \text{ m}^5 \\
&= 0.05708 \text{ cm}^5
\end{aligned}
$$

Choose Magnetics EE-43515 core, $\mu_r = 2300$
$K_g = 0.062458 \text{ cm}^5 \quad A_P = 1.264 \text{ cm}^4$
$H = 2.83\ cm, \quad W = 2.53 \text{ cm}, \quad L = 3.454 \text{ cm}$
$G = 1.93\ cm, \quad MPL = 6.9 \text{ cm}, \quad MLT = 6.6 \text{ cm}$
$A_m = 0.821\ cm^2, \quad W_a = 1.54 \text{ cm}^2, \quad At = 23.5 \text{ cm}^2$

$$\begin{aligned}
\text{Current density, } J &= \frac{P_{\text{tot}}}{\left(\frac{1}{0.3} * K_u * f_S * A_m * B_m\right)} \\
&= 248 * 10^4 \text{ A/m}^2 \\
&= 248\ A/\text{cm}^2 \\
\text{Primary winding, bare wire area } A_{w(b)} &= \frac{0.823}{248} \\
&= 3.3185 * 10^{-3} \text{ cm}^2 \\
\text{Secondary winding, bare wire area } A_{w(b)} &= \frac{6.0855}{248} \\
&= 24.54 * 10^{-3} \text{ cm}^2
\end{aligned}$$

Skin effect:

$$\begin{aligned}
\text{Current penetration depth at 50 KHz} &= \frac{6.62}{\sqrt{50,000}} \\
&= 0.0296 \text{ cm} \\
\text{Maximum bare wire area} &= \pi * 0.0296^2 \\
&= 0.002754 \text{ cm}^2
\end{aligned}$$

Choose:
Two strands of AWG 25 for the primary winding
One strand of AWG 25 for the reset winding
Twelve strands of AWG 24 for the secondary winding

$$\begin{aligned}
\text{Length of air gap, } l_g &= \frac{\mu_0 * N_p^2 A_m}{L_m} - \frac{MPL}{\mu_r} \\
&= 0.195086 \text{ cm} \\
\text{Fringing flux factor, } F &= 1 + \frac{l_g}{\sqrt{A_m}} * \ln\left(\frac{2 * G}{l_g}\right) \\
&= 1.64267
\end{aligned}$$

$$
\begin{aligned}
\text{Revised Number of primary turns, } N_p &= \sqrt{\frac{L_m * l_g}{\mu_0 * F * A_m}} \\
&= 94 \\
\text{Number of secondary turns, } N_s &= 6 \\
\text{Peak flux density} &= \frac{\mu_0 N_p I_p (d * T_S)}{l_g + l_m/\mu_r} \\
&= \frac{4\pi * 10^{-7} * 94 * 2.2345}{0.00198086} \\
&= 0.133 \text{ Tesla} \\
\text{Alternating flux density} &= \frac{\mu_0 N_p * \Delta I_P/2}{l_g + l_m/\mu_r} \\
&= \frac{4\pi * 10-7 * 94 * 0.8412}{0.00198086} \\
&= 0.0502 \text{ Tesla} \\
\text{Primary winding resistance} &= MLT * N_p * 531 \frac{\mu\Omega}{\text{cm}} * 10^{-6} \\
&= 6.6 * 94 * 531 * 10^{-6} \\
&= 0.3294\ \Omega \\
\text{Primary copper loss} &= 0.3294 * 0.823^2 \\
&= 0.223 \text{ Watts} \\
\text{Secondary winding resistance} &= 6.6 * 6 * 70.175 * 10^{-6} \\
&= 0.002779\ \Omega \\
\text{Secondary copper loss} &= 0.002779 * 6.0858^2 \\
&= 0.1029 \text{ Watts} \\
\text{Total copper loss} &= 0.3259 \text{ Watts} \\
\text{Regulation} &= \frac{0.3259}{36.6259} \\
&= 0.89.
\end{aligned}
$$

2.8.4 Flyback transformer

The flyback converter (discussed in Chapter 7) is the isolated version of the buck–boost converter. The isolation is provided by a two-winding transformer that functions more like an inductor. Energy flowing in from the input source is stored in the magnetizing inductance during an initial interval of the switching period, i.e., during $0 \leq t \leq d * T_S$ where T_S is the switching period and d is the duty ratio. When the excitation on the primary side is interrupted at the end of the charging interval, the stored energy is transferred to the load during the remainder of the switching period.

The design specifications are:

$$\begin{aligned}
\text{Input voltage, } V_i &= 36 \text{ V} \\
\text{Output voltage, } V_0 &= 4.8 \text{ V} \\
\text{Output current, } I_0 &= 4.8 \text{ A} \\
\text{Turns ratio, } N_p : N_s &= 5 \\
\text{Switching frequency, } f_S &= 50000 \text{ Hz} \\
\text{Peak core flux density, } B_m &= 0.25 \text{ Tesla} \\
\text{Window utilization factor, } K_u &= 0.35 \\
\text{Nominal duty ratio, } d &= 0.4 \\
\text{Efficiency, } \eta &= 0.8 \\
\text{Magnetizing inductance, } L_m &= 0.9 \text{ mH} \\
\text{Current density, } J &\simeq 200 \text{ A/cm}^2
\end{aligned}$$

The design procedure is as follows:

$$\begin{aligned}
\text{Average primary current, } I_{L0} &= \frac{N_s}{N_p} * \frac{V_0}{(1-D)} \\
&= \frac{4.8}{5 * 0.6} \\
&= 1.6 \text{ A} \\
\text{Peak-to-peak magnetizing current ripple, } \Delta I_L &= 0.2 * 1.6 \\
&= 0.32 \text{ A} \\
\text{Peak value of primary current, } I_{p,max} &= 1.6 + 0.16 = 1.76 \text{ A} \\
\text{Minimum primary current, } I_{p,min} &= 1.6 - 0.16 = 1.44 \text{ A} \\
\text{Required value of the magnetizing inductance, } L_m &= \frac{V_i * d * T_S}{\Delta I_L} \\
&= 0.9 \text{ mH} \\
\text{Primary current's rms value, } I_{p,\text{rms}} &= 1.6 * \sqrt{0.4} * \\
&\quad \sqrt{1 + \frac{(0.32/1.6)^2}{12}} \\
&= 1.013614 \text{ A}
\end{aligned}$$

$$\begin{aligned}
\text{Peak energy stored in } L_m &= \frac{1}{2} * 0.9 * 10^{-3} * 1.76^2 \\
&= 0.001394 \text{ Joules} \\
\text{Power delivered to load, } P_0 &= 4.8 * 4.8 = 23.04 \text{ Watts} \\
\text{Core geometry factor, } K_g &= \frac{\rho * (2 * 0.001394)^2}{P_0 * B_m^2 * \alpha} \\
&= \frac{1.724 * 10^{-8} * (0.002788)^2}{23.04 * 0.25^2 * 0.01} \\
&= 93.0487 * 10^{-13} \text{ m}^5 \\
&= 0.093049 \text{ cm}^5
\end{aligned}$$

Choose Magnetics MPP core MP-55548:

$$\begin{aligned}
K_g = 0.124886 \text{ cm}^5,\ A_P &= 2.00597 \text{ cm}^4 \\
OD = 3.366 \text{ cm},\ ID &= 1.946 \text{ cm} \\
MPL = 8.15 \text{ cm},\ MLT &= 4.348 \text{ cm} \\
A_m = 0.6768 \text{ cm}^2,\ W_a &= 2.9639 \text{ cm}^2 \\
At = 56.598 \text{ cm}^2,\ \mu_r &= 125 \\
A_L = 127\ mH/1000\, turns,\ Wt_{fe} &= 47\ grams \\
\text{Core loss, mWatts/gram} &= 3.91 * 10^{-3} * (f)^{1.28} * (B_{ac})^{2.14} \\
\text{Number of primary turns, } N_p &= 1000 * \sqrt{\frac{0.9}{127}} \simeq 85 \\
\text{Secondary turns, } N_s &= 17 \\
\text{Current density, } J &= \frac{2\mathcal{E}}{K_u * A_P * B_m} \\
&= 158.84 \text{ A/cm}^2.
\end{aligned}$$

The current density recommended for this core is low.
It is quite safe to use $J = 200$ A/cm^2.

$$\text{Primary winding bare wire area, } A_{w(b),p} = \frac{1.0136}{200} = 0.005068 \text{ cm}^2$$

Use 2 strands of AWG 23:

$$\begin{aligned}
\text{Specific resistance} &= 333\ \mu\Omega/\text{cm} \\
\text{Wire diameter with insulation} &= 0.0632 \text{ cm} \\
\text{Secondary winding bare wire area, } A_{w(b),s} &= \frac{5.068}{200} = 0.02534 \text{ cm}^2
\end{aligned}$$

$$
\begin{aligned}
\text{Number of strands of AWG 23}& \\
\text{wire for secondary winding} &= 10 \\
\text{Primary winding resistance} &= 4.348 * 85 * 333 * 10^{-6} \\
&= 0.12307\ \Omega \\
\text{Primary copper loss} &= 1.0136^2 * 0.12307 \\
&= 0.1264 \text{ Watts} \\
\text{Secondary winding resistance} &= 4.348 * 17 * 66.6 * 10^{-6} \\
&= 0.004923\ \Omega \\
\text{Secondary copper loss} = 5.068^2 * 0.004923& \\
&= 0.12645 \text{ Watts} \\
\text{Total copper loss} &= 0.2529 \text{ Watts} \\
\text{Regulation, } \alpha &= \frac{0.2529}{23.2929} \\
&= 0.01 = 1\%
\end{aligned}
$$

AWG 23 turns crossing the window area:

$$
\begin{aligned}
\text{Primary, } 2 * 85 &= 170 \\
\text{Secondary, } 10 * 17 &= 170 \\
\text{Total} &= 340 \\
\text{Total bare wire area in the window} &= 340 * 0.002508 \\
&= 0.8527 \text{ cm}^2 \\
\text{Window utilization factor, } K_u &= \frac{0.8527}{2.9639} \\
&= 0.288 \\
\text{Required relative permeability} &= \frac{MLP * B_m}{\mu_0 * N_p * I_{p,\max}} \\
&= \frac{8.15 * 10^{-2} * 0.25}{4\pi * 10^{-7} * 85 * 1.76} \\
&= 108.4 \text{ (less than 125)} \\
\text{Peak flux density, } B_m &= \frac{4\pi * 10^{-7} * 125 * 85 * 1.76}{8.15 * 10^{-2}} \\
&= 0.29 \text{ Tesla}
\end{aligned}
$$

$$
\begin{aligned}
\text{Alternating flux density, } B_{ac} &= \frac{4\pi * 10^{-7} * 125 * 85 * 0.16}{8.15 * 10^{-2}} \\
&= 0.0262 \text{ Tesla} \\
\text{Iron loss in mWatts/gram} &= 0.00391 * (50,000)^{1.28} \\
&\quad *(0.0262)^{2.14} \\
&= 0.00391 * 1.03438 * 10^{6} \\
&\quad *0.4123 * 10^{-3} \\
&= 1.6675 \\
\text{Iron loss, } P_{fe} &= 47 * 1.6675 \\
&= 0.0784 \text{ Watts} \\
\text{Total power loss} &= 0.3313 \text{ Watts} \\
\text{Power dissipated per } \mathrm{cm}^2 \text{ of surface area} &= \frac{0.3313}{56.598} \\
&= 0.00585 \text{ Watts/cm}^2 \\
\text{Estimated temperature rise} &= 450 * (0.00585)^{0.826} \\
&= 6.44\,^{\circ}\mathrm{C}.
\end{aligned}
$$

References

[1] W. H. Hayt, *Engineering Electromagnetics*, 5th Edition, 1988, McGraw-Hill Inc., New York.

[2] P. T. Krein, *Elements of Power Electronics,* International 2nd Ed., 1997, Oxford University Press.

[3] MIT EEStaff, *Magnetic Circuits and Transformers,* 1st Edition, 1965, MIT Press, Cambridge, Massachusetts.

[4] W. T. McLyman, *Transformer and Inductor Design Handbook,* 2nd Edition, 1988, Marcel Dekker Inc., New York.

[5] L. H. Dixon, *Magnetics Design Handbook,* Texas Instruments Literature SLUP132, 2001, Dallas, Texas.

[6] J. G. Kassakian, M. F. Schlecht, G. C. Verghese, *Principles of Power Electronics*, 1^{st}Ed., 2009, Pearson India.

[7] W. T. McLyman, *Magnetic Core Selection for Transformers and Inductors: A User's Guide to Practice and Specification,* 2nd Edition, 1997, Marcel Dekker Inc., New York.

3

Half-wave Converters

Half-wave, AC/DC converters have been in use since the days of vacuum tube electronics. They are the simplest among power electronic circuits and they admit analytic solutions. In this chapter, analysis of half-wave converters are presented and the simulation of these circuits is described.

3.1 Half-wave Converter with Resistive Load

Figure 3.1 shows the diode-based, half-wave converter circuit with sinusoidal excitation and a resistive load. The diode is forward biased for $\omega t > 0$ and it becomes conductive. Considering an ideal diode, the source voltage is applied across the load. For $\omega t > \pi$, the diode is reverse biased and it becomes non-conductive. The source voltage appears across the diode. Figure 3.2 shows the voltage waveforms across the components of the circuit.

The mean value of the load voltage is given by

$$V_{0,\text{mean}} = \frac{1}{2\pi}\int_0^{\pi} V_m \sin\omega t \, d\omega t = \frac{V_m}{2\pi}. \tag{3.1}$$

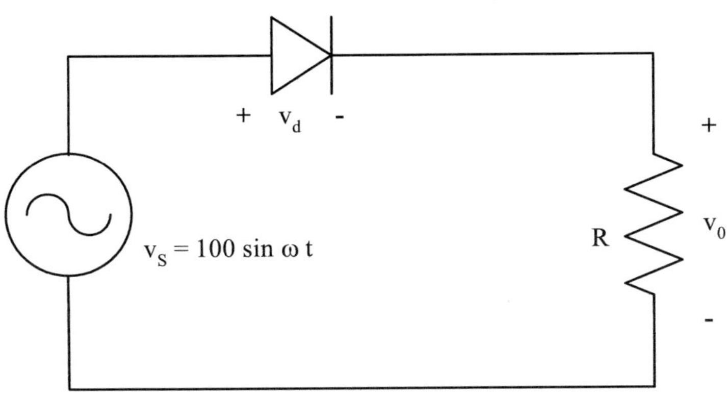

Figure 3.1 Half-wave converter, resistive load: circuit diagram.

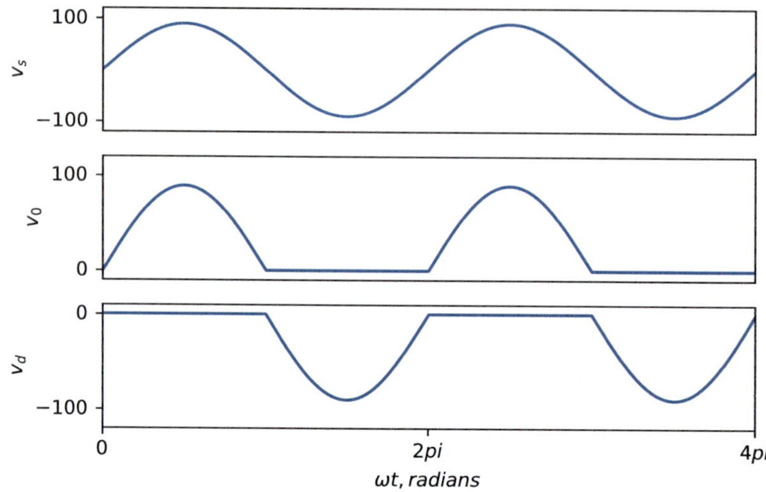

Figure 3.2 Half-wave converter, resistive load: response waveforms.

The RMS value of the load voltage is

$$V_{0,\mathrm{RMS}} = \sqrt{\frac{1}{2\pi}\int_0^{\pi} V_m^2 \sin^2 \omega t \, d\omega t} = \frac{V_m}{2}. \quad (3.2)$$

3.2 Half-wave Converter with Resistive-inductive Load

The converter is shown in Figure 3.3 with a RL load [1, 2]. It is assumed that the current i is zero immediately before $t = 0$. For $t > 0$, $v_s > 0$ and the diode becomes forward biased.

The load current is then governed by the equation

$$L\frac{di}{dt} + Ri = V_m \sin \omega t. \quad (3.3)$$

Applying the Laplace transform,

$$(sL + R)I(s) = \frac{\omega * V_m}{s^2 + \omega^2}. \quad (3.4)$$

From eqn (3.4), the transform of the current $I(s)$ is obtained as

$$I(s) = \frac{\omega L}{R^2 + (\omega L)^2} V_m \left\{ \frac{1}{s + R/L} - \frac{s}{s^2 + \omega^2} + \frac{R}{\omega L}\frac{\omega}{s^2 + \omega^2} \right\}. \quad (3.5)$$

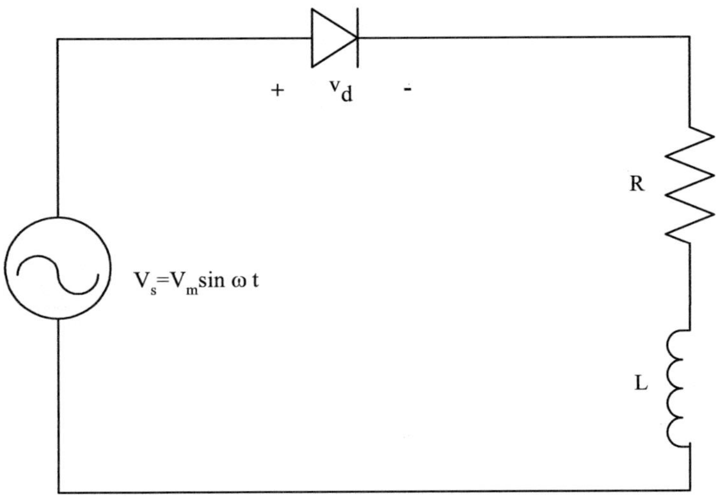

Figure 3.3 Half-wave converter, RL load: circuit diagram.

Putting $Z^2 = R^2 + (\omega L)^2$ and applying the inverse transform,

$$i(t) = \frac{V_m}{Z}\left\{\sin(\omega t - \theta) + \sin\theta \exp\left(-\frac{\omega t}{\omega L/R}\right)\right\}, \qquad (3.6)$$

where $\tan\theta = \frac{\omega L}{R}$. Note that $i(0) = 0$ and that there exists an angle β such that $i(\beta) = 0$. That is,

$$i(\beta) = \frac{V_m}{Z}\left\{\sin(\beta - \theta) + \sin(\theta)\exp\left(-\frac{\beta}{\omega L/R}\right)\right\} = 0. \qquad (3.7)$$

Eqn (3.7) is a nonlinear equation in β and its solution for β can only be obtained by numerical methods [1, 3].

3.2.1 Example 3.1

Consider a half-wave converter with the load $R = 10\ \Omega, L = 15$ mH. The source voltage is $120\ \text{V(rms)}, 60$ Hz. We determine the expression for the load current $i(\omega t)$ and calculate (a) the average load current, (b) the power dissipated by the load, and (c) the load power factor.

$$V_m = \sqrt{2}\cdot 120 = 169.7$$

$$\omega L = 5.655\Omega, \frac{\omega L}{R} = 0.5655$$

$$\begin{aligned}
\arctan\left(\frac{\omega L}{R}\right) &= 0.5146\,radians \\
Z &= \sqrt{10^2 + 5.655^2} = 11.488 \\
\cos\theta &= \frac{R}{Z} = 0.8705 \\
\sin\theta &= \frac{\omega L}{Z} = 0.4922 \\
i(\omega t) &= 14.772\left\{\sin(\omega t - 0.5146) + 0.4922\exp\left(-\frac{\omega t}{0.5655}\right)\right\}.
\end{aligned}$$

The angle β for which $i(\beta) = 0$ is obtained from the solution of the equation

$$\sin(\beta - 0.5146) + 0.4922\exp\left(-\frac{\beta}{0.5655}\right) = 0. \tag{3.8}$$

The solution is obtained by calling Program 3.01 with the command $ch3prog1()$. This program is an implementation of the Newton–Raphson procedure. At each iteration, the program calculates

$$\begin{aligned}
F(\beta) &= \sin(\beta - 0.5146) + 0.4922\exp\left(-\frac{\beta}{0.5655}\right) \\
dF(\beta) &= \cos(\beta - 0.5146) - 0.87003\exp\left(-\frac{\beta}{0.5655}\right).
\end{aligned}$$

The value of β is corrected as $\beta = \beta - F(\beta)/dF(\beta)$, and the program proceeds to the next iteration. After convergence, Program 3.01 prints:

```
F = 4.5624229969e - 05 beta = 3.65691177752
```

The average value of the load current and the power dissipated by the load may be obtained by running Program 3.02 with the command

```
I_mr(0.0, 3.657, 200)
```

In this program, $n = 200$ data points of the load current (or its square) are computed in the interval $\omega t = 0$ to $\omega t = \beta$. Trapezoidal integration is used to calculate the average and RMS values of the load current. Program 3.02 prints:

```
Imean = 5.05062667305Irms = 7.64461116347.
```

The mean and RMS values of the load voltage are obtained by making a substitution in Program 3.02, i.e., by putting F=169.7 sin(x). The values obtained are Vmean = 50.51V, Vrms = 86.02V. The power delivered to the load is P_0 = R*Irms*Irms = 584.4 watts. The power factor of the load is pf = P_0/(Vrms*Irms) =0.8887.

3.3 Half-wave Converter with a Generalized Load

A generalized DC load is the series connection of a resistor, an inductor (or a capacitor), and a direct voltage source. Figure 3.4 shows a half-wave converter with a generalized load. The converter is excited by the AC voltage source $V_m \sin \omega t$. The direct voltage source of the generalized load is V_{dc}.

In the circuit of Figure 3.4, the diode remains reverse biased until the moment t_α when $V_m \sin \omega t_\alpha = V_{\text{dc}}$. The analysis is simplified by putting $\omega t_\alpha = \alpha$ radians and setting the source phase angle to be α radians at $t = 0$. Then, the governing equation for the circuit current for $\omega t \geq \alpha$ is given by

$$L\frac{di}{dt} + Ri = \Big\{V_m \sin(\omega t + \alpha) - V_m \sin \alpha\Big\}u(t),\ u(t) = \text{unit-step function.} \tag{3.9}$$

Applying the Laplace transform to eqn (3.9), we get

$$(sL + R)I(s) = V_m\left\{\frac{\omega \cos \alpha}{s^2 + \omega^2} + \frac{s \sin \alpha}{s^2 + \omega^2} - \frac{\sin \alpha}{s}\right\}. \tag{3.10}$$

Putting $Z^2 = R^2 + (\omega L)^2$ and $\tan \theta = \frac{\omega L}{R}$, we get

$$\begin{aligned} I(s) &= \frac{V_m}{Z}\sin(\alpha - \theta)\frac{s}{s^2 + \omega^2} + \frac{V_m}{Z}\cos(\alpha - \theta)\frac{\omega}{s^2 + \omega^2} - \frac{V_m \sin \alpha}{sR} \\ &\quad - \left\{\frac{V_m}{Z}\sin(\alpha - \theta) - \frac{V_m \sin \alpha}{R}\right\}\frac{1}{s + \frac{R}{L}}. \end{aligned} \tag{3.11}$$

Taking the inverse transform of eqn (3.11), we get

$$\begin{aligned} i(\omega t) &= \frac{V_m}{Z}\sin(\omega t + \alpha - \theta) - \frac{V_m \sin \alpha}{R} \\ &\quad - \left\{\frac{V_m}{Z}\sin(\alpha - \theta) - \frac{V_m}{R}\sin \alpha\right\}e^{-\frac{\omega t}{\omega L/R}}. \end{aligned} \tag{3.12}$$

At $t = 0$, $i(0) = 0$. Besides, a phase angle $\omega t = \beta (> 0)$ exists such that

$$\begin{aligned} &\frac{V_m}{Z}\{\sin(\alpha + \beta - \theta) - \sin(\alpha - \theta)e^{-\beta/(\omega L/R)} \\ &\qquad + \frac{V_m}{R}\sin \alpha\, e^{-\beta/(\omega L/R) - 1)} = 0. \end{aligned} \tag{3.13}$$

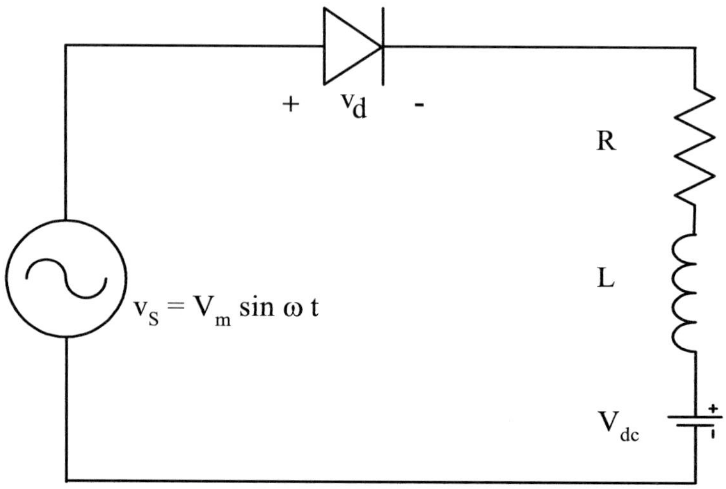

Figure 3.4 Half-wave converter, generalized load: circuit diagram.

The angle β is obtained by solving the nonlinear eqn (3.13). The diode conducts during the interval $\omega t = 0$ (i.e. the instant when $V_m \sin\alpha = V_{dc}$) to $\omega t = \beta$.

3.3.1 Example 3.2

Consider the half-wave converter supplying the generalized load $R = 0.5\,\Omega, L = 15\,\text{mH}, V_{dc} = 100$ V. The source is a 240 Vrms, 60 Hz alternating voltage. It is required to calculate (a) the power dissipated by the emf V_{dc}; (b) the power dissipated by the load resistance, (c) the power supplied by the alternating voltage source; and (c) the power factor of the half-wave converter presented to the source.

The peak source voltage $V_m = \sqrt{2} * 240 = 339.41$ V.

The diode starts conducting from $\omega t = \alpha$, where $V_m \sin\alpha = V_{dc} = 100$ V. That is, $\alpha = 17.135$ deg $= 0.299\,\text{radians}$.

$\omega L = 5.6549\ \Omega$, $Z = 5.677\ \Omega$, $\tan\theta = \omega L/R = 11.3097$.

$\theta = 84.947$ deg $= 1.4826\,\text{radians}$, $\sin\theta = 0.9961$, $\cos\theta = 0.08808$.

Substituting the numerical values,

$$i(\omega t) = 59.787\sin(\omega t - 1.1836) + 255.361e^{-0.08842\omega t} - 200. \quad (3.14)$$

The angle β is the solution of the nonlinear equation

$$59.787\sin(\beta - 1.1836) + 255.361e^{-0.08842*\beta} - 200 = 0. \quad (3.15)$$

Using Program 3.03, we get $\beta = 3.9777$ radians. Program 3.04 computes the mean and RMS values of the current: $I_{\text{mean}} = 22.217$ A, $I_{\text{rms}} = 32.72$ A.

Power dissipated by the load resistor $= 535.3$ watts.

Power delivered to $V_{\text{DC}} = I_{\text{mean}} * V_{\text{DC}} = 2221.7$ watts.

Total power delivered to the load $= 2757$ watts.

Putting $\text{F} = \text{V}_{\text{m}} \sin(\omega t + a)$ in Program 3.04, the RMS value of the load voltage is obtained as $\text{V}_{\text{rms}} = 188.5$ V.

The apparent power supplied to the load is $\text{V}_{\text{rms}} * \text{I}_{\text{rms}} = 6167.7$VA.

Therefore, the load power factor is $pf = 2757/6167.7 = 0.447$

3.4 Half-wave Converter with a Freewheeling Diode

A freewheeling diode is used to eliminate the sudden voltage spike that appears across an inductive load when its current is interrupted. Figure 3.5 shows the half-wave converter with the freewheeling diode D_2. The two diodes D_1 and D_2 do not conduct simultaneously. When one of the diodes conducts, the other diode is reverse biased by the source voltage. The diode D_1 conducts when the source voltage is of positive polarity. When the source voltage is of negative polarity, D_1 is reverse biased and the load current continues to circulate through the freewheeling diode D_2. The load voltage $v_0(\omega t)$ is, therefore, a series of positive, half-wave, sinusoidal pulses, and it is reduced to zero when D_2 conducts during the negative half-period of the source voltage.

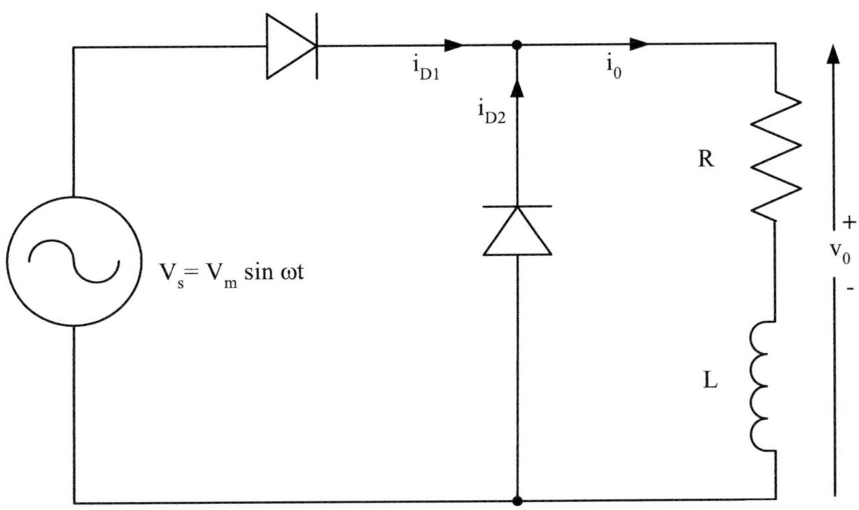

Figure 3.5 Half-wave converter with free wheeling diode: circuit diagram.

Applying Fourier analysis to the load voltage waveform [1, 2, 3], we obtain

$$v_0(\omega t) = \frac{V_m}{\pi} + \frac{V_m}{2}\sin(\omega t) - \sum_{h\,=\,2,\,4,\,6,\ldots}^{\infty} \frac{2V_m}{\pi(h^2-1)}\cos(h\omega t). \qquad (3.16)$$

The load current $i_0(\omega t)$ is given by a similar Fourier series and the current amplitude at each frequency may be calculated individually before summation. The diode current $i_{D1}(\omega t) = i_0(\omega t)$ for $2k\pi \leq (\omega t) < (2k+1)\pi,\ k = 0, 1, 2, \ldots$. Similarly, the diode current $i_{D2}(\omega t) = i_0(\omega t)$ for $(2k+1)\pi \leq (\omega t) < 2(k+1)\pi,\ k = 0, 1, 2, \ldots$.

3.4.1 Example 3.3

Consider a half-wave converter with a freewheeling diode connected to an RL load, $R = 2\ \Omega, L = 60$ mH. The source voltage is 120 Vrms, 60 Hz. It is required to determine the current waveform. The frequency components are calculated using eqn (3.16). The phasor notation is used in these calculations.

$$\begin{aligned}
I_0 &= \frac{120 * \sqrt{2}}{\pi * R} = 27\ A \\
I_1 &= \frac{60 + j0}{2 + j22.62} = 2.642\angle{-1.4826}\,radians \\
I_2 &= \frac{j25.46}{2 + j45.24} = 0.5622\angle 0.0442\,radians \\
i(\omega t) &= 27 + \sqrt{2}\Big\{2.642\sin(\omega t - 1.4826) - 0.5622\cos(2\omega t + 0.0442) - \ldots\Big\}.
\end{aligned}$$

The calculations are shown in Program 3.06. This program is called by the command $ch3fig306(esmax, freq, ls, rs, ncycle, n)$. The calling parameters are

$$\begin{aligned}
esmax &= 120 * \sqrt{2} = 169.7(V) \\
freq &= 60(\text{Hz}) \\
ls &= 0.06(\text{H}) \\
rs &= 2.0(\Omega) \\
ncycle &= 2\ (number\ of\ periods\ to\ be\ plotted) \\
n &= 512\ (number\ of\ data\ points\ per\ period).
\end{aligned}$$

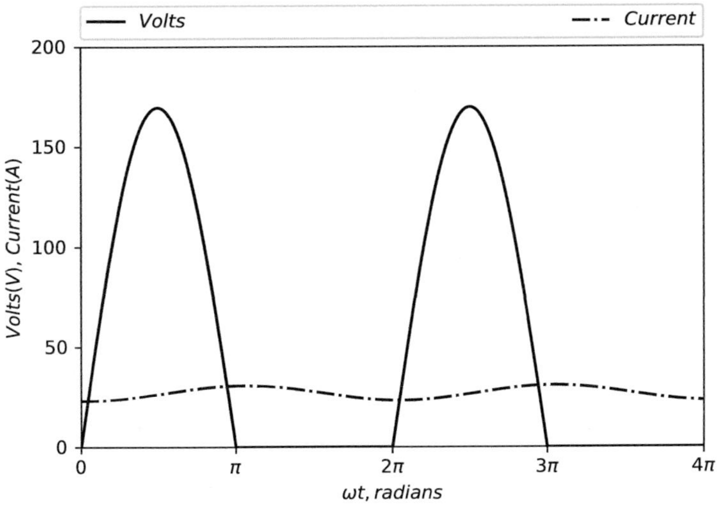

Figure 3.6 Half-wave converter, generalized load: response waveforms.

The results of the computation are

i0 = 27.008593842694637

Imax1 = 3.7366156098963432 theta1 = −1.48260625987938

Imax2 = −0.7952512238164975 theta2 = 0.044180937547767425

Imax4 = −0.07958335948028376 theta4 = 0.022101253937821405

Imax6 = −0.022741188050046025 theta6 = 0.014735502180955325

Figure 3.6 shows the response waveforms. Note that $i_{D1} = i_0$ and $i_{D2} = 0$ during the interval $0 \leq \omega t \leq \pi$. Similarly, $i_{D2} = i_0$ and $i_{D1} = 0$ during the interval $\pi \leq \omega t \leq 2\pi$. Thus, at $\omega t = \pi$, there is an instantaneous transfer of current from diode D_1 to diode D_2, given that the source is an ideal voltage source. (If the source was not ideal and it had some internal inductance, the transfer of current would not be instantaneous.) The amplitudes of the harmonic components of the load current decrease as the load inductance increases. A highly inductive load may therefore be represented by a constant current source.

3.5 Commutation

Commutation [1, 4] is the process by which current is transferred from one switching device to another when the converter is energized by a non-ideal

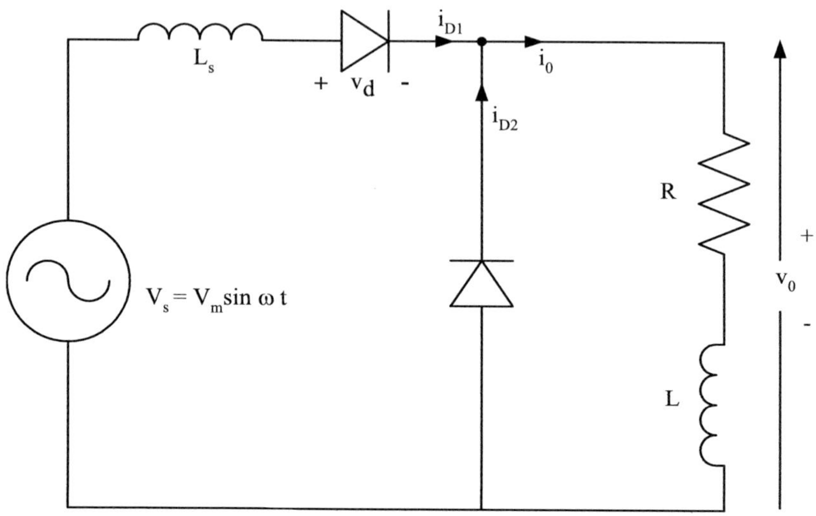

Figure 3.7 Half-wave converter, free-wheeling diode, non-ideal source: circuit diagram.

voltage source. Figure 3.7 shows a half-wave converter with a freewheeling diode. The alternating voltage source is not an ideal source and it has an internal inductance of L_s henries. The load is highly inductive and it is represented by a constant current source I_0.

The diode D_1 is reverse biased during the negative half-period of the source voltage and the load current I_0 circulates through the diode D_2. For $0 \leq \omega t \leq \pi$, the diode D_1 is forward biased, but the transfer of the load current from D_2 to D_1 is not instantaneous due to the presence of the source inductance. During the commutation interval $0 \leq \omega t \leq u$, the current i_{D1} increases from 0 to I_0 and the current i_{D2} decreases from I_0 to 0. The phase angle u is referred to as the commutation angle. Both diodes are in the conducting mode during the commutation interval and the load voltage is clamped to zero volts by the diode D_2.

During the commutation interval,

$$L_s \frac{di_{D1}}{dt} = V_m \sin(\omega t). \tag{3.17}$$

At $\omega t = 0$, $i_{D1} = 0$ and at $\omega t = u$, $i_{D1} = I_0$. Therefore, on integrating eqn (3.17),

$$\int_0^{I_0} di_{D1} = \frac{V_m}{L_s} \int_0^{\frac{u}{\omega}} \sin(\omega t)\, dt = \frac{V_m}{\omega L_s} \int_0^{u} \sin(\omega t)\, d\omega t. \tag{3.18}$$

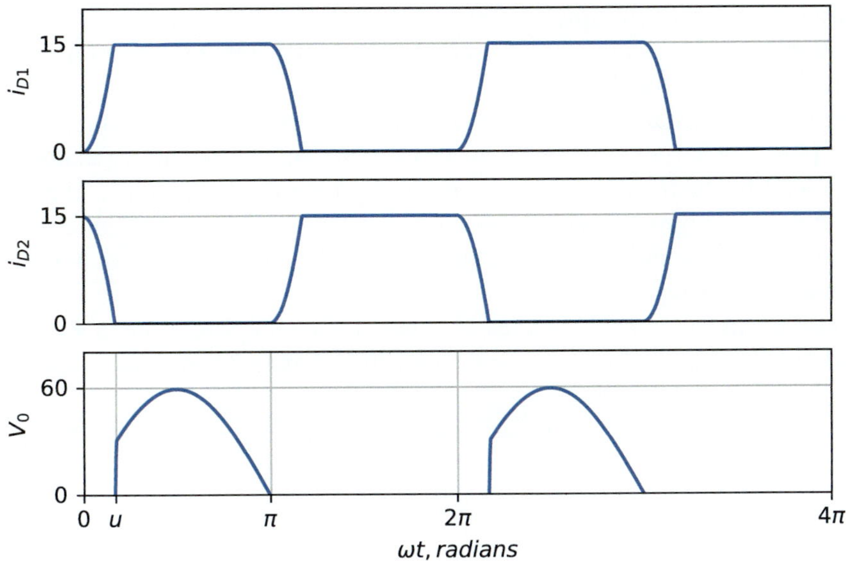

Figure 3.8 Half-wave converter of Figure (3.7): response waveforms.

That is,

$$\cos(u) = 1 - \frac{\omega L_s I_0}{V_m}. \tag{3.19}$$

Figure 3.8 shows the waveforms of the currents i_{D1}, i_{D2} and the load voltage v_0. Note that the load voltage is clamped to zero during the commutation interval. Therefore, the average load voltage is reduced by the commutation process. It is given by.

$$V_0 = \frac{1}{2\pi}\int_u^{\pi} V_m \sin(\omega t)\, d\omega t = \frac{V_m}{\pi}\left(1 - \frac{\omega L_s I_0}{2V_m}\right). \tag{3.20}$$

3.6 Simulation of Half-wave Converters

The simulation of power electronic circuits begins at an instant $t = 0$ when the state of the circuit is fully known. Using the state of the circuit and the excitations applied at $t = 0$, the state at $t = \Delta t$ is calculated. The time-step Δt should be sufficiently small. Now, with the state of the circuit and the excitations known at $t = \Delta t$, the state at $t = 2\Delta t$ is calculated. In this manner, the response of the circuit to a given excitation and with given initial conditions is determined. Since the power electronic circuit is nonlinear, the determination of the state at each time-step requires several iterations to converge to the solution.

The simulation starts with the mounting of the transient equivalent circuit for the power electronic converter. In this circuit, inductors and capacitors are replaced by their equivalents for the computation of transients. These equivalents are simply resistors with current sources connected in parallel. The current sources take into account the state of the inductor or capacitor in the previous time-step.

Consider an inductor L henries that has a time-varying voltage $v_L(t)$ applied to its terminals. Let $i_L(t)$ be the inductor current. Then,

$$L\frac{di_L}{dt} = v_L(t). \tag{3.21}$$

Putting $di_L \simeq i_L(t) - i_L(t - \Delta t)$ and $dt \simeq \Delta t$, we get

$$L\frac{i_L(t) - i_L(t - \Delta t)}{\Delta t} = v_L(t). \tag{3.22}$$

That is,

$$\begin{aligned} i_L(t) &= \frac{\Delta t}{L} v_L(t) + i_L(t - \Delta t) \\ &= gls * v_L(t) + pil, \end{aligned} \tag{3.23}$$

where $gls = \frac{\Delta t}{L}$ is the equivalent conductance of the inductor and $pil = i_L(t - \Delta t)$ is the known past history of the inductor current. Note that the equivalent conductance gls remains unchanged if the time-step Δt is constant. Eqn (3.23) leads to the transient equivalent for the inductor shown in Figure 3.9 (a).

Similarly, consider the capacitor C farads that has its terminal conditions given by $v_C(t)$, $i_C(t)$. From its governing equation $C\frac{dv_C}{dt} = i_C$, we get

$$\begin{aligned} i_C(t) &= \frac{C}{\Delta t}(v_C(t) - v_C(t - \Delta t)) \\ &= g_C * v_C(t) - g_C * pvc. \end{aligned} \tag{3.24}$$

The equivalent conductance of the capacitor for transient computations is $g_C = \frac{C}{\Delta t}$ and the known past history of the capacitor's voltage is $pvc = v_C(t - \Delta t)$. Eqn (3.24) suggests the equivalent for the capacitor shown in Figure 3.9 (b).

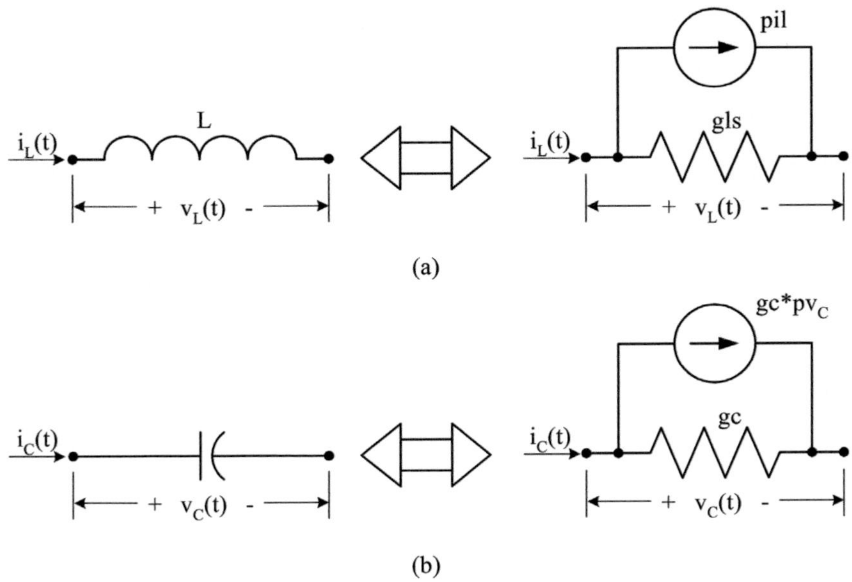

Figure 3.9 Transient equivalents for inductors and capacitors.

The nonlinear semiconductor devices remain in the transient equivalent circuit without any modification. Therefore, the resulting transient equivalent circuit that has to be solved at every time-step is a nonlinear resistive network with the applied excitations and the past-history current sources. We use a variant of the Newton–Raphson technique for solving this circuit [5]. In this technique, direct voltage sources are connected across every nonlinear semiconductor device. We refer to these voltage sources as probe voltages. Their magnitudes are estimates of the voltages that will appear across the nonlinear devices at the current time-step. If the estimates are exactly equal to the correct device voltages, then no current will flow into or out of the probe voltage sources. Thus, the currents flowing into the probe voltage sources are the error signals and their magnitudes are reduced to negligible levels over successive iterations by correcting the probe voltage magnitudes.

3.6.1 Half-wave converter with RL load

Consider the converter shown in Figure 3.10. We choose the total simulation time to be 20 periods of the frequency $f = 60$ Hz, and the number n of computed data points per period to be 512. The time-step Δt is given by $\Delta t = T/n = 1/(f * n)$.

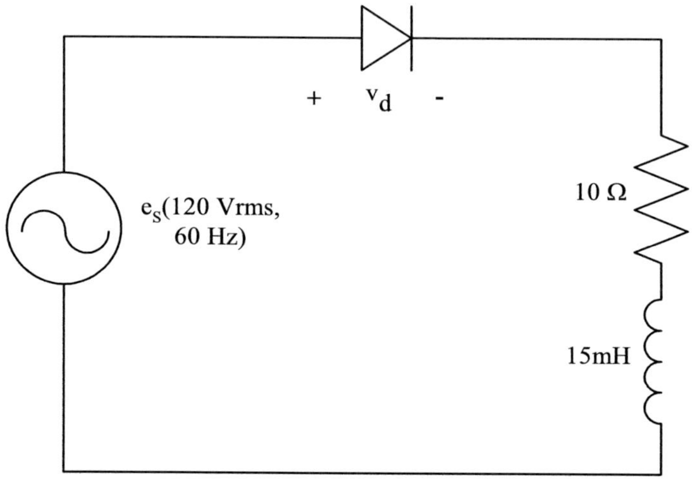

Figure 3.10 Half-wave converter, RL load: circuit diagram.

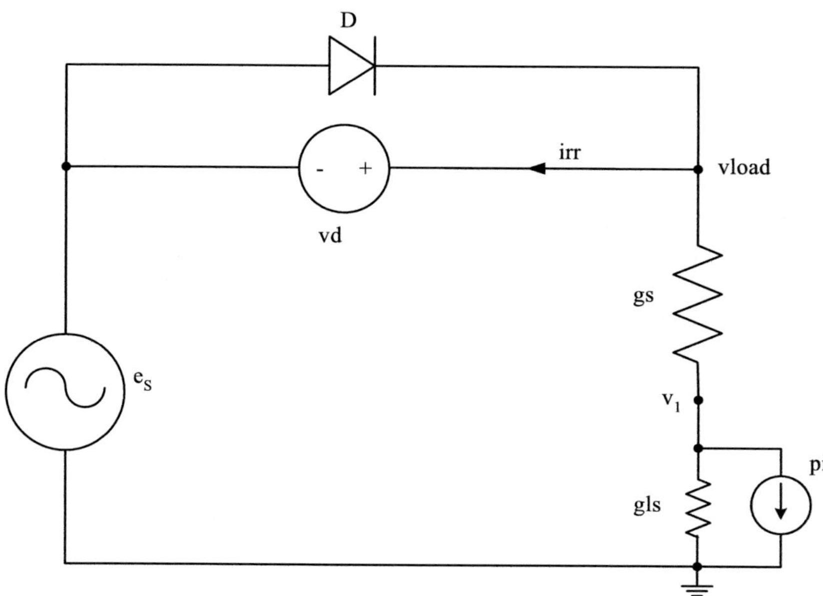

Figure 3.11 Half-wave converter, RL load: transient equivalent circuit.

The transient equivalent circuit is shown in Figure 3.11 and it is obtained by replacing the inductor by the parallel connection of its equivalent conductance $gls = \Delta t/L$ and the past-history current source $pil = i_L(t - \Delta t)$.

At each time-step, the instantaneous value of the source voltage and the past-history current are updated.

The iterative procedure at each time-step is as follows. A known, direct voltage source v_d is connected across the nonlinear diode and the circuit is analyzed. Note that v_d is the probe voltage whose magnitude will be adjusted over several iterations at each time-step. (In the first time-step, v_d may be assumed to be zero and in the subsequent time-steps, its starting value may be equal to the diode voltage at the preceding time-step.) The circuit equations are given by

$$g_s = 1/R \tag{3.25}$$
$$g_s * (v_1 - e_s - v_d) + gls * v_1 + pil = 0 \tag{3.26}$$
$$i_L = g_s * (e_s + v_d - v_1) \tag{3.27}$$
$$[i_d, g_d] = fdiode(v_d). \tag{3.28}$$

The current flowing into the voltage source v_d is

$$i_{rr} = -(i_d + i_L). \tag{3.29}$$

The current i_{rr} is considered to be an error and we seek the value of v_d for which $i_{rr} = 0$ (or below a specified tolerance).

The value of v_d is corrected using the circuit of Figure 3.12, which is referred to as the small–signal equivalent circuit. In this circuit, the voltage source e_s and the past–history current source pil have been removed. The voltage source v_d is replaced by an error current source $(-i_{rr})$, i.e., the error current is injected in the reverse direction. Finally, the diode is replaced by its conductance obtained by Eqn (3.28). The voltage δv_d that appears across the error current source is computed:

$$gstot = gs * gls/(gs + gls) \tag{3.30}$$
$$\delta v_d = i_{rr}/(gd + gstot). \tag{3.31}$$

Then, the corrected value of v_d is given by

$$(v_d)_{\text{corrected}} = v_d + \delta v_d. \tag{3.32}$$

With the corrected value of v_d and with the values of e_s and pil remaining unchanged, we proceed to the next iteration. After a few iterations, the error current is reduced to a negligible value and convergence to the solution is obtained. The circuit variables are stored for plotting the responses and the computation proceeds to the next time-step.

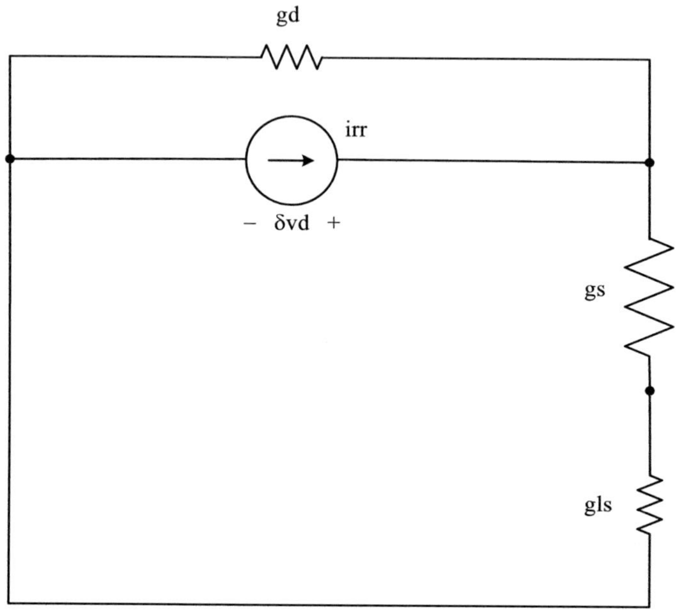

Figure 3.12 Half-wave converter, RL load: small–signal equivalent.

Program 3.13 simulates the circuit of Figure 3.10 and it is called by the command ch3fig13(esmax, freq, L, R, ncycle, n). The load variables v_{load}, i_L are shown plotted in Figure 3.13.

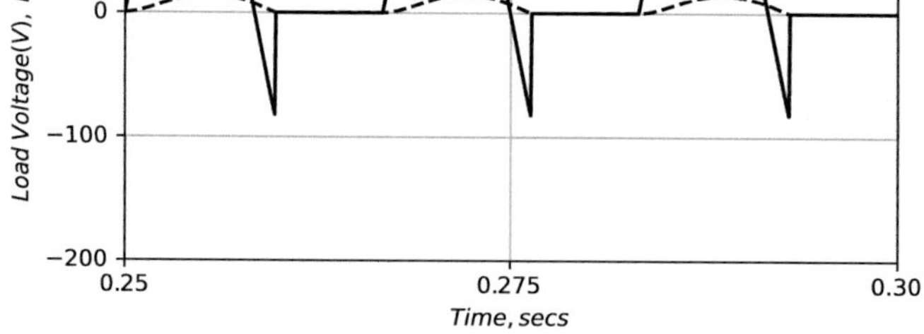

Figure 3.13 Half-wave converter, RL load: response waveforms.

3.6.2 Exercise

Figure 3.14 shows the half-wave converter connected to a generalized load. Program 3.15 simulates the circuit, and the load variables have been plotted as shown in Figure 3.15. We leave as an exercise the mounting of the transient and small-signal equivalent circuits and the verification of the program.

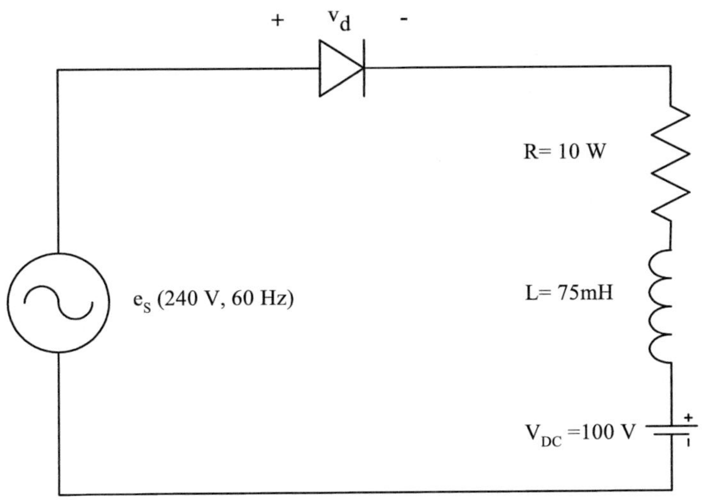

Figure 3.14 Half-wave converter, generalized load: circuit diagram.

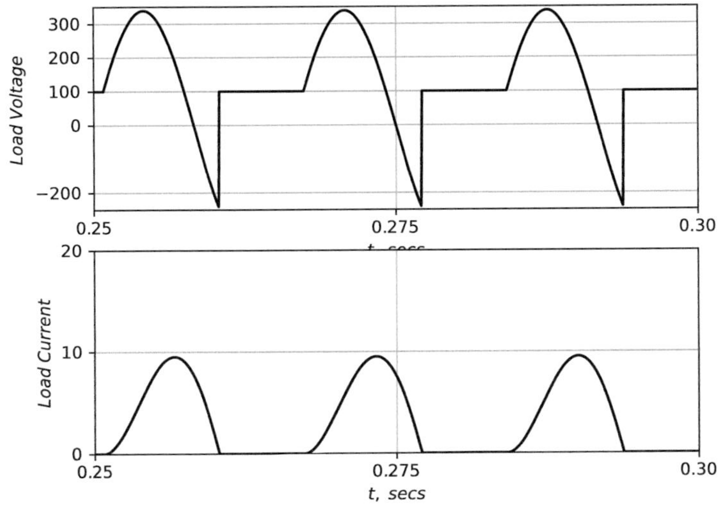

Figure 3.15 Half-wave converter with generalized load: response waveforms.

3.6.3 Simulation of the half-wave converter with a freewheeling diode

The converter is shown in Figure 3.16. The highly inductive load is represented by the constant current source of 5 A. The alternating voltage source is not ideal and it has an internal inductance of 1.5 mH.

The transient equivalent circuit of the converter is shown in Figure 3.17. The source inductance has been substituted by its equivalent. Note that the probe voltage sources vd_1 and vd_2 need to be connected across the diodes D_1 and D_2. However, it is more convenient to connect the source vd_1 to the input of the converter.

The equations governing the transient equivalent circuit of the converter are

$$i_L = gls * (e_s - vd_1) + pil \tag{3.33}$$

$$[id_1, gd_1] = fdiode(vd_2 - vd_1) \tag{3.34}$$

$$[id_2, gd_2] = fdiode(vd_2) \tag{3.35}$$

$$irr_1 = i_L + id_1 \tag{3.36}$$

$$irr_2 = -(id_1 + id_2 + I_0). \tag{3.37}$$

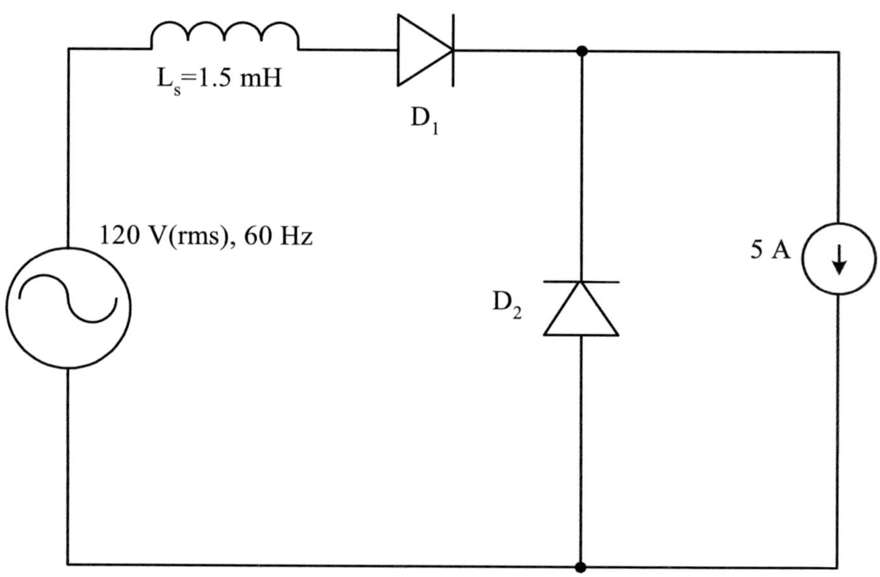

Figure 3.16 Half-wave converter with free wheeling diode: circuit diagram.

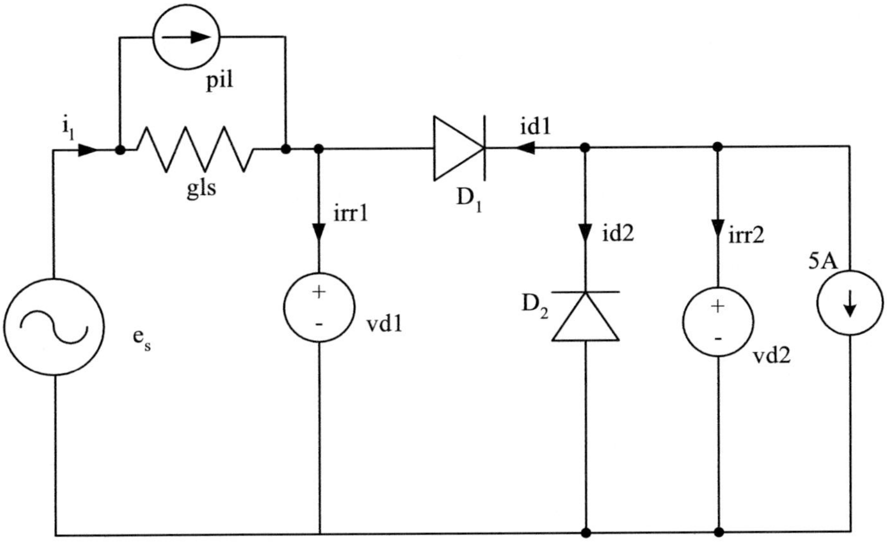

Figure 3.17 Half-wave converter with free wheeling diode: transient equivalent circuit.

Note that irr_1 and irr_2 are the error currents. The overall error may be defined as $irr = \sqrt{irr_1^2 + irr_2^2}$.

The corrections δvd_1 and δvd_2 are computed with the help of the small-signal equivalent circuit shown in Figure 3.18.

The equations are

$$gls * \delta vd_1 + gd_1 * (\delta vd_1 - \delta vd_2) = irr_1 \quad (3.38)$$

$$gd_2 * \delta vd_2 + gd_1 * (\delta vd_2 - \delta vd_1) = irr_2. \quad (3.39)$$

The magnitudes of the parallel voltage sources are corrected as given below:

$$vd_1 = vd_1 + \delta vd_1 \quad (3.40)$$

$$vd_2 = vd_2 + \delta vd_2. \quad (3.41)$$

We pass on to the next iteration without changing the magnitude of the alternating source voltage and the past–history current source pil. When the procedure converges after a few iterations, the variables to be plotted are saved and the past–history current source is updated. The computational procedure advances to the next time-step. Figure 3.19 shows the response waveforms and the simulation program is shown as Program 3.19.

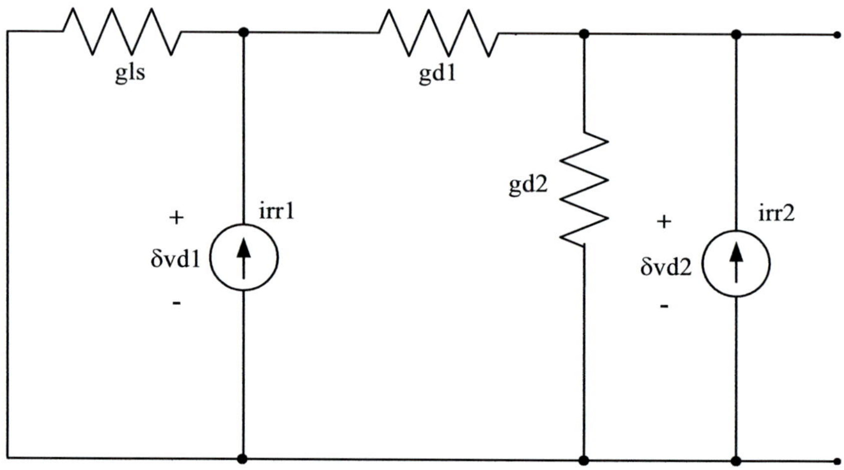

Figure 3.18 Half-wave converter with free wheeling diode: small-signal equivalent.

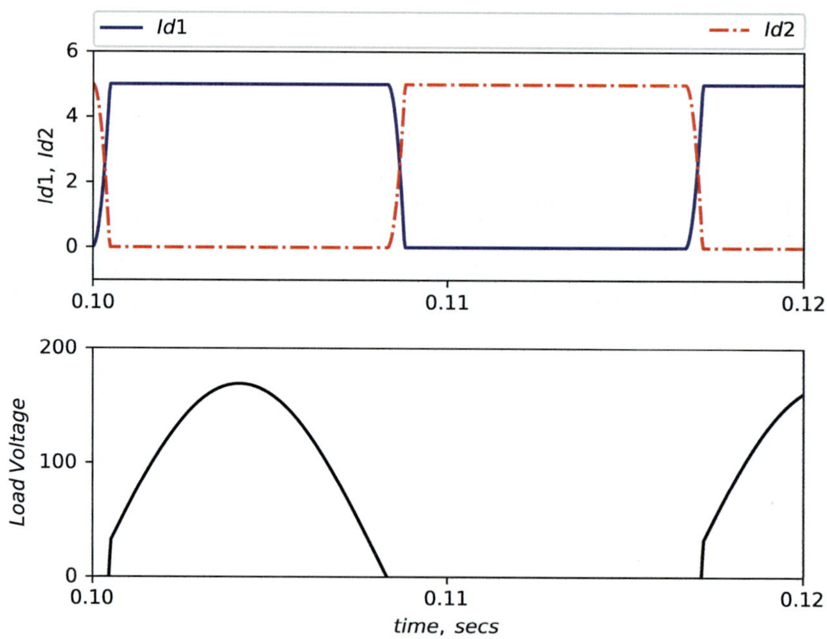

Figure 3.19 Half-wave converter with free wheeling diode: response waveforms.

3.6.4 Half-wave converter with capacitor filter

The half-wave converter with a capacitor connected to the output terminals is shown in Figure 3.20 [2, 3]. The function of the capacitor is to attenuate the ripple in the output DC voltage. The load connected to the converter is represented by the resistor R.

The transient equivalent circuit is shown in Figure 3.21. The probe voltage source v_d is connected across the diode in the simulation. As in the previous examples, the time-step $\Delta t = 1/(f * n)$, where f is the frequency of the source voltage and n is the number of computed data points per period. The equivalent conductance of the capacitor is $gc = C/\Delta t$, and the past-history current source for the capacitor is $gc * pvc = gc * v_C(t - \Delta t)$. It is assumed that the source voltage is switched on at $\omega t = 0$ with the capacitor uncharged.

The equations for the transient equivalent circuit are

$$v_C(t) = e_s(t) + v_d(t) \tag{3.42}$$

$$[i_d, g_d] = fdiode(v_C - e_s) \tag{3.43}$$

$$irr = -\left(i_d + gc(v_C - pvc) + \frac{1}{R} * v_C\right). \tag{3.44}$$

The current irr is the error signal. If $abs(irr)$ is greater than the specified tolerance $(erlim)$, then the probe voltage v_d has to be corrected. The correction δv_d is obtained from Figure 3.22, which is the small-signal equivalent.

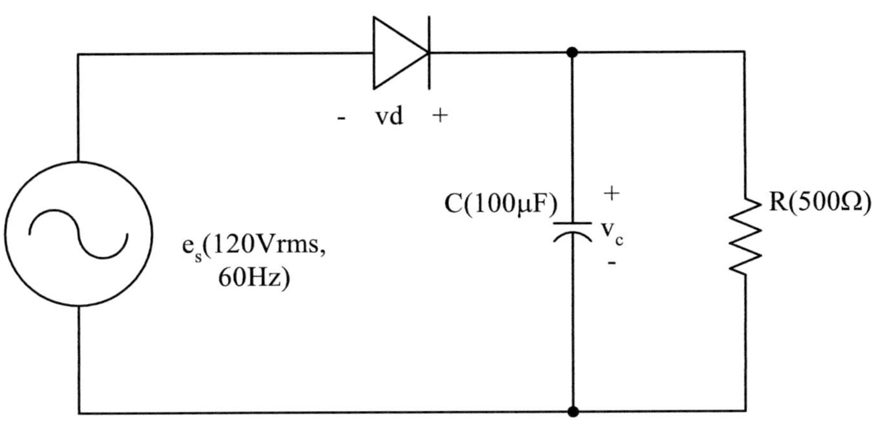

Figure 3.20 Half-wave converter with capacitor filter: circuit diagram.

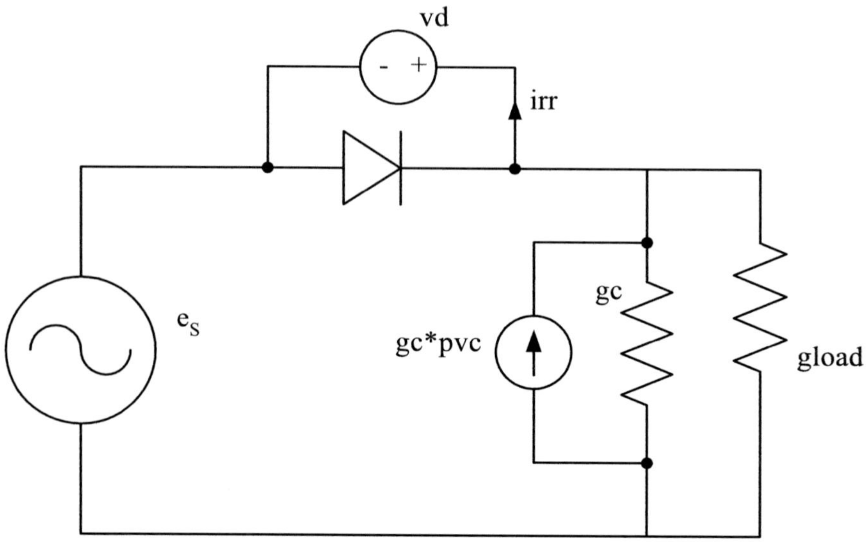

Figure 3.21 Half-wave converter with capacitor filter: transient equivalent circuit.

In this circuit, the probe voltage source has been replaced by the error current injected in the reverse direction and the diode has been substituted by its equivalent conductance at the operating point. The known sources such as the alternating voltage e_s and the past–history current source are removed. From Figure 3.22, the correction is obtained as

$$\delta v_d = irr/\left(g_d + gc + \frac{1}{R}\right). \tag{3.45}$$

The updated parallel voltage for the next iteration is $v_d = v_d + \delta v_d$.

Figure 3.23 shows the response waveforms plotted by Program 3.23. The capacitor current waveform shows that the capacitor is charged by short duration pulses when the diode is in the conducting state. The intervals in ωt during which the diode conducts are

$$(0, \alpha_0),\ (\alpha_1, 2\pi + \alpha_0),\ (2\pi + \alpha_1, 4\pi + \alpha_0)\ \ldots \tag{3.46}$$

On the other hand, the intervals of reverse bias during which the capacitor discharges through the load are

$$(\alpha_0, \alpha_1), (2\pi + \alpha_0, 2\pi + \alpha_1), (4\pi + \alpha_0, 4\pi + \alpha_1), \ldots \tag{3.47}$$

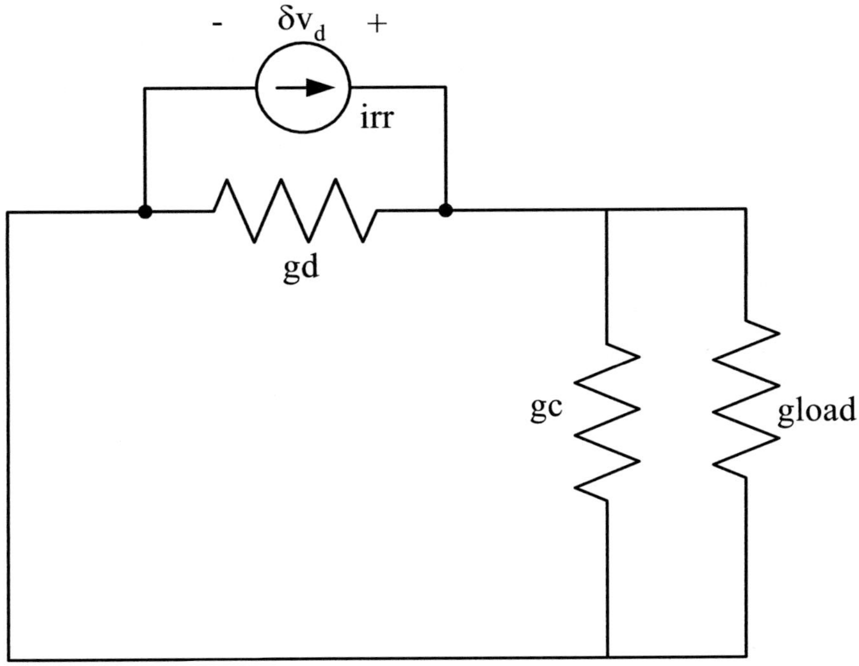

Figure 3.22 Half-wave converter with capacitor filter: small-signal equivalent.

The discussion above leads us to the conclusion that the output voltage $v_C(\omega t)$ is formed by two piecewise functions smoothly joined at $\omega t = \alpha_0,\ 2\pi + \alpha_0,\ 4\pi + \alpha_0, \ldots$ Thus,

$$\begin{aligned} v_C(\omega t) &= V_m \sin(\omega t),\ 0 \le \omega t \le \alpha_0 \\ &= V_m \sin\alpha_0\, e^{-(\omega t-\alpha_0)/\omega RC},\ \alpha_0 \le \omega t \le \alpha_1. \end{aligned} \tag{3.48}$$

Clearly, at $\omega t = \alpha_0$, the piecewise functions are equal to each other. Moreover, the transition from one function to the other is smooth. Therefore,

$$\frac{d}{d(\omega t)}(V_m \sin\omega t) = \frac{d}{d(\omega t)}\left(V_m \sin\alpha_0\, e^{-(\omega t-\alpha_0)/\omega RC}\right),\ \omega t = \alpha_0. \tag{3.49}$$

That is,

$$V_m \cos\alpha_0 = \frac{V_m \sin\alpha_0}{-\omega RC}. \tag{3.50}$$

Or, $\tan\alpha_0 = -\omega RC$ so that $\alpha_0 = \pi - \tan^{-1}(\omega RC)$.

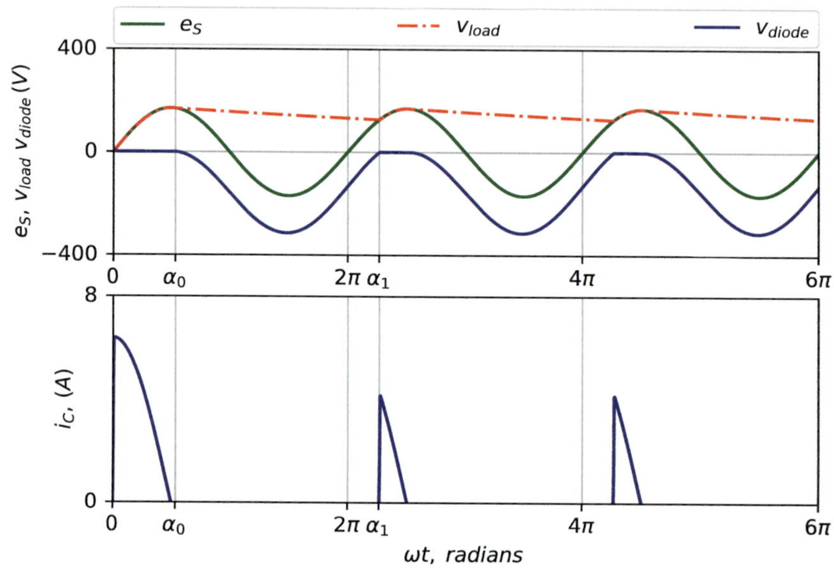

Figure 3.23 Half-wave converter with capacitor filter: response waveforms.

At $\omega t = \alpha_1$, the source voltage once again becomes equal to the capacitor voltage and the diode resumes conduction. That is,

$$V_m \sin \omega t = V_m \sin \alpha_0 \ e^{-(\omega t - \alpha_0)/\omega RC}, \ \omega t = \alpha_1. \tag{3.51}$$

Or,

$$\sin \alpha_1 - \sin \alpha_0 \ e^{-(\alpha_1 - \alpha_0)/\omega RC} = 0. \tag{3.52}$$

Eqn (3.52) may be solved numerically for α_1.

3.6.5 Example 3.3

Consider the converter shown in Figure 3.20. The variables are

$$\begin{aligned}
V_m &= 169.7 \ V \\
\omega RC &= 18.8496 \ radians \\
\alpha_0 &= \pi - \tan^{-1}(18.8496) \\
&= 1.6257 \ radians \\
F(\alpha_1) &= \sin \alpha_1 - \sin \alpha_0 \ e^{-(\alpha_1 - \alpha_0)/18.8496}
\end{aligned}$$

$$
\begin{aligned}
&= \sin\alpha_1 - 0.99849\, e^{-(\alpha_1 - 1.6257)/18.8496} \\
&= \sin\alpha_1 - 1.088428\, e^{-\alpha_1/18.8496} \\
dF(\alpha_1) &= \cos\alpha_1 + \frac{1.088428}{18.8496}\, e^{-\alpha_1/18.8496}.
\end{aligned}
$$

Numerical solution of the nonlinear equation $F(\alpha_1) = 0$ (using Program 3.24) gives $\alpha_1 = 7.125\ radians$.

Having obtained the angles α_0 and α_1, other circuit variables may be calculated. The first peak of the diode current is given by

$$
\begin{aligned}
C\frac{dv_C}{dt} &= \omega C V_m \\
&= 6.4\ \text{A}.
\end{aligned}
$$

The subsequent peaks are

$$
\begin{aligned}
i_{d,\text{peak}} &= \Big\{ i_C(\omega t) + i_R(\omega t) \Big\},\ \omega t = \alpha_1,\ 2\pi + \alpha_1, \ldots \\
&= \omega C V_m \cos\alpha_1\ +\ \frac{V_m \sin\alpha_1}{R} \\
&= 6.4 * 0.6661\ +\ 0.2531 \\
&= 4.516\ \text{A}.
\end{aligned}
$$

The peak-to-peak ripple in the output voltage is

$$
\Delta v_{C,p-p} = V_m * \Big(1\ -\ \sin\alpha_1\Big) = 43.13\ \text{V}.
$$

The ripple may also be calculated from the exponential decay of the capacitor voltage:

$$
\Delta v_{C,p-p} = V_m * \Big(1\ -\ \sin\alpha_0\, e^{-(\alpha_1 - \alpha_0)/\omega RC}\Big). \qquad (3.53)
$$

Approximating $\alpha_0 \simeq \pi/2$ and $(\alpha_1 - \alpha_0) \simeq 2\pi$, we get

$$
\begin{aligned}
\Delta v_{C,p-p} &= V_m(1\ -\ e^{-2\pi/\omega RC}) \\
&\simeq V_m \frac{2\pi}{\omega RC} \\
&= \frac{V_m}{fRC} \\
&= 56.6\ V.
\end{aligned}
$$

This is a conservative estimate of the calculated value of 43.13 V. The corresponding values obtained from Program 3.23 are

$$
\begin{aligned}
\alpha_0 &= 1.635 \text{ radians} \\
\alpha_1 &= 7.1177 \text{ radians} \\
\Delta v_{C,p-p} &= 43.447 \text{ V} \\
i_{d,\text{peak}} &= 6.3464 \text{ at } \omega t = 0 \\
&= 4.3719 \text{ A at } \omega t = \alpha_1.
\end{aligned}
$$

Note that Program 3.23 uses a diode model that is non-ideal, whereas the calculated values are based on an ideal diode.

3.7 Computational Programs

3.7.1 Program 3.01

```
#
def ch3prog1_F(x):
    """Function evaluation"""
    #
    import numpy as np
    #
    F = np.sin(x - 0.5146) + 0.492*np.exp(-x/0.5655)
    #
    return F
#
def ch3prog1_dF(x):
    """Derivative evaluation"""
    #
    import numpy as np
    #
    dF = np.cos(x - 0.5146) - 0.87003*np.exp(-x/0.5655)
    #
    return dF
#
def ch3prog1():
    """Newton Raphson procedure"""
    #
    #   calling command:
    #   ch3prog1()
    #
    from math import pi
    beta = pi
    epsilon = 0.001
    F = ch3prog1_F(beta)
    #
    while abs(F) >= epsilon:
        #
```

```
        dF = ch3prog1_dF(beta)
        beta -= F/dF
        F = ch3prog1_F(beta)
        #
    print('F = ', F, 'beta = ', beta)
```

3.7.2 Program 3.02

```
#
def ch3prog2_F(x):
    """Function evaluation"""
    #
    import numpy as np
    #
    F = 14.772*(np.sin(x - 0.5146) + 0.492*np.exp(-x/0.5655))
    #
    return F
#
def ch3prog2(b0, beta, n):
    """Mean and RMS evaluation"""
    #
    #   calling command:
    #   ch3prog2(0.0, 3.657, 200)
    #
    from math import pi
    import numpy as np
    #
    dteta = (beta - b0)/n
    teta = dteta*np.arange(n)
    #
    F_vect = []
    for j in range(n):
        F_vect.append(ch3prog2_F(teta[j]))
    #
    F_vect.append(ch3prog2_F(beta))
    #
    sumt = 0.0
    sumsq = 0.0
    for j in range(n):
        sumt += (dteta/2.0)*(F_vect[j] + F_vect[j+1])
        sumsq += (dteta/2.0)*(F_vect[j]**2 +F_vect[j+1]**2)
    #
    Imean = sumt/(2.0*pi)
    Irms = np.sqrt(sumsq/(2.0*pi))
    #
    print('Imean = ', Imean, 'Irms = ', Irms)
    #
```

3.7.3 Program 3.03

```
#
def ch3prog3_F(x):
    """Function evaluation"""
    #
    import numpy as np
    #
    F = 59.7869*np.sin(x - 1.1836) + 255.361*np.exp(-0.08842*x) -200.0
    #
    return F
#
def ch3prog3_dF(x):
    """Derivative evaluation"""
    #
    import numpy as np
    #
    dF = 59.7869*np.cos(x - 1.1836) - 22.579*np.exp(-0.08842*x)
    #
    return dF
#
def ch3prog3():
    """Newton Raphson procedure"""
    #
    from math import pi
    beta = pi
    epsilon = 0.0001
    F = ch3prog3_F(beta)
    #
    while abs(F) >= epsilon:
        #
        dF = ch3prog3_dF(beta)
        beta -= F/dF
        F = ch3prog3_F(beta)
        #
    print('F = ', F, 'beta = ', beta)
```

3.7.4 Program 3.04

```
#
def ch3prog4_F(x):
    """Function evaluation"""
    #
    #   calling command:
    #   I_mr3(0.0, 3.97774, 200)
    #
    import numpy as np
    #
    F = 59.7869*np.sin(x - 1.1836) + 255.361*np.exp(-0.08842*x) - 200.0
    #
    return F
#
```

```
def ch3prog4(b0, beta, n):
    """Mean and RMS evaluation"""
    #
    #   calling command:
    #   ch3prog4(0.0, 3.97774, 200)
    #
    from math import pi
    import numpy as np
    #
    dteta = (beta - b0)/n
    teta = dteta*np.arange(n)
    #
    F_vect = []
    for j in range(n):
        F_vect.append(ch3prog4_F(teta[j]))
    #
    F_vect.append(ch3prog4_F(beta))
    #
    sumt = 0.0
    sumsq = 0.0
    for j in range(n):
        sumt += (dteta/2.0)*(F_vect[j] + F_vect[j+1])
        sumsq += (dteta/2.0)*(F_vect[j]**2 +F_vect[j+1]**2)
    #
    Imean = sumt/(2.0*pi)
    Irms = np.sqrt(sumsq/(2.0*pi))
    #
    print('Imean = ', Imean, 'Irms = ', Irms)
    #
```

3.7.5 Program 3.06

```
#
def ch3fig06(esmax, freq, ls, rs, ncycle, n):
    """hw rectifier with free wheeling diode"""
    #
    #    calling command:

    #    ch3fig306(169.7, 60.0, 0.06, 2.0, 2, 512)
    #
    from math import pi
    import numpy as np
    import matplotlib.pyplot as plt
    import cmath as cm
    #
    N = ncycle*n
    dteta = 2.0*pi/n
    xl = 2.0*pi*freq*ls
    teta = dteta*np.arange(N)
    volt = []
    volt = esmax*np.sin(teta)
    i = []
    #
```

```
emax0 = esmax/pi
emax1 = esmax/2.0
emax2 = 2.0*esmax/(3.0*pi)
emax4 = 2.0*esmax/(15.0*pi)
emax6 = 2.0*esmax/(35.0*pi)
#
i0 = emax0/rs
I1 = complex(emax1, 0)/complex(rs, xl)
I2 = complex(0, emax2)/complex(rs, 2.0*xl)
I4 = complex(0, emax4)/complex(rs, 4.0*xl)
I6 = complex(0, emax6)/complex(rs, 6.0*xl)
Imax1 = abs(I1)
Imax2 = abs(I2)
Imax4 = abs(I4)
Imax6 = abs(I6)
tet1 = cm.phase(complex(I1))
tet2 = cm.phase(complex(I2))
tet4 = cm.phase(complex(I4))
tet6 = cm.phase(complex(I6))
#
print('i0 = ', i0)
print('Imax1 = ', Imax1, 'tet1 = ', tet1)
print('Imax2 = ', -Imax2, 'tet2 = ', tet2)
print('Imax4 = ', -Imax4, 'tet4 = ', tet4)
print('Imax6 = ', -Imax6, 'tet6 = ', tet6)
#
for j in range(N):
    #
    ang = teta[j]
    if volt[j] < 0.0:
        volt[j] = 0.0
    #
    i1 = Imax1*np.sin(ang + tet1)
    i2 = Imax2*np.sin(ang + tet2)
    i4 = Imax4*np.sin(ang + tet4)
    i6 = Imax6*np.sin(ang + tet6)
    curr = i0 + i1 - i2 - i4 - i6
    i.append(curr)
#
plt.plot(teta, volt, 'k-', label = '$Volts$')
plt.plot(teta, i, 'k-.', label = '$Current$')
plt.xlabel('$\omega t, radians$')
plt.xlim(0, 4*pi)
plt.xticks([0, pi, 2*pi, 3*pi, 4*pi], ['$0$', '$\pi$', '$2\pi$',     \
                                        '$3\pi$', '$4\pi$'])
plt.ylabel('$Volts (V)$, $Current (A)$')
plt.ylim(0, 200)
plt.yticks([0, 50, 100, 150, 200], ['$0$', '$50$', '$100$',           \
                                     '$150$', '$200$'])
#
plt.legend(bbox_to_anchor=(0., 1.02, 1, 0.102), loc=3, ncol=3,       \
                           mode="expand", borderaxespad=0.)
#
```

```
    plt.show()
    #
    #plt.draw
    #plt.savefig('Fig306.ps', dpi = 400, bbox_inches = 'tight')
    #
```

3.7.6 Program 3.13

```
#
def fdiode(vd):
    """ Diode characteristic.
        Returns diode current and its conductance"""
    glow = 0.1e-07
    ghigh = 0.1e+03
    Vr = -0.5
    #
    if vd >= Vr:
        id = glow*(vd - Vr)
        gd = glow
    else:
        id = ghigh*(vd - Vr)
        gd = ghigh
    #
    return id, gd
#
def ch3fig13_solver(esource, gs, gls, N):
    """Assumes esource is a list, fdiode is a function.
        Returns the lists iload and vload"""
    vd = 0.0
    pil = 0.0
    erlim = 0.1e-07
    gstot = gs*gls/(gs + gls)
    #
    iload = []
    vload = []
    #
    for j in range(N):
        #
        es = esource[j]
        #
        for its in range(1000):
            #
            v1 = (gs*(es + vd) - pil)/(gs + gls)
            il = gs*((es + vd) - v1)
            id, gd = fdiode(vd)
            irr = -(il + id)
            err = abs(irr)
            #
            if err < erlim:
                break
            else:
                dvd = irr/(gd + gstot)
```

```
                vd += dvd
            #
        pil = il
        iload.append(il)
        vload.append(es + vd)
        #
    return iload, vload
#
#
def ch3fig13(esmax, freq, ls, rs, ncycle, n):
    """ Main program, plots results"""
    #
    #ch3fig313(169.7, 60.0, 0.015, 10.0, 20, 512)
    #
    from math import pi
    import numpy as np
    import matplotlib.pyplot as plt
    #
    N = n*ncycle
    #
    delt = 1.0/(freq*n)
    dteta = 2.0*pi/n
    gs = 1.0/rs
    gls = delt/ls
    tk = []
    teta = []
    esource = []
    #
    for k in range(N):
        tk.append(k*delt)
        teta.append(k*dteta)
        esource.append(esmax*np.sin(k*dteta))
    #
    iload, vload = ch3fig13_solver(esource, gs, gls, N)
    #
    plt.plot(tk, iload, 'k--', label = 'Current')
    plt.plot(tk, vload, 'k-', label = 'Voltage')
    plt.xlabel('$Time, secs$')
    plt.ylabel('$Load\; Voltage (V),\; Load\; Current (A)$')
    plt.xlim(0.25, 0.30)
    plt.xticks([0.25, 0.275, 0.3], ['$0.25$', '$0.275$', '$0.30$'])
    plt.ylim(-200, 200)
    plt.yticks([-200, -100, 0, 100, 200], ['$-200$', '$-100$', '$0$',    \
                                           '$100$', '$200$'])
    plt.legend(bbox_to_anchor=(0., 1.02, 1., 0.102), loc=3, ncol=2,      \
                               mode="expand", borderaxespad=0.)
    plt.grid(True)
    #
    plt.show()
    #
    # plt.draw()
    # plt.savefig('Fig312.ps', dpi=400, bbox_inches='tight')
    #
```

3.7.7 Program 3.15

```
#
def fdiode(vd):
    """ Diode characteristic.
        Returns diode current and its conductance"""
    glow = 0.1e-07
    ghigh = 0.1e+03
    Vr = -0.5
    #
    if vd >= Vr:
        id = glow*(vd - Vr)
        gd = glow
    else:
        id = ghigh*(vd - Vr)
        gd = ghigh
    #
    return id, gd
#
def ch3fig15_solver(esource, gs, gls, Vdc, N):
    """Assumes esource is a list, fdiode is a function.
        Returns the list iload"""
    vd = 0.0
    pil = 0.0
    erlim = 0.1e-07
    gstot = gs*gls/(gs + gls)
    #
    iload = []
    vload = []
    #
    for j in range(N):
        #
        es = esource[j]
        #
        for its in range(1000):
            #
            v1 = (gs*(es + vd) + gls*Vdc - pil)/(gs + gls)
            il = gs*((es + vd) - v1)
            id, gd = fdiode(vd)
            irr = -(il + id)
            err = abs(irr)
            #
            if err < erlim:
                break
            else:
                dvd = irr/(gd + gstot)
                vd += dvd
        #
        pil = il
        iload.append(il)
        vload.append(es + vd)
        #
    return iload, vload
```

```
#
#
def ch3fig15(esrms, freq, ls, rs, Vdc, ncycle, n):
    """ Main program, plots results"""
    #
    #ch3fig15(240.0, 60.0, 0.075, 10.0, 100.0, 20, 512)
    #
    from math import pi
    import numpy as np
    import matplotlib.pyplot as plt
    #
    N = n*ncycle
    #
    delt = 1.0/(freq*n)
    gs = 1.0/rs
    gls = delt/ls
    dteta = 2.0*pi/n
    teta = dteta*np.arange(N)
    tk = delt*np.arange(N)
    esmax = np.sqrt(2)*esrms
    esource = esmax*np.sin(teta)
    #
    iload, vload = ch3fig15_solver(esource, gs, gls, Vdc, N)
    #
    plt.subplot(211)
    plt.plot(tk, vload, 'k-')
    plt.xlabel('$t,\;secs$')
    plt.xlim(0.25, 0.30)
    plt.xticks([0.25, 0.275, 0.30],     \
               ['$0.25$', '$0.275$', '$0.30$'])
    plt.ylabel('$Load\;Voltage$')
    plt.ylim(-250, 350)
    plt.yticks([-200, 0, 100, 200, 300],     \
               ['$-200$', '$0$', '$100$', '$200$', '$300$'])
    plt.grid(True)
    #
    plt.subplot(212)
    plt.plot(tk, iload, 'k-')
    plt.xlabel('$t,\;secs$')
    plt.xlim(0.25, 0.30)
    plt.xticks([0.25, 0.275, 0.30],     \
               ['$0.25$', '$0.275$', '$0.30$'])
    plt.ylabel('$Load\;Current$')
    plt.ylim(0, 20)
    plt.yticks([0, 10, 20], ['$0$', '$10$', '$20$'])
    plt.grid(True)
    #
    plt.show()
    #
    #plt.draw()
    #plt.savefig('Fig315.ps', dpi=400, bbox_inches='tight')
    #
```

3.7.8 Program 3.19

```
#
def fdiode(vd):
    """ Diode characteristic.
        Returns diode current and its conductance"""
    glow = 0.1e-07
    ghigh = 0.1e+03
    Vr = -0.5
    #
    if vd >= Vr:
        id = glow*(vd - Vr)
        gd = glow
    else:
        id = ghigh*(vd - Vr)
        gd = ghigh
    #
    return id, gd
#
def ch3fig19_solver(esource, gls, Iload, N):
    """Assumes esource is a list, fdiode is a function.
        Returns the list iload"""
    #
    import numpy as np
    #
    vd1 = 0.0
    vd2 = 0.0
    pil = 0.0
    erlim = 0.1e-07
    #
    id1 = []
    id2 = []
    vload = []
    #
    for j in range(N):
        #
        es = esource[j]
        #
        for its in range(1000):
            #
            il = gls*(es - vd1) + pil
            idiode1, gd1 = fdiode(vd2 - vd1)
            idiode2, gd2 = fdiode(vd2)
            irr1 = il + idiode1
            irr2 = -(Iload + idiode1 + idiode2)
            err = np.sqrt(irr1*irr1 + irr2*irr2)
            #
            if err < erlim:
                break
            else:
                gmat = np.array([[(gls + gd1), -gd1],
                                 [-gd1, (gd1 + gd2)]])
                bcol = np.array([irr1, irr2])
```

```
                xdv = np.linalg.solve(gmat, bcol)
                dvd1 = xdv[0]
                dvd2 = xdv[1]
                vd1 += dvd1
                vd2 += dvd2
            #
        pil = il
        id1.append(-idiode1)
        id2.append(-idiode2)
        vload.append(vd2)
        #
    return id1, id2, vload
#
#
def ch3fig19(esrms, freq, ls, Iload, ncycle, n):
    """ Main program, plots results"""
    #
    #ch3fig19(120.0, 60.0, 0.0015, 5, 20, 512)
    #
    from math import pi
    import numpy as np
    import matplotlib.pyplot as plt
    #
    N = n*ncycle
    #
    delt = 1.0/(freq*n)
    gls = delt/ls
    dteta = 2.0*pi/n
    teta = dteta*np.arange(N)
    tk = delt*np.arange(N)
    esmax = np.sqrt(2)*esrms
    esource = esmax*np.sin(teta)
    #
    id1, id2, vload = ch3fig19_solver(esource, gls, Iload, N)
    #
    plt.subplot(211)
    #
    plt.plot(tk, id1, 'b-', label = '$Id1$')
    plt.plot(tk, id2, 'r-.', label = '$Id2$')
    plt.xlim(0.1, 0.12)
    plt.xticks([0.10, 0.11, 0.12])
    plt.ylabel('$Id1,\;Id2$')
    plt.ylim(-1, 6)
    plt.yticks([0, 2, 4, 6], ['$0$', '$2$', '$4$', '$6$'])
    plt.legend(bbox_to_anchor=(0., 1.02, 1., .102), loc=3,        \
               ncol=2, mode="expand", borderaxespad=0.)
    plt.subplots_adjust(hspace=0.3)
    #
    plt.subplot(212)
    plt.plot(tk, vload, 'k-')
    plt.xlabel('$time,\;secs$')
    plt.xlim(0.1, 0.12)
    plt.xticks([0.10, 0.11, 0.12],['$0.10$', '$0.11$', '$0.12$'])
```

```
    plt.ylabel('$Load\;Voltage$')
    plt.ylim(0, 200)
    plt.yticks([0, 100, 200], ['$0$', '$100$', '$200$'])
    #
    plt.subplot(212)
    #
    #plt.show()
    #
    plt.draw()
    plt.savefig('Fig19.eps', dpi=400, bbox_inches='tight')
    #
```

3.7.9 Program 3.23

```
#
def fdiode(vd):
    """ Diode characteristic.
        Returns diode current and its conductance"""
    glow = 0.1e-07
    ghigh = 0.1e+03
    Vr = -0.5
    #
    if vd >= Vr:
        id = glow*(vd - Vr)
        gd = glow
    else:
        id = ghigh*(vd - Vr)
        gd = ghigh
    #
    return id, gd
#
def ch3fig23_solver(esource, gc, gload, N):
    """Assumes esource is a list, fdiode is a function.
        Returns the list iload"""
    #
    #import numpy as np
    #
    vd = 0.0
    pvc = 0.0
    erlim = 0.1e-07
    #
    ic = []
    vload = []
    vdiode = []
    #
    for j in range(N):
        #
        es = esource[j]
        #
        for its in range(1000):
            #
            vl = es + vd
```

```
            id, gd = fdiode(vd)
            irr = gc*pvc - (gc + gload)*vl -id
            err = abs(irr)
            #
            if err < erlim:
                break
            else:
                dvd = irr/(gd + gc + gload)
                vd += dvd
            #
        ccap = gc*(vl - pvc)
        ic.append(ccap)
        pvc = vl
        vload.append(vl)
        vdiode.append(-vd)
        #
    return ic, vload, vdiode
#
#
def ch3fig23(Erms, freq, C, R, ncycle, n):
    """ Main program, plots results"""
    #
    #  calling command:
    #  ch3fig23(120.0, 60.0, 0.1e-03, 500, 4, 512)
    #
    from math import pi
    import numpy as np
    import matplotlib.pyplot as plt
    #
    N = n*ncycle
    #
    delt = 1.0/(freq*n)
    gc = C/delt
    gload = 1.0/R
    dteta = 2.0*pi/n
    teta = dteta*np.arange(N)
    #tk = delt*np.arange(N)
    esmax = np.sqrt(2.0)*Erms
    esource = esmax*np.sin(teta)
    #
    ic, vload, vdiode = ch3fig23_solver(esource, gc, gload, N)
    #
    #   ang[0] is the instant when diode is reverse biased
    #   ang[1] is the instant when diode begins conducting again
    #
    k = 0
    ang = []
    ide = []
    delta_vc = []
    for j in range(N):
        if k > 1:
            break
        if vdiode[j]*vdiode[j+1] < 0:
```

```
            ang.append(teta[j])
            ide.append(ic[j+2]+vload[j+2]/R)
            delta_vc.append(esmax-vload[j])
            k += 1
    print('alpha0=     ',ang[0],'alpha1=     ',ang[1] )
    print('ide0=       ',ic[2],' ide1=       ',ide[1])
    print('dvc1=       ',delta_vc[1])
    #
    plt.subplot(211)
    plt.plot(teta, esource, 'g-', label = '$e_{S}$')
    plt.plot(teta, vload, 'r-.', label = '$v_{load}$')
    plt.plot(teta, vdiode, 'b-', label = '$v_{diode}$')
    #plt.xlabel('$\omega t,\;secs$')
    plt.xlim(0, 6*pi)
    plt.xticks([0, ang[0], 2*pi, ang[1], 4*pi, 6*pi], ['$0$',        \
      '$\\alpha_{0}$', '$2\pi$', '$\\alpha_{1}$', '$4\pi$', '$6\pi$'])
    plt.ylabel('$e_{S},\,v_{load}\,\,v_{diode}\,(V)$')
    plt.ylim(-400, 400)
    plt.yticks([-400, 0, 400], ['$-400$', '$0$', '$400$'])
    plt.grid(True)
    plt.legend(bbox_to_anchor=(0., 1.02, 1., .102), loc=3, ncol=3, \
               mode="expand", borderaxespad=0.)
    #
    plt.subplot(212)
    plt.plot(teta, ic, 'b-')
    plt.xlabel('$\omega t,\;radians$')
    plt.xlim(0, 6*pi)
    plt.xticks([0, ang[0], 2*pi, ang[1], 4*pi, 6*pi], ['$0$',          \
       '$\\alpha_{0}$', '$2\pi$', '$\\alpha_{1}$', '$4\pi$', '$6\pi$'])
    plt.ylabel('$i_{C},\,(A)$')
    plt.ylim(0, 8)
    plt.yticks([0, 8], ['$0$', '$8$'])
    plt.grid(True)
    #
    plt.show()
    #
    #plt.draw()
    #plt.savefig('Fig323.ps', dpi=400, bbox_inches='tight')
    #
```

3.7.10 Program 3.24

```
def ch3prog24_F(x):
    """Function evaluation"""
    #
    import numpy as np
    #
    F = np.sin(x) - 1.088428*np.exp(-x/18.8496)
    #
    return F
#
def ch3prog24_dF(x):
```

```
    """Derivative evaluation"""
    #
    import numpy as np
    #
    dF = np.cos(x) + 0.0577476*np.exp(-x/18.8496)
    #
    return dF
#
def ch3prog24():
    """Newton Raphson procedure"""
    #
    from math import pi
    alpha1 = 2*pi
    epsilon = 0.001
    F = ch3prog24_F(alpha1)
    #
    while abs(F) >= epsilon:
        dF = ch3prog24_dF(alpha1)
        alpha1 -= F/dF
        F = ch3prog24_F(alpha1)
    print('F = ', F, 'alpha1 = ', alpha1)
```

References

[1] N. Mohan, T. M. Undeland, W. P. Robbins, *Power Electronics: Converters, Applications, and Design*, 2^{nd}Ed., John Wiley & Sons Inc., USA, 1995.

[2] D. W. Hart, *Power Electronics*, McGraw Hill, New York, 2011.

[3] J. G. Kassakian, M. F. Schlecht, G. C. Verghese, *Principles of Power Electronics*, 1^{st}Ed., Pearson India, 2009.

[4] M. H. Rashid, *Power Electronics: Devices, Circuits and Applications*, 4^{th}Ed., Pearson India, Chennai, 2018.

[5] S. R. Naidu, A. M. N. Lima, *A new approach for the simulation of power electronic circuits*, IEEE Trans., on Circuits and Systems-1, vol.49, no.9, pp.1317-1324, 2002.

4

Full-wave AC–DC Converters

It has been observed in the half-wave converter that the current supplied by the alternating voltage source is a series of unidirectional pulses having a non-zero average value. Alternating voltage sources are not normally designed to supply currents that have a direct component. The full-wave AC/DC converter overcomes this drawback. The AC side input current to the converter consists of alternating pulses, the mean value of the current over a period of the alternating source voltage being zero. Another improvement in the performance of the full-wave converter is the smaller ripple in the output direct voltage. However, the benefits come at a cost; the full-wave converter requires more diodes than the half-wave converter.

The general form of the single-phase, full-wave, AC/DC converter is shown in Figure 4.1. It consists of the AC voltage source, the full-wave converter, and the DC load. The voltage source may not be ideal and may have an internal impedance (R_s, L_s). Other loads may be connected to the AC voltage source and, therefore, the voltage distortion at the point of common coupling (PCC) should be within limits. For this reason, an inductor $L_c \geq 5L_s$ is connected at the input of the converter. The diode bridge converts the alternating voltage of the AC source to a voltage waveform that has a significant direct component. At the output of the diode bridge, a filter (L_d, C_d) attenuates the ripple in the converted voltage. The DC load is connected at the output of the filter.

4.1 Simplified Full-wave Converter

Analytical, closed-form solution for the circuit of Figure 4.1 is not easily obtainable though its simulation is definitely possible. However, it is worthwhile simplifying the circuit and studying its conversion properties before working through the simulation of the general full-wave converter.

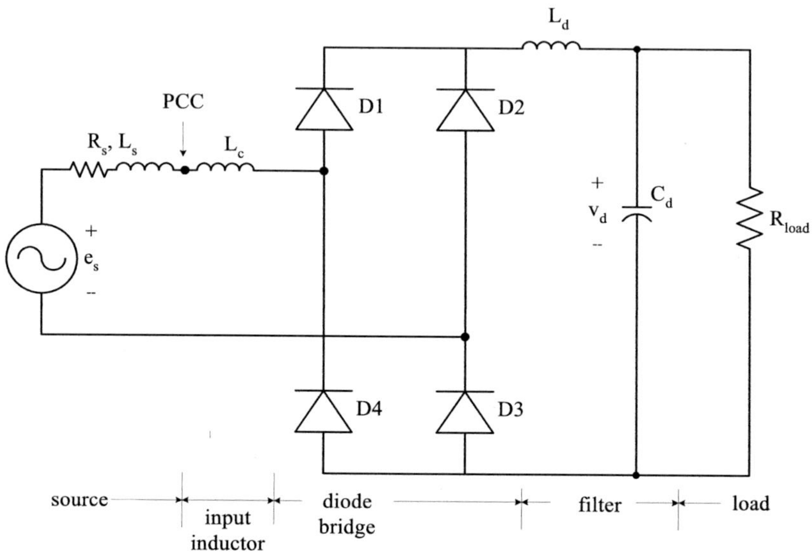

Figure 4.1 Single-phase, full-wave, AC/DC converter: circuit diagram.

A simplified form of the single-phase, full-wave AC/DC converter is shown in Figure 4.2. In this circuit, the alternating voltage source is the ideal source $v_s(\omega t) = V_m \sin \omega t = \sqrt{2}V_s \sin \omega t$. The load on the DC side is so highly inductive that it has been substituted by a constant current source of known amplitude I_0 amperes. All the diodes are assumed to be identical.

Consider the positive half-period of the alternating source voltage. During this interval, the diode D1 is driven into the conducting state and the diode D2 is reverse biased. (If D2 is assumed to be conducting, then $v_{AP} = v_{AB} + v_{BP} \simeq v_{AB}$. Thus, a forward bias is applied to D1.) Another argument to determine the states of D1 and D2 is to observe that their cathodes are tied together at the node P. The diode that conducts is the one that has a higher voltage applied to its anode. A similar reasoning will show that D3 conducts and D4 is reverse biased during the positive half-period of the source voltage. During the negative half-period of the source voltage, D1 and D3 are reverse biased and D2 and D4 are forward biased. On the basis of the discussion above, the voltage and current waveforms at the input and output terminals of the full-wave converter are shown in Figure 4.3 [1].

On the AC side, the current supplied by the alternating voltage source is an alternating current having a square waveform. This waveform possesses half-wave and odd symmetries and may therefore be represented by a Fourier

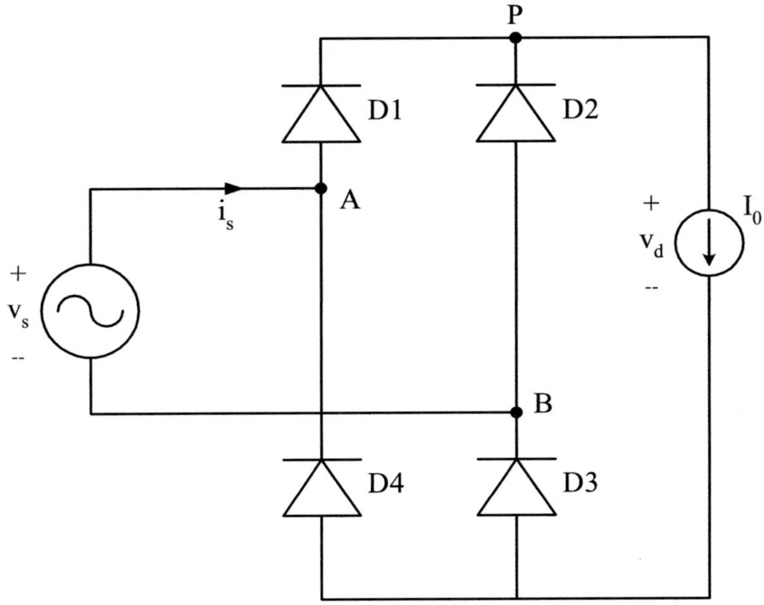

Figure 4.2 Simplified, full-wave AC/DC converter: circuit diagram.

series of the form

$$\begin{aligned} i_s(\omega t) &= b_1 \sin \omega t + b_3 \sin 3\omega t + b_5 \sin 5\omega t + \cdots \\ &= \sqrt{2}(I_{s1} \sin \omega t + I_{s3} \sin 3\omega t + I_{s5} \sin 5\omega t + \ldots). \end{aligned} \quad (4.1)$$

The amplitudes of the harmonic components are

$$\begin{aligned} b_j &= \frac{4}{\pi} \int_0^{\pi/2} I_0 \sin j\omega t d\omega t \\ &= \frac{4I_0}{\pi j}, \ j = 1, 3, 5, \ldots . \end{aligned} \quad (4.2)$$

In particular, $b_1 = \frac{4I_0}{\pi}$, and the RMS value of the fundamental frequency component is $I_{s1} = \frac{4I_0}{\pi\sqrt{2}}$. The RMS value of the current $i_s(\omega t)$ is given by

$$\begin{aligned} I_s &= \sqrt{\frac{1}{2\pi} \int_0^{2\pi} i_s^2(\omega t) d\omega t} \\ &= I_0. \end{aligned} \quad (4.3)$$

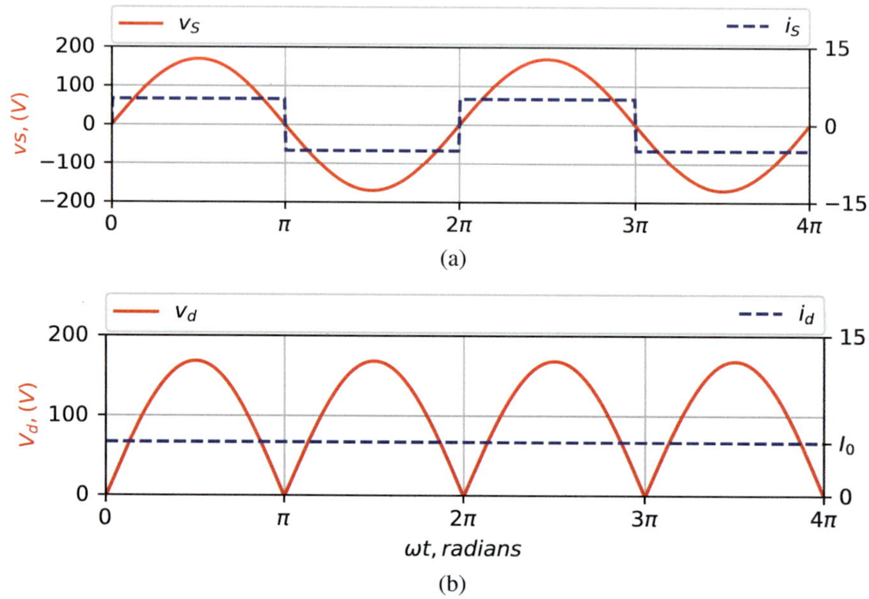

Figure 4.3 Simplified full-wave AC/DC converter: response waveforms.

On the DC side of the converter, the voltage waveform is a train of positive sine-wave pulses. The frequency of this waveform is twice the frequency of the input voltage waveform. The average output voltage is

$$\begin{aligned} V_{d0} &= \frac{1}{\pi}\int_0^{\pi} \sqrt{2}V_s \sin \omega t d\omega t \\ &= \frac{2\sqrt{2}}{\pi}V_s. \end{aligned} \tag{4.4}$$

The power supplied by the AC voltage source is

$$\begin{aligned} P_i &= V_s\frac{2\sqrt{2}}{\pi}I_0 \\ &= \frac{2\sqrt{2}}{\pi}V_sI_0. \end{aligned} \tag{4.5}$$

Note that the current harmonics do not carry any active power. The apparent input power is $S = V_sI_0$ and, finally, the power factor of the converter is

$$f_p = 2\sqrt{2}/\pi. \tag{4.6}$$

The power factor of the converter, as seen by the voltage source, is quite high in spite of the load on the DC side of the converter being highly inductive. This is due to the fact that the current on the DC side is a constant direct current and an inductance is a short circuit for this current. For this reason, no reactive power is supplied to the load.

The power factor of the converter as seen at its input terminals is high, but it is less than unity. This is due to the distortion in the current waveform, which is an alternating square wave and not a sine wave. The total harmonic distortion (THD) factor is given by

$$\text{THD} = \sqrt{(I_s/I_{s1})^2 - 1} = \sqrt{\frac{\pi^2}{8} - 1} = 0.4834. \quad (4.7)$$

Clearly, the distortion in the current waveform is unacceptable even though the converter's power factor is high.

4.2 Simplified Full-wave Converter with Non-ideal Voltage Source

We consider the same converter as in the previous section but with the alternating voltage source at the input having an internal inductance of L_s henries (Figure 4.4) [1]. The load on the DC side continues to be highly inductive and it is represented by the constant current source I_0 amperes. We assume that for $\omega t < 0$, the diodes D2 and D4 have been forward biased and the diodes D1 and D3 reverse biased. The load current has been circulating through the diodes D2 and D4 and also through the inductance L_s, in the direction from A to B.

For $\omega t > 0$, the alternating voltage source changes polarity and drives the diodes D1 and D3 into conduction. However, the instantaneous transfer of current from diode D2 to D1 (and from the diode D4 to D3) would imply an instantaneous change in the direction of the current flowing through the inductance L_s. Inductive currents cannot change instantaneously. Therefore, during an interval $0 \leq \omega t \leq u$, all the diodes conduct and current transients i_u are initiated in the loops shown in Figure 4.4. At the end of the interval, the load current is transferred from the diodes D2 and D4 to the diodes D1 and D3. This is the commutation process described in the previous chapter.

At an instant during the commutation interval, the current $i_{D1} = i_{D3} = i_u$ and $i_{D2} = i_{D4} = I_0 - i_u$. Applying Kirchhoff's law at the node A,

$$i_{D4} + i_s - i_{D1} = 0. \quad (4.8)$$

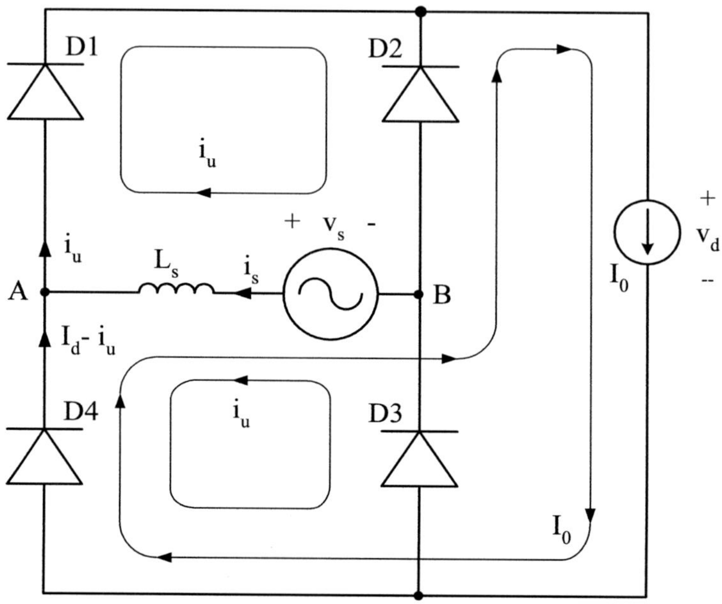

Figure 4.4 Simplified single-phase, full-wave AC/DC converter: non-ideal voltage source.

Therefore,

$$i_s = i_{D1} - i_{D4} = 2i_u - I_0. \tag{4.9}$$

At $\omega t = 0$, $i_u = 0$ and, therefore, $i_s = -I_0$. Similarly, at $\omega t = u$, $i_u = I_0$ and $i_s = I_0$. Further, during the commutation interval, all the diodes are conducting and, therefore, $v_{AB} = 0$. Thus,

$$L_s \frac{di_s}{dt} = \sqrt{2} V_s \sin \omega t. \tag{4.10}$$

Integrating the equation,

$$\int_{-I_0}^{I_0} di_s = \frac{\sqrt{2} V_s}{\omega L_s} \int_0^u \sin \omega t \, d\omega t. \tag{4.11}$$

That is,

$$\cos u = 1 - \frac{2\omega L_s I_0}{\sqrt{2} V_s}. \tag{4.12}$$

The load voltage and the AC side current waveforms are shown in Figure 4.5, plotted by Program 4.05.

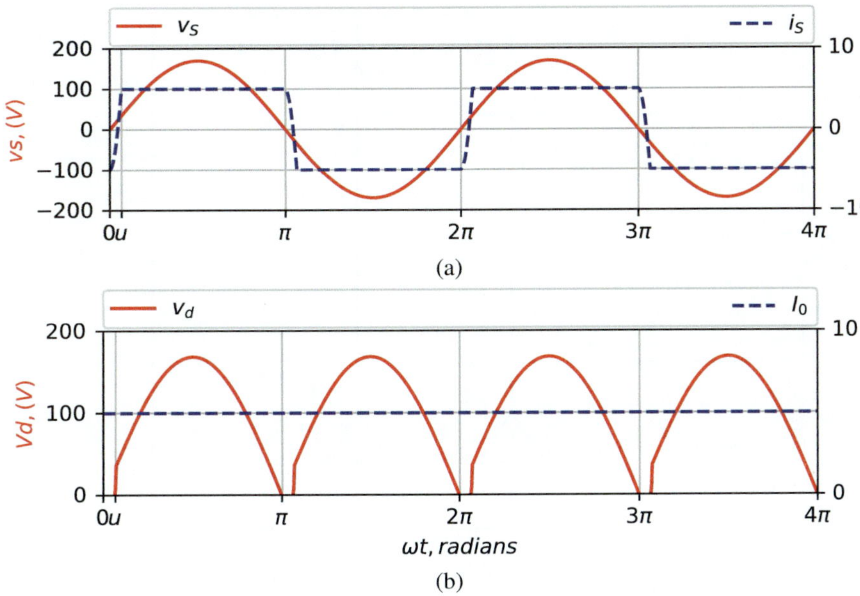

Figure 4.5 Simplified full-wave converter, non-ideal voltage source: response waveforms.

The output voltage is zero during the commutation interval because the diodes act as a short-circuit. The mean value of the output voltage is given by

$$\begin{aligned} V_d &= \frac{1}{\pi} \int_u^{\pi} \sqrt{2} V_s \sin \omega t \, d\omega t \\ &= \frac{\sqrt{2} V_s}{\pi} (1 + \cos u). \end{aligned} \tag{4.13}$$

4.3 Simulation of Single-phase, Full-wave Converters

4.3.1 Non-ideal alternating voltage source and highly capacitive load

The converter is shown in Figure 4.6 with the highly capacitive load represented by the direct voltage source V_d. The alternating voltage source ia given by $v_s(\omega t) = \sqrt{2} V_s \sin \omega t$. It has an internal impedance of L_s henries. For the useful functioning of the converter, it is necessary that $\sqrt{2} V_s > V_d$. The diodes are assumed to be identical.

The equivalent circuit for computing the converter's response is shown in Figure 4.7. The inductance has been substituted by its equivalent conductance

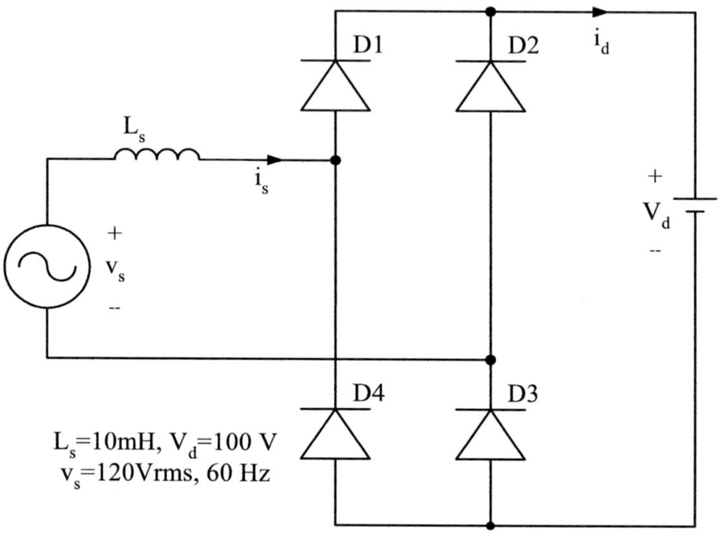

Figure 4.6 Full-wave converter, non-ideal voltage source, highly capacitive load: circuit diagram.

$gls = \Delta t/L_s$ and the past-history current source $pils$ is connected in parallel with the conductance. Probe voltages should be connected across each diode and their amplitudes be determined iteratively at each time-step such that no current flows into them. However, the same current flows at all instants in the diode pairs (D1 and D3) and (D2 and D4). Therefore, $v_{D1} = v_{D3}$ and $v_{D2} = v_{D4}$. There is the further constraint that $v_{D1} + v_{D4} = V_d$, which is the given data. The result is that a single probe voltage in parallel with any one of the diodes should be sufficient. The preference in the ensuing simulation is to place the probe voltage v_a at the input to the diode bridge.

On inspection of Figure 4.7, we obtain

$$v_{D1} + v_a + v_{D3} = V_d \tag{4.14}$$

$$v_{D2} - v_a + v_{D4} = V_d. \tag{4.15}$$

But $v_{D1} = v_{D3}$ and $v_{D2} = v_{D4}$. Thus,

$$v_{D1} = \frac{1}{2}(V_d - v_a) \tag{4.16}$$

$$v_{D2} = \frac{1}{2}(V_d + v_a). \tag{4.17}$$

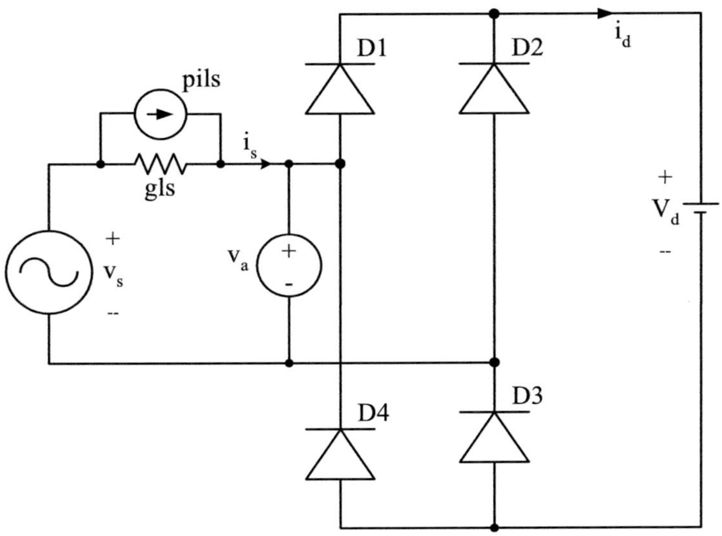

Figure 4.7 Full-wave converter, non-ideal voltage source, highly capacitive load: transient equivalent circuit.

Choosing a starting value for v_a, the variables of the circuit are computed as shown below:

$$ils = gls(v_s - v_a) + pils \tag{4.18}$$

$$id1, gd1 = fdiode(v_{D1}) \tag{4.19}$$

$$id2, gd2 = fdiode(v_{D2}) \tag{4.20}$$

$$id = -(id1 + id2). \tag{4.21}$$

The error current flowing into the probe voltage v_a is

$$i_{rr} = ils + id1 - id2. \tag{4.22}$$

At each time-step, it is required to find the amplitude of v_a such that i_{rr} is smaller than the specified tolerance. This is achieved by correcting the value of v_a over successive iterations.

The correction δv_a to be added to v_a is obtained from the small-signal equivalent circuit shown in Figure 4.8. In this circuit, the voltage sources v_s and V_d and the current source $pils$ are removed. The probe voltage v_a is replaced by the error current source i_{rr} with polarity reversed. The voltage developed across this current source is the correction to be added to the

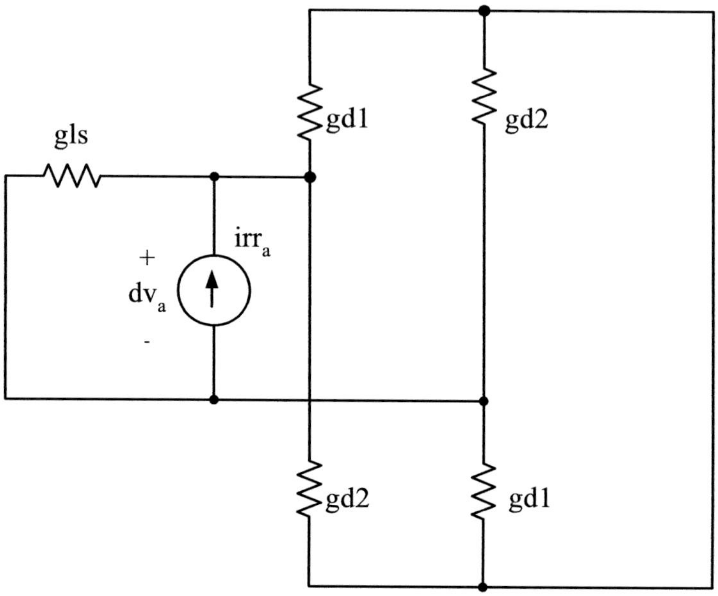

Figure 4.8 Full-wave converter, non-ideal voltage source, highly capacitive load: small-signal equivalent.

amplitude v_a. It is given by

$$\delta v_a = \frac{i_{rr}}{(gls + \frac{1}{2}(gd1 + gd2))}. \tag{4.23}$$

After convergence and before proceeding to the next time-step, the circuit variables that need to be plotted are stored in data files and the past-history current source $pils$ is updated.

The response waveforms for the circuit of Figure 4.6 are shown in Figure 4.9. There are some important conclusions to be noted in the waveforms:

- In each positive half-period, current begins to flow in the diodes D1 and D3, when the voltage v_s rises and becomes equal to V_d. In the negative half-period, the diodes D2 and D4 begin to conduct when $-v_s$ becomes equal to V_d.
- The average value of the inductor voltage v_{LS} (i.e., its direct component) is zero in the steady-state.

The code for simulating the circuit of Figure 4.6 is presented in Program 4.09.

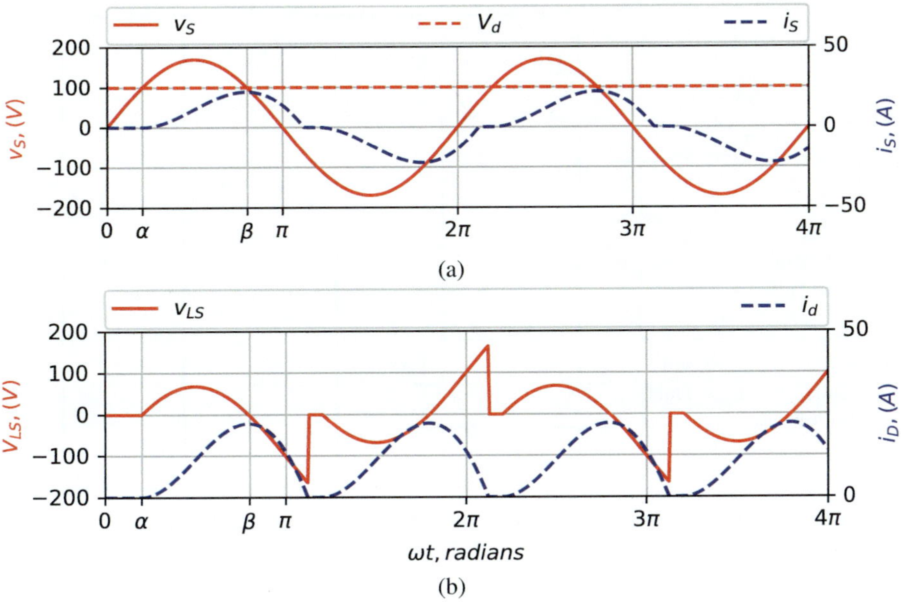

Figure 4.9 Full-wave converter, non-ideal voltage source, highly capacitive load: response waveforms.

4.3.2 The general, single-phase, full-wave AC/DC converter

The single-phase, full-wave converter to be simulated is shown in Figure 4.10. The objective of the simulation is to provide the following information:

- The waveform of the voltage $v_{\rm pcc}$ at the point of common coupling. The AC voltage source may be supplying other loads connected at the PCC. Therefore, the distortion in the waveform of $v_{\rm pcc}$ should be kept within limits.
- The waveform of the current $i_s(\omega t)$ on the AC side of the converter.
- The waveform of the voltage $v_c(\omega t)$ at the load terminals. The second harmonic ripple in this waveform should not be excessive.

Figure 4.11 shows the transient equivalent circuit of the converter. Probe voltages v_a and v_d are connected at the input and output terminals of the diode bridge and their amplitudes are determined such that negligible currents flow into them. Figure 4.11 shows the equivalent circuit of the converter. Note that the inductances L_s and L_c have been lumped together and that $gls = \Delta t/(L_s + L_c)$.

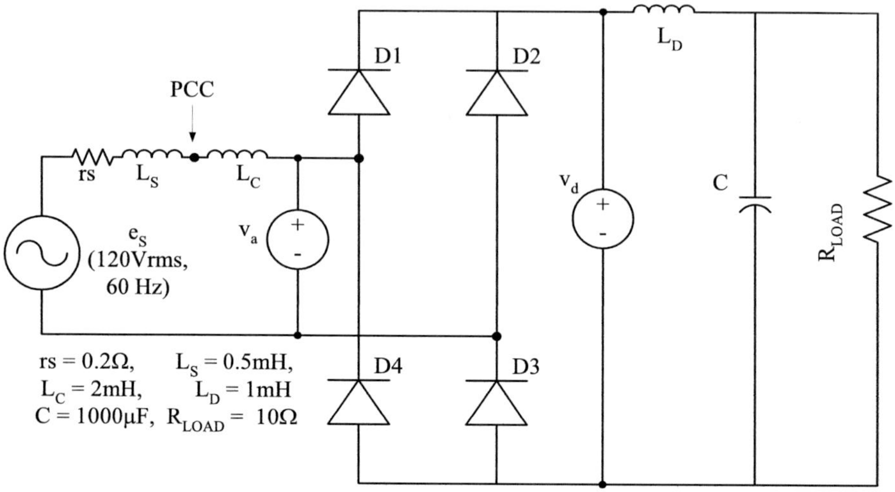

Figure 4.10 Single-phase, full-wave AC/DC converter, circuit diagram.

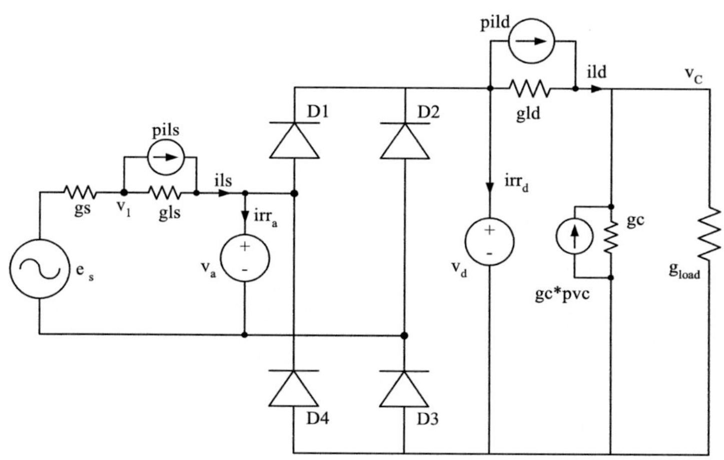

Figure 4.11 Single-phase, full-wave AC/DC converter: equivalent circuit.

Applying Kirchhoff's current law at the node with voltage v_1,

$$gs(v_1 - e_s) + gls(v_1 - v_a) + pils = 0. \tag{4.24}$$

From eqn 4.24, we get

$$v_1 = (gs * e_s + gls * v_a - pils)/(gs + gls). \tag{4.25}$$

Similarly, applying Kirchhoff's current law at the output terminal and re-arranging the equation, we obtain

$$v_c = (gld * v_d + pild + gc * pvc)/(gc + gload + gld). \tag{4.26}$$

Other equations that are obtained from the equivalent circuit are

$$\begin{aligned} vd1 &= 0.5(v_d - v_a) && (4.27)\\ vd2 &= 0.5(v_d + v_a) && (4.28)\\ id1, gd1 &= fdiode(vd1) && (4.29)\\ id2, gd2 &= fdiode(vd2) && (4.30)\\ irr_a &= ils + id1 - id2 && (4.31)\\ ild &= gld * (v_d - v_c) + pild && (4.32)\\ irr_d &= -(id1 + id2 + ild). && (4.33) \end{aligned}$$

Further,

$$\begin{aligned} v_{\mathrm{pcc}} &= v_1 + (v_a - v_1)\frac{L_s}{L_s + L_c}\\ &= \frac{L_c}{L_s + L_c} * v_1 + \frac{L_s}{L_s + L_c} * v_a\\ &= (1 - k_a) * v_1 + k_a * v_a, \text{ where } k_a = L_s/(L_s + L_c). \quad (4.34) \end{aligned}$$

The error currents irr_a and irr_d can take positive or negative amplitudes. Therefore, we define the simulation error as

$$err = \sqrt{irr_a^2 + irr_d^2}. \tag{4.35}$$

Convergence is obtained when $err < tolerance$. If $err > tolerance$, then v_a and v_d should be corrected by adding to them δv_a and δv_d, respectively. The corrections are calculated from the small-signal equivalent circuit shown in Figure 4.12.

In the small-signal equivalent circuit, all external sources have been removed and the probe voltages are substituted by the error current sources irr_a and irr_d with their polarities reversed. The diodes are replaced by the conductances obtained in eqn (4.29) and (4.30). Putting

$$\begin{aligned} gstot &= gs * gls/(gs + gls) && (4.36)\\ gdtot &= gld * (gc + gload)/(gld + gc + gload), && (4.37) \end{aligned}$$

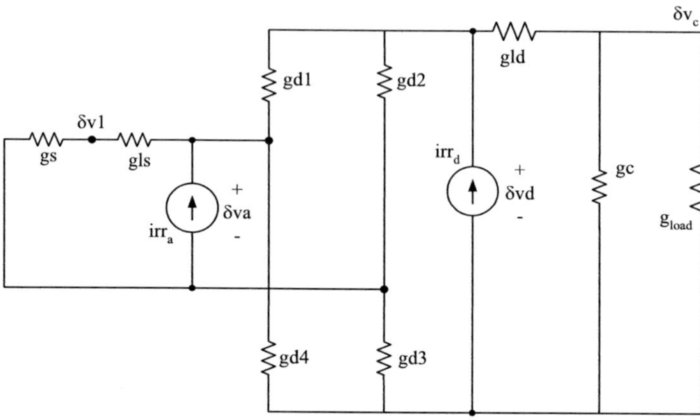

Figure 4.12 Single-phase, full-wave AC/DC converter: small-signal equivalent.

the following nodal equations are obtained:

$$gd2 * \frac{1}{2}(\delta v_a + \delta v_d) + gstot * \delta v_a - gd1 * \frac{1}{2}(\delta v_d - \delta v_a) = irr_a \quad (4.38)$$

$$gd1 * \frac{1}{2}(\delta v_d - \delta v_a) + gd2 * \frac{1}{2}(\delta v_a + \delta v_d) + gdtot * \delta v_d = irr_d. \quad (4.39)$$

The corrections δv_a and δv_d are obtained from the solution of eqn (4.38) and (4.39).

After convergence and before proceeding to the next time-step, the values of the variables to be plotted are stored and the AC source voltage and past-history terms are updated. The response waveforms are shown in Figure 4.13 and the code for generating the waveforms is presented in Program 4.13.

4.4 Three-phase, Uncontrolled AC/DC Converters

The three-phase, diode-based AC/DC converter is used for supplying large industrial DC loads [1, 2, 3]. The circuit diagram of a typical converter is shown in Figure 4.14. The three-phase AC source is assumed to be balanced but with an internal impedance R_s, L_s per phase.

The converter is connected to the terminals of the AC voltage source at the point of common coupling. An inductor L_c per phase is connected at the input side of the diode bridge to keep the distortion of the voltage waveforms at the PCC within limits. A filter L_d, C_d is inserted between the output terminals of the diode bridge and the DC load.

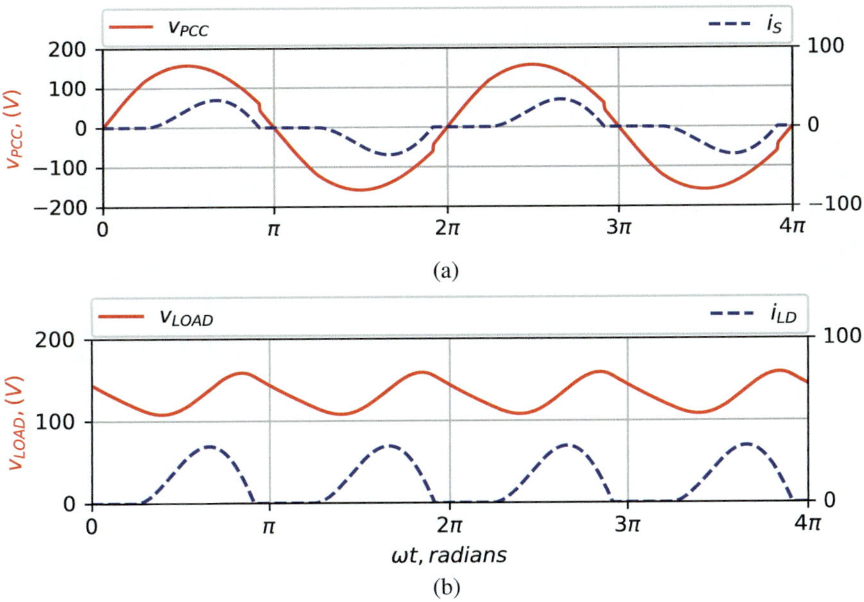

Figure 4.13 Single-phase, full-wave converter: response waveforms.

4.4.1 The simplified three-phase, AC/DC converter

The three-phase converter shown in Figure 4.14 cannot be easily analyzed. The simple form of the converter shown in Figure 4.15 is tractable and we shall study its input–output characteristics. The balanced, ideal three-phase voltage source is connected directly to the input of the converter. A highly inductive load represented by the (known) current source I_d is supplied by the converter.

The cathodes of the diodes D1, D3, and D5 are connected together at the terminal P. At any instant, only one of these diodes may be forward biased and it is the diode that has the highest voltage applied to its anode. Similarly, the diodes D2, D4, and D6 have their anodes connected to the terminal N and the diode that has the lowest voltage connected to its cathode is forward biased. Thus,

$$v_{Pn} = \max(es_a, es_b, es_c) \tag{4.40}$$

$$v_{Nn} = \min(es_a, es_b, es_c) \tag{4.41}$$

$$v_d = v_{Pn} - v_{Nn}. \tag{4.42}$$

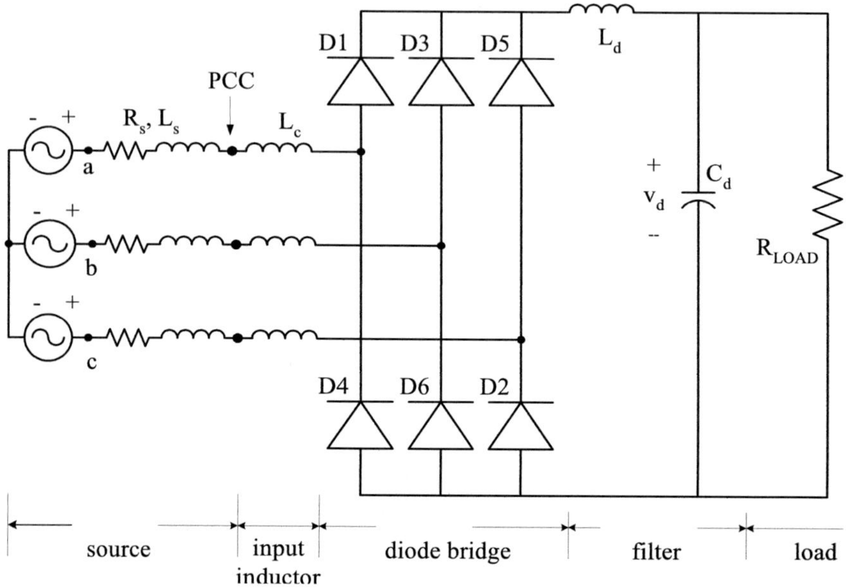

Figure 4.14 Three-phase, diode-based, AC/DC converter.

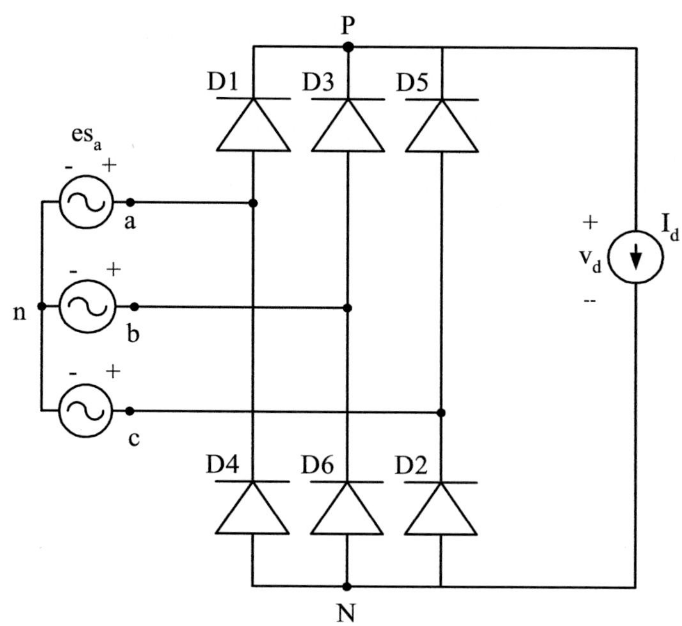

Figure 4.15 Simplified three-phase AC/DC converter.

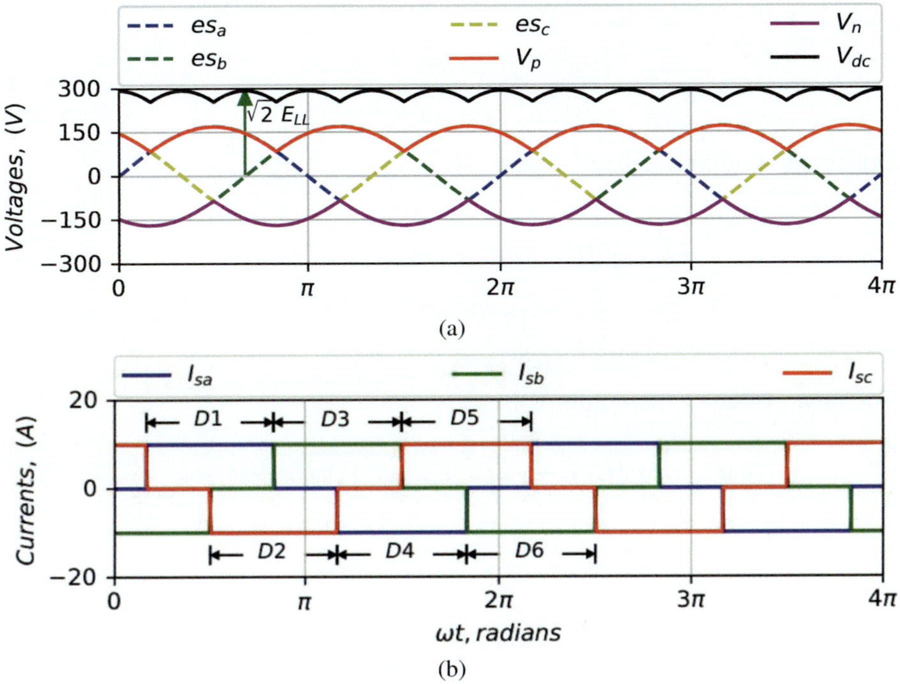

Figure 4.16 Simplified three-phase AC/DC converter: response waveforms.

The waveform of v_d is shown in Figure 4.16 (a) and it consists of six pulses per period of the fundamental frequency [1, 2]. For this reason, the converter is also referred to as the six-pulse converter. Each diode conducts for 120° per period.

The waveforms of the phase currents are shown in Figure 4.16 (b). The phase current i_{sa} is given by

$$\begin{aligned} i_{sa} &= I_d, \text{ when D1 conducts} \\ &= -I_d, \text{ when D4 conducts} \\ &= 0, \text{ when D1 and D4 are reverse biased.} \end{aligned}$$

In each period, the diodes conduct in the sequence D1-D6, D1-D2, D3-D2, D3-D4, D5-D4, and D5-D6. Commutation of the current from one diode to the next is instantaneous because the AC source is an ideal source.

From Figure 4.16 (a), we obtain

$$\begin{aligned} es_a &= \sqrt{2}E_{ph}\sin\omega t \\ es_b &= \sqrt{2}E_{ph}\sin(\omega t - 2\pi/3). \end{aligned}$$

During the interval $\frac{\pi}{6} \le \omega t \le \frac{\pi}{2}$,

$$\begin{aligned} v_d(\omega t) &= es_a - es_b \\ &= \sqrt{2}E_{ph}\{\sin\omega t - \sin(\omega t - 2\pi/3)\} \\ &= \sqrt{2}\sqrt{3}E_{ph}\sin(\omega t + \pi/6) \\ &= \sqrt{2}E_{LL}\sin(\omega t + \pi/6). \end{aligned}$$

The mean value of the output voltage is

$$\begin{aligned} V_{\mathrm{dc}} &= \sqrt{2}E_{LL}\frac{1}{\pi/3}\int_{\frac{\pi}{6}}^{\frac{\pi}{2}} \sin(\omega t + \pi/6)\, d\omega t \\ &= \frac{3\sqrt{2}}{\pi}E_{LL} \\ &= 1.35E_{LL}. \end{aligned} \tag{4.43}$$

The waveforms of the voltage and the current of phase-a are shown in Figure 4.17. The current is non-sinusoidal, but its waveform possesses odd and half-wave symmetries.

Applying Fourier analysis, we obtain

$$\begin{aligned} i_{sa}(\omega t) &= \frac{2\sqrt{3}}{\pi}I_d\{\sin\omega t + \frac{1}{5}\sin 5\omega t + \cdots\} \\ &= \sqrt{2}\{\frac{\sqrt{6}I_d}{\pi}\sin\omega t + \frac{\sqrt{6}I_d}{5\pi}\sin 5\omega t + \cdots\} \\ &= \sqrt{2}\{I_{s1,a}\sin\omega t + I_{s5,a}\sin 5\omega t + \cdots\}. \end{aligned}$$

The RMS value of $i_{sa}(\omega t)$ is

$$I_{sa} = \sqrt{\frac{1}{\pi}\int_{\frac{\pi}{6}}^{\frac{5\pi}{6}} I_d^2\, d\omega t} = \sqrt{\frac{2}{3}}I_d. \tag{4.44}$$

The fundamental component of $i_{sa}(\omega t)$ is in phase with the sinusoidal voltage $es_a(\omega t)$ and, therefore, the displacement power factor is $f_d = 1$. Then, the power factor of the load supplied by phase-a is

$$f_p = \frac{I_{s1,a}}{I_{sa}}f_d = \frac{3}{\pi}. \tag{4.45}$$

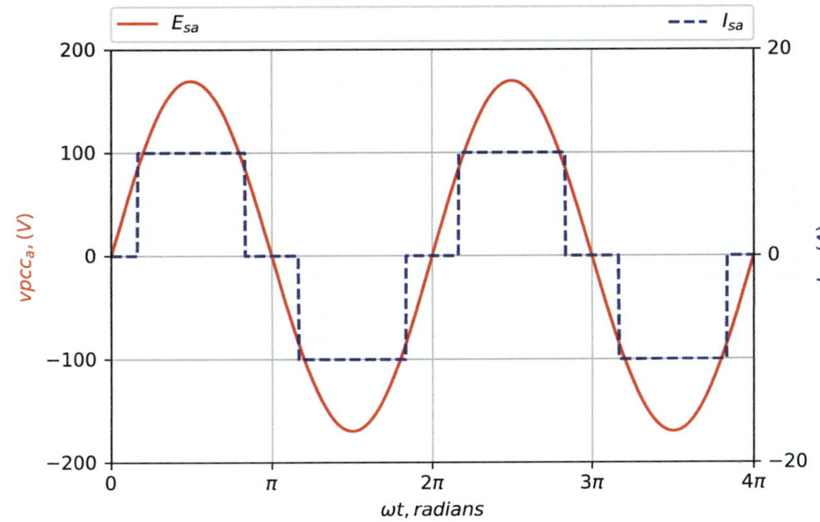

Figure 4.17 Phase-a: voltage and current waveforms.

4.4.2 Simplified AC/DC converter, non-ideal three-phase AC source

Consider the three-phase AC/DC converter shown in Figure 4.18. The three-phase AC source has an internal inductance of L_s henries per phase.

Thus, the transfer of current from one diode to the other is not instantaneous. Typically, the commutation of current from diode D5 to D1 starts at the instant when $es_a = es_c$ (Figure 4.19, plotted by Program 4.19), and a transient current i_u circulates in the closed loop comprising the diodes D5 and D1. Choosing $\omega t = 0$ when $es_a = es_c$, the transient current circulates during the interval $0 \leq \omega t \leq u$. During this interval, $i_{sa} = i_u$ and $i_{sc} = I_d - i_u$. At $\omega t = 0, i_u = 0$, and at $\omega t = u, i_u = I_d$.

Referring to Figure 4.20, which shows the circulating currents during the commutation interval, we obtain

$$\begin{aligned} es_a - es_c &= v_{La} - v_{Lc} \\ &= L_s \frac{di_u}{dt} - L_s \frac{d}{dt}(I_d - i_u) \\ &= 2L_s \frac{di_u}{dt}. \end{aligned} \tag{4.46}$$

But $es_a = \sqrt{2}E_{ph} \sin \omega t$ and $es_c = \sqrt{2}E_{ph} \sin(\omega t + 5\pi/3)$. Thus,

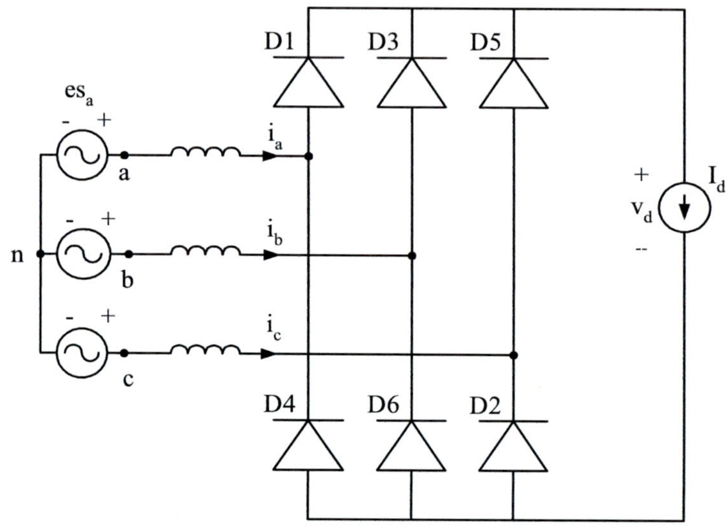

Figure 4.18 Three-phase AC/DC converter, non-ideal three-phase AC source.

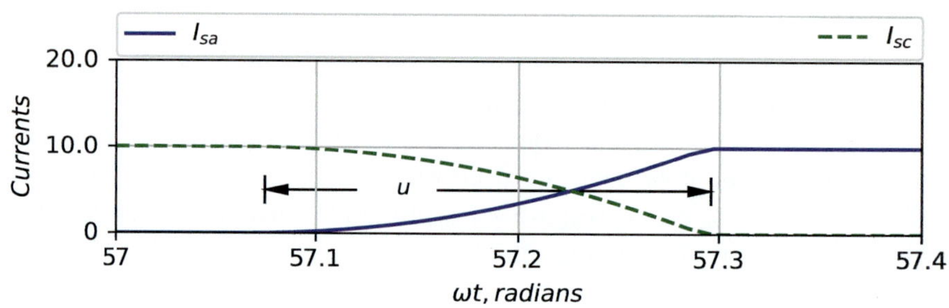

(a)

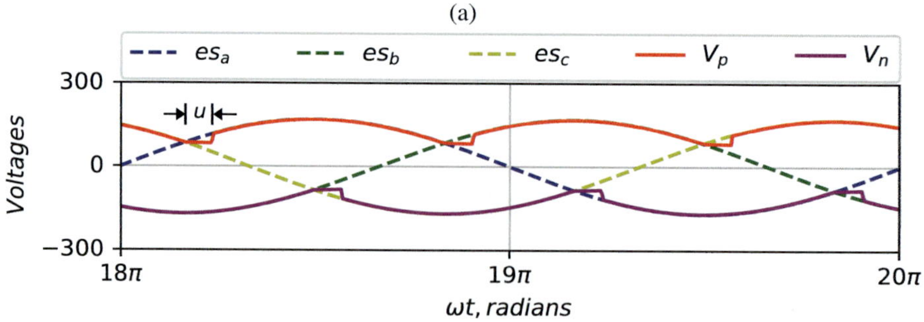

(b)

Figure 4.19 Three-phase AC/DC converter: current commutation.

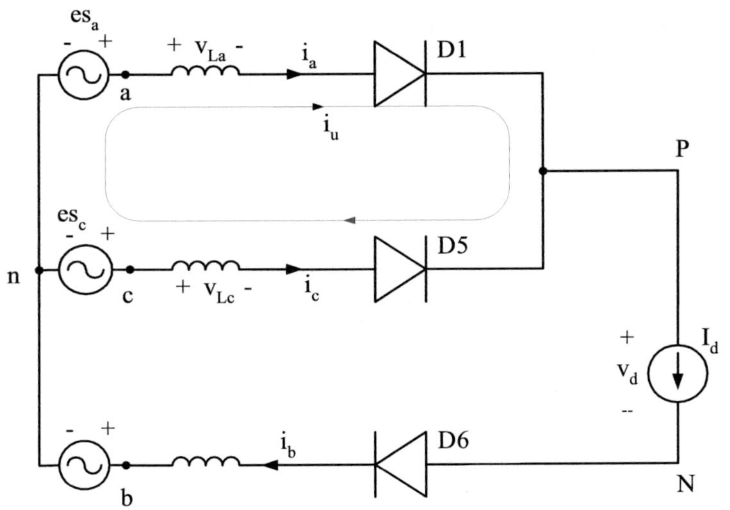

Figure 4.20 Current transfer from diode D5 to diode D1.

$$es_a - es_c = \sqrt{2}\frac{\sqrt{3}}{2}2E_{ph}\sin\omega t$$
$$= \sqrt{2}E_{LL}\sin\omega t. \tag{4.47}$$
$$\text{Therefore, } L_s\frac{di_u}{dt} = \frac{\sqrt{2}E_{LL}}{2}\sin\omega t. \tag{4.48}$$

Integrating eqn (4.48) between the limits $\omega t = 0$, $i_u = 0$ and $\omega t = u$, $i_u = I_d$, we obtain

$$\int_0^{I_d} di_u = \frac{\sqrt{2}E_{LL}}{2\omega L_s}\int_0^u \sin\omega t\, d\omega t.$$
$$\text{That is, } \cos u = 1 - \frac{2\omega L_s I_d}{\sqrt{2}E_{LL}}. \tag{4.49}$$

Referring once again to Figure 4.20, we obtain the following result:

$$v_{Pn} = es_a - L_s\frac{di_u}{dt}$$
$$= es_a - \frac{1}{2}(es_a - es_c)$$
$$= \frac{1}{2}(es_a + es_c). \tag{4.50}$$

During the commutation interval, the voltage of the positive terminal v_{Pn} does not follow the highest phase voltage. It is equal to the average of v_{an} and v_{cn}. This results in a loss in the direct output voltage and this loss occurs every $\pi/3$ radians. Thus,

$$\begin{aligned}\Delta V_d &= \frac{1}{\pi/3}\int_0^u (es_a - es_c)\, d\omega t \\ &= \frac{1}{\pi/3}\int_0^u L_s \frac{di_u}{dt}\, d\omega t \\ &= \frac{3\omega L_s}{\pi}\int_0^{I_d} di_u \\ &= \frac{3\omega L_s I_d}{\pi}.\end{aligned}$$

The average output voltage is given by

$$V_d = 1.35E_{LL} - \frac{3\omega L_s I_d}{\pi}. \tag{4.51}$$

4.4.3 Simulation of the three-phase AC/DC converter

A typical example of the three-phase AC/DC converter is shown in Figure 4.21. The variables of interest are the three-phase voltages at the *pcc*, the currents supplied by the AC source, the output voltage, and the current drawn by the load.

The transient equivalent of the converter is shown in Figure 4.22 [4]. Reactive elements have been substituted by their equivalents comprising conductances in parallel with their respective past-history current sources. Probe voltages vp, vn, vs_a, vs_b, and vs_c have been applied as shown in Figure 4.22. Note that there are six diodes that require six probe voltages. However, there are two constraints; the voltages across the three arms of the diode bridge are equal to each other. Therefore, only four probe voltages are necessary. In the simulation to follow, an extra probe voltage has been applied to make the nodal analysis easier.

Analysis of the transient equivalent circuit leads to the following equations:

1. Applying KCL to the PCC:

$$\begin{aligned}gs(v_{a1} - e_{sa}) + gls(v_{a1} - v_{sa}) + pisa &= 0 \\ gs(v_{b1} - e_{sb}) + gls(v_{b1} - v_{sb}) + pisb &= 0\end{aligned}$$

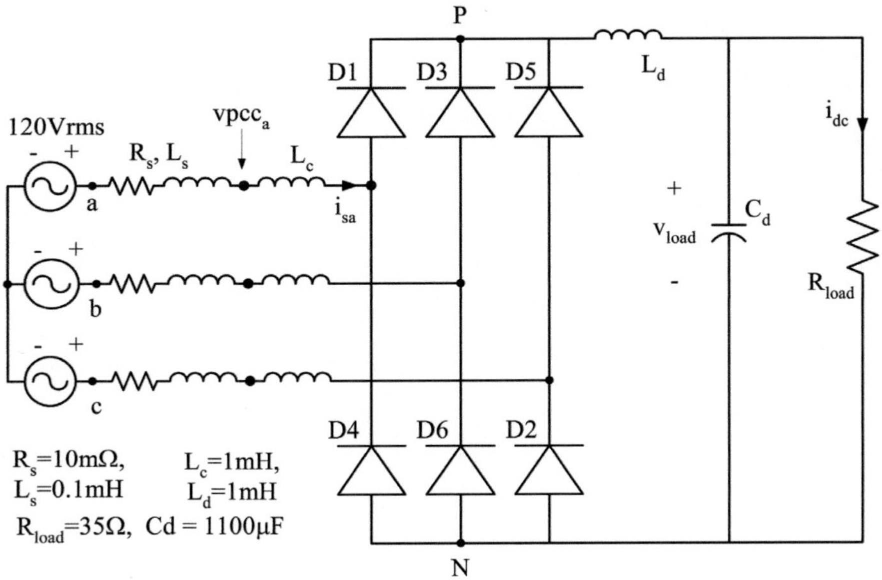

Figure 4.21 Typical three-phase AC/DC converter.

$$gs(v_{c1} - e_{sc}) + gls(v_{c1} - v_{sc}) + pisc = 0.$$

2. AC-side phase currents:

$$\begin{aligned} i_{sa} &= gs(e_{sa} - v_{a1}) \\ i_{sb} &= gs(e_{sb} - v_{b1}) \\ i_{sc} &= gs(e_{sc} - v_{c1}). \end{aligned}$$

3. Upper diode currents and conductances:

$$\begin{aligned} id1, gd1 &= fdiode(v_p - v_{sa}) \\ id3, gd3 &= fdiode(v_p - v_{sb}) \\ id5, gd5 &= fdiode(v_p - v_{sc}). \end{aligned}$$

4. Lower diode currents and conductances:

$$\begin{aligned} id4, gd4 &= fdiode(v_{sa} - v_n) \\ id6, gd6 &= fdiode(v_{sb} - v_n) \\ id2, gd2 &= fdiode(v_{sc} - v_n). \end{aligned}$$

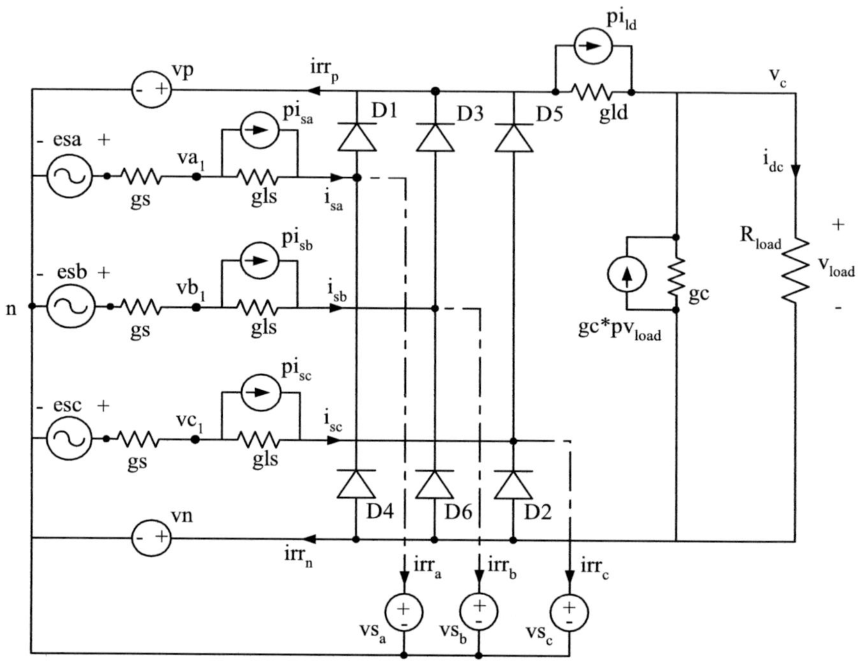

Figure 4.22 Three-phase AC/DC converter: transient equivalent circuit.

5. Apply KCL to the output node:

$$gld * \{v_c - (v_p - v_n)\} + (gc + gload) * v_c = pild + gc * pvc. \quad (4.52)$$

6. Error currents:

$$\begin{aligned} ild &= gld((v_p - v_n) - v_c) + pild \\ irr_a &= i_{sa} + id1 - id4 \\ irr_b &= i_{sb} + id3 - id6 \\ irr_c &= i_{sc} + id5 - id2 \\ irr_p &= -(id1 + id3 + id5 + ild) \\ irr_n &= id4 + id6 + id2 + ild. \end{aligned}$$

The small-signal equivalent circuit is set up as shown in Figure 4.23. Putting

$$\begin{aligned} gstot &= gs * gls/(gs + gls) \\ gdtot &= gld * (gc + gload)/(gld + gc + gload) \end{aligned}$$

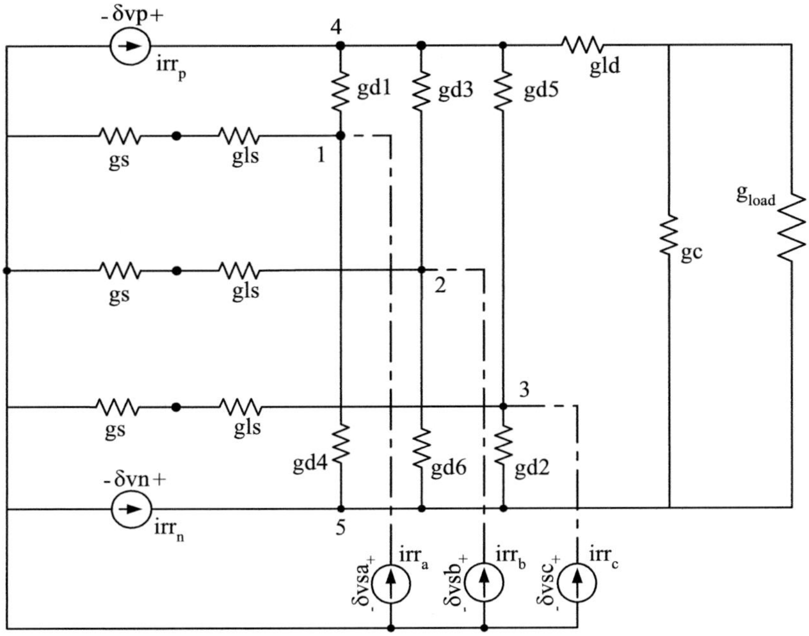

Figure 4.23 Three-phase AC/DC converter: small-signal equivalent.

and applying KCL to the nodes 1, 2, 3, 4, and 5, we obtain

$$gstot * \delta v_{sa} + gd1(\delta v_{sa} - \delta v_p) + gd4(\delta v_{sa} - \delta v_n) = irr_a \quad (4.53)$$

$$gstot * \delta v_{sb} + gd3(\delta v_{sb} - \delta v_p) + gd6(\delta v_{sb} - \delta v_n) = irr_b \quad (4.54)$$

$$gstot * \delta v_{sc} + gd5(\delta v_{sc} - \delta v_p) + gd2(\delta v_{sc} - \delta v_n) = irr_c \quad (4.55)$$

$$gd1(\delta v_p - \delta v_{sa}) + gd3(\delta v_p - \delta v_{sb}) + gd5(\delta v_p - \delta v_{sc}) + gdtot(\delta v_p - \delta v_n) = irr_p \quad (4.56)$$

$$gd4(\delta v_n - \delta v_{sa}) + gd6(\delta v_n - \delta v_{sb}) + gd2(\delta v_n - \delta v_{sc}) + gdtot(\delta v_n - \delta v_p) = irr_n. \quad (4.57)$$

The corrections to be added to the probe voltages are δv_{sa}, δv_{sb}, δv_{sc}, δv_p, and δv_n, and they are obtained by solving eqn (4.53)–(4.57).

The code for simulating the circuit of Figure 4.21 is Program 4.24 and the response waveforms are shown in Figure 4.24. A study of the waveforms leads to the following conclusions:

1. By choosing L_c to be $10L_s$, the phase voltage waveforms at the point of common coupling are almost free of distortion. However, a higher

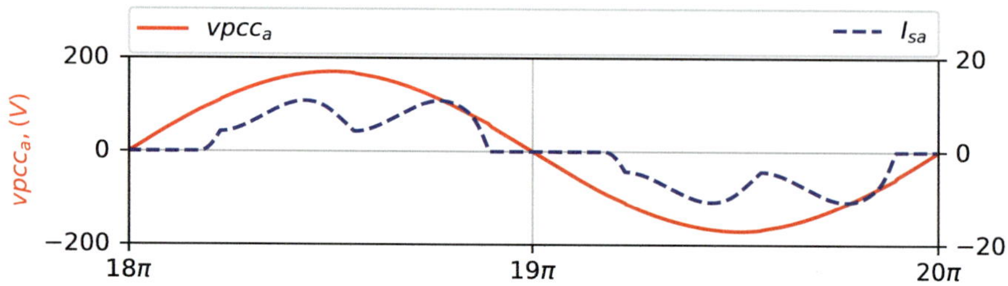

(a)

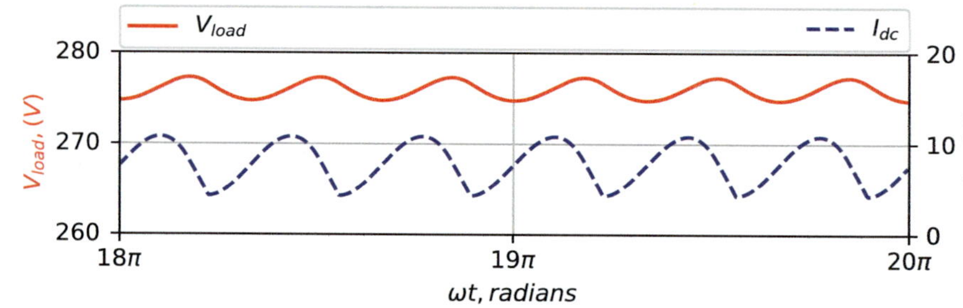

(b)

Figure 4.24 Three-phase AC/DC converter: response waveforms.

value of the inductance L_c gives rise to a larger loss in the direct output voltage.

2. A large value of the output filter capacitance attenuates the oscillations in the output voltage, but the converter may operate in the discontinuous conduction mode. Increasing the output filter inductance may shift the operation of the converter to the continuous conduction mode.

4.5 Solved Problems

1. Consider the single-phase, full-wave, AC/DC converter shown in Figure 4.4. Given that the AC source voltage is $120\,\mathrm{Vrms}$, $L_s = 0.001\,\mathrm{H}$ and $I_d = 5\,\mathrm{A}$, write the code for generating the response waveforms of Figure 4.5. (Ans: Program 4.05). The computed value of the commutation interval $u = 0.2086\,\mathrm{radians}$, as compared to the calculated value using eqn (4.12) of $0.2112\,\mathrm{radians}$.
2. Let the AC voltage source exciting the three-phase AC/DC converter of Figure 4.18 be $120\ \mathrm{Vrms}$ per phase. Let the source inductance be $10\,\mathrm{mH}$ per phase. Write the code to generate the response waveforms shown in Figure 4.19 (Ans: Program 4.19).

4.6 Computational Programs

4.6.1 Program 4.05

```
#
def fdiode(vd):
    """ Diode characteristic.
        Returns diode current and its conductance"""
    #
    glow = 0.1e-06
    ghigh = 0.1e+03
    Vr = -0.5
    #
    if vd >= Vr:
        id = glow*(vd - Vr)
        gd = glow
    else:
        id = ghigh*(vd - Vr)
        gd = ghigh
    #
    return id, gd
#
def ch4fig5_solver(esource, gls, Id, N):
    """Assumes esource is a list, fdiode is a function.
        Returns the list vdiode, iload"""
    #
    import numpy as np
    #
    vd = 0.0
    va = 0.0
    pils = -Id
    nsw = True
    erlim = 0.1e-07
    #
    isource = []
    vload = []
    iload = []
    #
    for j in range(N):
        #
        es = esource[j]
        #
        for its in range(1000):
            #
            vd1 = 0.5*(vd - va)
            vd2 = 0.5*(vd + va)
            id1, gd1 = fdiode(vd1)
            id2, gd2 = fdiode(vd2)
            ils = gls*(es - va) + pils
            irra = ils + id1 - id2
            irrd = -(id1 + id2 + Id)
            err = np.sqrt(irra**2 + irrd**2)
            #
```

```
            if err < erlim:
                break
            else:
                gmat = np.array([[(gls + (gd1 + gd2)/2.0), (gd2 - gd1)/2.0],
                                 [(gd2 - gd1)/2.0, (gd1 + gd2)/2.0]])
                bcol = np.array([irra, irrd])
                xdv = np.linalg.solve(gmat, bcol)
                dva = xdv[0]
                dvd = xdv[1]
                va += dva
                vd += dvd
        #
        if nsw == True and (ils - Id) >= 0.0:
            ku = j
            nsw = False
            #
            print('ku = ', ku)
            #
        pils = ils
        isource.append(ils)
        vload.append(vd)
        iload.append(Id)
        #
    return isource, iload, vload, ku
#
#
def ch4fig5(Erms, freq, ls, Id, ncycle, n):
    """ Main program, plots results
        Simplified single-phase, full-wave converter
        Non-ideal voltage source"""
    #
    #ch4fig5(120.0, 60.0, 0.001, 5.0, 2, 512)
    #
    from math import pi
    import numpy as np
    from mpl_toolkits.axes_grid1 import host_subplot
    import matplotlib.pyplot as plt
    plt.rcParams["figure.figsize"] = (6, 4)
    #
    N = n*ncycle
    #
    gls = 1.0/(freq*ls*n)
    dteta = 2.0*pi/n
    teta = dteta*np.arange(N)
    esmax = np.sqrt(2.0)*Erms
    esource = esmax*np.sin(teta)
    #
    isource, iload, vload, ku = ch4fig5_solver(esource, gls, Id, N)
    u = dteta*ku
    print('u = ', u)
    #
    ax1 = host_subplot(211)
    ax2 = ax1.twinx()
```

```
ax1.set_xlim([0, 4*pi])
ax1.set_xticks([0, u, pi, 2*pi, 3*pi, 4*pi])
ax1.set_xticklabels(['$0$', '$u$', '$\pi$', '$2\pi$', '$3\pi$', '$4\pi$'])
ax1.set_ylim([-200, 200])
ax1.set_yticks([-200, -100, 0, 100, 200])
ax1.set_yticklabels(['$-200$', '$-100$', '$0$', '$100$', '$200$'])
ax1.set_ylabel('$vs, (V)$')
ax1.text(6, -360, '$(a)$', size =10)
ax2.set_ylabel('$is, (A)$')
ax2.set_ylim([-10, 10])
ax2.set_yticks([-10, 0, 10])
ax2.set_yticklabels(['$-10$', '$0$', '$10$'])
p1, = ax1.plot(teta, esource, 'r', label = '$v_{S}$')
p2, = ax2.plot(teta, isource, 'b--', label = '$i_{S}$')
ax1.yaxis.get_label().set_color(p1.get_color())
ax2.yaxis.get_label().set_color(p2.get_color())
ax1.legend(bbox_to_anchor=(0., 1.02, 1., .102), loc=3, ncol=2, \
           mode="expand", borderaxespad=0.)
ax1.grid(True)
plt.subplots_adjust(hspace=0.8)
#
ax1 = host_subplot(212)
ax2 = ax1.twinx()
ax1.set_xlim([0, 4*pi])
ax1.set_xticks([0, u, pi, 2*pi, 3*pi, 4*pi])
ax1.set_xticklabels(['$0$', '$u$', '$\pi$', '$2\pi$', '$3\pi$', '$4\pi$'])
ax1.set_xlabel('$\omega t, radians$')
ax1.set_ylim([0, 200])
ax1.set_yticks([0, 100, 200])
ax1.set_yticklabels(['$0$', '$100$', '$200$'])
ax1.set_ylabel('$Vd, (V)$')
ax1.text(6, -120, '$(b)$', size =10)
ax2.set_ylabel('$id, (A)$')
ax2.set_ylim([0, 10])
ax2.set_yticks([0, 10])
ax2.set_yticklabels(['$0$', '$10$'])
p1, = ax1.plot(teta, vload, 'r', label = '$v_{d}$')
p2, = ax2.plot(teta, iload, 'b--', label = '$I_{0}$')
ax1.yaxis.get_label().set_color(p1.get_color())
ax2.yaxis.get_label().set_color(p2.get_color())
ax1.legend(bbox_to_anchor=(0., 1.02, 1., .102), loc=3, ncol=2, \
           mode="expand", borderaxespad=0.)
ax1.grid(True)
#
plt.show()
#
# plt.draw()
# plt.savefig('Fig405.ps', dpi=400, bbox_inches='tight')
#
```

4.6.2 Program 4.09

```
#
def fdiode(vd):
    """ Diode characteristic.
        Returns diode current and its conductance"""
    #
    glow = 0.1e-06
    ghigh = 0.1e+03
    Vr = -0.5
    #
    if vd >= Vr:
        id = glow*(vd - Vr)
        gd = glow
    else:
        id = ghigh*(vd - Vr)
        gd = ghigh
    #
    return id, gd
#
def ch4fig9_solver(esource, gls, Vd, N):
    """Assumes esource is a list, fdiode is a function.
        Returns the list iload, vload, vinduct"""
    va = 0.0
    pils = 0.0
    erlim = 0.1e-07
    #
    isource = []
    vinduct = []
    vload = []
    iload = []
    #
    for j in range(N):
        #
        es = esource[j]
        #
        for its in range(1000):
            #
            vls = es - va
            ils = gls*vls + pils
            vd1 = 0.5*(Vd - va)
            vd2 = 0.5*(Vd + va)
            id1, gd1 = fdiode(vd1)
            id2, gd2 = fdiode(vd2)
            ild = -(id1 + id2)
            irr = ils + id1 - id2
            err = abs(irr)
            #
            if err < erlim:
                break
            else:
                gdtot = 0.5*(gd1 + gd2)
                dva = irr/(gls + gdtot)
```

```
                va += dva
        #
        if its >= 1000:
            print('its = ', j, 'solver does not converge')
            break
        #
        isource.append(ils)
        vinduct.append(vls)
        iload.append(ild)
        vload.append(Vd)
        pils = ils
        #
    return isource, iload, vinduct, vload
#
#
def ch4fig9(Erms, freq, ls, Vd, ncycle, n):
    """ Main program, plots results
        Single-phase, full-wave converter
        non-ideal AC source, highly capacitive load"""
    #
    #calling command:
    #ch4fig9(120.0, 60.0, 0.01, 100.0, 2, 512)
    #
    from math import pi
    import numpy as np
    from mpl_toolkits.axes_grid1 import host_subplot
    import matplotlib.pyplot as plt
    plt.rcParams["figure.figsize"] = (6, 4)
    #
    N = n*ncycle
    #
    gls = 1.0/(freq*ls*n)
    dteta = 2.0*pi/n
    teta = dteta*np.arange(N)
    esmax = np.sqrt(2.0)*Erms
    esource = esmax*np.sin(teta)
    #
    isource, iload, vinduct, vload = ch4fig9_solver(esource, gls, Vd, N)
    #
    pdva1 = Vd
    pdva2 = Vd - esmax
    if pdva2 < 0.0:
        for j in range(n//2):
            if j <= n//4:
                dva1 = Vd - esource[j]
                if pdva1*dva1 < 0.0:
                    alpha = teta[j]
                    pdva1 = dva1
    #
            if j > n//4:
                dva2 = Vd - esource[j]
                if pdva2*dva2 < 0.0:
                    beta = teta[j]
```

```
                pdva2 = dva2
#
    print('alpha =', alpha, 'beta = ', beta)
#
n1 = 2*ncycle
#
ax1 = host_subplot(211)
ax2 = ax1.twinx()
ax1.set_xlim([0, n1*pi])
ax1.set_xticks([0, alpha, beta, pi, 2*pi, 3*pi, 4*pi])
ax1.set_xticklabels(['$0$', '$\\alpha $','$\\beta $','$\pi$',\
                  '$2\pi$', '$3\pi$', '$4\pi$'])
ax1.set_ylim([-200, 200])
ax1.set_yticks([-200, -100, 0, 100, 200])
ax1.set_yticklabels(['$-200$', '$-100$', '$0$', '$100$','$200$'])
ax1.set_ylabel('$v_{S}, (V)$')
ax1.text(6, -350, '$(a)$', size =10)
ax2.set_ylabel('$i_{S}, (A)$')
ax2.set_ylim([-50, 50])
ax2.set_yticks([-50, 0, 50])
ax2.set_yticklabels(['$-50$', '$0$', '$50$'])
p1, = ax1.plot(teta, esource, 'r', label = '$v_{S}$')
p2, = ax1.plot(teta, vload, 'r--', label = '$V_{d}$')
p3, = ax2.plot(teta, isource, 'b--', label = '$i_{S}$')
ax1.yaxis.get_label().set_color(p1.get_color())
ax2.yaxis.get_label().set_color(p3.get_color())
ax1.legend(bbox_to_anchor=(0., 1.02, 1., .102), loc=3, ncol=3, \
           mode="expand", borderaxespad=0.)
ax1.grid(True)
plt.subplots_adjust(hspace=0.8)
#
ax1 = host_subplot(212)
ax2 = ax1.twinx()
ax1.set_xlim([0, n1*pi])
ax1.set_xticks([0, alpha, beta, pi, 2*pi, 3*pi, 4*pi])
ax1.set_xticklabels(['$0$', '$\\alpha $','$\\beta $','$\pi$',\
                  '$2\pi$', '$3\pi$', '$4\pi$'])
ax1.set_xlabel('$\omega t, radians$')
ax1.set_ylim([-100, 200])
ax1.set_yticks([-200, -100, 0, 100, 200])
ax1.set_yticklabels(['$-200$', '$-100$', '$0$', '$100$', '$200$'])
ax1.set_ylabel('$V_{LS}, (V)$')
ax1.text(6, -440, '$(b)$', size =10)
ax2.set_ylabel('$i_{D}, (A)$')
ax2.set_ylim([0, 50])
ax2.set_yticks([0, 50])
ax2.set_yticklabels(['$0$', '$50$'])
p1, = ax1.plot(teta, vinduct, 'r', label = '$v_{LS}$')
p2, = ax2.plot(teta, iload, 'b--', label = '$i_{d}$')
ax1.yaxis.get_label().set_color(p1.get_color())
ax2.yaxis.get_label().set_color(p2.get_color())
ax1.legend(bbox_to_anchor=(0., 1.02, 1., .102), loc=3, ncol=2, \
           mode="expand", borderaxespad=0.)
```

```
    ax1.grid(True)
    #
    plt.show()
    #
    #plt.draw()
    #plt.savefig('fig409.ps', dpi=400, bbox_inches='tight')
    #
```

4.6.3 Program 4.13

```
#
def fdiode(vd):
    """ Diode characteristic.
        Returns diode current and its conductance"""
    #
    glow = 0.1e-06
    ghigh = 0.1e+03
    Vr = -0.5
    #
    if vd >= Vr:
        id = glow*(vd - Vr)
        gd = glow
    else:
        id = ghigh*(vd - Vr)
        gd = ghigh
    #
    return id, gd
#
def ch4fig13_solver(esource, gs, gls, gld, gc, gload, ka, N):
    """Assumes esource is a list, fdiode is a function.
        Returns the list iload"""
    #
    import numpy as np
    #
    k1 = 1.0 - ka
    gstot = gs*gls/(gs + gls)
    gdtot = gld*(gc + gload)/(gld + gc + gload)
    vd = 0.0
    va = 0.0
    pils = 0.0
    pild = 0.0
    pvc = 0.0
    erlim = 0.1e-07
    #
    isource = []
    vpcc = []
    cout = []
    vload = []
    #
    for j in range(N):
        #
        es = esource[j]
```

```
        #
        for its in range(1000):
            #
            v1 = (gs*es + gls*va - pils)/(gs + gls)
            ils = gs*(es - v1)
            vd1 = 0.5*(vd - va)
            vd2 = 0.5*(vd + va)
            id1, gd1 = fdiode(vd1)
            id2, gd2 = fdiode(vd2)
            irra = ils + id1 - id2
            vc = (gld*vd + pild + gc*pvc)/(gc + gload + gld)
            ild = gld*(vd - vc) + pild
            irrd = -(id1 + id2 + ild)
            err = np.sqrt(irra**2 + irrd**2)
            #
            if err < erlim:
                break
            else:
                gmat = np.array([[(gstot + (gd1 + gd2)/2.0), (gd2 - gd1)/2.0],
                                 [(gd2 - gd1)/2.0, (gdtot + (gd1 +gd2)/2.0)]])
                bcol = np.array([irra, irrd])
                xdv = np.linalg.solve(gmat, bcol)
                dva = xdv[0]
                dvd = xdv[1]
                va += dva
                vd += dvd
        #
        pils = ils
        pild = ild
        pvc = vc
        isource.append(ils)
        vpcc.append(k1*v1 + ka*va)
        cout.append(ild)
        vload.append(vc)
        #
    return isource, vpcc, cout, vload
#
#
def ch4fig13(Erms, freq, rs, ls, lc, ld, cd, rload, ncycle, n):
    """ Main program, plots results
        The general, single-phase, full-wave AC/DC converter"""
    #
    #calling command:
    #ch4fig13(120.0, 60.0, 0.2, 0.0005, 0.002, 0.001, 0.001, 10.0, 10, 512)
    #
    from math import pi
    import numpy as np
    from mpl_toolkits.axes_grid1 import host_subplot
    import matplotlib.pyplot as plt
    plt.rcParams["figure.figsize"] = (6, 4)
    #
    N = n*ncycle
    #
```

```
gs = 1.0/rs
gls = 1.0/(freq*n*(ls + lc))
gld = 1.0/(freq*n*ld)
gload = 1.0/rload
gc = freq*n*cd
ka = ls/(ls + lc)
dteta = 2.0*pi/n
teta = dteta*np.arange(N)
esmax = np.sqrt(2.0)*Erms
esource = esmax*np.sin(teta)
#
isource, vpcc, cout, vload = ch4fig13_solver(esource, gs, gls,\
                                gld, gc, gload, ka, N)
#
m0 = 2*(ncycle - 2)
m1 = 1 + m0
m2 = 1 + m1
m3 = 1 + m2
m4 = 1 + m3
#
ax1 = host_subplot(211)
ax2 = ax1.twinx()
ax1.set_xlim(m0*pi, m4*pi)
ax1.set_xticks([m0*pi, m1*pi, m2*pi, m3*pi, m4*pi])
ax1.set_xticklabels(['$0$', '$\pi$', '$2\pi$', '$3\pi$', '$4\pi$'])
ax1.set_ylim([-200, 200])
ax1.set_yticks([-200, -100, 0, 100, 200])
ax1.set_yticklabels(['$-200$', '$-100$', '$0$', '$100$', '$200$'])
ax1.set_ylabel('$v_{PCC}, (V)$')
ax1.text(m2*pi, -350, '$(a)$', size =10)
ax2.set_ylabel('$i_{S}, (A)$')
ax2.set_ylim([-100, 100])
ax2.set_yticks([-100, 0, 100])
ax2.set_yticklabels(['$-100$', '$0$', '$100$'])
p1, = ax1.plot(teta, vpcc, 'r', label = '$v_{PCC}$')
p2, = ax2.plot(teta, isource, 'b--', label = '$i_{S}$')
ax1.yaxis.get_label().set_color(p1.get_color())
ax2.yaxis.get_label().set_color(p2.get_color())
ax1.legend(bbox_to_anchor=(0., 1.02, 1., .102), loc=3, ncol=2, \
           mode="expand", borderaxespad=0.)
ax1.grid(True)
plt.subplots_adjust(hspace=0.8)
#
ax1 = host_subplot(212)
ax2 = ax1.twinx()
ax1.set_xlim(m0*pi, m4*pi)
ax1.set_xticks([m0*pi, m1*pi, m2*pi, m3*pi, m4*pi])
ax1.set_xticklabels(['$0$', '$\pi$', '$2\pi$', '$3\pi$', '$4\pi$'])
ax1.set_xlabel('$\omega t, radians$')
ax1.set_ylim([0, 200])
ax1.set_yticks([0, 100, 200])
ax1.set_yticklabels(['$0$', '$100$', '$200$'])
ax1.set_ylabel('$v_{LOAD}, (V)$')
```

```
    ax1.text(m2*pi, -120, '$(b)$', size =10)
    ax2.set_ylabel('$i_{LD}, (A)$')
    ax2.set_ylim([0, 100])
    ax2.set_yticks([0, 100])
    ax2.set_yticklabels(['$0$', '$100$'])
    p1, = ax1.plot(teta, vload, 'r', label = '$v_{LOAD}$')
    p2, = ax2.plot(teta, cout, 'b--', label = '$i_{LD}$')
    ax1.yaxis.get_label().set_color(p1.get_color())
    ax2.yaxis.get_label().set_color(p2.get_color())
    ax1.legend(bbox_to_anchor=(0., 1.02, 1., .102), loc=3, ncol=2, \
                mode="expand", borderaxespad=0.)
    ax1.grid(True)
    #
    plt.show()
    #
    #plt.draw()
    #plt.savefig('Fig413.ps', dpi=400, bbox_inches='tight')
    #
```

4.6.4 Program 4.19

```
#
def fdiode(vd):
    """ Diode characteristic.
        Returns diode current and its conductance"""
    #
    glow = 0.1e-06
    ghigh = 0.1e+03
    Vr = -0.5
    #
    if vd >= Vr:
        id = glow*(vd - Vr)
        gd = glow
    else:
        id = ghigh*(vd - Vr)
        gd = ghigh
    #
    return id, gd
#
def ch4fig19_solver(esource_a, esource_b, esource_c, gls, Id, N):
    """Assumes esource is a list, fdiode is a function.
        Returns the list iload
        Current transfer fron diode D5 to diode D1"""
    #
    import numpy as np
    #
    vp = 0.0
    vn = 0.0
    vsa = 0.0
    vsb = 0.0
    vsc = 0.0
    pcsa = 0.0
```

```
pcsb = 0.0
pcsc = 0.0
erlim = 0.1e-07
#
csa = []
csb = []
csc = []
v_P = []
v_N = []
v_D = []
#
for j in range(N):
    #
    esa = esource_a[j]
    esb = esource_b[j]
    esc = esource_c[j]
    #
    for its in range(1000):
        #
        ilsa = gls*(esa - vsa) + pcsa
        ilsb = gls*(esb - vsb) + pcsb
        ilsc = gls*(esc - vsc) + pcsc
        #
        cda1, gda1 = fdiode(vp - vsa)
        cda2, gda2 = fdiode(vsa - vn)
        cdb1, gdb1 = fdiode(vp - vsb)
        cdb2, gdb2 = fdiode(vsb - vn)
        cdc1, gdc1 = fdiode(vp - vsc)
        cdc2, gdc2 = fdiode(vsc - vn)
        #
        irra = ilsa + cda1 -  cda2
        irrb = ilsb + cdb1 -  cdb2
        irrc = ilsc + cdc1 -  cdc2
        irrp = -(cda1 + cdb1 + cdc1 + Id)
        irrn = cda2 + cdb2 + cdc2 + Id
        err = np.sqrt(irra**2 + irrb**2 + irrc**2 + irrp**2 + irrn**2)
        #
        if err < erlim:
            break
        else:
            gmat = np.array([[(gls+gda1+gda2), 0, 0, -gda1, -gda2],
                             [0, (gdb1+gdb2+gls), 0, -gdb1, -gdb2],
                             [0, 0, (gdc1+gdc2+ gls), -gdc1, -gdc2],
                             [-gda1, -gdb1, -gdc1, (gda1+gdb1+gdc1), 0],
                             [-gda2, -gdb2, -gdc2, 0, (gda2+gdb2+ gdc2)]])
            bcol = np.array([irra, irrb, irrc, irrp, irrn])
            xdv = np.linalg.solve(gmat, bcol)
            #
            dvsa = xdv[0]
            dvsb = xdv[1]
            dvsc = xdv[2]
            dvp = xdv[3]
            dvn = xdv[4]
```

```
                vsa += dvsa
                vsb += dvsb
                vsc += dvsc
                vp += dvp
                vn += dvn
        #
        pcsa = ilsa
        pcsb = ilsb
        pcsc = ilsc
        #
        csa.append(ilsa)
        csb.append(ilsb)
        csc.append(ilsc)
        v_P.append(vp)
        v_N.append(vn)
        v_D.append(vp - vn)
        #
    return csa, csb, csc, v_P, v_N, v_D
#
#
def ch4fig19(Erms, freq, ls, Id, ncycle, n):
    """ Main program, plots results
        Simplified 3-phase, AC/DC converter
        Non-ideal, 3-phase AC source"""
    #
    #calling command:
    #ch4fig19(120.0, 60.0, 0.1e-02, 10.0, 10, 512)
    #
    from math import pi
    import numpy as np
    import matplotlib.pyplot as plt
    #
    N = n*ncycle
    #
    gls = 1.0/(freq*n*ls)
    dteta = 2.0*pi/n
    phdif = 2.0*pi/3.0
    teta = dteta*np.arange(N)
    tetb = teta - phdif*np.ones(N)
    tetc = teta + phdif*np.ones(N)
    esmax = np.sqrt(2.0)*Erms
    esource_a = esmax*np.sin(teta)
    esource_b = esmax*np.sin(tetb)
    esource_c = esmax*np.sin(tetc)
    #
    csa, csb, csc, v_P, v_N, v_D = ch4fig19_solver(esource_a, esource_b, \
                                                   esource_c, gls, Id, N)
    #
    m0 = 2*(ncycle - 1)
    m1 = 1 + m0
    m2 = 1 + m1
    #
    plt.subplot(211)
```

```
plt.plot(teta, csa, 'b-', label = '$I_{sa}$')
plt.plot(teta, csc, 'g--', label = '$I_{sc}$')
plt.xlabel('$\omega t, radians$')
plt.xlim(57, 57.4)
plt.xticks([57, 57.1, 57.2, 57.3, 57.4], ['$57$', '$57.1$', '$57.2$', \
           '$57.3$', '$57.4$'])
plt.ylabel('$Currents$')
plt.ylim(0, 20)
plt.yticks([0, 10.0, 20.0], ['$0$', '$10.0$', '$20.0$'])
#
plt.plot((57.075, 57.075), (3.5, 6.5), linewidth = 1, color = 'k')
plt.arrow(57.12, 5, -0.034, 0, head_width = 1.0, head_length = 0.01, \
          fc = 'k', ec = 'k')
plt.text(57.14, 4.5, '$u$', family = 'monospace', fontsize = 9)
plt.arrow(57.16, 5, 0.125, 0, head_width = 1.0, head_length = 0.01, \
          fc = 'k', ec = 'k')
plt.plot((57.296, 57.296), (3.5, 6.5), linewidth = 1, color = 'k')
#
plt.text(57.19, -12, '$(a)$', size =10)


plt.legend(bbox_to_anchor=(0., 1.02, 1., .102), loc=3, ncol=2, \
           mode="expand", borderaxespad=0.)
plt.grid(True)
plt.subplots_adjust(hspace=1.0)
#
plt.subplot(212)
plt.plot(teta, esource_a, 'b--', label = '$es_{a}$')
plt.plot(teta, esource_b, 'g--', label = '$es_{b}$')
plt.plot(teta, esource_c, 'y--', label = '$es_{c}$')
plt.plot(teta, v_P, 'r-', label = '$V_{p}$')
plt.plot(teta, v_N, 'm-', label = '$V_{n}$')
plt.xlabel('$\omega t, radians$')
plt.xlim([m0*pi, m2*pi])
plt.xticks([m0*pi, m1*pi, m2*pi], ['$18\pi$', '$19\pi$', '$20\pi$'])
plt.ylabel('$Voltages$')
plt.ylim(-300, 300)
plt.yticks([-300, 0, 300], ['$-300$', '$0$', '$300$'])
#
plt.plot((57.075, 57.075), (150, 220), linewidth = 1, color = 'k')
plt.arrow(56.9, 185, 0.11, 0, head_width = 30.0, head_length = 0.05, \
          fc = 'k', ec = 'k')
plt.text(57.14, 170, '$u$', family = 'monospace', fontsize = 9)
plt.arrow(57.5, 185, -0.13, 0, head_width = 30.0, head_length = 0.05, \
          fc = 'k', ec = 'k')
plt.plot((57.296, 57.296), (150, 220), linewidth = 1, color = 'k')
#
plt.text(59.5, -650, '$(b)$', size =10)
plt.legend(bbox_to_anchor=(0., 1.02, 1., .102), loc=3, ncol=5, \
           mode="expand", borderaxespad=0.)
plt.grid(True)
#
plt.show()
```

```
    #
    # plt.draw()
    # plt.savefig('fig419.ps', dpi=400, bbox_inches='tight')
    #
```

4.6.5 Program 4.24

```
#
def fdiode(vd):
    """ Diode characteristic.
        Returns diode current and its conductance"""
    #
    glow = 0.1e-06
    ghigh = 0.1e+03
    Vr = -0.5
    #
    if vd >= Vr:
        id = glow*(vd - Vr)
        gd = glow
    else:
        id = ghigh*(vd - Vr)
        gd = ghigh
    #
    return id, gd
#
def ch4fig24_solver(esource_a, esource_b, esource_c, gs, gls, gld, gc, \
                    gload, ka, N):
    """Assumes esource is a list, fdiode is a function.
        Returns the list iload
        3-phase AC/DC converter"""
    #
    import numpy as np
    #
    k1 = 1.0 - ka
    gstot = gs*gls/(gs + gls)
    gdtot = gld*(gc + gload)/(gld + gc + gload)
    vp = 0.0
    vn = 0.0
    vsa = 0.0
    vsb = 0.0
    vsc = 0.0
    pcsa = 0.0
    pcsb = 0.0
    pcsc = 0.0
    pcld = 0.0
    pvc = 0.0
    erlim = 0.1e-07
    #
    csa = []
    csb = []
    csc = []
    vpcc_a = []
```

```
cout = []
vload = []
#
for j in range(N):
    #
    esa = esource_a[j]
    esb = esource_b[j]
    esc = esource_c[j]
    #
    for its in range(1000):
        #
        va1 = (gs*esa + gls*vsa - pcsa)/(gs + gls)
        vb1 = (gs*esb + gls*vsb - pcsb)/(gs + gls)
        vc1 = (gs*esc + gls*vsc - pcsc)/(gs + gls)
        #
        ilsa = gs*(esa - va1)
        ilsb = gs*(esb - vb1)
        ilsc = gs*(esc - vc1)
        #
        cda1, gda1 = fdiode(vp - vsa)
        cda2, gda2 = fdiode(vsa - vn)
        cdb1, gdb1 = fdiode(vp - vsb)
        cdb2, gdb2 = fdiode(vsb - vn)
        cdc1, gdc1 = fdiode(vp - vsc)
        cdc2, gdc2 = fdiode(vsc - vn)
        #
        vc = (gld*(vp - vn) + pcld + gc*pvc)/(gc + gload + gld)
        cld = gld*(vp - vn - vc) + pcld
        #
        irra = ilsa + cda1 -  cda2
        irrb = ilsb + cdb1 -  cdb2
        irrc = ilsc + cdc1 -  cdc2
        irrp = -(cda1 + cdb1 + cdc1 + cld)
        irrn = cda2 + cdb2 + cdc2 + cld
        err = np.sqrt(irra**2 + irrb**2 + irrc**2 + irrp**2 + irrn**2)
        #
        if err < erlim:
            break
        else:
            gmat = np.array([[(gstot+gda1+gda2), 0, 0, -gda1, -gda2],
                             [0, (gdb1+gdb2+gstot), 0, -gdb1, -gdb2],
                             [0, 0, (gdc1+gdc2+gstot), -gdc1, -gdc2],
                             [-gda1, -gdb1, -gdc1, (gda1+gdb1+gdc1+gdtot),  \
                                                                -gdtot],
                             [-gda2, -gdb2, -gdc2, -gdtot, (                \
                                          gda2+gdb2+gdc2+gdtot)]])
            bcol = np.array([irra, irrb, irrc, irrp, irrn])
            xdv = np.linalg.solve(gmat, bcol)
            #
            dvsa = xdv[0]
            dvsb = xdv[1]
            dvsc = xdv[2]
            dvp = xdv[3]
```

```
                dvn = xdv[4]
                vsa += dvsa
                vsb += dvsb
                vsc += dvsc
                vp += dvp
                vn += dvn
        #
        pcsa = ilsa
        pcsb = ilsb
        pcsc = ilsc
        pcld = cld
        pvc = vc
        #
        csa.append(ilsa)
        csb.append(ilsb)
        csc.append(ilsc)
        vpcc_a.append(k1*va1 + ka*vsa)
        cout.append(cld)
        vload.append(vc)
        #
    return csa, csb, csc, vpcc_a, cout, vload
#
#
def ch4fig24(Erms, freq, rs, ls, lc, ld, cd, rload, ncycle, n):
    """ Main program, plots results"""
    #
    #calling command:
    #ch4fig24(120.0, 60.0, 0.1e-01, 0.1e-03, 0.1e-02, 0.1e-02,
    #                                    0.11e-02, 35.0, 10, 512)
    #
    from math import pi
    import numpy as np
    from mpl_toolkits.axes_grid1 import host_subplot
    import matplotlib.pyplot as plt
    plt.rcParams["figure.figsize"] = (6, 4)
    #
    N = n*ncycle
    #
    gs = 1.0/rs
    gls = 1.0/(freq*n*(ls + lc))
    gld = 1.0/(freq*n*ld)
    gload = 1.0/rload
    gc = freq*n*cd
    ka = ls/(ls + lc)
    dteta = 2.0*pi/n
    phdif = 2.0*pi/3.0
    teta = dteta*np.arange(N)
    tetb = teta - phdif*np.ones(N)
    tetc = teta + phdif*np.ones(N)
    esmax = np.sqrt(2.0)*Erms
    esource_a = esmax*np.sin(teta)
    esource_b = esmax*np.sin(tetb)
    esource_c = esmax*np.sin(tetc)
```

```
#
csa, csb, csc, vpcc_a, cout, vload = ch4fig24_solver        \
    (esource_a, esource_b, esource_c, gs, gls, gld,          \
     gc, gload, ka, N)
#
m0 = 2*(ncycle - 1)
m1 = 1 + m0
m2 = 1 + m1
#
ax1 = host_subplot(211)
ax2 = ax1.twinx()
ax1.set_xlim([m0*pi, m2*pi])
ax1.set_xticks([m0*pi, m1*pi, m2*pi])
ax1.set_xticklabels(['$18\pi$', '$19\pi$', '$20\pi$'])
ax1.set_ylim([-200, 200])
ax1.set_yticks([-200, 0, 200])
ax1.set_yticklabels(['$-200$', '$0$', '$200$'])
ax1.set_ylabel('$vpcc_{a}, (V)$')
ax1.text(59.5, -350, '$(a)$', size =10)
ax2.set_ylabel('$I_{sa}, (A)$')
ax2.set_ylim([-20, 20])
ax2.set_yticks([-20, 0, 20])
ax2.set_yticklabels(['$-20$', '$0$', '$20$'])
p1, = ax1.plot(teta, vpcc_a, 'r', label = '$vpcc_a$')
p2, = ax2.plot(teta, csa, 'b--', label = '$I_{sa}$')
ax1.yaxis.get_label().set_color(p1.get_color())
ax2.yaxis.get_label().set_color(p2.get_color())
ax1.legend(bbox_to_anchor=(0., 1.02, 1., .102), loc=3, ncol=2, \
           mode="expand", borderaxespad=0.)
ax1.grid(True)
plt.subplots_adjust(hspace=0.8)
#
ax3 = host_subplot(212)
ax4 = ax3.twinx()
ax3.set_xlim([m0*pi, m2*pi])
ax3.set_xticks([m0*pi, m1*pi, m2*pi])
ax3.set_xticklabels(['$18\pi$', '$19\pi$', '$20\pi$'])
ax3.set_ylim(260, 280)
ax3.set_yticks([260, 270, 280])
ax3.set_yticklabels(['$260$','$270$', '$280$'])
ax3.set_xlabel('$\omega t, radians$')
ax3.set_ylabel('$V_{load}, (V)$')
ax3.text(59.5, 248, '$(b)$', size =10)
ax4.set_ylabel('$i_{dc}, (A)$')
ax4.set_ylim([0, 20])
ax4.set_yticks([0, 10, 20])
ax4.set_yticklabels(['$0$', '$10$', '$20$'])
p3, = ax3.plot(teta, vload, 'r', label = '$V_{load}$')
p4, = ax4.plot(teta, cout, 'b--', label = '$I_{dc}$')
ax3.yaxis.get_label().set_color(p3.get_color())
ax4.yaxis.get_label().set_color(p4.get_color())
ax3.legend(bbox_to_anchor=(0., 1.02, 1., .102), loc=3, ncol=2, \
           mode="expand", borderaxespad=0.)
```

```
ax3.grid(True)
#
plt.show()
#
#plt.draw()
#plt.savefig('fig424.ps', dpi=400, bbox_inches='tight')
#
```

References

[1] N. Mohan, T. M. Undeland, W. P. Robbins, *Power Electronics: Converters, Applications, and Design*, 2^{nd}Ed., John Wiley & Sons Inc., USA, 1995.

[2] D. W. Hart, *Power Electronics*, McGraw Hill, New York, 2011.

[3] M. H. Rashid, *Power Electronics: Devices, Circuits and Applications*, 4^{th}Ed., Pearson India, Chennai, 2018.

[4] S. R. Naidu, A. M. N. Lima, *A new approach for the simulation of power electronic circuits*, IEEE Trans., on Circuits and Systems-1, vol.49, no.9, pp.1317-1324, 2002.

5

Controlled Thyristor-based AC–DC Converters

In many applications, it is necessary to introduce control functions in the operation of AC/DC converters. A typical situation arises when the amplitudes of the input alternating voltages are fluctuating and it is necessary to maintain the output load current equal to a reference amplitude. One of the earliest forms of the controlled AC/DC converter is the bidirectional, thyristor-based converter, which is capable of controlling the flow of power in both directions, i.e., power flow from the AC side to the DC side or vice versa.

The thyristor begins to conduct when it is forward biased and a positive voltage pulse is applied between the gate and the cathode. The thyristor is forward biased when its cathode to anode voltage is negative (which is the same as the anode to cathode voltage being positive). Once conducting, it remains in this state and its anode current is controlled by the external circuit. When the amplitude of the thyristor's current falls below the holding current, the thyristor enters the non-conducting or blocking state and it remains in this state until it subsequently becomes forward biased and a gating pulse is applied once again.

Many thyristor-based converter circuits were proposed at the start of the power electronics revolution. The invention of more versatile controlled switches (such as the IGBT) has now relegated the thyristor mainly to high-power HVDC transmission. A few well-known examples of thyristor-based converters are discussed in this chapter and the simulation procedure is described in [1, 2].

5.1 The Basic Thyristor Circuit

Figure 5.1 shows the basic thyristor circuit consisting of an AC voltage source, a thyristor with the $R_s\,C_s$ snubber circuit in parallel, and the inductive load represented by $R_d\,L_d$. A trigger pulse is applied to the gate in every

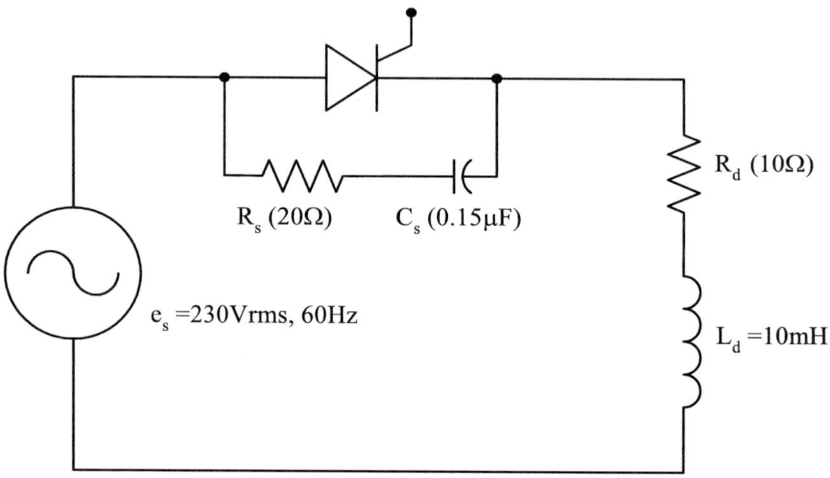

Figure 5.1 Thyristor-based, half-wave AC/DC converter.

period after a delay of α radians from the positive zero-crossing of the AC voltage waveform. The trigger signal should be of sufficient duration so that the anode current, which starts rising from zero, soon exceeds the latching current and the thyristor is turned on. The gate loses control of the anode current and, therefore, the trigger signal is removed.

5.1.1 Simulation of the basic circuit

The values of the parameters used in the simulation are shown in Figure 5.1. In the transient equivalent of the basic circuit shown in Figure 5.2, the capacitor and inductor have been substituted by their equivalents and a probe voltage v_n has been applied. The past-history terms associated with the capacitor and the inductor are pvc and pid. The thyristor's characteristic is represented by the function $fthy$, which requires as its parameters the cathode to anode voltage and the conducting state of the thyristor. If the thyristor is known to be in the conducting state, then the parameter $nst = 1$. The parameter is set to -1 if the thyristor is in the non-conducting state.

From the transient equivalent circuit, the error current irr_n flowing into the probe voltage is calculated as shown below:

$$v_c = v_1 - v_n$$
$$gc * v_c + gsnb * (v_1 - e_s) = gc * pvc$$

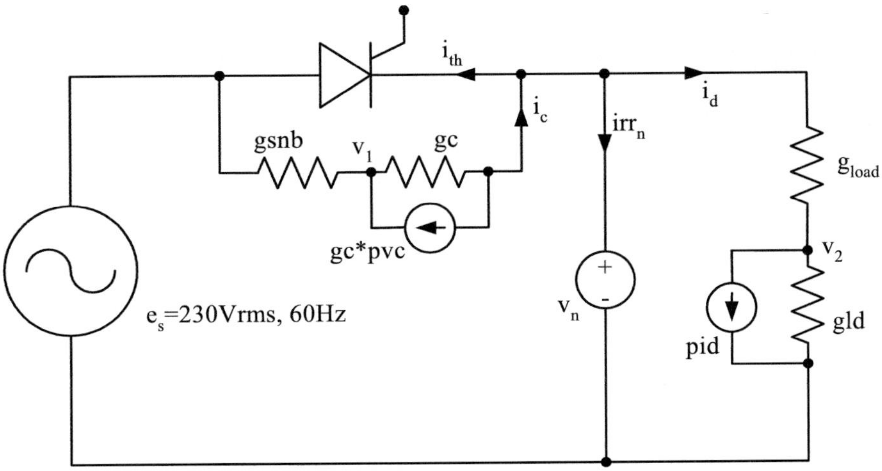

Figure 5.2 Thyristor-based, half-wave AC/DC converter: transient equivalent circuit.

$$gld * v_2 + gload * (v_2 - v_n) + pid = 0$$
$$i_{thy}, g_{thy} = fthy(v_n - e_s, nst)$$
$$i_d = gload * (v_n - v_2)$$
$$i_c = gc * v_c - gc * pvc$$
$$irr_n = i_c - i_{thy} - i_d.$$

The correction δv_n to be added to v_n is obtained from Figure 5.3. From this figure,

$$gdtot = gload * gld/(gload + gld)$$
$$gstot = gsnb * gc/(gsnb + gc)$$
$$\delta v_n = irr_n/(gthy + gdtot + gstot).$$

After a few iterations, the two-step procedure converges and the circuit variables are stored for plotting. Before proceeding to the next time-step, the past-history terms and the indicator of the thyristor's conducting state are updated.

Figure 5.4 (plotted by Program 5.04) shows the response waveforms for the basic thyristor circuit. The conduction state for the thyristor starts after a delay of α radians from the positive zero-crossing of the source voltage and it remains in this state until $\omega t = \beta$ when the thyristor current falls below

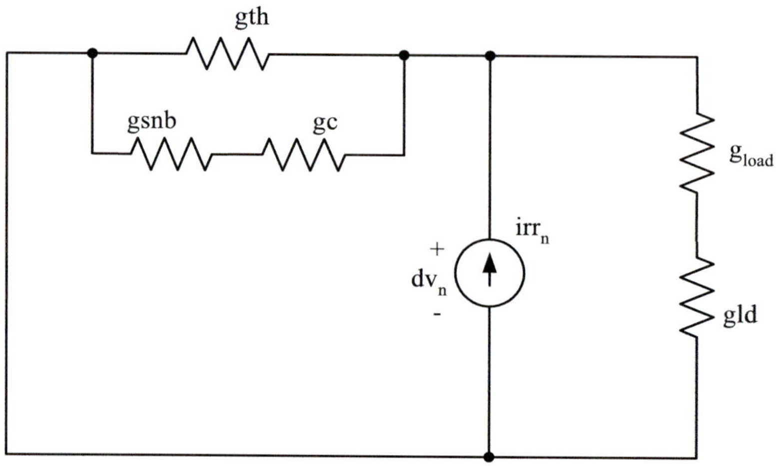

Figure 5.3 Thyristor-based, half-wave AC/DC converter: small-signal equivalent circuit.

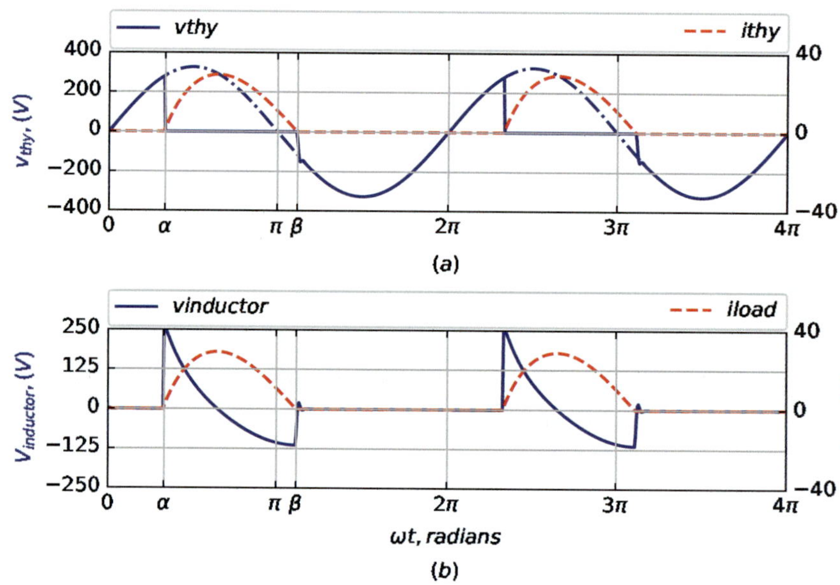

Figure 5.4 Thyristor-based, half-wave AC/DC converter: response waveforms.

the holding current. (The angle β may only be determined numerically.) The inductor voltage is non-zero when the thyristor is conducting and its average value over a period is zero.

5.2 Exercise 5.1

Consider the circuit shown in Figure 5.5. The response waveforms are shown in Figure 5.6. Write the code for generating the waveforms. (Ans: Program 5.06)

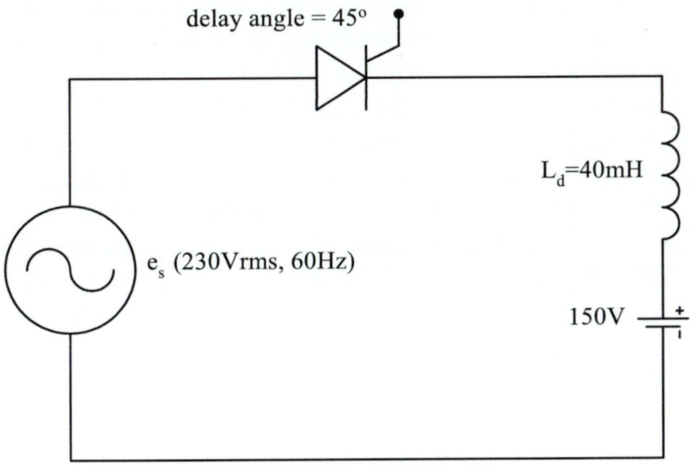

Figure 5.5 AC/DC converter with inductive load and a back emf.

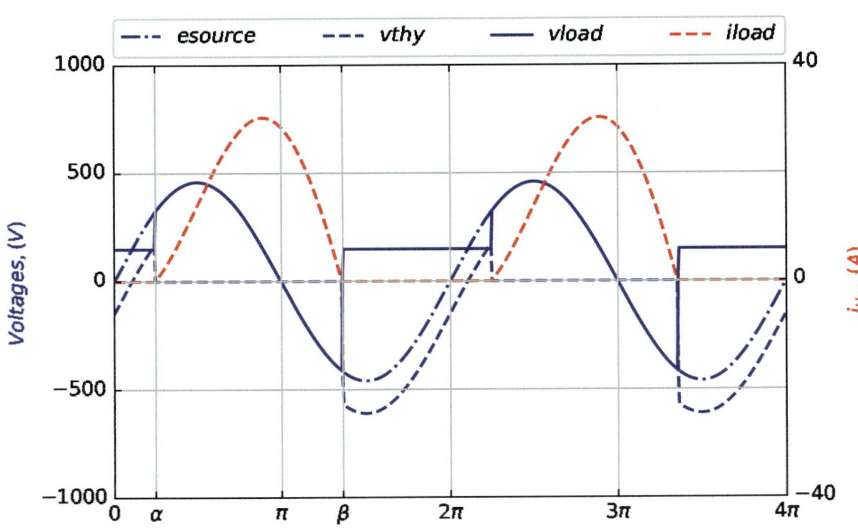

Figure 5.6 Converter with inductive load and back emf: response waveforms.

5.3 Single-phase, Full-wave, Controlled AC/DC Converter

Figure 5.7 shows the thyristor-based, single-phase, full-wave converter. The alternating voltage source is an ideal source and the load is a highly inductive load approximated by the current source I_d. The thyristors T_1, T_3 are triggered at the same instant and they are assumed to conduct together. Similarly, thyristors T_2, T_4 are triggered together and they conduct simultaneously.

Suppose the thyristors T_1, T_3 are triggered at $\omega t = 0$ and the thyristors T_2, T_4 are triggered at $\omega t = \pi$. Then the converter will function precisely like a diode-based converter. For this reason, the instants $\omega t = 0$ and $\omega t = \pi$ are known as the instants of natural conduction. Delays in triggering are measured from these instants. Thus, if the delay is α radians, then the thyristors T_1, T_3 are triggered at $\omega t = \alpha$ and the thyristors T_2, T_4 are triggered at $\omega t = \alpha + \pi$. The thyristors T_1, T_3 are forward biased in the interval $0 \leq \omega t \leq \alpha$, but they remain in the forward blocking state since trigger pulses have not been applied to their gates. Therefore, the thyristors T_2, T_4 that were conducting before $\omega t = 0$ remain conducting. At $\omega t = \alpha$ the thyristors T_1, T_3 are triggered and the currents flowing through T_2, T_4

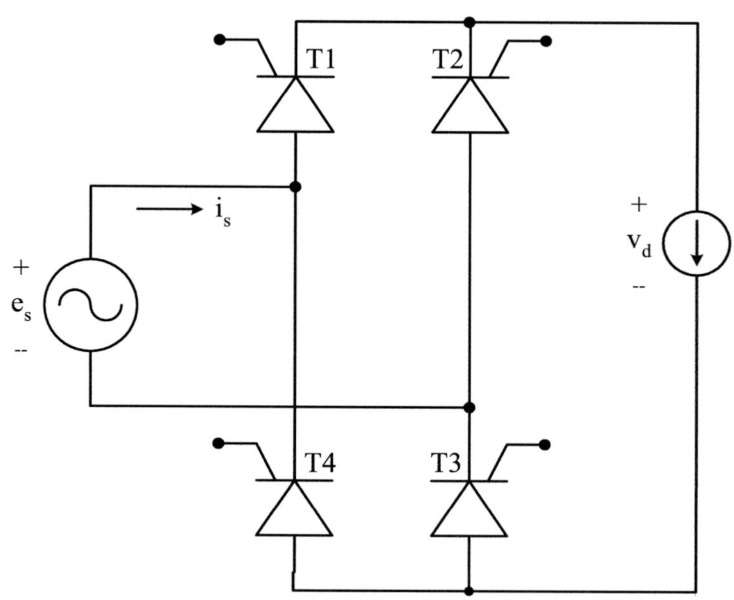

Figure 5.7 Simplified, full-wave, controlled AC/DC converter.

are instantaneously transferred to T_1, T_3. Instantaneous commutation of the current takes place because the alternating voltage source is ideal.

Figure 5.8 (plotted by Program 5.08) shows the waveforms of the input current and the load voltage. The input current is a phase-shifted square wave. The load voltage waveform has a direct component and a significant second harmonic. The average value of the load voltage is given by

$$\begin{aligned} V_d &= \frac{1}{\pi}\int_{\alpha}^{\pi+\alpha} \sqrt{2}E_s \sin\omega t \, d\omega t \\ &= \frac{2\sqrt{2}}{\pi} E_s \cos\alpha \\ &= 0.9E_s \cos\alpha. \end{aligned} \tag{5.1}$$

The input current waveform is delayed by α radians with respect to the alternating source voltage waveform. Thus, the input current may be expressed as

$$i_s(\omega t) = \sqrt{2}\,(I_{s1}\sin(\omega t - \alpha) + I_{s3}\sin 3(\omega t - \alpha) + I_{s5}\sin 5(\omega t - \alpha) + \cdots), \tag{5.2}$$

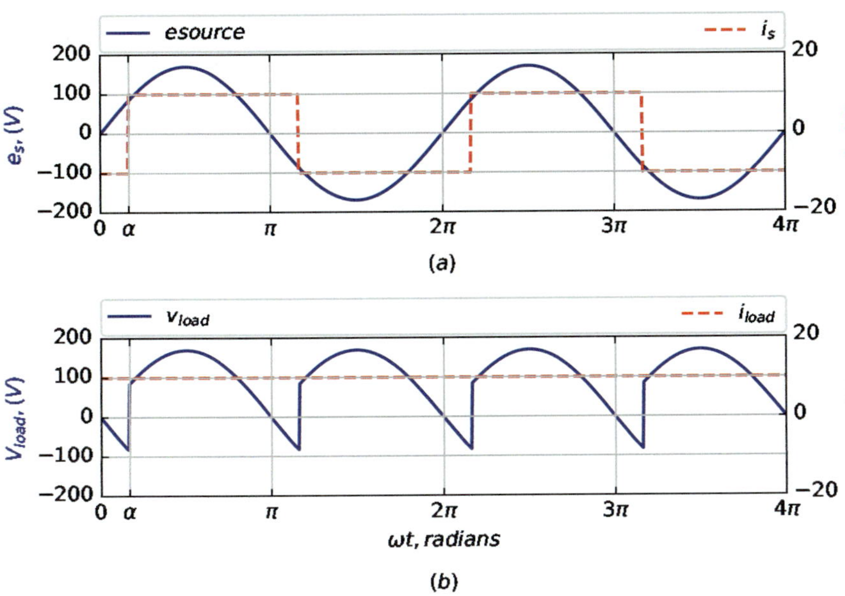

Figure 5.8 Simplified, full-wave, controlled AC/DC converter: response waveforms.

where $I_{sk} = \frac{2\sqrt{2}}{k\pi} I_d$, $k = 1, 3, 5, \ldots$. The RMS value of the input current is

$$I_s = \sqrt{\frac{1}{2\pi}\int_0^{2\pi} (i_s(\omega t))^2 \, d\omega t} = I_d. \tag{5.3}$$

Therefore, the power factor of the converter is

$$\begin{aligned} f_p &= \frac{I_{s1}}{I_s} f_d \\ &= \frac{2\sqrt{2}}{\pi} I_d * \frac{1}{I_d} \cos\alpha \\ &= 0.9\cos\alpha. \end{aligned} \tag{5.4}$$

The THD factor for the input current waveform is the same as for the diode-based, full-wave converter, which is 0.4843. The active power supplied by the voltage source to the converter is

$$\begin{aligned} P &= E_s I_{s1} \cos\alpha \\ &= 0.9 E_s I_d \cos\alpha. \end{aligned} \tag{5.5}$$

The reactive power supplied to the converter is

$$\begin{aligned} Q &= E_s I_{s1} \sin\alpha \\ &= 0.9 E_s I_d \sin\alpha. \end{aligned} \tag{5.6}$$

5.3.1 Effect of the source inductance of a non-ideal AC source

The source inductance will not permit the instantaneous transfer of current from one set of thyristors to the other. A finite interval of commutation of u radians is required for the transfer of current and during this interval, both sets of thyristors will be in the conducting state. Thus, $v_d(\omega t) = 0$ during the intervals $\alpha \le \omega t \le \alpha + u$ and $\pi + \alpha \le \omega t \le \pi + \alpha + u$. Figure 5.9 shows the commutation process during the interval $\alpha \le \omega t \le \alpha + u$, when transient currents $i_u(\omega t)$ circulate in the two loops [1, 2].

Consider the commutation of the current from the thyristors T_2T_4 to T_1T_3 at $\omega t = \alpha$. The transient currents satisfy the conditions $i_u(\alpha) = 0$ and $i_u(\alpha + u) = I_d$. The source current is given by $i_s(\omega t) = 2i_u(\omega t) - I_d$. Thus, $i_s(\alpha) = -I_d$ and $i_s(\alpha + u) = I_d$. During commutation

$$\begin{aligned} L_s \frac{di_s}{dt} &= e_s(\omega t) \\ &= \sqrt{2} E_s \sin\omega t. \end{aligned} \tag{5.7}$$

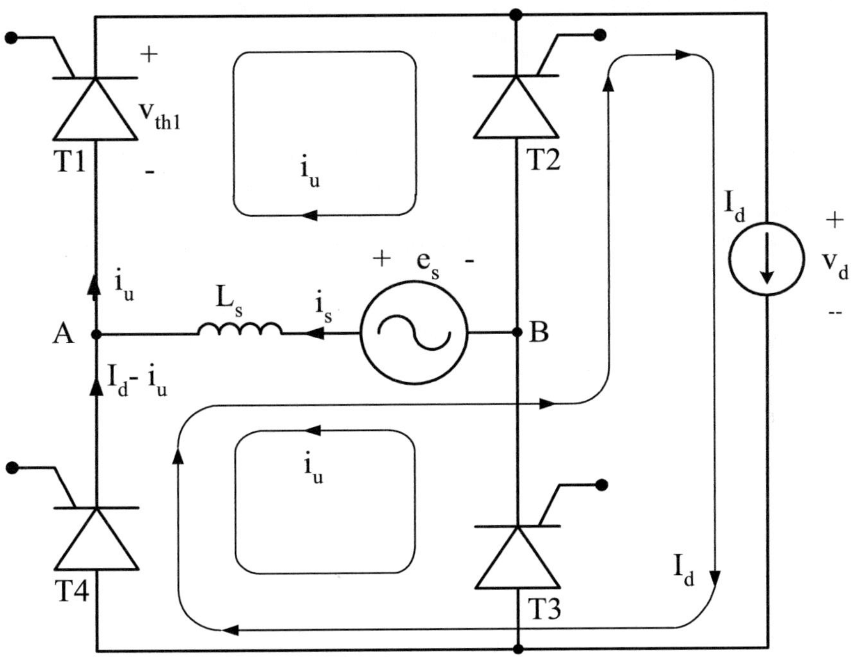

Figure 5.9 Single-phase, full-wave, controlled AC/DC converter: commutation process.

Therefore,

$$di_s = \frac{\sqrt{2}E_s}{\omega L_s} \sin \omega t \, d\omega t. \tag{5.8}$$

Integration yields

$$\int_{-I_d}^{I_d} di_s = \frac{\sqrt{2}E_s}{\omega L_s} \int_{\alpha}^{\alpha+u} \sin \omega t \, d\omega t. \tag{5.9}$$

That is,

$$\cos(\alpha + u) = \cos \alpha - \frac{2\omega L_s I_d}{\sqrt{2}E_s}. \tag{5.10}$$

The average value of the load voltage is given by

$$\begin{aligned} V_d &= \frac{1}{\pi} \int_{\alpha+u}^{\alpha+\pi} \sqrt{2}E_s \sin \omega t \, d\omega t \\ &= \frac{\sqrt{2}E_s}{\pi} \big(\cos(\alpha + u) + \cos \alpha \big). \end{aligned} \tag{5.11}$$

Using eqn (5.10) and (5.11), the average load voltage may be expressed as

$$V_d = \frac{2\sqrt{2}}{\pi} E_s \cos\alpha - \frac{2\omega L_s I_d}{\pi}. \tag{5.12}$$

Figure 5.10 (plotted by Program 5.10) shows the waveforms of the input current, the source voltage, and the load voltage. The zero-crossings of the input current are delayed by $(\alpha + u/2)$ radians. The fundamental component of the input current is solely responsible for transferring power to the load. If the RMS value of the fundamental component is I_{s1}, then

$$P = E_s I_{s1} \cos(\alpha + u/2) = V_d I_d. \tag{5.13}$$

Therefore,

$$I_{s1} = \frac{V_d I_d}{E_s \cos(\alpha + u/2)}. \tag{5.14}$$

Note that P is negative for $\alpha > \pi/2$.

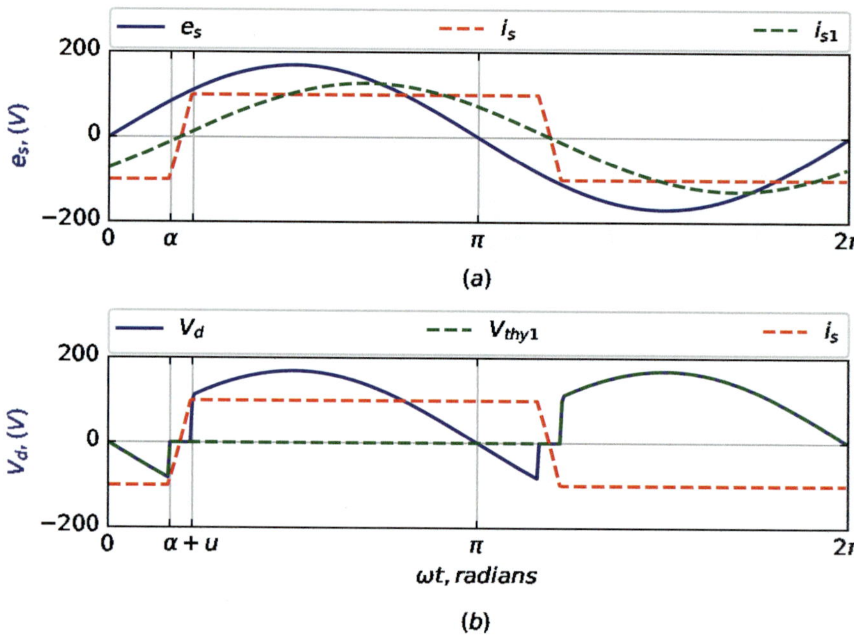

Figure 5.10 Single-phase, full-wave, controlled AC/DC converter with non-ideal AC source: response waveforms.

5.3.2 Inverter mode of operation for $\alpha > \pi/2$

The source voltage waveform and the input current waveform for $\alpha > \pi/2$ are shown in Figure 5.11. The output voltage waveform $V_d(\omega t)$ has an average value that is negative. Thus, the power flow is from the DC side to the AC side and the converter is operating in the inverter mode. This mode of operation is possible only when there is a source of energy on the DC side.

Note that the thyristor T_1 conducts during the interval $\alpha \leq \omega t \leq \pi+\alpha+u$ (see Figure 5.11). At $\omega t = \pi + \alpha + u$, it ceases to conduct and its cathode to anode voltage V_{thy1} jumps to a positive value. Thereafter, the thyristor remains reverse biased during the interval $\gamma = \{\pi+\alpha+u \leq \omega t \leq 2\pi\}$, where γ is defined as the extinction angle. It is a matter of concern that the extinction angle is quite small when the converter is operating in the inverter mode. When γ becomes less than the limit specified by the manufacturer, there is the danger that the thyristor could spontaneously jump into the conduction state as soon as it becomes forward biased for $\omega t > 2\pi$.

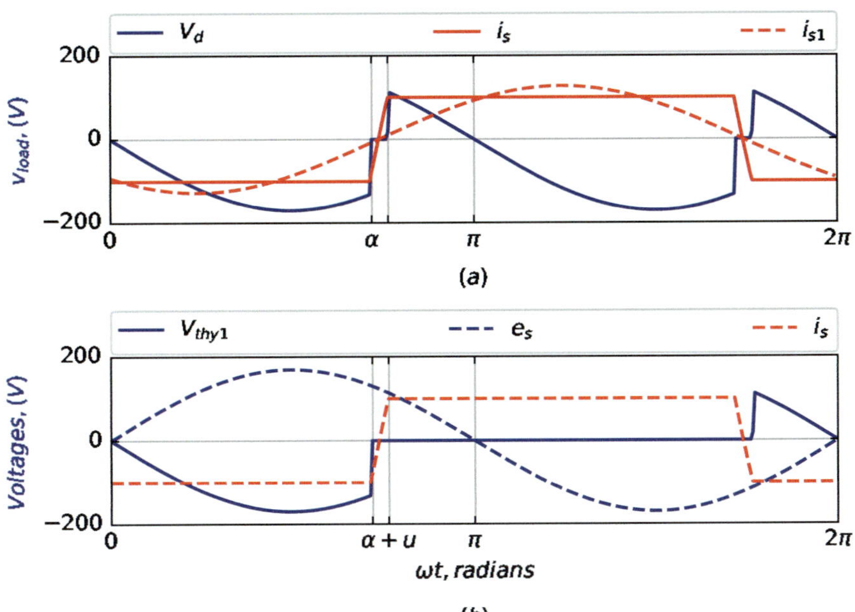

Figure 5.11 Single-phase, full-wave, controlled AC/DC converter, $\alpha > \pi/2$: response waveforms.

5.4 Three-phase, Controlled, AC/DC Converters

The three-phase, controlled AC/DC converter in its idealized form is shown in Figure 5.12 [1–4]. The AC source is an ideal source and the highly inductive load has been represented by the constant current source I_d. In each period of the AC source frequency, the thyristors are triggered in the same sequence as their reference numbers. The constant load current circulates through one thyristor of the upper group (T_1, T_3, T_5) and another of the lower group (T_4, T_6, T_2). If positive currents are continuously injected into the gates of the thyristors, they will function like diodes and conduct current whenever they are forward biased. The set of instants when the thyristors begin to conduct are the instants of natural conduction. These instants are shown numbered 1-6 in Figure 5.13(a), plotted by Program 5.13. Taking the instant of positive zero-crossing of es_a as the origin, the instant of natural conduction of thyristor T_1 is $\omega t = \pi/6$.

In the normal functioning of the controlled converter, each thyristor is triggered at the delayed instant that is α radians after the corresponding instant of natural conduction. Thus, thyristor T_1 will begin conducting at $\omega t = \alpha + \pi/6$ and stop conducting at $\omega t = \alpha + 5\pi/6$ radians. The load current circulates through the pair $(T_1 T_6)$ during the interval $\alpha + \pi/6 \leq \omega t \leq \alpha + \pi/2$

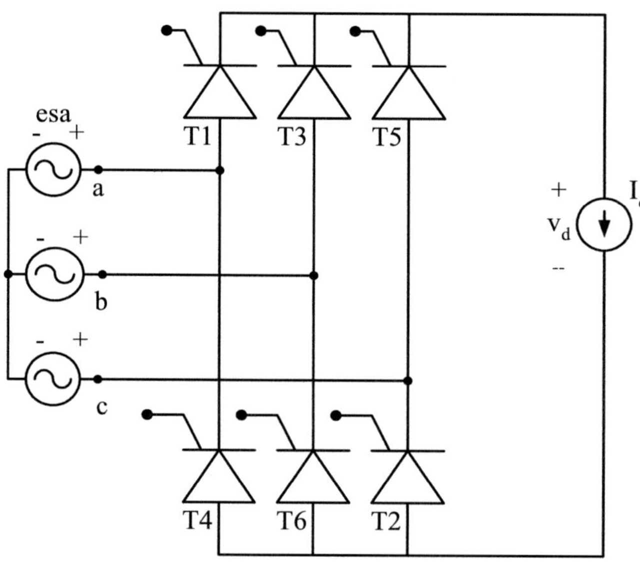

Figure 5.12 Three-phase, idealized, controlled AC/DC converter.

and through (T_1T_2) during the interval $\alpha + \pi/2 \le \omega t \le \alpha + 5\pi/6$. Other conducting intervals and the conducting thyristors are given in Table 5.1. The last column of this table indicates that the output voltage waveform consists of six pulses per period of the alternating input voltage.

The average output voltage is given by

$$V_d = \frac{1}{\pi/3}\int_{\alpha+\pi/6}^{\alpha+\pi/2} (es_a - es_b)\, d\omega t$$

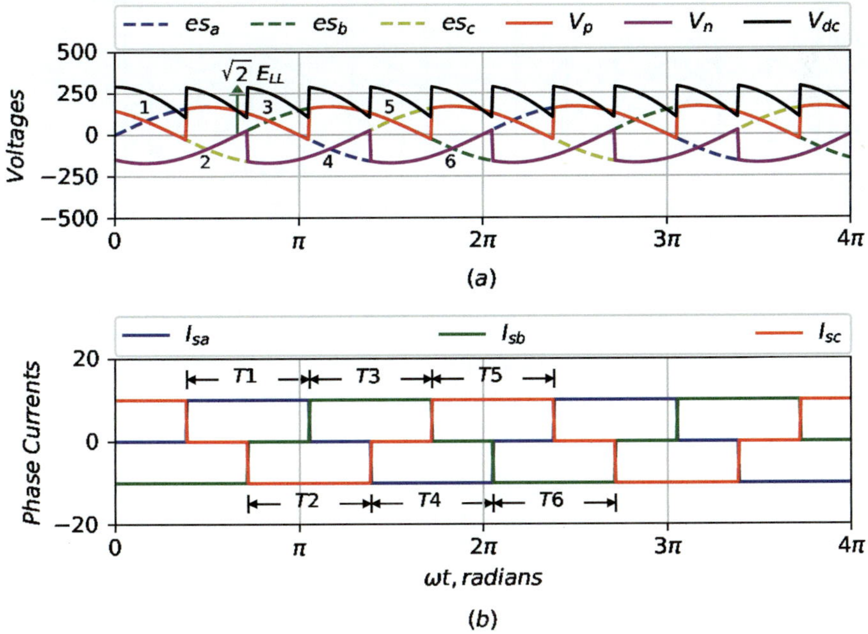

Figure 5.13 Three-phase, idealized, controlled AC/DC converter: response waveforms.

Table 5.1 Conduction pattern for the three-phase controlled converter.

Interval	Conducting thyristor	v_{Pn}	v_{Nn}	v_d
$\alpha + \pi/6 \le \omega t \le \alpha + \pi/2$	T_1, T_6	es_a	es_b	$es_a - es_b$
$\alpha + \pi/2 \le \omega t \le \alpha + 5\pi/6$	T_1, T_2	es_a	es_c	$es_a - es_c$
$\alpha + 5\pi/6 \le \omega t \le \alpha + 7\pi/6$	T_3, T_2	es_b	es_c	$es_b - es_c$
$\alpha + 7\pi/6 \le \omega t \le \alpha + 3\pi/2$	T_3, T_4	es_b	es_a	$es_b - es_a$
$\alpha + 3\pi/2 \le \omega t \le \alpha + 11\pi/6$	T_5, T_4	es_c	es_a	$es_c - es_a$
$\alpha + 11\pi/6 \le \omega t \le \alpha + 2\pi + \pi/6$	T_5, T_6	es_c	es_b	$es_c - es_b$

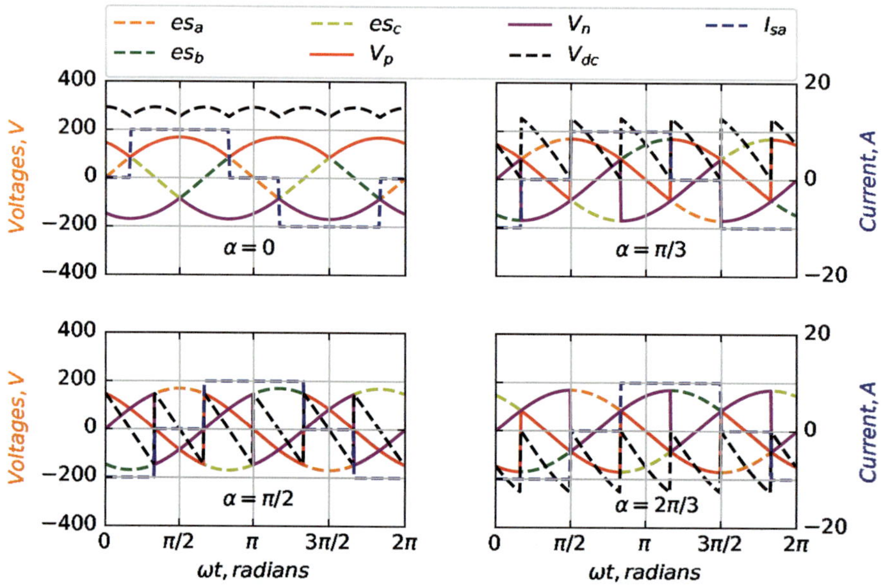

Figure 5.14 Input current, output voltage as functions of the delay angle.

$$
\begin{aligned}
&= \frac{3}{\pi}\sqrt{2}E_{LL}\int_{\alpha+\pi/6}^{\alpha+\pi/2+\alpha} \sin(\omega t + \pi/6)\, d\omega t \\
&= \frac{3}{\pi}\sqrt{2}E_{LL}\cos\alpha. \qquad (5.15)
\end{aligned}
$$

Therefore, the loss in the output voltage due to the delayed triggering is

$$
\begin{aligned}
\Delta V_d &= \frac{3}{\pi}\sqrt{2}E_{LL} - \frac{3}{\pi}\sqrt{2}E_{LL}\cos\alpha \\
&= \frac{3}{\pi}\sqrt{2}E_{LL}(1 - \cos\alpha). \qquad (5.16)
\end{aligned}
$$

Figure 5.14 shows the voltage and current waveforms for several values of the delay angle [1, 2].

5.4.1 AC side phase currents

Figure 5.15 shows the phase-a current. It consists of alternating rectangular pulses, shifted in phase by α radians and the RMS value is given by $I_{sa} = I_d\sqrt{2}/\sqrt{3}$. This current may be represented by

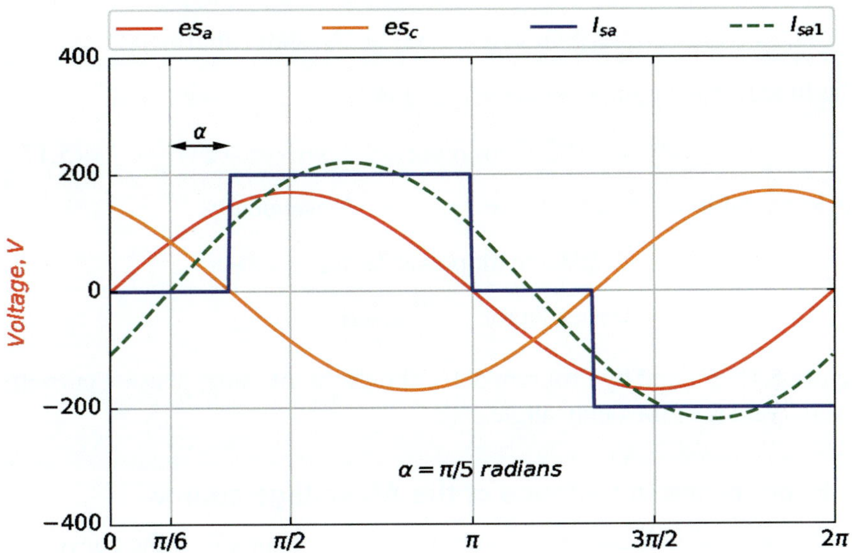

Figure 5.15 Three-phase, controlled AC/DC converter: phase-a current waveform.

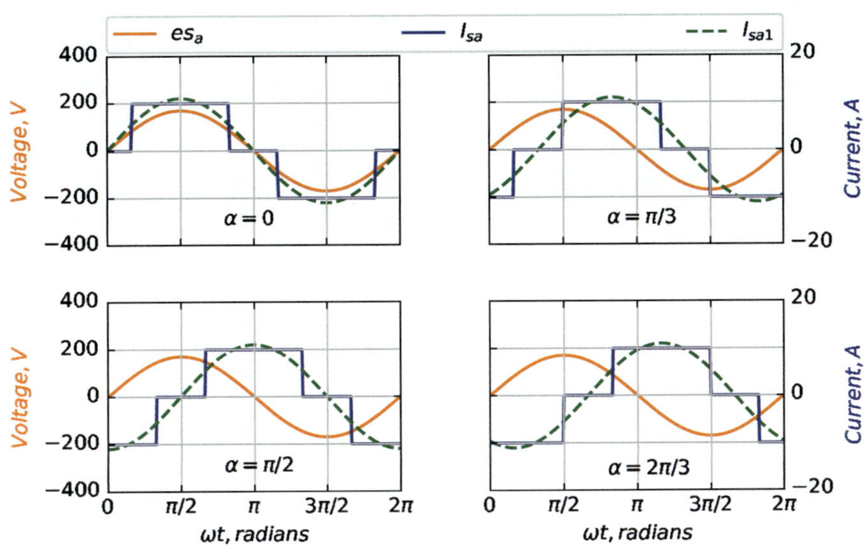

Figure 5.16 Three-phase, controlled AC/DC converter: AC side, phase-a current waveform as a function of the delay angle.

$$i_{sa}(\omega t) = \sqrt{2} * \left(I_{s1}\sin(\omega t-\alpha) - I_{s5}\sin 5(\omega t-\alpha) - I_{s7}\sin 7(\omega t-\alpha) + I_{s11}\sin 11(\omega t-\alpha) + I_{s13}\sin 13(\omega t-\alpha) - \cdots\right).$$

The fundamental component of $i_{sa}(\omega t)$ is

$$i_{sa1}(\omega t) = \sqrt{2}I_{s1}(\cos\alpha\sin\omega t - \sin\alpha\cos\omega t) \tag{5.17}$$

and its RMS value is given by $I_{s1} = I_d\sqrt{6}/\pi$. Consequently,

$$f_d = \text{displacement power factor} = \cos\alpha$$
$$f_p = \text{power factor} = \frac{3}{\pi}\cos\alpha.$$

Figure 5.16, plotted by Program 5.16, shows the AC side, phase-a current for several values of the delay angle.

5.4.2 Effect of the inductance of the AC voltage source

The presence of an inductive reactance at the input terminals of the converter should be considered for two reasons: (1) an ideal AC voltage source does not exist and a practical voltage source has an internal inductance; (2) the converter distorts the voltage waveform at the point of common coupling on the AC side. Deep notches will be observed in the line-to-line voltage waveforms at the input terminals and this is caused by the simultaneous conduction of three thyristors during the commutation process. Some national standards make it mandatory to insert an inductive reactance of at least 5% at the input terminals.

Consider the converter shown in Figure 5.17. The inductance L_s is the sum of the internal inductance of the voltage source and any inductance inserted at the input terminals of the converter. Suppose that T_5 and T_6 were conducting before T_1 is triggered at $\omega t = \alpha + \pi/6$. (Note that the delay angle is measured from $\omega t = \pi/6$, which is the instant of natural conduction for T_1.) When T_1 is triggered, a transient current is initiated in the loop comprising the phases "a" and "c" (Figures 5.18) and the load current commutates from T_5 to T_1 over the interval $\alpha + \pi/6 \leq \omega t \leq \alpha + u + \pi/6$. (The commutation interval is u radians.)

From Figure 5.18,

$$v_{Pn} = es_a - L_s\frac{di_a}{dt} \tag{5.18}$$

$$v_{Pn} = es_c - L_s\frac{di_c}{dt}. \tag{5.19}$$

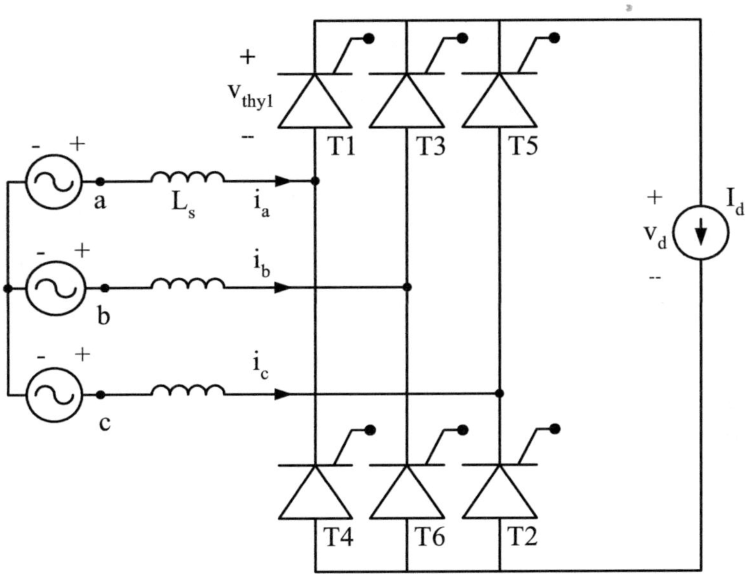

Figure 5.17 Three-phase, controlled AC/DC converter, non-ideal AC voltage source.

Therefore,

$$2v_{Pn} = es_a + es_c - L_s \frac{d}{dt}(i_a + i_c). \tag{5.20}$$

But, $i_a + i_c = I_d =$ constant load current. Therefore,

$$\frac{d}{dt}(i_a + i_c) = \frac{dI_d}{dt} = 0. \tag{5.21}$$

Thus,

$$v_{Pn} = \frac{1}{2}(es_a + es_c). \tag{5.22}$$

There is a dip in the voltage v_{Pn} during the commutation interval. Substituting eqn (5.22) in eqn (5.18),

$$L_s \frac{di_a}{dt} = \frac{1}{2}(es_a - es_c) = \frac{\sqrt{2}E_{LL}}{2} \sin(\omega t - \pi/6). \tag{5.23}$$

Integrating,

$$\int_0^{I_d} di_a = \frac{\sqrt{2}V_{LL}}{2\omega L_s} \int_{\alpha+\pi/6}^{\alpha+u+\pi/6} \sin(\omega t - \pi/6)\, d\omega t. \tag{5.24}$$

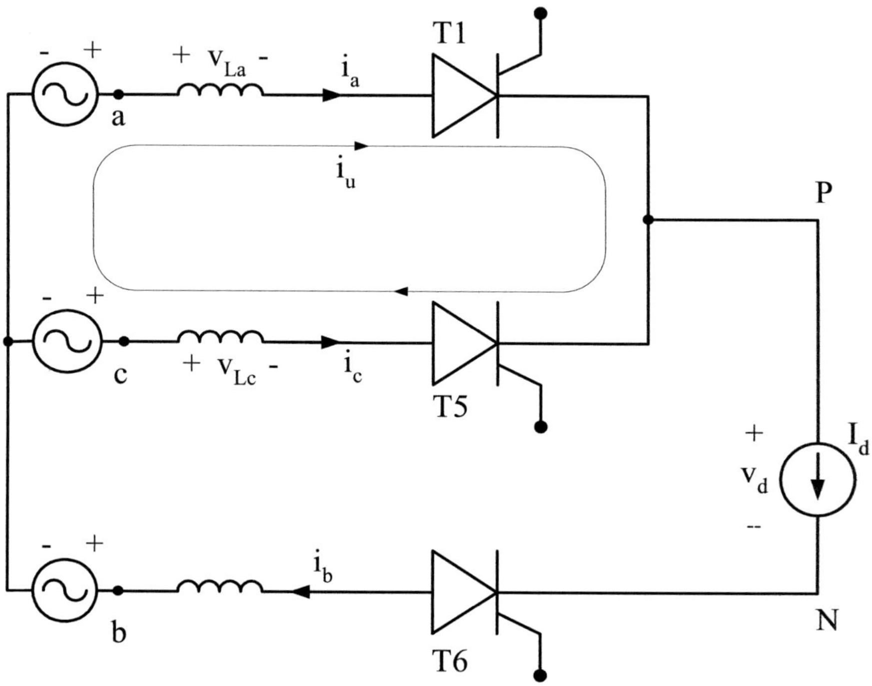

Figure 5.18 Three-phase, controlled AC/DC converter: commutation of current, phase-c to phase-a.

Or,

$$I_d = \frac{\sqrt{2}E_{LL}}{2\omega L_s}\big(\cos\alpha - \cos(\alpha+u)\big). \tag{5.25}$$

Rewriting the equation,

$$\cos(\alpha+u) = \cos\alpha - \frac{2\omega L_s I_d}{\sqrt{2}E_{LL}}. \tag{5.26}$$

The average output voltage is given by

$$\begin{aligned} V_d &= \frac{1}{\pi/3}\left[\int_{\alpha+\pi/6}^{\alpha+\pi/2}(es_a - es_b)d\omega t - \int_{\alpha+\pi/6}^{\alpha+u+\pi/6}(es_a - v_{Pn})d\omega t\right] \\ &= \frac{3}{\pi}\left[\sqrt{2}E_{LL}\cos\alpha - \int_{\alpha+\pi/6}^{\alpha+u+\pi/6} L_s\frac{di_a}{dt}d\omega t\right] \end{aligned}$$

$$= \frac{3}{\pi}\left[\sqrt{2}E_{LL}\cos\alpha - \int_0^{I_d} \omega L_s\, di_a\right]$$
$$= 1.35E_{LL}\cos\alpha - R_c I_d. \tag{5.27}$$

where $R_c = 3\omega L_s/\pi$.

Consider a three-phase system with $E_{LL} = 207.85$ V, $L_s = 5$ mH, $I_d = 10$ A. Let the delay angle $\alpha = \pi/6$. Then $\cos\alpha = 0.866$. Substituting in eqn (5.26),

$$\begin{aligned}\cos(\alpha + u) &= 0.866 - \frac{2 * 376.99 * 5 * 10^{-3} * 10}{\sqrt{2} * 207.85}\\ &= 0.7378\\ \text{Thus, } (\alpha + \text{u}) &= 0.7410\,\text{radians},\\ u &= 0.2174\,\text{radians}.\end{aligned}$$

Figure 5.19 (plotted by Program 5.19) shows the voltage waveforms of the DC side positive and negative terminals. The commutation of current, phase-c to phase-a, is also shown in this figure.

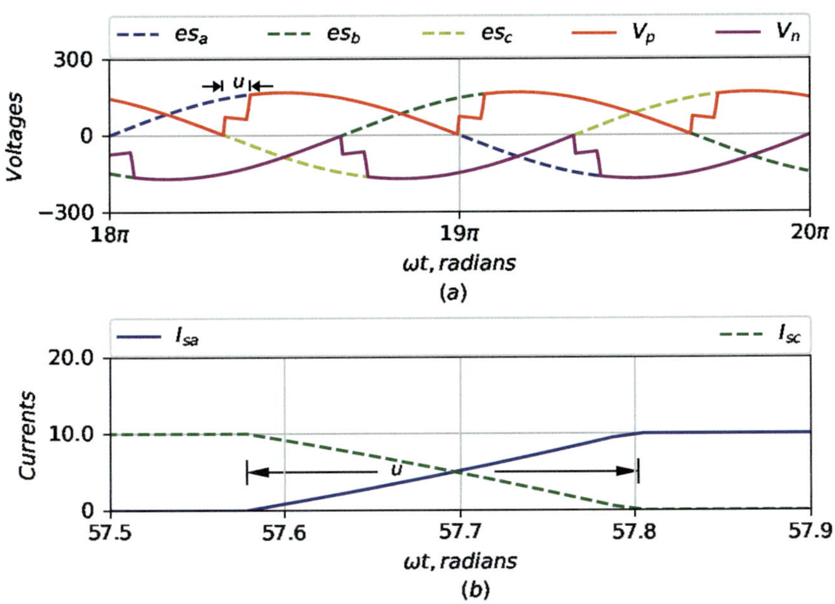

Figure 5.19 Three-phase, controlled AC/DC converter's response waveforms: commutation of the current from phase "c" to phase "a."

5.4.3 AC side currents

The voltage es_a and the current i_{sa} of phase-a are shown in Figure 5.20. The waveform of i_{sa} consists of alternating trapezoids and its fundamental frequency component i_{sa1} is phase shifted from es_a by $(\alpha + u/2)$ radians. For $\pi/2 \leq \alpha \leq \pi$, the phase delay of i_{sa1} from the voltage es_a, as shown in Figure 5.21 (plotted by Program 5.21), is greater than $\pi/2$ radians. Consequently, the alternating voltage source receives power from the constant current source on the DC side and converter operates in the inverter mode. Also shown in Figure 5.21 is the cathode to anode voltage $v_{thy5}(\omega t)$. After a delay of $(\alpha + u + \pi/6)$ radians from the positive zero-crossing of the phase voltage es_a, the voltage v_{thy5} jumps to a positive value and remains positive during the extinction angle γ that is given by $\gamma = \pi - (\alpha + u)$. For $\omega t > 7\pi/6$ radians from the positive zero-crossing of the voltage es_a, the voltage v_{thy5} is less than zero and, therefore, T_5 is forward biased. The thyristor T_5 may conduct spontaneously if it is forward biased after an extinction angle that is less than the minimum specified by the manufacturer.

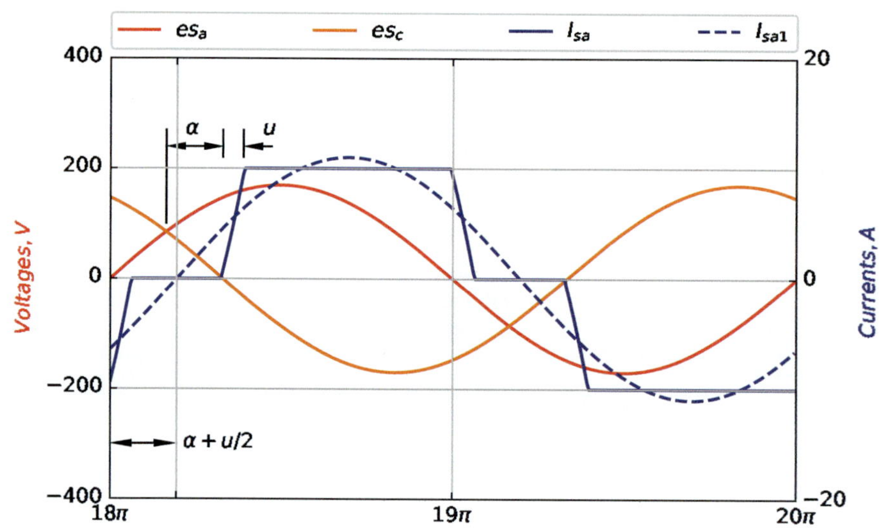

Figure 5.20 Three-phase, controlled AC/DC converter, non-ideal AC voltage source: AC side currents.

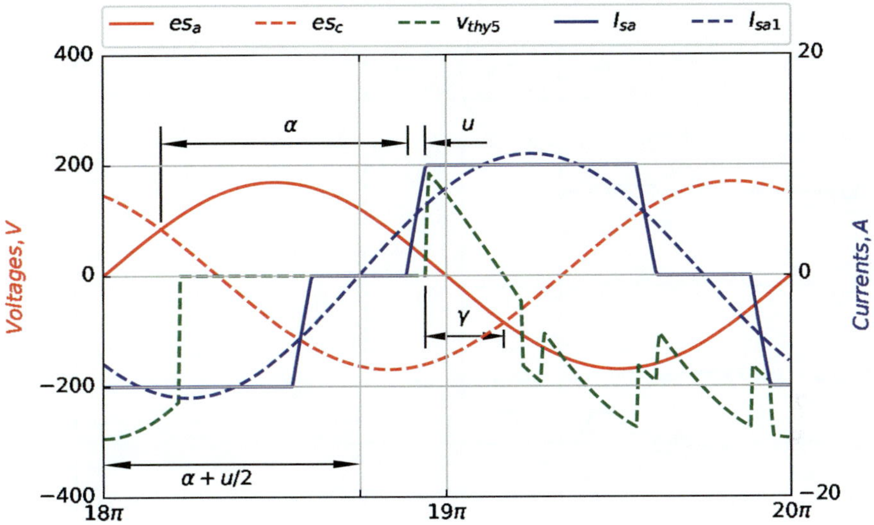

Figure 5.21 Three-phase, controlled AC/DC converter: inverter mode of operation.

5.5 Simulation of the three-phase, Controlled, AC/DC Converter

A practical example of the three-phase, controlled, AC/DC converter is shown in Figure 5.22. (It is a scaled version of a converter operating in the transmission voltage levels [5].) Figure 5.23 shows the transient equivalent circuit that includes the five probe voltages $v_p, v_{sa}, v_{sb}, v_{sc}, v_n$ [6]. The simulation technique is, therefore, identical to that utilized for the diode-based, three-phase, AC/DC converter. The only modification required is that the characteristic function of the diode should be replaced by the corresponding characteristic function of the thyristor. The input parameters for the thyristor characteristic are the cathode to anode voltage and the indicator of the conducting state, nst. A thyristor that enters the blocked state remains blocked until its triggering instant in the next period. The output variables of the thyristor characteristic function are the cathode to anode current and the conductance. A negative current indicates that the thyristor is in the conducting state and, therefore, the conductance will be very low.

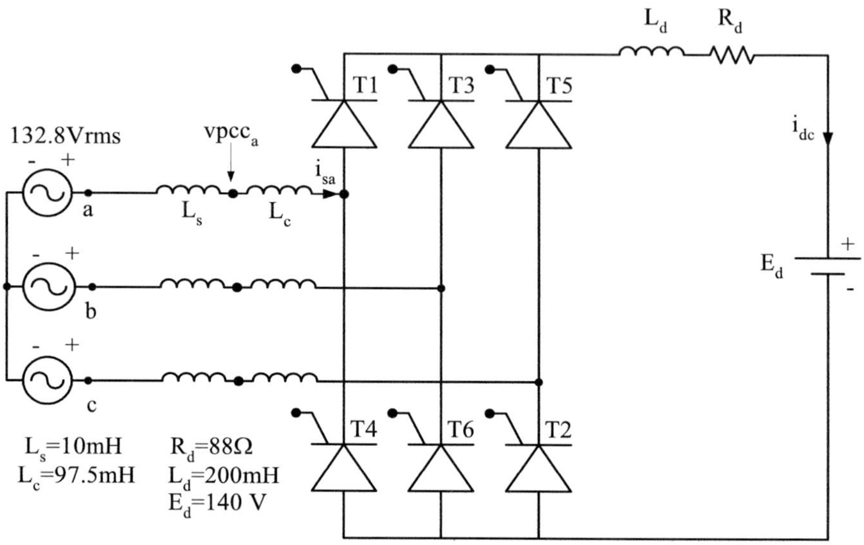

Figure 5.22 A practical three-phase, controlled AC/DC converter.

It is convenient to choose 360 time-steps per period so that delay angles specified in degrees correspond to the same number of time-steps. The instants of natural conduction of the thyristors T_1 to T_6 are the time-steps $30, 90, 150, 210, 270$, and 330. Given the delay as n_α time-steps, the triggering instants of the thyristors are obtained by adding n_α to the corresponding instants of natural conduction. Each thyristor is forced to remain in the conducting state for two time-steps following the triggering instant. The thyristor current is monitored at the end of each time-step and, if it is negative (see Figure 5.23), then it is maintained in the conducting state for the next time-step. If the thyristor current is positive at the end of a time-step, then it has entered the blocked state and the parameter nst is set to -1 for the next time-step. Note that the state of any thyristor cannot be changed within the iteration loop.

In the iteration loop of the simulation, the error currents irr_a, irr_b, irr_c, irr_p, irr_n flowing into the probe voltages are calculated. The square-root of the sum of the squares of the error currents is the overall error. If it is greater than the specified tolerance, then the corrections to be added to the probe voltages are calculated using the small-signal equivalent shown in Figure 5.24. After convergence and before proceeding to the next time-step,

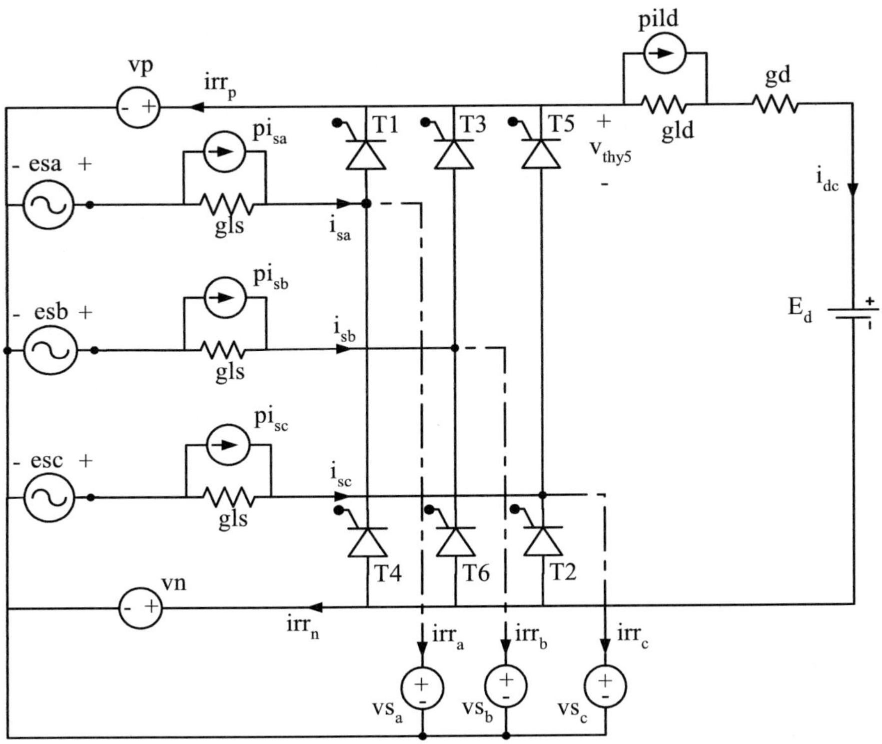

Figure 5.23 Three-phase, controlled AC/DC converter: transient equivalent circuit.

the variables to be plotted are stored and the past-history terms are updated. Figure 5.25 (plotted by Program 5.25) presents the results of the simulation.

5.5.1 Feedback control of the three-phase, AC/DC converter

It is essential to describe briefly the control of the three-phase, thyristor-based converter. Suppose that the objective is to control the average value of the load current I_d to be equal to a reference value I_{dref}. Then the load current is measured and supplied as the input to the control system, which will calculate the error and the new trigger delay for the next period. It is expected that the updated trigger delay will reduce the error.

Applying Kirchhoff's voltage law to the output loop,

$$v_d(\omega t) - E_d = L_d \frac{d}{dt} i_d(\omega t) + R_d i_d(\omega t). \tag{5.28}$$

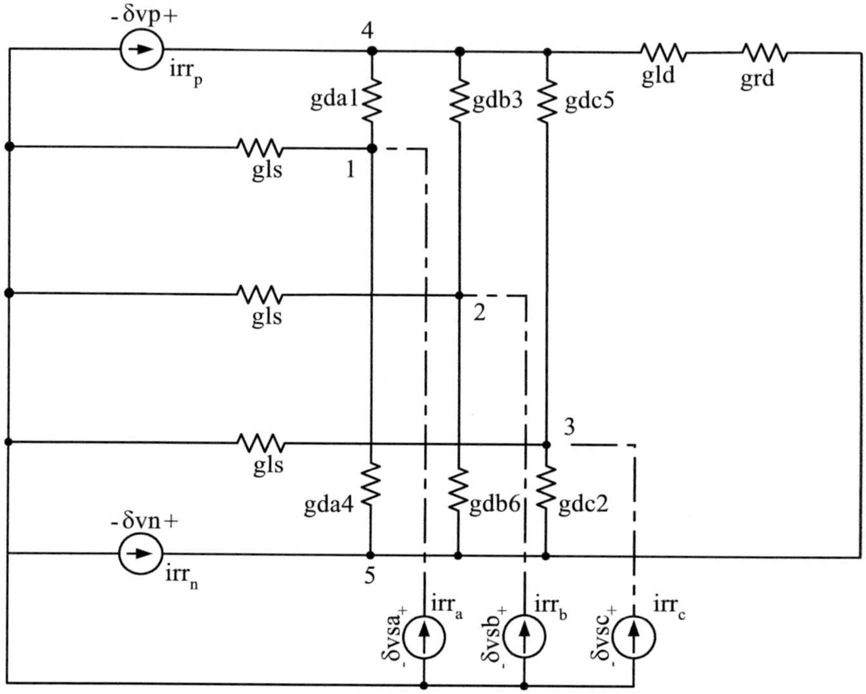

Figure 5.24 Three-phase, controlled AC/DC converter: small-signal equivalent circuit.

The load current consists of a direct component, the sixth harmonic, and multiples of the sixth harmonic. However, the harmonic components may be neglected due to the high inductance L_d. Therefore, the load current is assumed to be a direct current with possible step changes in its amplitude. Under these assumptions, eqn (5.28) may be written as

$$(1.35V_{LL}\cos\alpha - R_c I_d) - E_{dc} = L_d \frac{d}{dt} I_d + R_d I_d, \tag{5.29}$$

where $R_c = 3\omega L_s/\pi$. Substituting the parameter values,

$$1.35 * 230 * \cos\alpha - 38.7 * I_d - 140 = 0.2\frac{dI_d}{dt} + 88 * I_d. \tag{5.30}$$

Applying the Laplace transform and simplifying,

$$\cos\alpha - 0.4509 = \frac{0.2s + 126.7}{310.5} I_d(s). \tag{5.31}$$

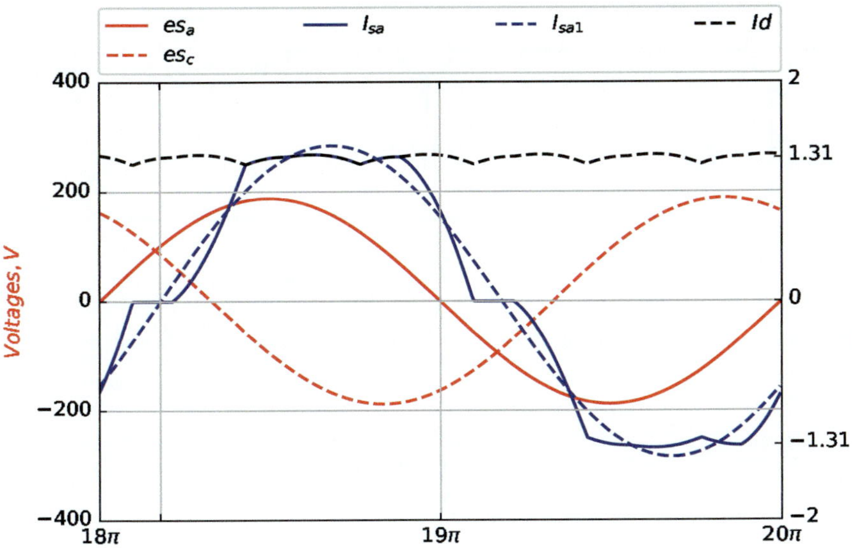

Figure 5.25 Three-phase, controlled AC/DC converter: response waveforms.

Normalizing all currents by dividing by $I_{d\max} = 1.414$A, which is the value of I_d for $\alpha = 0$,

$$\frac{I_d}{I_{d\max}} = \frac{310.5/(0.2 * 1.414)}{s + 633.5} * (\cos\alpha - 0.4509). \qquad (5.32)$$

The control system suggested by eqn (5.32) is shown in Figure 5.26. The controller is a simple integrator with gain K. The controller is chosen so that the system is critically damped:

$$1097.78 * K = \left(\frac{1}{2} * 633.5\right)^2$$
$$K = 91.39.$$

The constant 0.4509 that is subtracted from $\cos\alpha$ at the plant's input is treated as a disturbance by the controller and its effect at the output is rejected. From the proposed control system, we get

$$\frac{d\cos\alpha}{dt} = K \cdot ierr$$

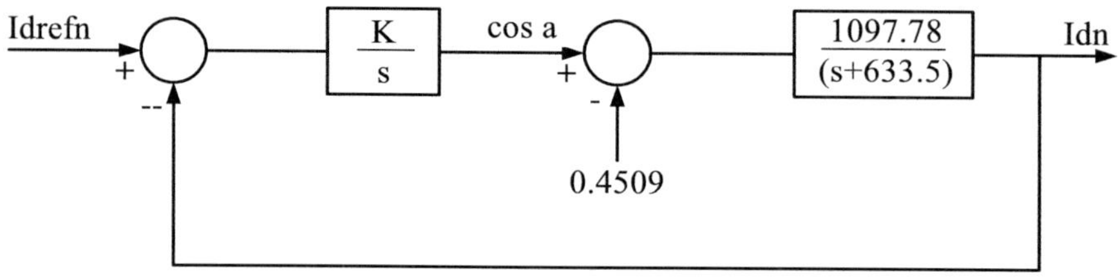

Figure 5.26 Three-phase, controlled AC/DC converter: feedback control system.

$$\text{Or, } \Delta(\cos\alpha) = K \cdot ierr \cdot \Delta t$$
$$= \left(\frac{K}{f \cdot n}\right) \cdot ierr$$
$$= K_{ierr} \cdot ierr.$$

At the beginning of each time-step, the current i_d is monitored and its average value I_d estimated. For this purpose, the load current is expressed as

$$i_d(\omega t) = I_d + I_{c6}\cos 6\omega t + I_{s6}\sin 6\omega t. \tag{5.33}$$

The amplitudes I_d, I_{c6}, and I_{s6} are estimated using the recursive least square (RLS) estimator. The RLS estimator uses the current sample i_d and the estimate of the average value in the previous time-step, to arrive at the current estimate of the average value I_d. The RLS estimator is given below without the underlying theory [7]:

- Put $X = [0 0 0]$
- Put $P = 100 * \begin{bmatrix} 1 0 0 \\ 0 1 0 \\ 0 0 1 \end{bmatrix}$.

At each time-step:

- Form the row vector $H = [1 \quad \cos 6(n\omega\Delta t) \quad \sin 6(n\omega\Delta t)]$
- HT = transpose of H
- The scalar $r = 1 + H * P * HT$
- $Kr = \frac{1}{r} * (P * HT)$
- $kf = i_d(n\omega\Delta t) - H * X$

- $X = X + kf * Kr = [\, I_d \, I_{c6} \, I_{s6} \,]$
- $I_d = X\,[0]$.

Update the matrix P for the next time-step:

- $dP = Kr * H * P$
- $P = \frac{1}{ff} * (P - dP), \; ff = \text{forgetting factor} = 0.98.$

The code for simulating the converter over a time span of 15 periods, including its control, is given in Program 5.27. The simulation starts with the control system disabled and the delay angle set to 10°. The average load current settles at 1.31 A. After five periods of simulation, the control system is enabled and the reference current is set at 1.2 A. At the start of the 10th period, the reference current is again changed to 0.9 A. Figure 5.27 shows the results of the simulation and the functioning of the feedback control. The control appears to be slow mainly because the delay angle is changed only once at the beginning of each period. Moreover, the controller is a simple integrator. A PI controller with proper pole placement is likely to speed up the control.

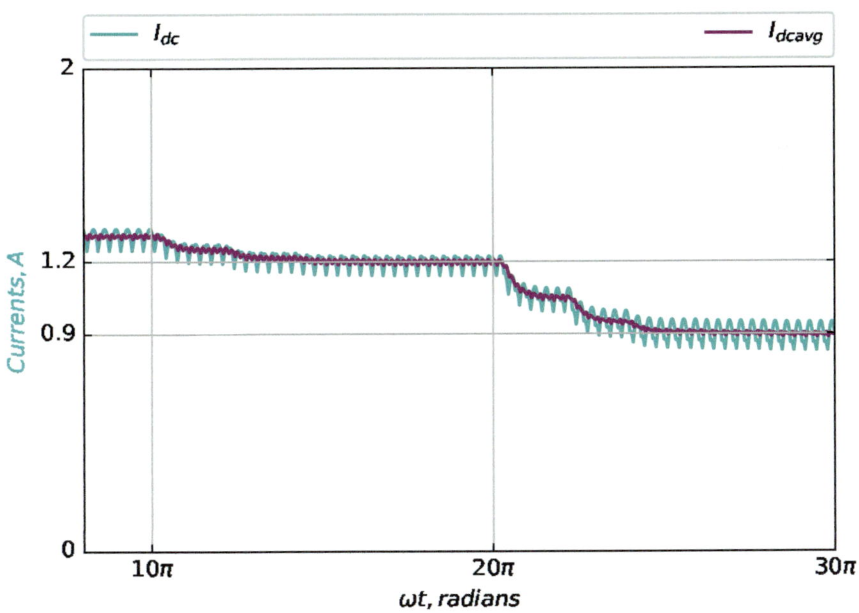

Figure 5.27 Three-phase, controlled AC/DC converter with feedback control: average load current.

5.6 Computational Programs

5.6.1 Program 5.04

```
#
def fthy(vth, nsw):
    """ Thyristor characteristic
        nsw: trigger signal;
        vth: thyristor cathode to anode voltage
        Returns thyristor current and conductance"""
    #
    glow = 0.1e-07
    ghigh = 0.1e+03
    #
    if nsw == -1:
        gdth = glow
    if nsw == 1:
        if vth < 0.0:
            gdth = ghigh
        else:
            gdth = glow
    #
    idth = gdth*vth
    #
    return idth, gdth
#
def ch5fig4_solver(esource, gsnb, gc, gload, gld, nalpha, ncycle, n):
    #
    vn = 0.1
    pvc = 0.0
    pid = 0.0
    ihold = -0.01
    erlim = 0.1e-07
    nst = -1
    #
    gstot = gsnb*gc/(gsnb + gc)
    gdtot = gload*gld/(gload + gld)
    #
    ithyristor = []
    vthyristor = []
    iload = []
    vinductor = []
    #
    for k in range(ncycle):
        #
        nk = n*k
        #
        for j in range(n):
            #
            i = nk + j
            es = esource[i]
            #
            if j >= nalpha and j < (nalpha + 5):
```

```
            nst = 1
        #
        for its in range(1000):
            #
            v1 = (gsnb*es + gc*vn + gc*pvc)/(gc + gsnb)
            v2 = (gload*vn - pid)/(gld + gload)
            #
            ith, gth = fthy(vn - es, nst)
            id = gload*(vn - v2)
            ic = gc*(v1 - vn) - gc*pvc
            irrn = ic - ith -id
            err = abs(irrn)
            #
            if err < erlim:
                break
            else:
                dvn = irrn/(gth + gstot + gdtot)
                vn += dvn
        #
        if its >= 999:
            print('time-step= ', i, 'solver does not converge')
            break
        #
        ithyristor.append(-ith)
        vthyristor.append(es - vn)
        iload.append(id)
        vinductor.append(v2)
        #
        pid = id
        pvc = v1 - vn
        if nst == 1 and ith > ihold:
            nst = -1
            if k == 0:
                nbeta = j
        #
    return ithyristor, vthyristor, iload, vinductor, nbeta
#
def ch5fig4(esmax, freq, rsnb, cb, rload, ld, nalpha, ncycle, n):
    """Main program, plots results
       Basic thyristor-based, half-wave AC/DC converter"""
    #
    #Calling command:
    #ch5fig4(325.3, 60.0, 20.0, 0.15e-06, 10.0, 0.01, 60, 2, 360)
    #
    from math import pi
    import numpy as np
    from mpl_toolkits.axes_grid1 import host_subplot
    import matplotlib.pyplot as plt
    #
    N = ncycle*n
    delt = 1.0/(freq*n)
    dteta = 2.0*pi/n
    alpha = nalpha*dteta
```

```
teta = dteta*np.arange(N)
esource = esmax*np.sin(teta)
#
gsnb = 1.0/rsnb
gc = cb/delt
gload = 1.0/rload
gld = delt/ld
#
ithy, vthy, iload, vinductor, nbeta = ch5fig4_solver(esource, gsnb,  \
                                  gc, gload, gld, nalpha, ncycle, n)
#
bt = nbeta*dteta
n1 = 2*ncycle
#
ax1 = host_subplot(211)
ax2 = ax1.twinx()
ax1.set_xlim([0, n1*pi])
ax1.set_xticks([0, alpha, pi, bt, 2*pi, 3*pi, 4*pi])
ax1.set_xticklabels(['$0$', '$\\alpha$', '$\pi$', '$\\beta$',        \
                                   '$2\pi$', '$3\pi$', '$4\pi$'])
ax1.set_ylim([-400, 400])
ax1.set_yticks([-400, -200, 0, 200, 400])
ax1.set_yticklabels(['$-400$', '$-200$', '$0$', '$200$','$400$'])
ax1.set_ylabel('$v_{thy}, (V)$')
ax1.text(6, -700, '$(a)$', size =10)
ax2.set_ylabel('$i_{thy}, (A)$')
ax2.set_ylim([-40, 40])
ax2.set_yticks([-40, 0, 40])
ax2.set_yticklabels(['$-40$', '$0$', '$40$'])
p1, = ax1.plot(teta, vthy, 'b-', label = '$v_{thy}$')
p2, = ax2.plot(teta, ithy, 'r--', label = '$i_{thy}$')
ax1.yaxis.get_label().set_color(p1.get_color())
ax2.yaxis.get_label().set_color(p2.get_color())
ax1.legend(bbox_to_anchor=(0., 1.02, 1., .102), loc=3, ncol=2,      \
           mode="expand", borderaxespad=0.)
ax1.grid(True)
plt.subplots_adjust(hspace=0.75)
#
ax1 = host_subplot(212)
ax2 = ax1.twinx()
ax1.set_xlim([0, n1*pi])
ax1.set_xticks([0, alpha, pi, bt, 2*pi, 3*pi, 4*pi])
ax1.set_xticklabels(['$0$', '$\\alpha$', '$\pi$', '$\\beta$',      \
                               '$2\pi$', '$3\pi$', '$4\pi$'])
ax1.set_ylim([-250, 250])
ax1.set_yticks([-250, -125, 0, 125, 250])
ax1.set_yticklabels(['$-250$', '$-125$', '$0$', '$125$', '$250$'])
ax1.set_xlabel('$\omega t, radians$')
ax1.set_ylabel('$V_{L}, (V)$')
ax1.text(6, -530, '$(b)$', size =10)
ax2.set_ylabel('$i_{load}, (A)$')
ax2.set_ylim([-40, 40])
ax2.set_yticks([-40, 0, 40])
```

```
    ax2.set_yticklabels(['$-40$','$0$', '$40$'])
    p1, = ax1.plot(teta, vinductor, 'b-', label = '$v_{L}$')
    p2, = ax2.plot(teta, iload, 'r--', label = '$i_{load}$')
    ax1.yaxis.get_label().set_color(p1.get_color())
    ax2.yaxis.get_label().set_color(p2.get_color())
    ax1.legend(bbox_to_anchor=(0., 1.02, 1., .102), loc=3, ncol=2,      \
                                mode="expand", borderaxespad=0.)
    ax1.grid(True)
    #
    plt.show()
    #
    # plt.draw()
    # plt.savefig('Fig504.ps', dpi=400, bbox_inches='tight')
    #
```

5.6.2 Program 5.06

```
#
def fthy(vth, nsw):
    """  Thyristor characteristic
         nsw: trigger signal
         vth: thyristor cathode to anode voltage;
         Returns thyristor current and conductance"""
    #
    glow = 0.1e-07
    ghigh = 0.1e+03
    #
    if nsw == -1:
        gdth = glow
    if nsw == 1:
        if vth < 0.0:
            gdth = ghigh
        else:
            gdth = glow
    #
    idth = gdth*vth
    #
    return idth, gdth
#
def ch5fig6_solver(esource, gld, Ed, nalpha, ncycle, n):
    #
    vn = 0.1
    pid = 0.0
    ihold = -0.01
    erlim = 0.1e-07
    nst = -1
    #
    ithyristor = []
    vthyristor = []
    iload = []
    vload = []
    #
```

```
    for k in range(ncycle):
        #
        nk = n*k
        #
        for j in range(n):
            #
            i = nk + j
            es = esource[i]
            #
            if j >= nalpha and j < (nalpha + 5):
                nst = 1
            #
            for its in range(1000):
                #
                ith, gth = fthy(vn - es, nst)
                id = gld*(vn - Ed) + pid
                irrn = -(ith + id)
                err = abs(irrn)
                #
                if err < erlim:
                    break
                else:
                    dvn = irrn/(gth + gld)
                    vn += dvn
            #
            if its >= 999:
                print('time-step= ', i, 'solver does not converge')
                break
            #
            ithyristor.append(-ith)
            vthyristor.append(es - vn)
            iload.append(id)
            vload.append(vn)
            #
            pid = id
            if nst == 1 and ith > ihold:
                nst = -1
                if k == 0:
                    nbeta = j
            #
    return ithyristor, vthyristor, iload, vload, nbeta
#
def ch5fig6(Erms, freq, ld, Ed, nalpha, ncycle, n):
    """Main program, plots results
       Thyristor-based AC/DC converter with
       inductive load and a back emf"""
    #
    #Calling command:
    #ch5fig6(325.3, 60.0, 0.04, 150.0, 45, 2, 360)
    #
    from math import pi
    import numpy as np
    from mpl_toolkits.axes_grid1 import host_subplot
```

```
import matplotlib.pyplot as plt
#
N = ncycle*n
delt = 1.0/(freq*n)
dteta = 2.0*pi/n
alpha = nalpha*dteta
teta = dteta*np.arange(N)
esmax = Erms*np.sqrt(2.0)
esource = esmax*np.sin(teta)

#
gld = delt/ld
#
ithy, vthy, iload, vload, nbeta = ch5fig6_solver(esource, gld, Ed,   \
                                            nalpha, ncycle, n)
#
bt = nbeta*dteta
n1 = 2*ncycle
#
ax1 = host_subplot(111)
ax2 = ax1.twinx()
ax1.set_xlim([0, n1*pi])
ax1.set_xticks([0, alpha, pi, bt, 2*pi, 3*pi, 4*pi])
ax1.set_xticklabels(['$0$', '$\\alpha$', '$\pi$', '$\\beta$',         \
                                '$2\pi$', '$3\pi$', '$4\pi$'])
ax1.set_ylim([-1000, 1000])
ax1.set_yticks([-1000, -500, 0, 500, 1000])
ax1.set_yticklabels(['$-1000$', '$-500$', '$0$', '$500$','$1000$'])
ax1.set_ylabel('$Voltages, (V)$')
ax2.set_ylabel('$i_{load}, (A)$')
ax2.set_ylim([-40, 40])
ax2.set_yticks([-40, 0, 40])
ax2.set_yticklabels(['$-40$', '$0$', '$40$'])
p1, = ax1.plot(teta, esource, 'b-.', label ='$e_{S}$')
p2, = ax1.plot(teta, vthy, 'b--', label = '$v_{thy}$')
p3, = ax1.plot(teta, vload, 'b-', label = '$v_{load}$')
p4, = ax2.plot(teta, iload, 'r--', label = '$i_{load}$')
ax1.yaxis.get_label().set_color(p1.get_color())
ax2.yaxis.get_label().set_color(p4.get_color())
ax1.legend(bbox_to_anchor=(0., 1.02, 1., .102), loc=3, ncol=4,     \
                            mode="expand", borderaxespad=0.)
ax1.grid(True)
#
plt.show()
#
# plt.draw()
# plt.savefig('Fig506.ps', dpi=400, bbox_inches='tight')
#
```

5.6.3 Program 5.08

```
#
def fthy(vth, nsw):
    """Thyristor characteristic
       vth: thyristor cathode to anode voltage;
       nsw: trigger signal
       Returns thyristor current and conductance"""
    #
    glow = 0.1e-07
    ghigh = 0.1e+03
    #
    if nsw == -1:
        gdth = glow
    if nsw == 1:
        if vth < 0.0:
            gdth = ghigh
        else:
            gdth = glow
    #
    idth = gdth*vth
    #
    return idth, gdth
#
def ch5fig8_solver(esource, Id, nalpha, ncycle, n):
    #
    vd = 0.1
    ihold = -0.01
    erlim = 0.1e-07
    ntrig1 = nalpha
    ntrig2 = 180 + nalpha
    nst1 = -1
    nst2 = 1
    #
    isource = []
    iload = []
    vload = []
    #
    for k in range(ncycle):
        #
        nk = n*k
        #
        for j in range(n):
            #
            i = nk + j
            es = esource[i]
            #
            if j >= ntrig1 and j < (ntrig1 + 5):
                nst1 = 1
            if j >= ntrig2 and j < (ntrig2 + 5):
                nst2 = 1
            #
            for its in range(1000):
```

```
                #
                vth1 = 0.5*(vd - es)
                vth2 = 0.5*(vd + es)
                ith1, gth1 = fthy(vth1, nst1)
                ith2, gth2 = fthy(vth2, nst2)
                irrd = -(ith1 + ith2 +Id)
                err = abs(irrd)
                #
                if err < erlim:
                    break
                else:
                    gtot = 0.5*(gth1 + gth2)
                    dvd = irrd/gtot
                    vd += dvd
            #
            if its >= 999:
                print('time-step= ', i, 'solver does not converge')
                break
            #
            isource.append(ith2 - ith1)
            iload.append(Id)
            vload.append(vd)
            #
            if nst1 == 1 and ith1 > ihold:
                nst1 = -1
            if nst2 == 1 and ith2 > ihold:
                nst2 = -1
            #
    return isource, iload, vload
#
def ch5fig8(Erms, freq, Id, nalpha, ncycle, n):
    """Main program, plots results
       Thyristor-based, single-phase, full-wave, controlled
       AC/DC converter"""
    #
    #Calling command:
    #ch5fig8(120.0, 60.0, 10.0, 30, 2, 360)
    #
    from math import pi
    import numpy as np
    from mpl_toolkits.axes_grid1 import host_subplot
    import matplotlib.pyplot as plt
    #
    dteta = 2.0*pi/n
    alpha = nalpha*dteta
    N = ncycle*n
    teta = dteta*np.arange(N)
    esmax = Erms*np.sqrt(2.0)
    esource = esmax*np.sin(teta)
    #
    isource, iload, vload = ch5fig8_solver(esource, Id,           \
                                            nalpha, ncycle, n)
    #
```

```
n1 = 2*ncycle
#
ax1 = host_subplot(211)
ax2 = ax1.twinx()
ax1.set_xlim([0, n1*pi])
ax1.set_xticks([0, alpha, pi, 2*pi, 3*pi, 4*pi])
ax1.set_xticklabels(['$0$', '$\\alpha$', '$\pi$', '$2\pi$',    \
                                        '$3\pi$', '$4\pi$'])
ax1.set_ylim([-200, 200])
ax1.set_yticks([-200, -100, 0, 100, 200])
ax1.set_yticklabels(['$-200$', '$-100$', '$0$', '$100$','$200$'])
ax1.set_ylabel('$e_{s}, (V)$')
ax1.text(6, -350, '$(a)$', size =10)
ax2.set_ylabel('$i_{s}, (A)$')
ax2.set_ylim([-20, 20])
ax2.set_yticks([-20, 0, 20])
ax2.set_yticklabels(['$-20$', '$0$', '$20$'])
p1, = ax1.plot(teta, esource, 'r', label = '$e_{S}$')
p2, = ax2.plot(teta, isource, 'b--', label = '$i_{S}$')
ax1.yaxis.get_label().set_color(p1.get_color())
ax2.yaxis.get_label().set_color(p2.get_color())
ax1.legend(bbox_to_anchor=(0., 1.02, 1., .102), loc=3, ncol=2,   \
                                mode="expand", borderaxespad=0.)
ax1.grid(True)
plt.subplots_adjust(hspace=0.8)
#
ax1 = host_subplot(212)
ax2 = ax1.twinx()
ax1.set_xlim([0, n1*pi])
ax1.set_xticks([0, alpha, pi, 2*pi, 3*pi, 4*pi])
ax1.set_xticklabels(['$0$', '$\\alpha$', '$\pi$', '$2\pi$',      \
                                          '$3\pi$', '$4\pi$'])
ax1.set_ylim([-200, 200])
ax1.set_yticks([-200, -100, 0, 100, 200])
ax1.set_yticklabels(['$-200$', '$-100$', '$0$', '$100$', '$200$'])
ax1.set_xlabel('$\omega t, radians$')
ax1.set_ylabel('$V_{load}, (V)$')
ax1.text(6, -440, '$(b)$', size =10)
ax2.set_ylabel('$I_{load}, (A)$')
ax2.set_ylim([-20, 20])
ax2.set_yticks([-20, 0, 20])
ax2.set_yticklabels(['$-20$','$0$', '$20$'])
p1, = ax1.plot(teta, vload, 'r', label = '$v_{load}$')
p2, = ax2.plot(teta, iload, 'b--', label = '$i_{load}$')
ax1.yaxis.get_label().set_color(p1.get_color())
ax2.yaxis.get_label().set_color(p2.get_color())
ax1.legend(bbox_to_anchor=(0., 1.02, 1., .102), loc=3, ncol=2,   \
                              mode="expand", borderaxespad=0.)
ax1.grid(True)
#
plt.show()
#
# plt.draw()
```

```
    # plt.savefig('Fig508.ps', dpi=400, bbox_inches='tight')
    #
```

5.6.4 Program 5.10

```
#
def fthy(vth, nsw):
    """  Thyristor characteristic
         vth: thyristor cathode to anode voltage;
         nsw: trigger signal
         Returns thyristor current and conductance"""
    #
    glow = 0.1e-07
    ghigh = 0.1e+03
    #
    if nsw == -1:
        gdth = glow
    if nsw == 1:
        if vth < 0.0:
            gdth = ghigh
        else:
            gdth = glow
    #
    idth = gdth*vth
    #
    return idth, gdth
#
def ch5fig10_solver(esource, gls, Id, nalpha, ncycle, n):
    #
    import numpy as np
    #
    vd = 0.1
    va = 0.1
    ihold = -0.001
    erlim = 0.1e-07
    ntrig1 = nalpha
    ntrig2 = nalpha + n/2
    nst1 = -1
    nst2 = 1
    nrise = True
    pils = -Id
    #
    isource = []
    iload = []
    vload = []
    vthyristor1 = []
    #
    for k in range(ncycle):
        #
        for j in range(n):
            #
            es = esource[j]
```

```
    #
    if j >= ntrig1 and j < (ntrig1 + 5):
        nst1 = 1
    if j >= ntrig2 and j < (ntrig2 + 5):
        nst2 = 1
    #
    for its in range(1000):
        #
        vth1 = 0.5*(vd - va)
        vth2 = 0.5*(vd + va)
        ith1, gth1 = fthy(vth1, nst1)
        ith2, gth2 = fthy(vth2, nst2)
        ils = gls*(es - va) + pils
        irra = ils + ith1 - ith2
        irrd = -(ith1 + ith2 + Id)
        err = np.sqrt(irra**2 + irrd**2)
        #
        if err < erlim:
            break
        else:
            gsum = 0.5*(gth1 + gth2)
            gdif = 0.5*(gth2 - gth1)
            gmat = np.array([[(gls + gsum), gdif],
                              [gdif, gsum]])
            bcol = np.array([irra, irrd])
            xdv = np.linalg.solve(gmat, bcol)
            dva = xdv[0]
            dvd = xdv[1]
            va = va + dva
            vd = vd + dvd
    #
    if its >= 999:
        print('time-step= ', (k*n + j), 'solver does    \
                                       not converge')
        break
    #
    if k == (ncycle - 1):
        isource.append(ils)
        iload.append(Id)
        vload.append(vd)
        vthyristor1.append(vth1)
        #
        if j > nalpha:
            if nrise == True and (ils - Id) >= 0.0:
                ku = j
                nrise = False
        #
    pils = ils
    #
    if nst1 == 1 and ith1 > ihold:
        nst1 = -1
    if nst2 == 1 and ith2 > ihold:
        nst2 = -1
```

```
        #
    return isource, iload, vload, vthyristor1, ku
#
def ch5fig10(Erms, freq, ls, Id, nalpha, ncycle, n):
    """Main program, plots results
       Single-phase, full-wave, controlled, AC/DC converter
       non-ideal AC source"""
    #
    #Calling command:
    #ch5fig10(120.0, 60.0, 0.005, 5.0, 30, 2, 360)
    #
    from math import pi
    import numpy as np
    from mpl_toolkits.axes_grid1 import host_subplot
    import matplotlib.pyplot as plt
    #
    dteta = 2.0*pi/n
    alpha = nalpha*dteta
    delt = 1.0/(freq*n)
    gls = delt/ls
    teta = dteta*np.arange(n)
    esmax = Erms*np.sqrt(2.0)
    esource = esmax*np.sin(teta)
    #
    isource, iload, vload, vthyristor1, ku = \
                    ch5fig10_solver(esource, gls, Id,      \
                                         nalpha, ncycle, n)
    u = ku*dteta
    #
    is_freq = np.fft.fft(isource)
    z = is_freq[1]
    a2 = (z.real)/(n/2.0)
    b2 = -(z.imag)/(n/2.0)
    is1 = a2*np.cos(teta) + b2*np.sin(teta)
    is1_peak = np.sqrt(a2**2 + b2**2)
    print('u = ', u, 'is1_peak = ', is1_peak)
    #
    ax1 = host_subplot(211)
    ax2 = ax1.twinx()
    ax1.set_xlim([0, 2*pi])
    ax1.set_xticks([0, alpha, u, pi, 2*pi])
    ax1.set_xticklabels(['$0$', '$\\alpha$', '$u$',     \
                                        '$\pi$', '$2\pi$'])
    ax1.set_ylim([-200, 200])
    ax1.set_yticks([-200, 0, 200])
    ax1.set_yticklabels(['$-200$', '$0$', '$200$'])
    ax1.set_ylabel('$e_{s}, (V)$')
    ax1.text(3, -350, '$(a)$', size =10)
    ax2.set_ylabel('$i_{s}, (A)$')
    ax2.set_ylim([-10, 10])
    ax2.set_yticks([-10, 0, 10])
    ax2.set_yticklabels(['$-10$', '$0$', '$10$'])
    p1, = ax1.plot(teta, esource, 'r', label = '$e_{s}$')
```

```
    p2, = ax2.plot(teta, isource, 'b--', label = '$i_{s}$')
    p3, = ax2.plot(teta, is1, 'g--', label = '$i_{s1}$')
    ax1.yaxis.get_label().set_color(p1.get_color())
    ax2.yaxis.get_label().set_color(p2.get_color())
    ax1.legend(bbox_to_anchor=(0., 1.02, 1., .102), loc=3, ncol=3, \
                                   mode="expand", borderaxespad=0.)
    ax1.grid(True)
    plt.subplots_adjust(hspace=0.8)
    #
    ax1 = host_subplot(212)
    ax2 = ax1.twinx()
    ax1.set_xlim([0, 2*pi])
    ax1.set_xticks([0, alpha, u, pi, 2*pi])
    ax1.set_xticklabels(['$0$', '$\\alpha$', '$u$', '$\pi$', '$2\pi$'])
    ax1.set_ylim([-200, 200])
    ax1.set_yticks([-200, 0, 200])
    ax1.set_yticklabels(['$-200$', '$0$', '$200$'])
    ax1.set_xlabel('$\omega t, radians$')
    ax1.set_ylabel('$V_{d}, (V)$')
    ax1.text(3, -440, '$(b)$', size =10)
    ax2.set_ylabel('$I_{d}, (A)$')
    ax2.set_ylim([-10, 10])
    ax2.set_yticks([-10, 0, 10])
    ax2.set_yticklabels(['$-10$','$0$', '$10$'])
    p1, = ax1.plot(teta, vload, 'r', label = '$V_{d}$')
    p2, = ax2.plot(teta, isource, 'b--', label = '$i_{s}$')
    p3, = ax1.plot(teta, vthyristor1, 'm--', label = '$V_{thy1}$')
    ax1.yaxis.get_label().set_color(p1.get_color())
    ax2.yaxis.get_label().set_color(p2.get_color())
    ax1.legend(bbox_to_anchor=(0., 1.02, 1., .102), loc=3, ncol=3, \
                                   mode="expand", borderaxespad=0.)
    ax1.grid(True)
    #
    plt.show()
    #
    # plt.draw()
    # plt.savefig('Fig510.ps', dpi=400, bbox_inches='tight')
    #
```

5.6.5 Program 5.13

```
#
def fthy(vth, nsw):
    """  Thyristor characteristic
         vth: thyristor cathode to anode voltage;
         nsw: trigger signal
         Returns thyristor current and conductance"""
    #
    glow = 0.1e-07
    ghigh = 0.1e+03
    #
    if nsw == -1:
```

```
        gdth = glow
    if nsw == 1:
        if vth < 0.0:
            gdth = ghigh
        else:
            gdth = glow
    #
    idth = gdth*vth
    #
    return idth, gdth
#
def ch5fig15_solver(esource_a, esource_b, esource_c, Id,        \
                                            nalpha, ncycle, n):
    """Assumes esource is a list, fdiode is a function.
        Returns the list iload"""
    #
    import numpy as np
    #
    vp = 0.0
    vn = 0.0
    erlim = 0.1e-07
    ihold = -0.001
    #
    ntrig1 = 30 + nalpha
    ntrig2 = 90 + nalpha
    ntrig3 = 150 + nalpha
    ntrig4 = 210 + nalpha
    ntrig5 = 270 + nalpha
    ntrig6 = 330 + nalpha
    #
    if ntrig1 >= n:
        ntrig1 = ntrig1 - n
    if ntrig2 >= n:
        ntrig2 = ntrig2 - n
    if ntrig3 >= n:
        ntrig3 = ntrig3 - n
    if ntrig4 >= n:
        ntrig4 = ntrig4 - n
    if ntrig5 >= n:
        ntrig5 = ntrig5 - n
    if ntrig6 >= n:
        ntrig6 = ntrig6 - n
    #
    nst1 = -1
    nst2 = -1
    nst3 = -1
    nst4 = -1
    nst5 = 1
    nst6 = 1
    #
    csa = []
    csb = []
    csc = []
```

```
v_P = []
v_N = []
v_D = []
#
for k in range(ncycle):
    #
    for j in range(n):
        #
        esa = esource_a[j]
        esb = esource_b[j]
        esc = esource_c[j]
        #
        if j >= ntrig1 and j < (ntrig1 + 5):
            nst1 = 1
        if j >= ntrig2 and j < (ntrig2 + 5):
            nst2 = 1
        if j >= ntrig3 and j < (ntrig3 + 5):
            nst3 = 1
        if j >= ntrig4 and j < (ntrig4 + 5):
            nst4 = 1
        if j >= ntrig5 and j < (ntrig5 + 5):
            nst5 = 1
        if j >= ntrig6 and j < (ntrig6 + 5):
            nst6 = 1
        #
        for its in range(1000):
            #
            cd1, gd1 = fthy(vp - esa, nst1)
            cd4, gd4 = fthy(esa - vn, nst4)
            cd3, gd3 = fthy(vp - esb, nst3)
            cd6, gd6 = fthy(esb - vn, nst6)
            cd5, gd5 = fthy(vp - esc, nst5)
            cd2, gd2 = fthy(esc - vn, nst2)
            #
            ilsa = cd4 - cd1
            ilsb = cd6 - cd3
            ilsc = cd2 - cd5
            #
            irrp = -(cd1 + cd3 + cd5 + Id)
            irrn = cd2 + cd6 + cd4 + Id
            err = np.sqrt(irrp**2 + irrn**2)
            #
            if err < erlim:
                break
            else:
                dvp = irrp/(gd1 + gd3 + gd5)
                dvn = irrn/(gd2 + gd4 + gd6)
                vp += dvp
                vn += dvn
                vd = vp - vn
        #
        if its >= 999:
            print('time-step= ', (k*n + j), 'solver does not converge')
```

```
                break
            #
            if k == (ncycle - 1):
                csa.append(ilsa)
                csb.append(ilsb)
                csc.append(ilsc)
                v_P.append(vp)
                v_N.append(vn)
                v_D.append(vd)
            #
            if nst1 == 1 and cd1 > ihold:
                nst1 = -1
            if nst2 == 1 and cd2 > ihold:
                nst2 = -1
            if nst3 == 1 and cd3 > ihold:
                nst3 = -1
            if nst4 == 1 and cd4 > ihold:
                nst4 = -1
            if nst5 == 1 and cd5 > ihold:
                nst5 = -1
            if nst6 == 1 and cd6 > ihold:
                nst6 = -1
        #
    return csa, csb, csc, v_P, v_N, v_D
#
#
def ch5fig15(Ell, freq, Id, nalpha, ncycle, n):
    """ Main program, plots results
        3-phase, controlled, AC/DC converter
        AC-side, phase-a, current waveform"""
    #
    #calling command:
    #ch5fig15(207.85, 60.0, 10.0, 30, 5, 360)
    #
    from math import pi
    import numpy as np
    import matplotlib.pyplot as plt
    from mpl_toolkits.axes_grid1 import host_subplot
    #
    dteta = 2.0*pi/n
    phdif = 2.0*pi/3.0
    teta = dteta*np.arange(n)
    tetb = teta - phdif*np.ones(n)
    tetc = teta + phdif*np.ones(n)
    esmax = np.sqrt(2.0)*Ell/np.sqrt(3.0)
    esource_a = esmax*np.sin(teta)
    esource_b = esmax*np.sin(tetb)
    esource_c = esmax*np.sin(tetc)
    #
    csa, csb, csc, v_P, v_N, v_D = ch5fig15_solver(esource_a, esource_b, \
                                        esource_c, Id, nalpha, ncycle, n)
    #
    cs_freq = np.fft.fft(csa)
```

```
z = cs_freq[1]
a2 = (2.0/n)*(z.real)
b2 = -(2.0/n)*(z.imag)
cs1 = a2*np.cos(teta) + b2*np.sin(teta)
pcs1 = cs1[0]
for i in range(n):
    if pcs1*cs1[i] <= 0.0:
        ndelay = i
        break
    pcs1 = cs1[i]
alpha = dteta*(ndelay - 0.5)
#
ax1 = host_subplot(1, 1, 1)
ax2 = ax1.twinx()
ax1.set_xlim(0, 2*pi)
ax1.set_xticks([0, pi/6.0, alpha, pi/6.0+alpha, pi/2.0, pi,      \
                                        3*pi/2.0, 2*pi])
ax1.set_xticklabels(['$0$', '$\ $', '$\\alpha$', '$\ $',         \
                    '$\pi /2$', '$\pi$', '$3\pi /2$', '$2\pi$'])
ax1.set_ylim([-400, 400])
ax1.set_yticks([-400, -200, 0, 200, 400])
ax1.set_yticklabels(['$-400$', '$-200$', '$0$', '$200$', '$400$'])
ax1.arrow(pi/6+alpha/2, 250, -alpha/4, 0, fc='k', ec='k',        \
                        head_width=10, head_length=0.2*alpha)
ax1.text(pi/6+0.4*alpha, 260, '$\\alpha$', family='monospace',  \
                                                fontsize=9)
ax1.arrow(pi/6+alpha/2, 250, alpha/4, 0, fc='k', ec='k',         \
                        head_width=10, head_length=0.2*alpha)
ax1.text(2.5, -320, '$\\alpha = \pi /5\ radians $', size =10)
ax1.set_ylabel('$Voltage, V$')
ax2.set_ylim(-20, 20)
ax2.set_yticks([-20, 0, 20])
ax2.set_yticklabels(['$-20$', '$0$', '$20$'])
ax2.set_ylabel('$Current, A$')
p1, = ax1.plot(teta, esource_a, 'r-', label = '$es_{a}$')
p2, = ax1.plot(teta, esource_c, 'C1-', label = '$es_{c}$')
p3, = ax2.plot(teta, csa[0:n], 'b', label = '$I_{sa}$')
p4, = ax2.plot(teta, cs1, 'g--', label = '$I_{sa1}$')
ax1.yaxis.get_label().set_color(p1.get_color())
ax2.yaxis.get_label().set_color(p3.get_color())
ax1.legend(bbox_to_anchor=(0., 1.02, 1., .102), loc=3, ncol=4,  \
                        mode="expand", borderaxespad=0.)
ax1.grid(True)
#
plt.show()
#
# plt.draw()
# plt.savefig('Fig515.ps', dpi=400, bbox_inches='tight')
#
```

5.6.6 Program 5.16

```
#
def fthy(vth, nsw):
    """  Thyristor characteristic
         nsw: trigger signal;
         vth: thyristor cathode to anode voltage
         Returns thyristor current and conductance"""
    #
    glow = 0.1e-07
    ghigh = 0.1e+03
    #
    if nsw == -1:
        gdth = glow
    if nsw == 1:
        if vth < 0.0:
            gdth = ghigh
        else:
            gdth = glow
    #
    idth = gdth*vth
    #
    return idth, gdth
#
def ch5fig16_solver(esource_a, esource_b, esource_c, Id,       \
                                            nlph, ncycle, n):
    """Assumes esource is a list, fdiode is a function.
        Returns the list iload"""
    #
    import numpy as np
    #
    vp = 0.0
    vn = 0.0
    erlim = 0.1e-07
    ihold = -0.001
    #
    csa = []
    #
    for i in range(len(nlph)):
        #
        nalpha = nlph[i]
        #
        ntrig1 = 30 + nalpha
        ntrig2 = 90 + nalpha
        ntrig3 = 150 + nalpha
        ntrig4 = 210 + nalpha
        ntrig5 = 270 + nalpha
        ntrig6 = 330 + nalpha
        #
        if ntrig1 >= n:
            ntrig1 = ntrig1 - n
        if ntrig2 >= n:
            ntrig2 = ntrig2 - n
```

```
if ntrig3 >= n:
    ntrig3 = ntrig3 - n
if ntrig4 >= n:
    ntrig4 = ntrig4 - n
if ntrig5 >= n:
    ntrig5 = ntrig5 - n
if ntrig6 >= n:
    ntrig6 = ntrig6 - n
#
nst1 = -1
nst2 = -1
nst3 = -1
nst4 = -1
nst5 = 1
nst6 = 1
#
for k in range(ncycle):
    #
    for j in range(n):
        #
        esa = esource_a[j]
        esb = esource_b[j]
        esc = esource_c[j]
        #
        if j >= ntrig1 and j < (ntrig1 + 5):
            nst1 = 1
        if j >= ntrig2 and j < (ntrig2 + 5):
            nst2 = 1
        if j >= ntrig3 and j < (ntrig3 + 5):
            nst3 = 1
        if j >= ntrig4 and j < (ntrig4 + 5):
            nst4 = 1
        if j >= ntrig5 and j < (ntrig5 + 5):
            nst5 = 1
        if j >= ntrig6 and j < (ntrig6 + 5):
            nst6 = 1
        #
        for its in range(1000):
            #
            cd1, gd1 = fthy(vp - esa, nst1)
            cd4, gd4 = fthy(esa - vn, nst4)
            cd3, gd3 = fthy(vp - esb, nst3)
            cd6, gd6 = fthy(esb - vn, nst6)
            cd5, gd5 = fthy(vp - esc, nst5)
            cd2, gd2 = fthy(esc - vn, nst2)
            #
            ilsa = cd4 - cd1
            #
            irrp = -(cd1 + cd3 + cd5 + Id)
            irrn = cd2 + cd6 + cd4 + Id
            err = np.sqrt(irrp**2 + irrn**2)
            #
            if err < erlim:
```

```
                        break
                    else:
                        dvp = irrp/(gd1 + gd3 + gd5)
                        dvn = irrn/(gd2 + gd4 + gd6)
                        vp += dvp
                        vn += dvn
                        vd = vp - vn
                    #
                if k == (ncycle - 1):
                    csa.append(ilsa)
                #
                if its >= 999:
                    print('time-step= ', (k*n + j), 'solver does not converge')
                    break
                #
                if nst1 == 1 and cd1 > ihold:
                    nst1 = -1
                if nst2 == 1 and cd2 > ihold:
                    nst2 = -1
                if nst3 == 1 and cd3 > ihold:
                    nst3 = -1
                if nst4 == 1 and cd4 > ihold:
                    nst4 = -1
                if nst5 == 1 and cd5 > ihold:
                    nst5 = -1
                if nst6 == 1 and cd6 > ihold:
                    nst6 = -1
        #
    return csa
#
#
def ch5fig16(Ell, freq, Id, ncycle, n):
    """ Main program, plots results"""
    #
    #calling command:
    #ch5fig16(207.85, 60.0, 10.0, 2, 360)
    #
    from math import pi
    import numpy as np
    import matplotlib.pyplot as plt
    from mpl_toolkits.axes_grid1 import host_subplot
    #
    dteta = 2.0*pi/n
    nlph = [0, 60, 90, 120]
    #
    phdif = 2.0*pi/3.0
    teta = dteta*np.arange(n)
    tetb = teta - phdif*np.ones(n)
    tetc = teta + phdif*np.ones(n)
    esmax = np.sqrt(2.0)*Ell/np.sqrt(3.0)
    esource_a = esmax*np.sin(teta)
    esource_b = esmax*np.sin(tetb)
    esource_c = esmax*np.sin(tetc)
```

```
#
csa = ch5fig16_solver(esource_a, esource_b, \
                                esource_c, Id, nlph, ncycle, n)
#
cs_freq = np.fft.fft(csa[0:n])
z = cs_freq[1]
a2 = (2.0/n)*(z.real)
b2 = -(2.0/n)*(z.imag)
csa1 = a2*np.cos(teta) + b2*np.sin(teta)
#
ax1 = host_subplot(2, 2, 1)
ax2 = ax1.twinx()
ax1.set_xlim(0, 2*pi)
ax1.set_xticks([0, pi/2.0, pi, 3*pi/2.0, 2*pi])
ax1.set_xticklabels(['', '', '', '', ''])
ax1.set_ylim([-400, 400])
ax1.set_yticks([-400, -200, 0, 200, 400])
ax1.set_yticklabels(['$-400$', '$-200$', '$0$', '$200$', '$400$'])
p1, = ax1.plot(teta, esource_a, 'C1-', label = '$es_{a}$')
ax1.text(2.5, -320, '$\\alpha = 0$', size =10)
ax1.set_ylabel('$Voltage, V$')
ax2.set_ylim(-20, 20)
ax2.set_yticks([-20, 0, 20])
ax2.set_yticklabels(['', '', ''])
p2, = ax2.plot(teta, csa[0:n], 'b-', label = '$I_{sa}$')
p3, = ax2.plot(teta, csa1, 'g--', label = '$I_{sa1}$')
ax1.legend(bbox_to_anchor=(0., 1.02, 2.3, .102), loc=3, ncol=3,     \
                                    mode="expand", borderaxespad=0.)
ax1.yaxis.get_label().set_color(p1.get_color())
ax2.yaxis.get_label().set_color(p2.get_color())
ax1.grid(True)
plt.subplots_adjust(wspace = 0.3, hspace = 0.3)
#
cs_freq = np.fft.fft(csa[n:2*n])
z = cs_freq[1]
a2 = (2.0/n)*(z.real)
b2 = -(2.0/n)*(z.imag)
csa2 = a2*np.cos(teta) + b2*np.sin(teta)
#
ax1 = host_subplot(2, 2, 2)
ax2 = ax1.twinx()
ax1.set_xlim(0, 2*pi)
ax1.set_xticks([0, pi/2.0, pi, 3*pi/2.0, 2*pi])
ax1.set_xticklabels(['', '', '', '', ''])
ax1.set_ylim([-400, 400])
ax1.set_yticks([-400, -200, 0, 200, 400])
ax1.set_yticklabels(['', '', '', '', ''])
p1, = ax1.plot(teta, esource_a, 'C1-', label = '_nolegend_')
ax1.text(2.5, -320, '$\\alpha = \pi /3$', size =10)
ax2.set_ylabel('$Current, A$')
ax2.set_ylim(-20, 20)
ax2.set_yticks([-20, 0, 20])
ax2.set_yticklabels(['$-20$', '$0$', '$20$'])
```

```
p2, = ax2.plot(teta, csa[n:2*n], 'b-', label = '_nolegend_')
p3, = ax2.plot(teta, csa2, 'g--', label = '_nolegend_')
ax1.legend(bbox_to_anchor=(0., 1.02, 1., .102), loc=3, ncol=2,      \
                                 mode="expand", borderaxespad=0.)
ax1.yaxis.get_label().set_color(p1.get_color())
ax2.yaxis.get_label().set_color(p2.get_color())
ax1.grid(True)
plt.subplots_adjust(wspace = 0.3, hspace = 0.3)
#
cs_freq = np.fft.fft(csa[2*n:3*n])
z = cs_freq[1]
a2 = (2.0/n)*(z.real)
b2 = -(2.0/n)*(z.imag)
csa3 = a2*np.cos(teta) + b2*np.sin(teta)
#
ax1 = host_subplot(2, 2, 3)
ax2 = ax1.twinx()
ax1.set_xlim(0, 2*pi)
ax1.set_xticks([0, pi/2.0, pi, 3*pi/2.0, 2*pi])
ax1.set_xticklabels(['$0$', '$\pi /2$', '$\pi$',                \
                                   '$3\pi /2$', '$2\pi$'])
ax1.set_xlabel('$\omega t, radians$')
ax1.set_ylim([-400, 400])
ax1.set_yticks([-400, -200, 0, 200, 400])
ax1.set_yticklabels(['$-400$', '$-200$', '$0$', '$200$', '$400$'])
ax1.text(2.5, -320, '$\\alpha = \pi /2$', size = 10)
ax1.set_ylabel('$Voltage, V$')
ax2.set_ylim(-20, 20)
ax2.set_yticks([-20, 0, 20])
ax2.set_yticklabels(['', '', ''])
p1, = ax1.plot(teta, esource_a, 'C1-', label = '_nolegend_')
p2, = ax2.plot(teta, csa[2*n:3*n], 'b-', label = '_nolegend_')
p3, = ax2.plot(teta, csa3, 'g--', label = '_nolegend_')
ax1.yaxis.get_label().set_color(p1.get_color())
ax2.yaxis.get_label().set_color(p2.get_color())
ax1.grid(True)
plt.subplots_adjust(wspace = 0.3, hspace = 0.3)
#
cs_freq = np.fft.fft(csa[3*n:4*n])
z = cs_freq[1]
a2 = (2.0/n)*(z.real)
b2 = -(2.0/n)*(z.imag)
csa4 = a2*np.cos(teta) + b2*np.sin(teta)
#
ax1 = host_subplot(2, 2, 4)
ax2 = ax1.twinx()
ax1.set_xlim(0, 2*pi)
ax1.set_xticks([0, pi/2.0, pi, 3*pi/2.0, 2*pi])
ax1.set_xticklabels(['$0$', '$\pi /2$', '$\pi$',          \
                                '$3\pi /2$', '$2\pi$'])
ax1.set_xlabel('$\omega t, radians$')
ax1.set_ylim([-400, 400])
ax1.set_yticks([-400, -200, 0, 200, 400])
```

```
    ax1.set_yticklabels(['', '', '', '', ''])
    ax1.text(2.5, -340, '$\\alpha = 2\pi /3$', size =10)
    ax2.set_ylabel('$Current, A$')
    ax2.set_ylim(-20, 20)
    ax2.set_yticks([-20, 0, 20])
    ax2.set_yticklabels(['$-20$', '$0$', '$20$'])
    p1, = ax1.plot(teta, esource_a, 'C1-', label = '_nolegend_')
    p2, = ax2.plot(teta, csa[3*n:4*n], 'b-', label = '_nolegend_')
    p3, = ax2.plot(teta, csa4, 'g--', label = '_nolegend_')
    ax1.yaxis.get_label().set_color(p1.get_color())
    ax2.yaxis.get_label().set_color(p2.get_color())
    ax1.grid(True)
    plt.subplots_adjust(wspace = 0.3, hspace = 0.3)
    #
    plt.show()
    #
    #plt.draw()
    #plt.savefig('Fig516.ps', dpi=400, bbox_inches='tight')
    #
```

5.6.7 Program 5.19

```
#
def fthy(vth, nsw):
    """  Thyristor characteristic
         vth: thyristor cathode to anode voltage;
         nsw: trigger signal
         Returns thyristor current and conductance"""
    #
    glow = 0.1e-07
    ghigh = 0.1e+03
    #
    if nsw == -1:
        gdth = glow
    if nsw == 1:
        if vth < 0.0:
            gdth = ghigh
        else:
            gdth = glow
    #
    idth = gdth*vth
    #
    return idth, gdth
#
#
def ch5fig19_solver(esource_a, esource_b, esource_c, \
                                        gls, Id, nalpha, ncycle, n):
    """Assumes esource is a list, fdiode is a function.
       Returns the list iload"""
    #
    import numpy as np
    #
```

```
vp = 0.0
vn = 0.0
vsa = 0.0
vsb = 0.0
vsc = 0.0
pcsa = 0.0
pcsb = 0.0
pcsc = 0.0
erlim = 0.1e-07
ihold = -0.001
#
ntrig1 = 30 + nalpha
ntrig2 = 90 + nalpha
ntrig3 = 150 + nalpha
ntrig4 = 210 + nalpha
ntrig5 = 270 + nalpha
ntrig6 = 330 + nalpha
#
if ntrig1 >= n:
    ntrig1 = ntrig1 - n
if ntrig2 >= n:
    ntrig2 = ntrig2 - n
if ntrig3 >= n:
    ntrig3 = ntrig3 - n
if ntrig4 >= n:
    ntrig4 = ntrig4 - n
if ntrig5 >= n:
    ntrig5 = ntrig5 - n
if ntrig6 >= n:
    ntrig6 = ntrig6 - n
#
nst1 = -1
nst2 = -1
nst3 = -1
nst4 = -1
nst5 = 1
nst6 = 1
#
csa = []
csb = []
csc = []
v_P = []
v_N = []
v_D = []
#
for k in range(ncycle):
    #
    for j in range(n):
        #
        esa = esource_a[j]
        esb = esource_b[j]
        esc = esource_c[j]
        #
```

```
        if j >= ntrig1 and j < (ntrig1 + 2):
            nst1 = 1
        if j >= ntrig2 and j < (ntrig2 + 2):
            nst2 = 1
        if j >= ntrig3 and j < (ntrig3 + 2):
            nst3 = 1
        if j >= ntrig4 and j < (ntrig4 + 2):
            nst4 = 1
        if j >= ntrig5 and j < (ntrig5 + 2):
            nst5 = 1
        if j >= ntrig6 and j < (ntrig6 + 2):
            nst6 = 1
        #
        for its in range(1000):
            #
            ilsa = gls*(esa - vsa) + pcsa
            ilsb = gls*(esb - vsb) + pcsb
            ilsc = gls*(esc - vsc) + pcsc
            #
            cda1, gda1 = fthy((vp - vsa), nst1)
            cda4, gda4 = fthy((vsa - vn), nst4)
            cdb3, gdb3 = fthy((vp - vsb), nst3)
            cdb6, gdb6 = fthy((vsb - vn), nst6)
            cdc5, gdc5 = fthy((vp - vsc), nst5)
            cdc2, gdc2 = fthy((vsc - vn), nst2)
            #
            irra = ilsa + cda1 -  cda4
            irrb = ilsb + cdb3 -  cdb6
            irrc = ilsc + cdc5 -  cdc2
            irrp = -(cda1 + cdb3 + cdc5 + Id)
            irrn = cda4 + cdb6 + cdc2 + Id
            err = np.sqrt(irra**2 + irrb**2 + irrc**2 +        \
                                      irrp**2 + irrn**2)
            #
            if err < erlim:
                break
            else:
                gmat = np.array([[(gls+gda1+gda4), 0, 0, -gda1, -gda4],
                              [0, (gdb3+gdb6+gls), 0, -gdb3, -gdb6],
                              [0, 0, (gdc5+gdc2+gls), -gdc5, -gdc2],
                              [-gda1, -gdb3, -gdc5, (gda1+gdb3+gdc5), 0],
                              [-gda4, -gdb6, -gdc2, 0,            \
                                          (gda4+gdb6+ gdc2)]])
                bcol = np.array([irra, irrb, irrc, irrp, irrn])
                xdv = np.linalg.solve(gmat, bcol)
                #
                dvsa = xdv[0]
                dvsb = xdv[1]
                dvsc = xdv[2]
                dvp = xdv[3]
                dvn = xdv[4]
                vsa += dvsa
                vsb += dvsb
```

```
                    vsc += dvsc
                    vp += dvp
                    vn += dvn
            #
            if its >= 999:
                print('time-step= ', (k*n + j), 'solver does not converge')
                break
            #
            pcsa = ilsa
            pcsb = ilsb
            pcsc = ilsc
            #
            if k == ncycle - 1:
                csa.append(ilsa)
                csb.append(ilsb)
                csc.append(ilsc)
                v_P.append(vp)
                v_N.append(vn)
                v_D.append(vp - vn)
            #
            if nst1 == 1 and cda1 > ihold:
                nst1 = -1
            if nst2 == 1 and cdc2 > ihold:
                nst2 = -1
            if nst3 == 1 and cdb3 > ihold:
                nst3 = -1
            if nst4 == 1 and cda4 > ihold:
                nst4 = -1
            if nst5 == 1 and cdc5 > ihold:
                nst5 = -1
            if nst6 == 1 and cdb6 > ihold:
                nst6 = -1
        #
    return csa, csb, csc, v_P, v_N, v_D
#
#
def ch5fig19(Erms, freq, ls, Id, nalpha, ncycle, n):
    """ Main program, plots results
        3-phase, controlled, AC/DC converter
        Commutation of current from phase-c to phase-a"""
    #
    #calling command:
    #ch5fig19(120.0, 60.0, 0.5e-02, 10.0, 30, 10, 360)
    #
    from math import pi
    import numpy as np
    import matplotlib.pyplot as plt
    #
    gls = 1.0/(freq*n*ls)
    dteta = 2.0*pi/n
    phdif = 2.0*pi/3.0
    teta = dteta*np.arange(n)
    tetb = teta - phdif*np.ones(n)
```

```
tetc = teta + phdif*np.ones(n)
esmax = np.sqrt(2.0)*Erms
esource_a = esmax*np.sin(teta)
esource_b = esmax*np.sin(tetb)
esource_c = esmax*np.sin(tetc)
#
csa, csb, csc, v_P, v_N, v_D = ch5fig19_solver(esource_a, esource_b, \
                        esource_c, gls, Id, nalpha, ncycle, n)
#
m0 = 2*(ncycle - 1)
m1 = 1 + m0
m2 = 1 + m1
ang = teta + (m0*pi)*np.ones(n)
#
plt.subplot(211)
plt.xlabel('$\omega t, radians$')
plt.xlim([m0*pi, m2*pi])
plt.xticks([m0*pi, m1*pi, m2*pi], ['$18\pi$', '$19\pi$', '$20\pi$'])
plt.ylabel('$Voltages$')
plt.ylim(-300, 300)
plt.yticks([-300, 0, 300], ['$-300$', '$0$', '$300$'])
plt.plot(ang, esource_a, 'b--', label = '$es_{a}$')
plt.plot(ang, esource_b, 'g--', label = '$es_{b}$')
plt.plot(ang, esource_c, 'y--', label = '$es_{c}$')
plt.plot(ang, v_P, 'r-', label = '$V_{p}$')
plt.plot(ang, v_N, 'm-', label = '$V_{n}$')
#
plt.plot((57.57, 57.57), (180, 215), linewidth = 1, color = 'k')
plt.arrow(57.39, 195, 0.11, 0, head_width = 30.0,               \
            head_length = 0.05, fc = 'k', ec = 'k')
plt.text(57.664, 190, '$u$', family = 'monospace', fontsize = 9)
plt.arrow(58.03, 195, -0.13, 0, head_width = 30.0,              \
            head_length = 0.05, fc = 'k', ec = 'k')
plt.plot((57.82, 57.82), (180, 215), linewidth = 1, color = 'k')
plt.text(59.5, -650, '$(a)$', size =10)
plt.legend(bbox_to_anchor=(0., 1.01, 1., .101), loc=3, ncol=5, \
                        mode="expand", borderaxespad=0.)
plt.grid(True)
plt.subplots_adjust(hspace=0.95)
#
plt.subplot(212)
plt.xlabel('$\omega t, radians$')
plt.ylabel('$Currents$')
plt.xlim(57.5, 57.9)
plt.xticks([57.5, 57.6, 57.7, 57.8, 57.9], ['$57.5$',          \
              '$57.6$', '$57.7$', '$57.8$', '$57.9$'])
plt.ylim(0, 20)
plt.yticks([0, 10.0, 20.0], ['$0$', '$10.0$', '$20.0$'])
#
plt.plot((57.5785, 57.5785), (3.5, 6.5), linewidth = 1, color = 'k')
plt.plot((57.802, 57.802), (3.5, 6.5), linewidth = 1, color = 'k')
plt.arrow(57.652, 5, -0.0615, 0, head_width = 1.0,              \
            head_length = 0.01, fc = 'k', ec = 'k')
```

```
    plt.text(57.66, 4.5, '$u$', family = 'monospace', fontsize = 9)
    plt.arrow(57.72, 5, 0.070, 0, head_width = 1.0,                \
                      head_length = 0.01, fc = 'k', ec = 'k')
    plt.plot(ang, csa, 'b-', label = '$I_{sa}$')
    plt.plot(ang, csc, 'g--', label = '$I_{sc}$')
    plt.text(57.7, -11.5, '$(b)$', size =10)
    plt.legend(bbox_to_anchor=(0., 1.01, 1., .101), loc=3, ncol=2, \
                                   mode="expand", borderaxespad=0.)
    plt.grid(True)
    #
    plt.show()
    #
    # plt.draw()
    # plt.savefig('Fig519.ps', dpi=400, bbox_inches='tight')
    #
```

5.6.8 Program 5.21

```
#
def fthy(vth, nsw):
    """  Thyristor characteristiv
         vth: thyristor cathode to anode voltage;
         nsw: trigger signal
         Returns thyristor current and conductance"""
    #
    glow = 0.1e-07
    ghigh = 0.1e+03
    #
    if nsw == -1:
        gdth = glow
    if nsw == 1:
        if vth < 0.0:
            gdth = ghigh
        else:
            gdth = glow
    #
    idth = gdth*vth
    #
    return idth, gdth
#
#
def ch5fig21_solver(esource_a, esource_b, esource_c, \
                             gls, Id, nalpha, ncycle, n):
    """Assumes esource is a list, fdiode is a function.
        Returns the lists csa, vthy5"""
    #
    import numpy as np
    #
    vp = 0.0
    vn = 0.0
    vsa = 0.0
    vsb = 0.0
    vsc = 0.0
```

```
pcsa = 0.0
pcsb = 0.0
pcsc = 0.0
erlim = 0.1e-07
ihold = -0.001
#
ntrig1 = 30 + nalpha
ntrig2 = 90 + nalpha
ntrig3 = 150 + nalpha
ntrig4 = 210 + nalpha
ntrig5 = 270 + nalpha
ntrig6 = 330 + nalpha
#
if ntrig1 >= n:
    ntrig1 = ntrig1 - n
if ntrig2 >= n:
    ntrig2 = ntrig2 - n
if ntrig3 >= n:
    ntrig3 = ntrig3 - n
if ntrig4 >= n:
    ntrig4 = ntrig4 - n
if ntrig5 >= n:
    ntrig5 = ntrig5 - n
if ntrig6 >= n:
    ntrig6 = ntrig6 - n
#
nst1 = -1
nst2 = -1
nst3 = -1
nst4 = -1
nst5 = 1
nst6 = 1
#
csa = []
vthy5 = []
#
for k in range(ncycle):
    #
    for j in range(n):
        #
        esa = esource_a[j]
        esb = esource_b[j]
        esc = esource_c[j]
        #
        if j >= ntrig1 and j < (ntrig1 + 2):
            nst1 = 1
        if j >= ntrig2 and j < (ntrig2 + 2):
            nst2 = 1
        if j >= ntrig3 and j < (ntrig3 + 2):
            nst3 = 1
        if j >= ntrig4 and j < (ntrig4 + 2):
            nst4 = 1
        if j >= ntrig5 and j < (ntrig5 + 2):
```

```
        nst5 = 1
    if j >= ntrig6 and j < (ntrig6 + 2):
        nst6 = 1
    #
    for its in range(1000):
        #
        ilsa = gls*(esa - vsa) + pcsa
        ilsb = gls*(esb - vsb) + pcsb
        ilsc = gls*(esc - vsc) + pcsc
        #
        cda1, gda1 = fthy((vp - vsa), nst1)
        cda4, gda4 = fthy((vsa - vn), nst4)
        cdb3, gdb3 = fthy((vp - vsb), nst3)
        cdb6, gdb6 = fthy((vsb - vn), nst6)
        cdc5, gdc5 = fthy((vp - vsc), nst5)
        cdc2, gdc2 = fthy((vsc - vn), nst2)
        #
        irra = ilsa + cda1 -  cda4
        irrb = ilsb + cdb3 -  cdb6
        irrc = ilsc + cdc5 -  cdc2
        irrp = -(cda1 + cdb3 + cdc5 + Id)
        irrn = cda4 + cdb6 + cdc2 + Id
        err = np.sqrt(irra**2 + irrb**2 + irrc**2 + irrp**2 + irrn**2)
        #
        if err < erlim:
            break
        else:
            gmat = np.array([[(gls+gda1+gda4), 0, 0, -gda1, -gda4],
                         [0, (gdb3+gdb6+gls), 0, -gdb3, -gdb6],
                         [0, 0, (gdc5+gdc2+gls), -gdc5, -gdc2],
                         [-gda1, -gdb3, -gdc5, (gda1+gdb3+gdc5), 0],
                         [-gda4, -gdb6, -gdc2, 0, (gda4+gdb6+ gdc2)]])
            bcol = np.array([irra, irrb, irrc, irrp, irrn])
            xdv = np.linalg.solve(gmat, bcol)
            #
            dvsa = xdv[0]
            dvsb = xdv[1]
            dvsc = xdv[2]
            dvp = xdv[3]
            dvn = xdv[4]
            vsa += dvsa
            vsb += dvsb
            vsc += dvsc
            vp += dvp
            vn += dvn
    #
    if its >= 999:
        print('time-step= ', (k*n + j), 'solver does not converge')
        break
    #
    pcsa = ilsa
    pcsb = ilsb
    pcsc = ilsc
```

```
            #
            if k == ncycle - 1:
                csa.append(ilsa)
                vthy5.append(vp - vsc)
            #
            if nst1 == 1 and cda1 > ihold:
                nst1 = -1
            if nst2 == 1 and cdc2 > ihold:
                nst2 = -1
            if nst3 == 1 and cdb3 > ihold:
                nst3 = -1
            if nst4 == 1 and cda4 > ihold:
                nst4 = -1
            if nst5 == 1 and cdc5 > ihold:
                nst5 = -1
            if nst6 == 1 and cdb6 > ihold:
                nst6 = -1
        #
    return csa, vthy5
#
#
def ch5fig21(Erms, freq, ls, Id, nalpha, ncycle, n):
    """ Main program, plots results
        3-phase, controlled, AC/DC converter
        Inverter mode of operation"""
    #
    #calling command:
    #ch5fig21(120.0, 60.0, 0.5e-02, 10.0, 130, 10, 360)
    #
    from math import pi
    import numpy as np
    import matplotlib.pyplot as plt
    from mpl_toolkits.axes_grid1 import host_subplot
    #
    gls = 1.0/(freq*n*ls)
    dteta = 2.0*pi/n
    phdif = 2.0*pi/3.0
    teta = dteta*np.arange(n)
    tetb = teta - phdif*np.ones(n)
    tetc = teta + phdif*np.ones(n)
    esmax = np.sqrt(2.0)*Erms
    esource_a = esmax*np.sin(teta)
    esource_b = esmax*np.sin(tetb)
    esource_c = esmax*np.sin(tetc)
    #
    csa, vthy5 = ch5fig21_solver(esource_a, esource_b,              \
                    esource_c, gls, Id, nalpha, ncycle, n)
    #
    m0 = 2*(ncycle - 1)
    m1 = 1 + m0
    m2 = 1 + m1
    ang = teta + (m0*pi)*np.ones(n)
    #
```

```
cs_freq = np.fft.fft(csa)
z = cs_freq[1]
a2 = (2.0/n)*(z.real)
b2 = -(2.0/n)*(z.imag)
cs1 = a2*np.cos(teta) + b2*np.sin(teta)
pcs1 = cs1[0]
for i in range(n):
    if pcs1*cs1[i] <= 0.0:
        ndelay = i
        break
    pcs1 = cs1[i]
delay = dteta*(ndelay - 0.5)
alpha1 = delay + m0*pi
alpha = nalpha*dteta
u = 2.0*(alpha1 - m0*pi - alpha)
beta1 = m0*pi + pi/6.0
beta2 = beta1 + alpha
beta3 = beta2 + u
#
dgm = pi - alpha - u
gm1 = beta1 + alpha + u
gm2 = beta1 + pi
#
ax1 = host_subplot(1, 1, 1)
ax2 = ax1.twinx()
ax1.set_xlim([m0*pi, m2*pi])
ax1.set_xticks([m0*pi, alpha1, m1*pi, m2*pi])
ax1.set_xticklabels(['$18\pi$', '$\ $', '$19\pi$', '$20\pi$'])
ax1.set_ylim([-400, 400])
ax1.set_yticks([-400, -200, 0, 200, 400])
ax1.set_yticklabels(['$-400$', '$-200$', '$0$', '$200$', '$400$'])
ax1.set_ylabel('$Voltages, V$')
#
plt.plot((beta1, beta1), (100, 270), linewidth = 1, color = 'k')
plt.plot((beta2, beta2), (220, 270), linewidth = 1, color = 'k')
plt.plot((beta3, beta3), (220, 270), linewidth = 1, color = 'k')
#
ax1.arrow(beta1 + 0.5*alpha, 240, -0.4*alpha, 0, fc='k', ec='k',   \
                        head_width=10, head_length=0.075*alpha)
ax1.arrow(beta2 - 0.5*alpha, 240, 0.4*alpha, 0, fc='k', ec='k',    \
                        head_width=10, head_length=0.075*alpha)
ax1.text(beta1 + 0.5*alpha, 260, '$\\alpha$', family='monospace',  \
                                                      fontsize=10)
#
ax1.arrow(beta3 + 0.25*alpha, 240, -0.15*alpha, 0, fc='k', ec='k', \
                       head_width=10, head_length=0.075*alpha)
ax1.text(beta3 + 0.15*alpha, 260, '$u$', family='monospace',       \
                                                     fontsize=10)
#
ax1.arrow(m0*pi + 0.5*delay, -340, -0.4*delay, 0, fc='k', ec='k',  \
                       head_width=10, head_length=0.075*delay)
ax1.arrow(m0*pi + 0.5*delay, -340, 0.4*delay, 0, fc='k', ec='k',   \
                       head_width=10, head_length=0.075*delay)
```

```
    ax1.text(m0*pi + 0.3*delay, -380, '$\\alpha +u/2$',                \
                                 family='monospace', fontsize=10)
    #
    plt.plot((gm1, gm1), (-20, -130), linewidth = 1, color = 'k')
    plt.plot((gm2, gm2), (-100, -130), linewidth = 1, color = 'k')
    #
    ax1.arrow(gm1 + 0.5*dgm, -110, -0.25*dgm, 0, fc='k', ec='k',       \
                         head_width=10, head_length=0.20*dgm)
    ax1.arrow(gm1 + 0.5*dgm, -110, 0.25*dgm, 0, fc='k', ec='k',        \
                         head_width=10, head_length=0.20*dgm)
    ax1.text(gm1 + 0.4*dgm, -85, '$\\gamma$',                          \
                               family='monospace', fontsize=10)
    #
    ax2.set_ylim(-20, 20)
    ax2.set_yticks([-20, 0, 20])
    ax2.set_yticklabels(['$-20$', '$0$', '$20$'])
    ax2.set_ylabel('$Currents, A$')
    p1, = ax1.plot(ang, esource_a, 'r-', label = '$es_{a}$')
    p2, = ax1.plot(ang, esource_c, 'r--', label = '$es_{c}$')
    p3, = ax1.plot(ang, vthy5, 'C1--', label = '$v_{thy5}$')
    p4, = ax2.plot(ang, csa, 'b-', label = '$I_{sa}$')
    p5, = ax2.plot(ang, cs1, 'b--', label = '$I_{sa1}$')
    ax1.yaxis.get_label().set_color(p1.get_color())
    ax2.yaxis.get_label().set_color(p4.get_color())
    ax1.legend(bbox_to_anchor=(0., 1.02, 1., .102), loc=3, ncol=5,     \
                                mode="expand", borderaxespad=0.)
    ax1.grid(True)
    #
    plt.show()
    #
    # plt.draw()
    # plt.savefig('Fig521.ps', dpi=400, bbox_inches='tight')
    #
```

5.6.9 Program 5.25

```
#
def fthy(vth, nsw):
    """  Thyristor characteristic
         vth: thyristor cathode to anode voltage;
         nsw: trigger signal;
         Returns thyristor current and conductance"""
    #
    glow = 0.1e-07
    ghigh = 0.1e+03
    #
    if nsw == -1:
        gdth = glow
    if nsw == 1:
        if vth < 0.0:
            gdth = ghigh
        else:
```

```
            gdth = glow
    #
    idth = gdth*vth
    #
    return idth, gdth
#
#
def ch5fig25_solver(esource_a, esource_b, esource_c, \
                                        gls, gld, grd, Ed, nalpha, ncycle, n):
    """Assumes esource_a,b,c are lists, fdiode is a function.
        Returns the lists csa, iload"""
    #
    import numpy as np
    #
    vp = 0.0
    vn = 0.0
    vsa = 0.0
    vsb = 0.0
    vsc = 0.0
    pcsa = 0.0
    pcsb = 0.0
    pcsc = 0.0
    pild = 0.0
    gdtot = grd*gld/(grd + gld)
    erlim = 0.1e-07
    ihold = -0.001
    #
    #
    nt1 = round(n/12)
    nt2 = 3*nt1
    nt3 = 5*nt1
    nt4 = 7*nt1
    nt5 = 9*nt1
    nt6 = 11*nt1
    #
    ntrig1 = nt1 + nalpha
    ntrig2 = nt2 + nalpha
    ntrig3 = nt3 + nalpha
    ntrig4 = nt4 + nalpha
    ntrig5 = nt5 + nalpha
    ntrig6 = nt6 + nalpha
    #
    if ntrig1 >= n:
        ntrig1 = ntrig1 - n
    if ntrig2 >= n:
        ntrig2 = ntrig2 - n
    if ntrig3 >= n:
        ntrig3 = ntrig3 - n
    if ntrig4 >= n:
        ntrig4 = ntrig4 - n
    if ntrig5 >= n:
        ntrig5 = ntrig5 - n
    if ntrig6 >= n:
```

```
        ntrig6 = ntrig6 - n
    #
    nst1 = -1
    nst2 = -1
    nst3 = -1
    nst4 = -1
    nst5 = 1        # intializing T5 and
    nst6 = 1        # T6 to conduction mode
    #
    csa = []
    Id = []
    #
    for k in range(ncycle):
        #
        for j in range(n):
            #
            esa = esource_a[j]
            esb = esource_b[j]
            esc = esource_c[j]
            #
            if j >= ntrig1 and j < (ntrig1 + 2):
                nst1 = 1
            if j >= ntrig2 and j < (ntrig2 + 2):
                nst2 = 1
            if j >= ntrig3 and j < (ntrig3 + 2):
                nst3 = 1
            if j >= ntrig4 and j < (ntrig4 + 2):
                nst4 = 1
            if j >= ntrig5 and j < (ntrig5 + 2):
                nst5 = 1
            if j >= ntrig6 and j < (ntrig6 + 2):
                nst6 = 1
            #
            for its in range(1000):
                #
                ilsa = gls*(esa - vsa) + pcsa
                ilsb = gls*(esb - vsb) + pcsb
                ilsc = gls*(esc - vsc) + pcsc
                #
                vd2 = (grd*Ed + gld*(vp - vn) + pild)/(grd + gld)
                ild = gld*(vp - vn - vd2) + pild
                #
                cda1, gda1 = fthy((vp - vsa), nst1)
                cda4, gda4 = fthy((vsa - vn), nst4)
                cdb3, gdb3 = fthy((vp - vsb), nst3)
                cdb6, gdb6 = fthy((vsb - vn), nst6)
                cdc5, gdc5 = fthy((vp - vsc), nst5)
                cdc2, gdc2 = fthy((vsc - vn), nst2)
                #
                irra = ilsa + cda1 -  cda4
                irrb = ilsb + cdb3 -  cdb6
                irrc = ilsc + cdc5 -  cdc2
                irrp = -(cda1 + cdb3 + cdc5 + ild)
```

```
        irrn = cda4 + cdb6 + cdc2 + ild
        err = np.sqrt(irra**2 + irrb**2 + irrc**2 + irrp**2 + irrn**2)
        #
        if err < erlim:
            break
        else:
            gmat = np.array([[(gls+gda1+gda4), 0, 0, -gda1, -gda4],
              [0, (gdb3+gdb6+gls), 0, -gdb3, -gdb6],
              [0, 0, (gdc5+gdc2+gls), -gdc5, -gdc2],
              [-gda1, -gdb3, -gdc5, (gda1+gdb3+gdc5+gdtot), -gdtot],
              [-gda4, -gdb6, -gdc2, -gdtot, (gda4+gdb6+ gdc2+gdtot)]])
            bcol = np.array([irra, irrb, irrc, irrp, irrn])
            xdv = np.linalg.solve(gmat, bcol)
            #
            dvsa = xdv[0]
            dvsb = xdv[1]
            dvsc = xdv[2]
            dvp = xdv[3]
            dvn = xdv[4]
            vsa += dvsa
            vsb += dvsb
            vsc += dvsc
            vp += dvp
            vn += dvn
    #
    if its >= 999:
        print('time-step= ', (k*n + j), 'solver does not converge')
        break
    #
    pcsa = ilsa
    pcsb = ilsb
    pcsc = ilsc
    pild = ild
    #
    if k == ncycle - 1:
        csa.append(ilsa)
        Id.append(ild)
    #
    # check to see if thyristors are turning off
    #
    if nst1 == 1 and cda1 > ihold:
        nst1 = -1
    if nst2 == 1 and cdc2 > ihold:
        nst2 = -1
    if nst3 == 1 and cdb3 > ihold:
        nst3 = -1
    if nst4 == 1 and cda4 > ihold:
        nst4 = -1
    if nst5 == 1 and cdc5 > ihold:
        nst5 = -1
    if nst6 == 1 and cdb6 > ihold:
        nst6 = -1
#
```

```
    #
    #
    return csa, Id
#
#
def ch5fig25(Ell, freq, ls, ld, rd, Ed, nalpha, ncycle, n):
    """ Main program, plots results
        3-phase, controlled, AC/DC converter"""
    #
    #calling command:
    #ch5fig25(230.0, 60.0, 0.1075, 0.2, 88.0, 140.0, 10, 10, 360)
    #
    from math import pi
    import numpy as np
    import matplotlib.pyplot as plt
    from mpl_toolkits.axes_grid1 import host_subplot
    #
    dteta = 2.0*pi/n
    delt = 1.0/(freq*n)
    gls = delt/ls
    gld = delt/ld
    grd = 1.0/rd
    phdif = 2.0*pi/3.0
    teta = dteta*np.arange(n)
    tetb = teta - phdif*np.ones(n)
    tetc = teta + phdif*np.ones(n)
    esmax = np.sqrt(2.0/3.0)*Ell
    esource_a = esmax*np.sin(teta)
    esource_b = esmax*np.sin(tetb)
    esource_c = esmax*np.sin(tetc)
    #
    csa, Id = ch5fig25_solver(esource_a, esource_b, \
                              esource_c, gls, gld, grd, Ed, nalpha, ncycle, n)
    #
    Idmean = np.mean(Id)
    print('Idmean = ', Idmean, 'nalpha = ', nalpha)
    #
    m0 = 2*(ncycle - 1)
    m1 = 1 + m0
    m2 = 1 + m1
    ang = teta + (m0*pi)*np.ones(n)
    #
    #   waveform of the fundamental frequency component of csa
    #
    cs_freq = np.fft.fft(csa)
    z = cs_freq[1]
    a2 = (2.0/n)*(z.real)
    b2 = -(2.0/n)*(z.imag)
    cs1 = a2*np.cos(teta) + b2*np.sin(teta)
    #
    pcs1 = cs1[0]
    for i in range(n):
        if pcs1*cs1[i] <= 0.0:
```

```
            ndelay = i
            break
        pcs1 = cs1[i]
    delay = dteta*(ndelay - 0.5)
    alpha1 = delay + m0*pi
    #
    ax1 = host_subplot(1, 1, 1)
    ax2 = ax1.twinx()
    ax1.set_xlim([m0*pi, m2*pi])
    ax1.set_xticks([m0*pi, alpha1, m1*pi, m2*pi])
    ax1.set_xticklabels(['$18\pi$', '$\ $', '$19\pi$', '$20\pi$'])
    ax1.set_ylim([-400, 400])
    ax1.set_yticks([-400, -200, 0, 200, 400])
    ax1.set_yticklabels(['$-400$', '$-200$', '$0$', '$200$', '$400$'])
    ax1.set_ylabel('$Voltages, V$')
    ax2.set_ylim(-2, 2)
    ax2.set_yticks([-2, -1.31, 0, 1.31, 2])
    ax2.set_yticklabels(['$-2$', '$-1.31$', '$0$', '$1.31$', '$2$'])
    ax2.set_ylabel('$Currents, A$')
    p1, = ax1.plot(ang, esource_a, 'r', label = '$es_{a}$')
    p2, = ax1.plot(ang, esource_c, 'r--', label = '$es_{c}$')
    p3, = ax2.plot(ang, csa, 'b', label = '$I_{sa}$')
    p4, = ax2.plot(ang, cs1, 'b--', label = '$I_{sa1}$')
    p5, = ax2.plot(ang, Id, 'k--', label = '$Id$')
    ax1.yaxis.get_label().set_color(p1.get_color())
    ax2.yaxis.get_label().set_color(p3.get_color())
    ax1.legend(bbox_to_anchor=(0., 1.02, 1., .102), loc=3, ncol=5,      \
                                   mode="expand", borderaxespad=0.)
    ax1.grid(True)
    #
    plt.show()
    #plt.savefig('Fig525.ps', dpi=400, bbox_inches='tight')
    #
```

5.6.10 Program 5.27

```
#
def fthy(vth, nsw):
    """  Thyristor characteristic
         vth: thyristor cathode to anode voltage;
         nsw: trigger signal;
         Returns thyristor current and conductance"""
    #
    glow = 0.1e-07
    ghigh = 0.1e+03
    #
    if nsw == -1:
        gdth = glow
    if nsw == 1:
        if vth < 0.0:
            gdth = ghigh
        else:
```

```
            gdth = glow
    #
    idth = gdth*vth
    #
    return idth, gdth
#
#
def ch5fig27_solver(esource_a, esource_b, esource_c, \
                    gls, gld, grd, Ed, Idref, teta, ncycle, n):
    """Assumes esource_a,b,c are lists, fthy is a function.
       Returns the list iload"""
    #
    import numpy as np
    from math import pi
    #
    vp = 0.0
    vn = 0.0
    vsa = 0.0
    vsb = 0.0
    vsc = 0.0
    pcsa = 0.0
    pcsb = 0.0
    pcsc = 0.0
    pild = 0.0
    gdtot = grd*gld/(grd + gld)
    erlim = 0.1e-07
    ihold = -0.001
    #
    ff = 1.0/0.98
    X = np.matrix([ [0], [0], [0] ])
    P = np.matrix(100.0*np.eye(3))
    cs6 = np.cos(6.0*teta)
    sn6 = np.sin(6.0*teta)
    #
    nalpha = round(n/36)
    kierr = 91.3938/(60.0*n)
    Idhat = 1.4142
    Idrefn = Idref/Idhat
    pcsalpha = np.cos(2.0*pi*nalpha/n)
    print('pcsalpha = ', pcsalpha, 'kierr = ', kierr,       \
                                        'nalpha = ', nalpha)
    print('')
    #
    Idc = []
    Idcavg = []
    ang = []
    m0 = ncycle/3
    m1 = 2*m0
    #
    nt1 = round(n/12)
    nt2 = 3*nt1
    nt3 = 5*nt1
    nt4 = 7*nt1
```

```
nt5 = 9*nt1
nt6 = 11*nt1
#
nst1 = -1
nst2 = -1
nst3 = -1
nst4 = -1
nst5 = 1
nst6 = 1
#
for k in range(ncycle + 1):
    #
    ntrig1 = nt1 + nalpha
    ntrig2 = nt2 + nalpha
    ntrig3 = nt3 + nalpha
    ntrig4 = nt4 + nalpha
    ntrig5 = nt5 + nalpha
    ntrig6 = nt6 + nalpha
    #
    if ntrig1 >= n:
        ntrig1 = ntrig1 - n
    if ntrig2 >= n:
        ntrig2 = ntrig2 - n
    if ntrig3 >= n:
        ntrig3 = ntrig3 - n
    if ntrig4 >= n:
        ntrig4 = ntrig4 - n
    if ntrig5 >= n:
        ntrig5 = ntrig5 - n
    if ntrig6 >= n:
        ntrig6 = ntrig6 - n
    #
    if k >= (m0 - 1):
        ang0 = 2.0*k*pi
    if k >= (m0 - 1) and k < (m1 - 1):
        Idrefn = Idref/Idhat
    if k >= (m1 - 1):
        Idrefn = 0.75*Idref/Idhat
        #

    #
    for j in range(n):
        #
        esa = esource_a[j]
        esb = esource_b[j]
        esc = esource_c[j]
        #
        if j >= ntrig1 and j < (ntrig1 + 4):
            nst1 = 1
        if j >= ntrig2 and j < (ntrig2 + 4):
            nst2 = 1
        if j >= ntrig3 and j < (ntrig3 + 4):
            nst3 = 1
```

```
        if j >= ntrig4 and j < (ntrig4 + 4):
            nst4 = 1
        if j >= ntrig5 and j < (ntrig5 + 4):
            nst5 = 1
        if j >= ntrig6 and j < (ntrig6 + 4):
            nst6 = 1
        #
        for its in range(1000):
            #
            ilsa = gls*(esa - vsa) + pcsa
            ilsb = gls*(esb - vsb) + pcsb
            ilsc = gls*(esc - vsc) + pcsc
            #
            vd2 = (grd*Ed + gld*(vp - vn) + pild)/(grd + gld)
            ild = gld*(vp - vn - vd2) + pild
            #
            cda1, gda1 = fthy((vp - vsa), nst1)
            cda4, gda4 = fthy((vsa - vn), nst4)
            cdb3, gdb3 = fthy((vp - vsb), nst3)
            cdb6, gdb6 = fthy((vsb - vn), nst6)
            cdc5, gdc5 = fthy((vp - vsc), nst5)
            cdc2, gdc2 = fthy((vsc - vn), nst2)
            #
            irra = ilsa + cda1 -  cda4
            irrb = ilsb + cdb3 -  cdb6
            irrc = ilsc + cdc5 -  cdc2
            irrp = -(cda1 + cdb3 + cdc5 + ild)
            irrn = cda4 + cdb6 + cdc2 + ild
            err = np.sqrt(irra**2 + irrb**2 + irrc**2 + irrp**2 +    \
                                                            irrn**2)
            #
            if err < erlim:
                break
            else:
                gmat = np.array([[(gls+gda1+gda4), 0, 0, -gda1, -gda4],
                  [0, (gdb3+gdb6+gls), 0, -gdb3, -gdb6],
                  [0, 0, (gdc5+gdc2+gls), -gdc5, -gdc2],
                  [-gda1, -gdb3, -gdc5, (gda1+gdb3+gdc5+gdtot), -gdtot],
                  [-gda4, -gdb6, -gdc2, -gdtot,                         \
                                      (gda4+gdb6+ gdc2+gdtot)]])
                bcol = np.array([irra, irrb, irrc, irrp, irrn])
                xdv = np.linalg.solve(gmat, bcol)
                #
                dvsa = xdv[0]
                dvsb = xdv[1]
                dvsc = xdv[2]
                dvp = xdv[3]
                dvn = xdv[4]
                vsa += dvsa
                vsb += dvsb
                vsc += dvsc
                vp += dvp
                vn += dvn
```

```
#
if its >= 999:
    print('time-step= ', (k*n + j), 'solver does not converge')
    break
#
pcsa = ilsa
pcsb = ilsb
pcsc = ilsc
pild = ild
#
H = np.matrix([ 1.0, cs6[j], sn6[j] ])
HT = np.matrix([ [1.0], [cs6[j]], [sn6[j]] ])
rsum = np.dot(H, np.dot(P, HT))
r = 1.0 +float(rsum)
Kr = (1.0/r)*np.dot(P, HT)
kf = ild - float(np.dot(H, X))
X = X + kf*Kr
Id0 = X.item(0,0)
dP = np.dot(Kr, np.dot(H, P))
P = ff*(P - dP)
#
if k >= (m0 - 1):
    Idn = Id0/Idhat
    ierr = Idrefn - Idn
    #
    dcsalpha = kierr*ierr
    #
    #   Scale down the correction by (at least) a factor of 5
    #   to facilitate convergence
    #
    csalpha = pcsalpha + 0.2*dcsalpha
    if csalpha > 0.999:
        csalpha = 0.999
    if csalpha <= 0.001:
        csalpha = 0.001
    nalpha = round(n*np.arccos(csalpha)/(2.0*pi))
    if j == 1:
        print('pcsalpha = ', pcsalpha, 'dcsalpha = ', \
                                dcsalpha, 'csalpha = ', csalpha)
        print('ierr =', ierr, 'Id0 = ', Id0, 'nalpha = ', nalpha)
    pcsalpha = csalpha
    #
    Idc.append(ild)
    Idcavg.append(Id0)
    ang.append(ang0 + teta[j])
#
if nst1 == 1 and cda1 > ihold:
    nst1 = -1
if nst2 == 1 and cdc2 > ihold:
    nst2 = -1
if nst3 == 1 and cdb3 > ihold:
    nst3 = -1
if nst4 == 1 and cda4 > ihold:
```

```
                nst4 = -1
            if nst5 == 1 and cdc5 > ihold:
                nst5 = -1
            if nst6 == 1 and cdb6 > ihold:
                nst6 = -1
        #
    return Idc, Idcavg, ang
#
#
def ch5fig27(Ell, freq, ls, ld, rd, Ed, Idref, ncycle, n):
    """ Main program, plots results
        3-phase, controlled, AC/DC converter
        with feedback control"""
    #
    #calling command:
    #ch5fig27(230.0, 60.0, 0.1075, 0.2, 88.0, 140.0, 1.2, 15, 1800)
    #
    from math import pi
    import numpy as np
    import matplotlib.pyplot as plt
    from mpl_toolkits.axes_grid1 import host_subplot
    #
    dteta = 2.0*pi/n
    delt = 1.0/(freq*n)
    gls = delt/ls
    gld = delt/ld
    grd = 1.0/rd
    phdif = 2.0*pi/3.0
    teta = dteta*np.arange(n)
    tetb = teta - phdif*np.ones(n)
    tetc = teta + phdif*np.ones(n)
    esmax = np.sqrt(2.0/3.0)*Ell
    esource_a = esmax*np.sin(teta)
    esource_b = esmax*np.sin(tetb)
    esource_c = esmax*np.sin(tetc)
    #
    Idc, Idcavg, ang = ch5fig27_solver(esource_a, esource_b, esource_c, \
                                gls, gld, grd, Ed, Idref, teta, ncycle, n)
    #
    m0 = 2*ncycle/3
    m1 = 2*m0
    m2 = 3*m0
    #
    ax1 = host_subplot(1, 1, 1)
    ax1.set_xlim([(m0 - 2)*pi, m2*pi])
    ax1.set_xticks([m0*pi, m1*pi, m2*pi])
    ax1.set_xticklabels(['$10\pi$', '$20\pi$', '$30\pi$'])
    ax1.set_xlabel('$\omega t, radians$')
    ax1.set_ylim([0, 2])
    ax1.set_yticks([0, 0.9, 1.2, 2])
    ax1.set_yticklabels(['$0$', '$0.9$', '$1.2$', '$2$'])
    ax1.set_ylabel('$Currents, A$')
    p1, = ax1.plot(ang, Idc, 'c-', label = '$I_{dc}$')
```

```
p2, = ax1.plot(ang, Idcavg, 'm-', label = '$I_{dcavg}$')
ax1.yaxis.get_label().set_color(p1.get_color())
ax1.legend(bbox_to_anchor=(0., 1.02, 1., .102), loc=3, ncol=2,    \
                                mode="expand", borderaxespad=0.)
ax1.grid(True)
#
plt.show()
#
#plt.draw()
#plt.savefig('Fig527.ps', dpi=400, bbox_inches='tight')
#
```

References

[1] N. Mohan, T. M. Undeland, W. P. Robbins, *Power Electronics: Converters, Applications, and Design*, 2^{nd}Ed., John Wiley & Sons Inc., USA, 1995.

[2] M. H. Rashid, *Power Electronics: Devices, Circuits and Applications*, 4^{th}Ed., Pearson India, Chennai, 2018.

[3] D. W. Hart, *Power Electronics*, McGraw Hill, New York, 2011.

[4] J. G. Kassakian, M. F. Schlecht, G. C. Verghese, *Principles of Power Electronics*, 1^{st}Ed., 2009, Pearson India.

[5] K. G. Fehrle, R. H. Lasseter, *Simulation of control systems and application to HVDC converters*, IEEE Course Text 81EH0173-5-PWR, Course Organizer A. G. Phadke, Piscataway, New Jersey.

[6] S. R. Naidu, A. M. N. Lima, *A new approach for the simulation of power electronic circuits*, IEEE Trans., on Circuits and Systems-1, vol.49, no.9, pp.1317-1324, 2002.

[7] K. J. Astrom, B. Wittenmark, *Computer-Controlled Systems: Theory and Design*, Dover Publications, Inc., Mineola, New York, 2011.

6

Non-isolated DC–DC Converters

The absence of a physical DC transformer was probably one of the reasons for developing switch-mode DC–DC converters. Over the past several decades, the power electronics community has seen significant efforts to design and deploy DC–DC converters. The three basic converter circuits in common use are the buck (step-down) converter, the boost (step-up) converter, and the buck−boost (up-down) converter. There are several variants of the DC–DC converter such as the non-isolated converter, the transformer isolated converter, and many resonant versions. This chapter is focused on the study of non-isolated DC–DC converters.

6.1 The Non-isolated Buck Converter

This is the simplest of the switch-mode, DC–DC converters. The direct output voltage is less than and of the same polarity as the direct input voltage. Figure 6.1 shows the basic circuit of the converter's power stage, ignoring parasitic effects. The switch SW is usually an N-channel MOSFET and the freewheeling diode is a Schottky diode. The inductor L and the capacitor C form the output filter. During normal operation, SW is repeatedly turned on for t_{ON} seconds and turned off for t_{OFF} seconds. Thus, the switching period is $T_{\mathrm{S}} = t_{\mathrm{ON}} + t_{\mathrm{OFF}}$ and the switching frequency is $f_{\mathrm{S}} = 1/T_{\mathrm{S}}$. The ratio $D = t_{\mathrm{ON}}/T_{\mathrm{S}}$ is the duty ratio.

The output voltage of the converter v_0 is less than the input voltage. Therefore, the inductor current rises linearly when the switch is turned on and it decays linearly when the switch is turned off. In the steady-state, the converter can operate in the continuous conduction mode (CCM) or in the discontinuous conduction mode (DCM). In CCM, the inductor current flows continuously during the entire switching period. The converter passes through two states per switching period: the ON state (SW on and diode reverse biased) and the OFF state (SW off and diode forward biased). When operating

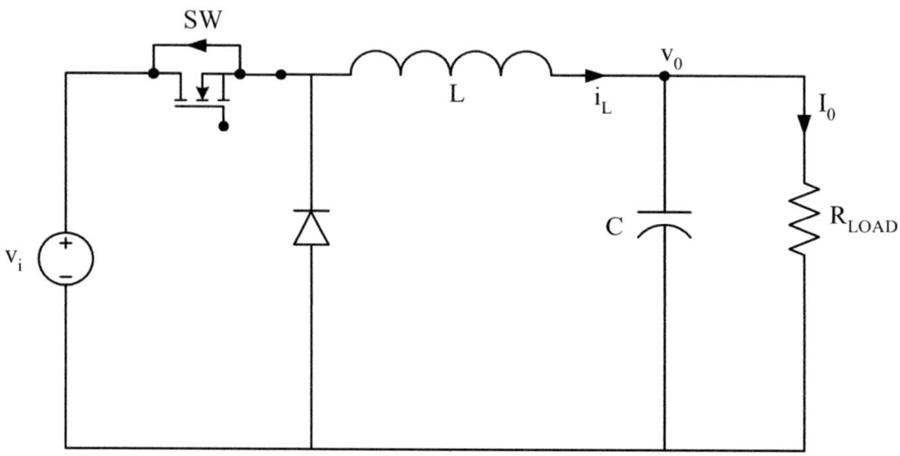

Figure 6.1 Power stage of the basic buck converter.

in DCM, the inductor current rises from zero to a peak value and then decays back to zero. The current remains at zero for the rest of the switching period. The converter passes through three states: the ON state and the OFF state (as in CCM), and a third IDLE state (SW off and the diode reverse biased). The buck converter may be designed to operate in either of the two modes. However, it is desirable that the converter operates in one mode only over its expected load conditions.

6.1.1 Steady-state analysis of the basic buck converter

Consider the steady-state operation of the buck converter in the continuous conduction mode [1]. Average values of the circuit variables are represented by upper case letters and their instantaneous values by lower case letters. Figure 6.2 shows the inductor voltage and the inductor current. The input voltage is the direct voltage V_i and the output voltage is considered to be the direct voltage V_O. The voltage applied to the inductor is $V_i - V_O$ when the switch is turned on and $-V_O$ when the switch is turned off. In the steady-state, the average inductor voltage over a period of the switching frequency is zero. Therefore,

$$D * T_S(V_i - V_O) - (1 - D) * T_S V_O = 0. \tag{6.1}$$

Simplifying,

$$V_O = DV_i. \tag{6.2}$$

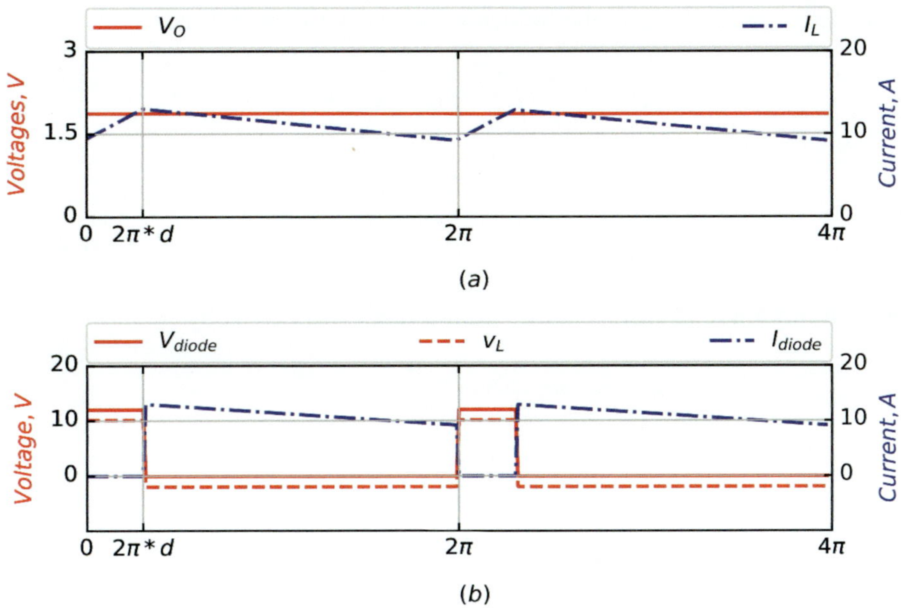

Figure 6.2 Buck converter: current and voltage waveforms.

The inductor current is given by

$$i_L = i_C + v_O/R_{\text{LOAD}}. \tag{6.3}$$

Taking the average of the inductor current over a switching period and noting that the average capacitive current is zero, we get

$$I_L = V_O/R_{\text{LOAD}} = I_O. \tag{6.4}$$

The switch current is equal to the inductor current when the switch is turned on and the diode current is equal to the inductor current when the switch is turned off (Figure 6.2 (b)). The average switch current (which is also the average input current) is given by

$$I_{\text{SW}} = \frac{1}{T_{\text{S}}} \int_0^{DT_{\text{S}}} i_L \, dt = DI_L. \tag{6.5}$$

Thus, we obtain $V_i I_{\text{SW}} = V_O I_O$, which is true since there are no losses in the basic buck converter.

The peak−peak ripple in the inductor current is obtained from

$$\begin{aligned}\Delta I &= \frac{1}{L}\int_0^{DT_S}(V_i - V_0)dt \\ &= \frac{DT_s}{L}(V_i - V_0) \\ &= (1-D)T_S\frac{V_0}{L}. \end{aligned} \tag{6.6}$$

The minimum load current for operation in CCM is given by

$$\begin{aligned} I_{0,\min} &= \frac{1}{2}\Delta I_L \\ &= \frac{1}{2}\frac{(1-D)T_S V_0}{L}. \end{aligned} \tag{6.7}$$

The output voltage V_0 is a constant direct voltage only if the capacitance C is infinite. For a finite capacitance, a ripple is superposed on the output voltage. As a first approximation, we assume that the capacitive current is the ripple component of the inductor current (6.2 a). Then, the peak−peak ripple in the output voltage is given by

$$\begin{aligned} \Delta V_0 &= \frac{1}{C}\int_{t_1}^{t_2} i_C(t)dt \\ &= \frac{1}{C}\left(\frac{1}{2}\frac{T_S}{2}\frac{\Delta I_L}{2}\right) \\ &= \frac{V_0 T_S(1-D)}{8}\cdot\frac{T_S}{LC} \\ &= \frac{V_0(1-D)}{8}4\pi^2\frac{f_c^2}{f_S^2}. \end{aligned}$$

Therefore,

$$\frac{\Delta V_0}{V_0} = \frac{\pi^2(1-D)}{2}\frac{f_c^2}{f_S^2}. \tag{6.8}$$

where $f_C = \frac{1}{2\pi\sqrt{LC}}$.

It is usual to design the filter such that the peak−peak ripple component of the inductor current is 20%−40% of the load current and the peak−peak output voltage ripple does not exceed 1% of the output voltage.

6.1.2 Inclusion of losses

Assume that the buck converter is operating in CCM. Losses in the circuit components are modeled by inserting resistors appropriately. The voltage drop in the switch is represented by a resistance R_{SW}. A voltage source V_{DE} in series with the ON resistance R_D is a reasonably accurate representation of the freewheeling diode when it is forward biased. Copper losses in the inductor and the dielectric losses in the capacitor are taken into account by the resistors R_L and R_C, respectively. (It is usual to refer to these resistors as the equivalent series resistances (ESRs). Likewise, ESL would refer to the equivalent series inductance of the capacitor.)

Figure 6.3 shows the converter circuit after inserting the resistors.

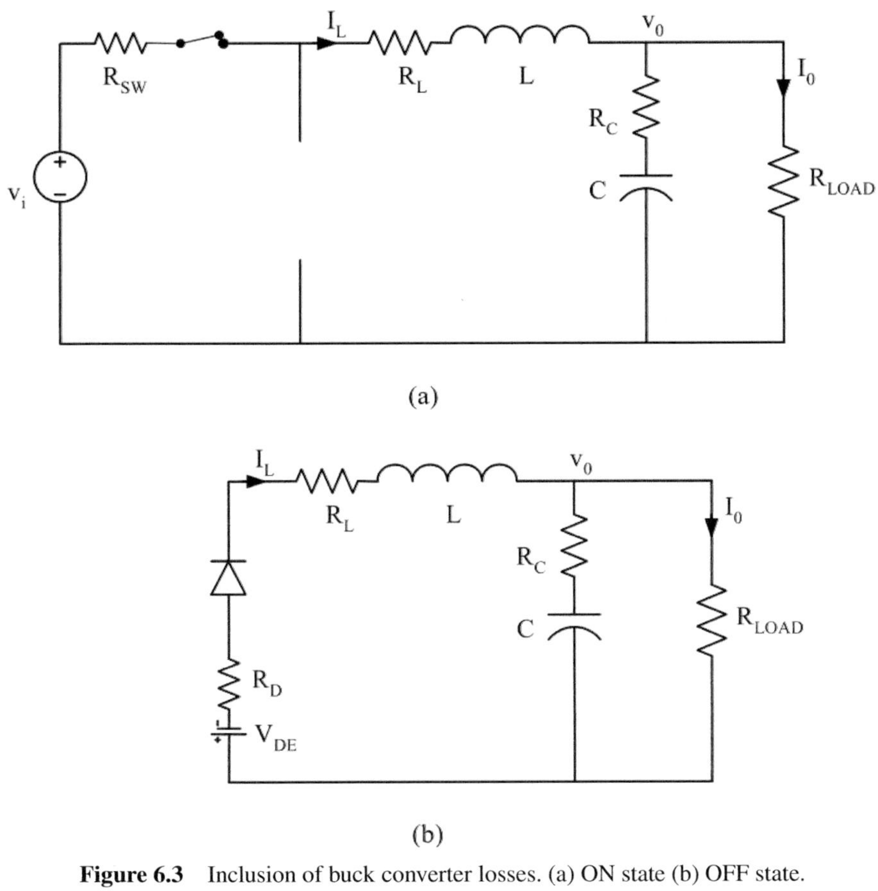

Figure 6.3 Inclusion of buck converter losses. (a) ON state (b) OFF state.

In order to obtain the output/input voltage ratio, the circuit equations are written separately for the two operating states of the converter. In the ON state,

$$\begin{aligned} v_L &= v_i - v_0 - (R_{\mathrm{SW}} + R_L) * i_L \\ i_C &= i_L - v_0/R_{LOAD}. \end{aligned}$$

In the OFF state,

$$\begin{aligned} v_L &= -v_0 - V_{DE} - (R_D + R_L) * i_L \\ i_C &= i_L - v_0/R_{LOAD}. \end{aligned}$$

Taking the average value of the variables and considering that the converter operates in the two states for fractions D and $(1-D)$ of each switching period, we get

$$I_C = I_0 - V_0/R_{LOAD} \tag{6.9}$$

$$\begin{aligned} V_L &= D(V_i - V_0 - I_0(R_{\mathrm{SW}} + R_L)) + \\ &\quad (1-D)(-V_0 - V_{DE} - (R_L + R_D)I_0). \end{aligned} \tag{6.10}$$

In the steady-state, the average values $V_L = 0$ and $I_C = 0$. Setting the right-hand sides of eqn (6.9) and (6.10) to zero and simplifying the equations, we get

$$I_0 = \frac{V_0}{R_{LOAD}} \tag{6.11}$$

$$V_0 = \frac{D * V_i - (1-D) * V_{DE}}{1 + \frac{R_L + D*R_{\mathrm{SW}} + (1-D)*R_D}{R_{LOAD}}}. \tag{6.12}$$

6.1.3 Selection of the power stage's components

Consider a 12 V/1.8 V buck converter as an example [2, 3]. It is given that the minimum and maximum input voltages are 10.5 V and 13.5 V. The maximum load I_0 is 12 A and the switching frequency f_{S} is 625 kHz. Assuming that the converter's efficiency is $\eta = 0.9$, an estimate of the duty ratio is

$$D = \frac{V_0}{\eta V_{i,\max}} = 0.148. \tag{6.13}$$

The minimum inductance is calculated based on the acceptable ripple current. The inductor ripple current ΔI_L is the peak–peak change in the inductor current during the ON time of the power switch. Then

$$L\frac{\Delta I_L}{t_{\text{ON}}} = V_{L,\text{ON}} = V_i - V_{\text{SW,ON}} - V_0. \tag{6.14}$$

Therefore,

$$L = \frac{t_{\text{ON}}(V_i - V_{\text{SW,ON}} - V_0)}{\Delta I_L}, \tag{6.15}$$

where $t_{\text{ON}} = D/f_{\text{S}}$ and $V_{\text{SW,ON}}$ is the voltage drop due to the switch resistance, when turned on. The recommended value for ΔI_L is $0.2-0.4$ of I_L. Choosing $\Delta I_L = 0.3I_L$, the minimum value for the inductance is 0.67 μH. The inductance is chosen to be 0.68 μH. Choice of a lower inductance would imply fewer turns of a conductor having a larger cross section. Consequently, the resistive losses will be lower and the saturation current will be higher. Additionally, the response to a load transient will be faster. On the other hand, a larger inductance results in a lower current ripple and, hence, a lower inductor core loss. The lower inductor current ripple will also lead to lower RMS value for the capacitive current.

Common capacitor types used in low-voltage, DC–DC converters are aluminum electrolytic, tantalum, and ceramic capacitors. Other types have also been introduced into the market, such as the solid polymer aluminum (SPA) capacitor, aluminum with organic electrolyte (OS-CON) capacitor, and tantalum with organic electrolyte capacitor (POSCAP). Ceramic and tantalum capacitors have high volumetric capacitance (i.e., $\mu\text{F} * \text{volts}/\text{volume}$). Ceramic capacitors have the least ESR and ESL, followed by SPA and POSCAP capacitors. Conventional aluminum electrolytic capacitors have high ESR and low ripple current rating. The low ESR ceramic capacitors have the highest cost per Farad.

The capacitance required to limit the voltage ripple to 1% of the output voltage is

$$C = \frac{\Delta I_L}{8 f_{\text{S}} \Delta V_0} = 4\ \mu\text{F}. \tag{6.16}$$

Often, the selection of the output capacitor is dictated by the response to a load transient. Considering a ΔI_0 drop in the load current (and without considering the presence of any feedback control), an estimate for the minimum capacitance to limit the overshoot in the output voltage is obtained by putting $\Delta I_0 = 3\%$ of I_0 and $V_{\text{overshoot}} = 1\%$ of V_0. That is, $\Delta I_0 = 0.36$ A and $V_{\text{overshoot}} = 0.018$ V. Then,

$$C = \frac{\frac{1}{2}\Delta I_0^2 L}{V_{\text{overshoot}}^2} = 136\ \mu\text{F}. \tag{6.17}$$

Any capacitor value above the minimum value may be used, provided that the feedback compensation is designed accordingly. For example, a 470 μF, 15 mΩ ESR aluminum electrolytic capacitor is chosen for bulk storage. Two 10 μF, 1 mΩ ESR ceramic capacitors are connected in parallel to damp the voltage spikes in the output due to step changes in the load current.

Either a P-channel or an N-channel MOSFET may be used as the power switch. N-channel MOSFETs have a lower ON resistance and they are relatively less expensive. On the other hand, P-channel MOSFETs have simpler gate drive requirements. Their disadvantages are the higher ON resistance, slower switching times, and higher cost.

The free-wheeling diode conducts when the power switch is turned off and it provides a path for the inductor current. It is usually a Schottky diode. The breakdown voltage of the diode should be greater than the maximum input voltage with some margin for voltage spikes. The current rating should be at least twice the maximum output current. (Power dissipation may require the current rating to be much higher.) The power dissipated by the diode is given by

$$P_D = (1 - D) * I_0 V_{\mathrm{FD}}, \tag{6.18}$$

where V_{FD} is the forward voltage drop. The diode's junction temperature may be estimated as

$$T_{\mathrm{JD}} = T_A + R_{\mathrm{JA}} * P_D, \tag{6.19}$$

where T_A is the ambient temperature and R_{JA} is the junction to ambient thermal resistance.

Finally, the power stage of the buck converter has the following parameter values:

$$\begin{aligned} v_i &= 12\ \mathrm{V},\ \ \mathrm{v_0} = 1.8\ \mathrm{V},\ \ \text{load } \mathrm{I_0} = 12\ \mathrm{A} \\ L &= 0.68\ \mu\mathrm{F},\ \ \mathrm{C} = 490\ \mu\mathrm{F},\ \ (\mathrm{ESR})\ \mathrm{R_C} = 0.015\ \Omega. \end{aligned}$$

6.1.4 Simulation of the buck converter

The power stage of the buck converter is shown in Figure 6.4. The transient simulation is performed over $ncycle$ periods and n is the number of time-steps in each period. The closure of the switch extends over nc time-steps, where $nc = \mathrm{round(D * n)}$.

The transient equivalent circuit of the power stage is shown in Figure 6.5. The probe voltages are v_D and v_0. The past-history current sources for the inductor and the capacitor are pil and $gcp * pvc$, respectively. During each

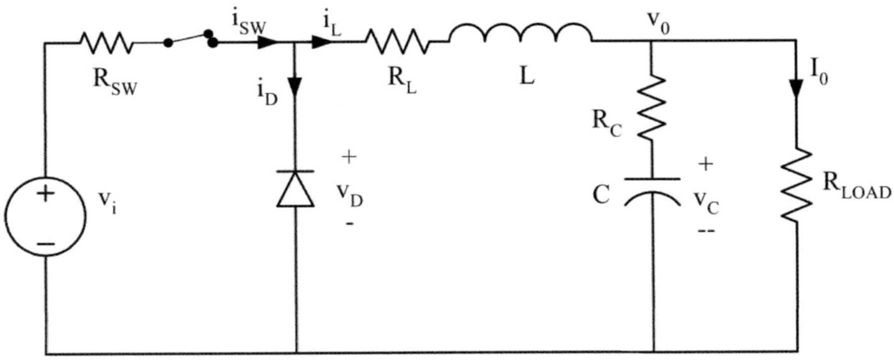

$R_{SW} = 10\text{m}\Omega$, $L = 0.68\mu\text{H}$, $v_i = 12$ V
$R_L = 80\text{m}\Omega$, $C = 490\mu\text{F}$, $f_s = 625$ kHz
$R_C = 15\text{m}\Omega$, $R_{LOAD} = 0.15\Omega$, $d = 0.22$

Figure 6.4 Buck converter: circuit diagram.

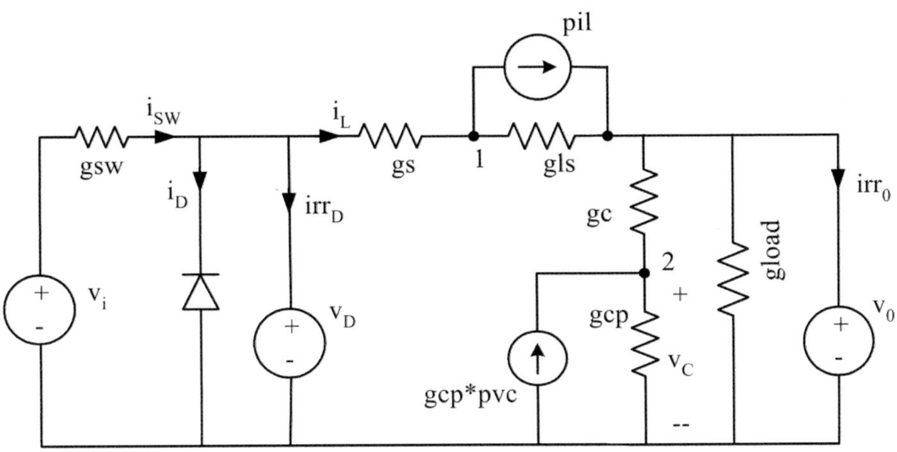

Figure 6.5 Buck converter: transient equivalent circuit.

switching period, the conductance of the switch is $gsw = 100\,\text{S}$ for $0 \leq j \leq nc$ and $gsw = 10^{-7}\,\text{S}$ for $nc \leq j \leq n$.

Applying KCL at node 1, we get

$$gs * (v_D - v_1) - pil - gls * (v_1 - v_0) = 0. \tag{6.20}$$

Therefore,

$$v_1 = \frac{gs * v_D + gls * v_0 - pil}{gs + gls}. \tag{6.21}$$

Applying KCL at node 2,

$$gc * (v_C - v_0) + gcp * v_C = gcp * pvc. \tag{6.22}$$

Therefore,

$$v_C = \frac{gc * v_0 + gcp * pvc}{gc + gcp}. \tag{6.23}$$

Other equations that follow are

$$\begin{aligned} i_L &= gs * (v_D - v_1) && (6.24) \\ i_{SW} &= gsw * (v_i - v_D) && (6.25) \\ i_C &= gc * (v_0 - v_C) && (6.26) \\ i_D, gd &= fdiode(v_D). && (6.27) \end{aligned}$$

From eqn (6.21)–(6.23), the error currents flowing into the probe voltages are computed as follows:

$$\begin{aligned} irr_D &= i_{SW} - i_D - i_L && (6.28) \\ irr_0 &= i_L - i_{LOAD} - i_C. && (6.29) \end{aligned}$$

The error currents are progressively reduced by adding corrections δv_D, δv_0 to the probe voltages. The small-signal equivalent circuit shown in Figure 6.6 provides the following equations for the corrections:

$$\begin{bmatrix} gd + gsw + gser & -gser \\ -gser & gpar + gser + gload \end{bmatrix} \begin{bmatrix} \delta v_D \\ \delta v_0 \end{bmatrix} = \begin{bmatrix} irr_D \\ irr_0 \end{bmatrix}. \tag{6.30}$$

In eqn (6.30),

$$\begin{aligned} gser &= gs * gls/(gs + gls) \\ gpar &= gc * gcp/(gc + gcp). \end{aligned}$$

After a few iterations, the procedure converges. The variables are stored for plotting purposes, the past-history terms are updated, and we proceed to the next time step. Figure 6.7 shows the waveforms of the inductor current and the output voltage. The code for simulating the buck converter is presented in Program 6.07.

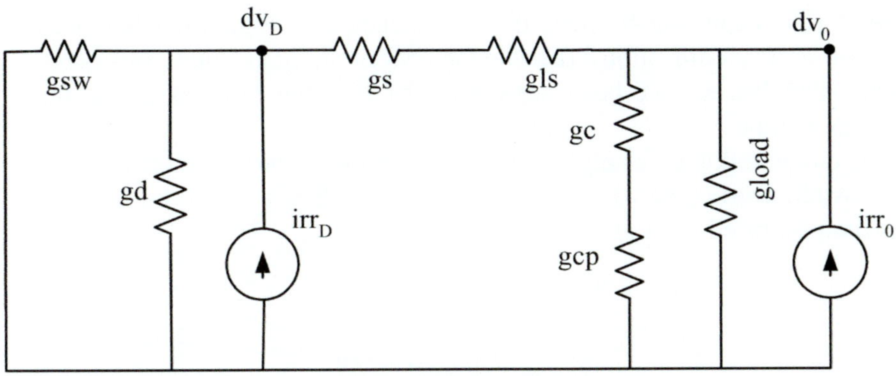

Figure 6.6 Buck converter: small-signal equivalent circuit.

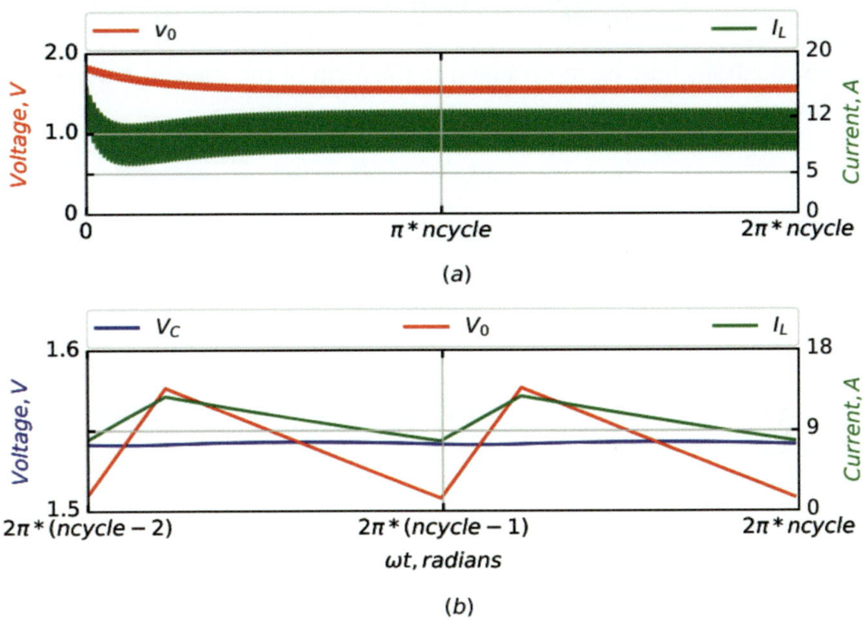

Figure 6.7 Buck converter: response waveforms.

6.1.5 Buck converter model in CCM

The model we seek is the representation of the converter by mathematical equations [4]. The converter's variables have an average behavior over each switching period and the model should describe the variation of the average from period to period. Let T_S be the switching period and d be the duty ratio.

In the development that follows, the nomenclature of Figure 6.1 has been used with some modifications: (a) only the ESR of the capacitor is considered and all other losses have been neglected; (b) R is the load resistor at the converter's output.

The circuit topology changes within each period according to the position of the switch. During the jth period and for $(j-1)TS \leq \mathrm{t} \leq (\mathrm{j}-1++\mathrm{d})\mathrm{T_S}$, the switch is closed and the circuit equations are

$$
\begin{aligned}
L\frac{di_L}{dt} &= v_i - v_0 \\
i_L &= \frac{v_0 - v_C}{R_C} + \frac{v_0}{R} \\
C\frac{dv_C}{dt} &= \frac{v_0 - v_c}{R_C}.
\end{aligned}
$$

These equations may be re-arranged in the form of the state equations:

$$
\begin{aligned}
L\frac{di_L}{dt} &= v_i - R_p i_L - \frac{R}{R+R_C}v_C && (6.31) \\
\text{where } \mathrm{R_p} &= R\,R_C/(R+R_C) \\
C\frac{dv_C}{dt} &= \frac{R}{R+R_C}i_L - \frac{1}{R+R_C}v_C && (6.32) \\
C\frac{dv_C}{dt} &= \frac{v_0 - v_c}{R_C}. && (6.33)
\end{aligned}
$$

Eqn (6.33) is not a state equation but is a useful, auxiliary equation.

During the interval $(j-1+d)T_S \leq t \leq j\,T_S$, the switch is open and the circuit equations are

$$
\begin{aligned}
L\frac{di_L}{dt} &= -v_0 \\
i_L &= \frac{v_0 - v_C}{R_C} + \frac{v_0}{R} \\
C\frac{dv_C}{dt} &= \frac{v_0 - v_c}{R_C}.
\end{aligned}
$$

The state equations that result from a re-arrangement of these equations are:

$$
L\frac{di_L}{dt} = -\frac{R\,R_C}{R+R_C}i_L - \frac{R}{R+R_C}v_C \qquad (6.34)
$$

$$C\frac{dv_C}{dt} = \frac{R}{R+R_C}i_L - \frac{1}{R+R_C}v_C \tag{6.35}$$

$$C\frac{dv_C}{dt} = \frac{v_0 - v_c}{R_C}. \tag{6.36}$$

Taking the average of the RHS of eqn (6.31)–(6.36), we get

$$L\frac{d\langle i_L\rangle}{dt} = \langle d\rangle\langle v_i\rangle - R_p\langle i_L\rangle - \frac{R}{R+R_C}\langle v_C\rangle, \tag{6.37}$$

$$C\frac{d\langle v_C\rangle}{dt} = \frac{R}{R+R_C}\langle i_L\rangle - \frac{1}{R+R_C}\langle v_C\rangle \tag{6.38}$$

$$R_C\,C\frac{d\langle v_C\rangle}{dt} = \langle v_0\rangle - \langle v_C\rangle. \tag{6.39}$$

The notation $\langle x\rangle$ represents the average of the variable x during the jth switching period. The average $\langle x\rangle$ consists of a constant (or direct) component X and a periodic, time-varying component $\tilde{x}$. Thus, $\langle x\rangle = X + \tilde{x}$.

Putting $\langle i_L\rangle = I_L + \tilde{i}_L$ and $\langle v_C\rangle = V_C + \tilde{v}_C$, eqn (6.37)–(6.39) are written as

$$L\frac{d\tilde{i}_L}{dt} = (D+\tilde{d})(V_i+\tilde{v}_i) - R_p(I_L+\tilde{i}_L) - \frac{R}{R+R_C}(V_C+\tilde{v}_C) \tag{6.40}$$

$$C\frac{d\tilde{v}_C}{dt} = \frac{R}{R+R_C}(I_L+\tilde{i}_L) - \frac{1}{R+R_C}(V_C+\tilde{v}_C) \tag{6.41}$$

$$R_C C\frac{d\tilde{v}_C}{dt} = (V_0+\tilde{v}_0) - (V_C) + \tilde{v}_C). \tag{6.42}$$

(Note that $\tilde{i}_L$ is not the inductor current ripple. It is the variation in the average value of i_L from period to period.)

Eqn (6.40) is nonlinear due to the product $(D+\tilde{d})(V_i+\tilde{v}_i)$. Assuming that the periodic, time-varying components are much smaller than the direct components, the product may be linearized as

$$(D+\tilde{d})(V_i+\tilde{v}_i) = DV_i + V_i\tilde{d} + D\tilde{v}_i. \tag{6.43}$$

Further, neglecting variations in the input voltage $\tilde{v}_i$, the product is simplified as

$$(D+\tilde{d})(V_i+\tilde{v}_i) = DV_i + V_i\tilde{d}. \tag{6.44}$$

Then, eqn (6.40)–(6.42) may be written as

$$\begin{aligned} L\frac{d\tilde{i}_L}{dt} &= \left(DV_i - R_pI_L - \frac{R}{R+R_C}V_C\right) + \left(V_i\tilde{d} - R_p\tilde{i}_L\right. \\ &\quad \left. - \frac{R}{R+R_C}\tilde{v}_C\right) \qquad (6.45) \\ C\frac{d\tilde{v}_C}{dt} &= \left(\frac{R}{R+R_C}I_L - \frac{1}{R+R_C}V_C\right) + \left(\frac{R}{R+R_C}\tilde{i}_L\right. \\ &\quad \left. - \frac{1}{R+R_C}\tilde{v}_C\right) \qquad (6.46) \\ R_C\,C\frac{d\tilde{v}_C}{dt} &= (V_0 - V_C) + (\tilde{v}_0 - \tilde{v}_C). \qquad (6.47) \end{aligned}$$

The LHS of eqn (6.45)–(6.47) are periodic and time-varying without any direct component. Therefore, we put

$$\begin{aligned} DV_i &= R_pI_L + \frac{R}{R+R_C}V_C \\ \frac{R}{R+R_C}I_L &= \frac{1}{R+R_C}V_C \\ V_0 &= V_C. \end{aligned}$$

Simplifying, we get the steady-state solution presented in Section 6.1:

$$DV_i = V_C,\ V_C = RI_L,\ V_0 = V_C. \qquad (6.48)$$

Finally, the smal-signal equations are

$$\begin{aligned} L\frac{d\tilde{i}_L}{dt} &= V_i\tilde{d} - R_p\tilde{i}_L - \frac{R}{R+R_C}\tilde{v}_C \qquad (6.49) \\ C\frac{d\tilde{v}_C}{dt} &= \frac{R}{R+R_C}\tilde{i}_L - \frac{1}{R+R_C}\tilde{v}_C \qquad (6.50) \\ R_CC\frac{d\tilde{v}_C}{dt} &= \tilde{v}_0 - \tilde{v}_C. \qquad (6.51) \end{aligned}$$

Eqn (6.49)–(6.51) are the linearized equations for the buck converter in the neighborhood of an operating point.

6.1.6 Control to output transfer function (CCM)

Applying the Laplace transform to eqn (6.49)–(6.51), we get

$$sL\tilde{i}_L(s) = V_i\tilde{d}(s) - R_p\tilde{i}_L(s) - \frac{R}{R+R_C}\tilde{v}_C(s) \tag{6.52}$$

$$sC\,\tilde{v}_C(s) = \frac{R}{R+R_C}\tilde{i}_L(s) - \frac{1}{R+R_C}\tilde{v}_C(s) \tag{6.53}$$

$$R_C\,C\tilde{v}_C(s) = \tilde{v}_O(s) - \tilde{v}_C(s). \tag{6.54}$$

Eliminating $\tilde{i}_L(s)$ in eqn (6.52) and (6.53) and using eqn (6.54), we get

$$\begin{aligned} \tilde{v}_O(s) &= \frac{V_i(1+sCR_C)}{s^2LC\frac{R+R_C}{R} + s\left(\frac{L}{R}+CR_C\right)+1}\tilde{d}(s) \\ &\simeq \frac{V_i(1+sCR_C)}{s^2LC + s\left(\frac{L}{R}+CR_C\right)+1}\tilde{d}(s) \\ &= G(s)\cdot\tilde{d}(s). \end{aligned} \tag{6.55}$$

The control to output transfer function has a zero at $\omega_{\text{ESR}} = 1/(R_C\,C)$ and two poles. (It is usual to consider the poles as a double pole at $\omega_{LC} = 1/\sqrt{LC}$.)

6.1.7 Feedback control of the buck converter in CCM

Disturbances such as fluctuations in the input voltage or the load will cause variations in the output voltage. For this reason, a feedback control system is necessary, which takes as its input the output voltage of the power stage and will stabilize it by appropriately adjusting the duty ratio. Figure 6.8 shows a typical feedback control system [5].

The control system consists of an error amplifier and compensating network, which is either a Type II(PI) or a Type III(PID) compensator. The Type III(PID) compensator has the extra branch R_3C_3 added to the Type II(PI) compensator. A reference voltage V_{REF} is connected to the non-inverting input of the error amplifier. The PWM generator compares the output v_e of the error amplifier to a periodic saw-tooth voltage. The peak–peak amplitude of the saw-tooth is V_{RAMP} and its frequency is precisely the switching frequency f_S. The PWM voltage output of the generator has the duty ratio set by the ratio v_e/V_{RAMP} (as shown in Figure 6.8).

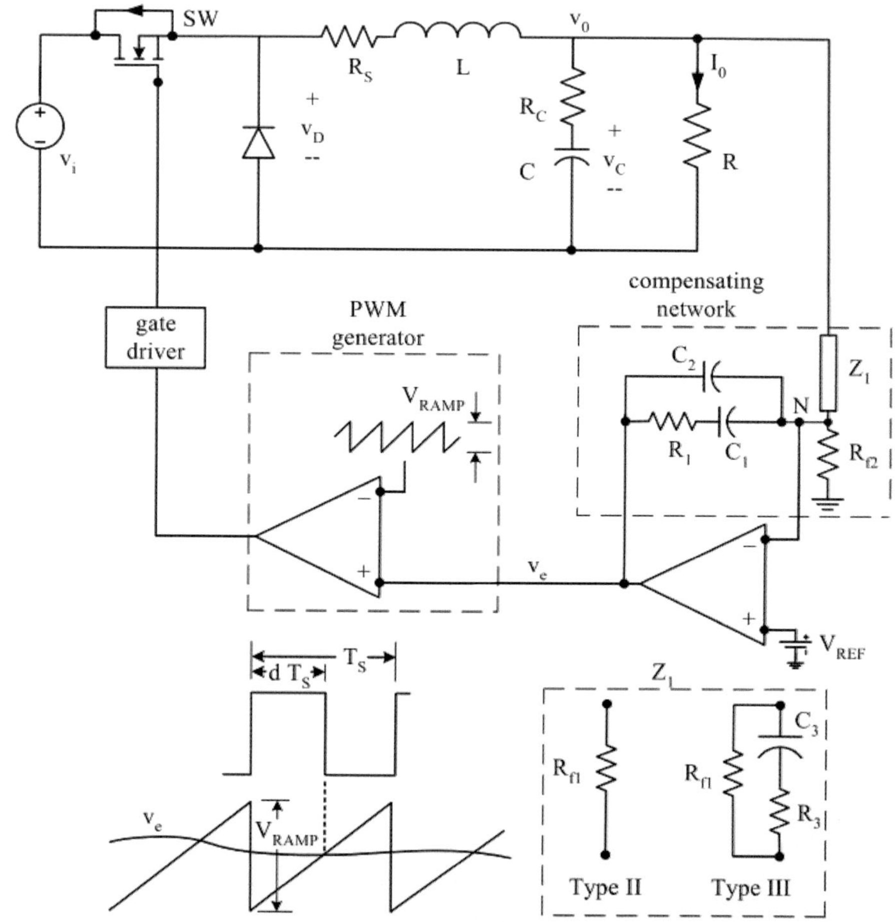

Figure 6.8 Buck converter: power stage and feedback control system.

The design procedure requires the consideration of the following angular frequencies:

- ω_{ESR}, the frequency of the zero of the power stage. This frequency is mainly a characteristic of the filter capacitor. It is low for aluminum electrolytic capacitors and high for ceramic capacitors.
- $\omega_{LC} = 1/\sqrt{LC}$ is the location of the double poles of the power stage (neglecting the losses). It is the least of the frequencies being considered.

- ω_0 is the zero crossover frequency of the loop gain. The response of the system is fast when the crossover frequency is high. On the other hand, the crossover frequency should be low enough to attenuate the switching noise.

The type of compensator that is chosen will depend on the relative locations of the frequencies:

- If $\omega_{LC} < \omega_{\mathrm{ESR}} < \omega_0 < \omega_{\mathrm{S}}/2$: choose the Type II compensator.
- If $\omega_{LC} < \omega_0 < \omega_{\mathrm{ESR}} < \omega_{\mathrm{S}}/2$: choose the Type IIIA compensator.
- If $\omega_{LC} < \omega_0 < \omega_{\mathrm{S}}/2 < \omega_{\mathrm{ESR}}$: choose the Type IIIB compensator.

The Type IIIA and Type IIIB compensators have the same circuit topology. It is usual to choose $\omega_0 = \omega_{\mathrm{S}}/10$ for the Type II compensator, $\omega_0 = \omega_{\mathrm{S}}/8$ for the Type IIIA compensator, and $\omega_0 = \omega_{\mathrm{S}}/6$ for the Type IIIB compensator.

The transfer function of the Type II compensating network will be derived as follows. Assuming a very high gain for the error amplifier and applying KCL at the node N, we get

$$\frac{V_{\mathrm{REF}} - v_0}{R_{f1}} + \frac{V_{\mathrm{REF}}}{R_{f2}} + (V_{\mathrm{REF}} - v_e)\left\{sC_2 + \frac{sC_1}{1 + sC_1R_1}\right\} = 0. \quad (6.56)$$

Assuming that $C_1 \gg C_2$ and putting $\mu = \frac{R_{f2}}{R_{f1}+R_{f2}}$, we get

$$(V_{\mathrm{REF}} - \mu v_0) = \mu(v_e - V_{\mathrm{REF}})\frac{sC_1R_{f1}(1 + sC_2R_1)}{(1 + sC_1R_1)}. \quad (6.57)$$

Suppose that there is a drop in the output voltage from the steady-state value V_0 to $V_0 - \tilde{v}_0$. Let the resulting overshoot in v_e be $\tilde{v}_e$. Putting $v_e = V_e + \tilde{v}_e$, choosing $\mu V_0 = V_{\mathrm{REF}}$ and noting that in the steady-state $V_e = V_{\mathrm{REF}}$, we get

$$\tilde{v}_e = \frac{1 + sC_1R_1}{sC_1R_{f1}(1 + sC_2R_1)}\tilde{v}_0 = -H(s) * \tilde{v}_0. \quad (6.58)$$

where the negative transfer function emphasizes the opposing polarities of $\tilde{v}_0$ and $\tilde{v}_e$. The Type II(PI) compensating network is characterized by a zero at $\omega_{Z1} = \frac{1}{R_1C_1}$, a pole at the origin and a second pole at $\omega_{P2} = \frac{1}{R_1C_2}$.

In a similar manner, it may be shown that the transfer function of the Type III compensating network is given by

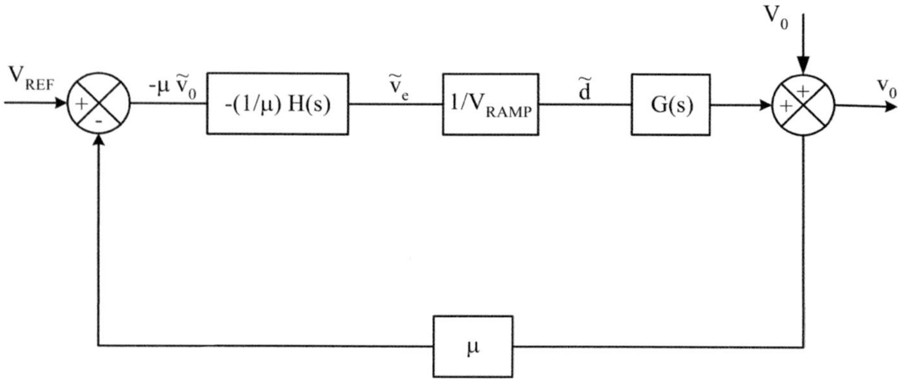

Figure 6.9 Buck converter's feedback control system.

$$H(s) = -\frac{(1 + sR_1C_1)\{1 + sC_3(R_{f1} + R_3)\}}{sR_{f1}C_1(1 + sR_1C_2)(1 + sR_{f3}C_3)}. \tag{6.59}$$

The two zeros and three poles of $H(s)$ are

$$\begin{aligned} \omega_{Z1} &= 1/(R_1C_1), \quad \omega_{Z2} = 1/\{C_3(R_{f1} + R_3)\} \\ \omega_{P1} &= 0, \quad \omega_{P2} = 1/(R_3C_3), \quad \omega_{P3} = 1/(R_1C_2). \end{aligned}$$

The block diagram of the power stage and the feedback control is shown in Figure 6.9. The loop gain of the system is $H(s)G(s)$.

6.1.8 Placement of the poles and zeros of the compensating network

Design of the compensating network essentially reduces to the placement of the poles and zeros of its transfer function in the complex frequency plane. The principal objectives of the design are that (a) the crossover frequency of the loop gain should be adequately located, (b) the slope of the loop gain at the crossover frequency should be about −20 dB/decade, and (c) the phase margin should be greater than 45°.

At the crossover frequency, the loop gain of the Type II compensating network is given by

$$20\log\left|H(s)G(s)\right| \simeq 20\log\left|\frac{V_i}{V_{RAMP}}\frac{1}{R_{f1}}\frac{sC_1R_1}{sC_1}\frac{sCR_C}{s^2LC}\right| = 0. \tag{6.60}$$

Therefore,

$$\left| \frac{V_i}{V_{RAMP}} \frac{R_1}{R_{f1}} \frac{CR_C}{j\omega_0/(\omega_{LC})^2} \right| = 1. \tag{6.61}$$

That is,

$$R_1 = R_{f1} \frac{V_{RAMP}}{V_i} \frac{\omega_0\, \omega_{\text{ESR}}}{(\omega_{LC})^2}. \tag{6.62}$$

In a similar manner, the loop gain at the crossover frequency for the Type III compensating network is given by

$$\begin{aligned} |H(s)G(s)| &\simeq \left| \frac{(1 + j\omega_0 R_1 C_1)(1 + j\omega_0 R_{f1} C_3)}{j\omega_0 R_{f1} C_1 (1 + j\omega_0 R_1 C_2)(1 + j\omega_0 R_3 C_3)} * \right. \\ &\qquad \left. \frac{V_i}{V_{RAMP}} \frac{1 + j\omega_0 R_C C}{(j\omega_0)^2 LC} \right| \\ &= 1. \end{aligned}$$

Therefore,

$$\frac{(\omega_0 R_1 C_1)(\omega_0 R_{f1} C_3)}{(\omega_0 R_{f1} C_1)(1)(1)} \frac{V_i}{V_{RAMP}} \frac{1}{\omega_0^2 LC} = 1. \tag{6.63}$$

That is,

$$R_1 = \frac{V_{RAMP}}{V_i} \frac{\omega_0 LC}{C_3}. \tag{6.64}$$

Several techniques exist based on the root locus or Bode diagrams for placing the poles and zeros of the compensating network. However, the practical rules given below are likely to give reasonably good results.

6.1.8.1 Type II compensator's pole-zero placement

Choose R_{f1} to be a few kΩs and then calculate R_1 using eqn (6.62). The value of R_{f2} is obtained from

$$\frac{R_{f2}}{R_{f1} + R_{f2}} = \frac{V_{\text{REF}}}{V_0}. \tag{6.65}$$

Finally, choose $\omega_{Z1} = \frac{1}{R_1 C_1} = 0.75\omega_{LC}$ and $\omega_{P2} = \frac{1}{R_1 C_2} = \omega_S/2$. It is now possible to calculate the values of the compensating network's elements. If the calculated values are not available, the nearest available values should be used. It may be necessary to pass through several iterations before finalizing the compensating network.

6.1.8.2 Pole-zero placement for the Type IIIA compensator

Choose $\omega_0 = \omega_S/8$ and a few nF for C_3. Place the zeros of the compensator at $\omega_{Z1} = \frac{1}{R_1C_1} = 0.75\omega_{LC}$ and $\omega_{Z2} = \frac{1}{C_3(R_3+R_{f1})} = \omega_{LC}$. The two poles are located at $\omega_{P2} = \frac{1}{R_3C_3} = \omega_{\text{ESR}}$ and $\omega_{P3} = \frac{1}{R_1C_2} = \omega_S/2$. It is now possible to calculate the network elements.

6.1.8.3 Pole-zero placement for the Type IIIB compensator

Choose $\omega_0 = \omega_S/6$ and a few nF for C_3. The pole zero pair ω_{Z2}, ω_{P2} are placed so that they form a lead compensator that gives a maximum phase lead of $\theta = 70°$ at the crossover frequency. Thus,

$$\omega_{Z2} = \omega_0\sqrt{\frac{1-\sin\theta}{1+\sin\theta}}, \quad \omega_{P2} = \omega_0\sqrt{\frac{1+\sin\theta}{1-\sin\theta}}. \tag{6.66}$$

Finally, place ω_{Z1} at $0.5\omega_{Z2}$ and ω_{P3} at $\omega_S/2$.

6.1.9 Type II compensator design example

The buck converter that is considered is the one shown in Figure 6.4. In the steady-state, the converter attenuates the 12 V input voltage to 1.8 V output. The duty ratio at the operating point is 0.22.

The frequencies to be considered in the design of the control system are:

$$\begin{aligned}
\omega_S &= 2\pi * 625{,}000 = 3.927e+06 \text{ rads/sec} \\
\omega_0 &= \omega_S/10 = 392{,}700 \text{ rads/sec} \\
\omega_{LC} &= \frac{1}{\sqrt{LC}} = 54783 \text{ rads/sec} \\
\omega_{\text{ESR}} &= \frac{1}{R_C C} = 136054 \text{ rads/sec.}
\end{aligned}$$

Observe that $\omega_{LC} < \omega_{\text{ESR}} < \omega_0 < \omega_S/2$ and, therefore, a Type II compensating network is recommended in the control loop. The error amplifier has the reference voltage $V_{\text{REF}} = 0.7$ V connected to its non-inverting input. The saw-toothed voltage signal applied to the inverting input of the PWM generator has a frequency of 625 kHz and its peak voltage $V_{RAMP} = 1.8$ V.

From the relation $\frac{V_{\text{REF}}}{V_0} = \frac{R_{f2}}{R_{f1}+R_{f2}}$, we get $R_{f2} = \frac{0.7}{1.1}R_{f1}$. Choosing $R_{f1} = 1200\,\Omega$ leads to $R_{f2} = 764\,\Omega$. From eqn (6.62), we get $R_1 = 3204\,\Omega$. Choose

$$\omega_{Z1} = 0.75 * \omega_{LC} = 41087 \text{ rads/sec}$$

$$\omega_{P2} = 0.5 * \omega_S = 1.9635e + 06 \text{ rads/sec.}$$

The capacitances C_1 and C_2 are given by

$$C_1 = \frac{1}{R_1 \omega_{Z1}} = 7.6 \text{ nF}$$
$$C_2 = \frac{1}{R_1 \omega_{P2}} = 0.16 \text{ nF.}$$

Figure 6.10 (plotted by Program 6.10) shows the Bode diagrams for the transfer functions of the power stage, the compensating network, and the loop gain. The code for plotting the diagrams is presented in Program 6.10.

We proceed with the simulation of the buck converter with the feedback control included. It is necessary to insert the equations for the transient equivalent circuit of the error amplifier (Figure 6.11) in the computational program. Applying Kirchhoff's laws to the circuit of Figure 6.11, we obtain

$$\begin{bmatrix} g1 + gc2 & g1 \\ g1 & g1 + gc1 \end{bmatrix} \begin{bmatrix} v_e \\ v_{C1} \end{bmatrix} = \begin{bmatrix} u_A \\ u_B \end{bmatrix}, \quad (6.67)$$

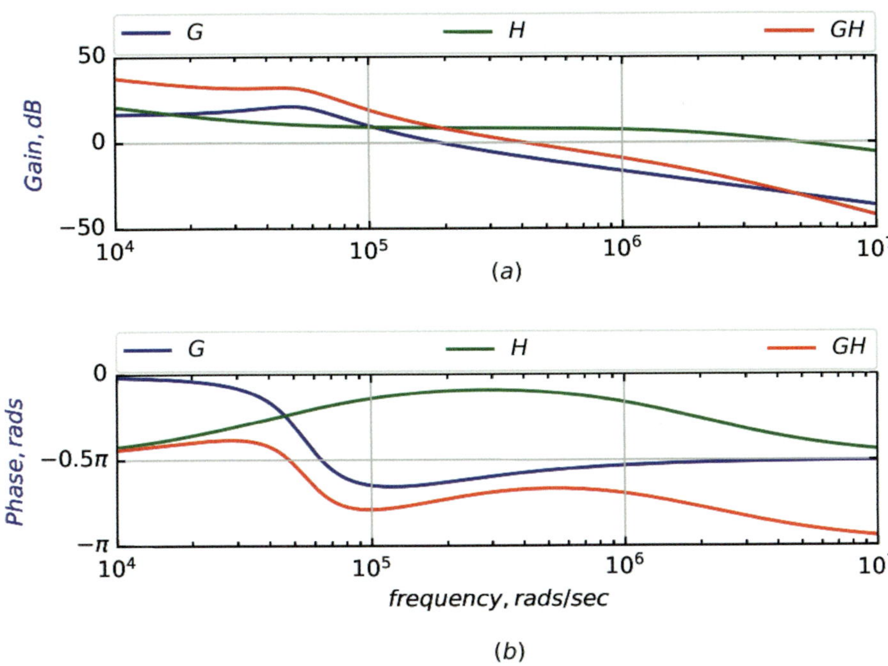

Figure 6.10 Buck converter with voltage mode control: Bode plots.

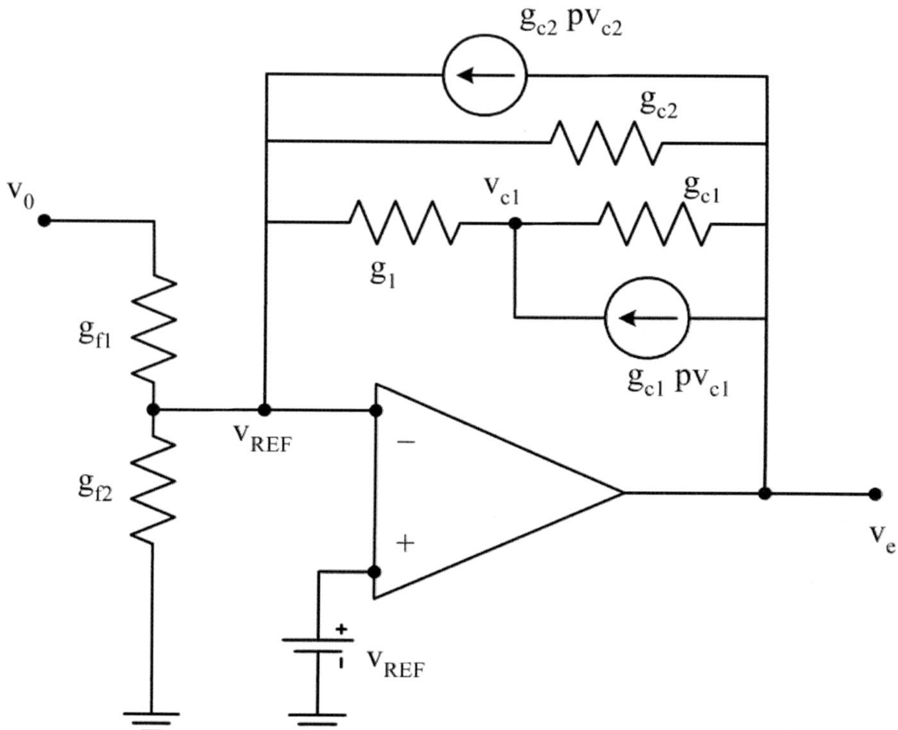

Figure 6.11 Transient equivalent circuit of the error amplifier.

where

$$\begin{bmatrix} u_A \\ u_B \end{bmatrix} = \begin{bmatrix} V_{\mathrm{REF}}(gf1 + gf2 + gc2 + g1) - gf1 * v_0 - gc2 * pvc2 \\ g1 * V_{\mathrm{REF}} + gc1 * pvc1 \end{bmatrix}. \tag{6.68}$$

The past-history terms $gc1 * pvc1$ and $gc2 * pvc2$ are updated at the end of every time-step by $gc1 * vc1$ and $gc2 * (V_{\mathrm{REF}} - v_e)$.

The duty ratio is maintained at a fixed value during each period of the computation and the output of the error amplifier is stored at every time-step. At the end of the current period, the average of the error amplifier output over the period is calculated and the new duty ratio to be used in the next period is obtained. Figure 6.12, plotted by Program 6.12, shows the waveforms obtained from a simulation over 200 periods. After 100 periods, a 30% reduction in the load is introduced. Observe the overshoot in the output voltage due to the load transient. However, the output voltage returns to the set value after a few periods.

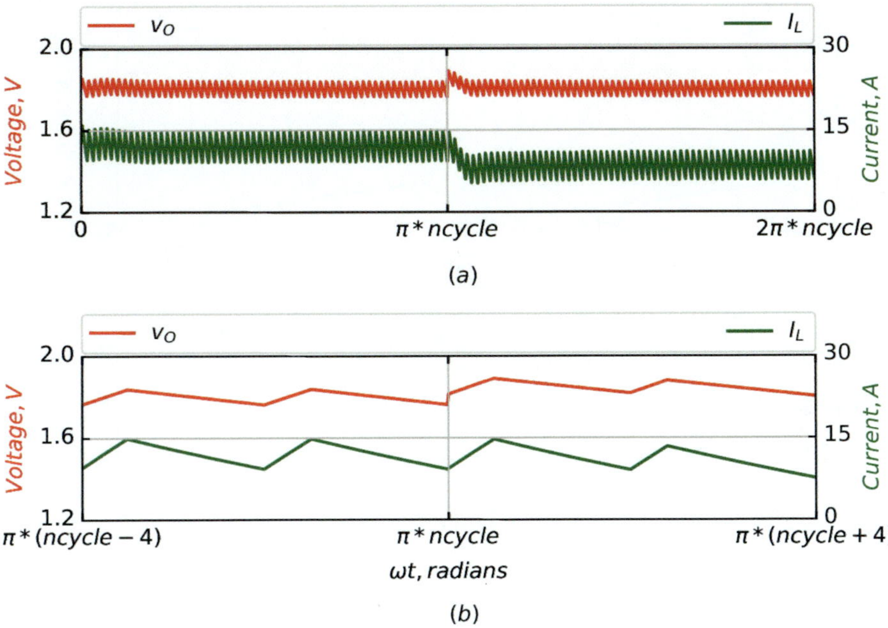

Figure 6.12 Buck converter, voltage mode control: transient response.

6.1.10 The buck converter operating in DCM

When operating in CCM, the peak–peak ripple in the inductor current given by eqn (6.6) does not depend on the load. On the other hand, the average value of the inductor current is, to a first approximation, equal to the load current. Therefore, there exists a critical value of inductance L_{cr} for which the load current is equal to half of the peak–peak ripple in the inductor load current. That is,

$$I_{0,cr} = \frac{1}{2}\Delta I_L = \frac{V_0 T_s}{2\,L_{cr}}(1 - D). \tag{6.69}$$

Thus,

$$L_{cr} = \frac{R_0}{2\,f_S}(1 - D). \tag{6.70}$$

For load currents less than the critical value, the converter operates in DCM.

Figure 6.13 shows the typical inductor voltage and current waveforms when the converter is operating in DCM. There are three distinct states of operation:

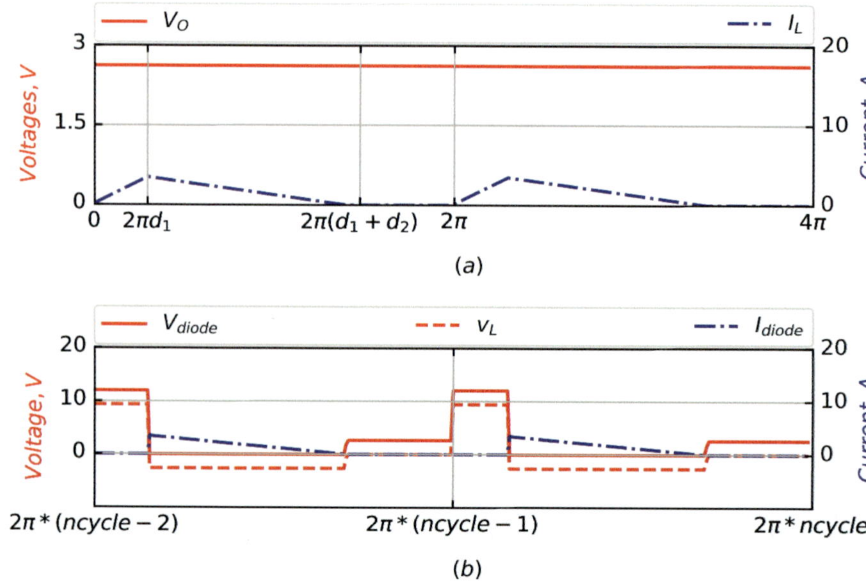

Figure 6.13 Buck converter in DCM: response waveforms.

(a) the ON state for $0 \leq t \leq d_1 T_S$, switch S conducting and diode reverse biased;
(b) the OFF state for $d_1 T_S \leq t \leq (d_1 + d_2) T_S$, S blocked and diode forward biased;
(c) IDLE state for $(d_1 + d_2) T_S \leq t \leq T_S$, S blocked and diode reverse biased.

The voltage ratio in this mode of operation is obtained as follows:

- The average value of the inductor voltage over a switching period is zero. Therefore, $\frac{V_O}{V_i} = \frac{d_1}{d_1+d_2}$.
- The average value of the inductor current is equal to the load current. That is, $I_O = \frac{1}{2}\frac{d_1}{d_1+d_2}\Delta I_L$.
- The inductor current ripple $\Delta I_L = \frac{1}{L}\int_0^{d_1 T_S}(V_i - V_O)\,dt = \frac{d_1 T_S}{L}(V_i - V_O)$.

Putting $I_O = V_O/R$ and $K = \frac{2L}{R_O T_S}$, we get

$$\frac{V_O}{V_i} = \frac{2}{1+\sqrt{1+\frac{4K}{d_1^2}}}. \tag{6.71}$$

6.1.10.1 Mathematical model for the buck converter operating in DCM

We follow the same procedure that was used for determining the converter's model in CCM. However, the topology of the power stage changes thrice in each period. Therefore, the state equations are written for each of the three topologies before they are averaged.

During the jth period and in the interval $jT_S \leq t < (j+d_1)T_S$, the power switch is conducting and we obtain

$$\begin{aligned} L\frac{di_L}{dt} &= v_i - v_0 \\ i_L &= C\frac{dv_C}{dt} + \frac{v_0}{R} = \frac{v_0 - v_C}{R_C} + \frac{v_0}{R}. \end{aligned}$$

Re-arranging in the form of state equations, we obtain

$$L\frac{di_L}{dt} = -R_p i_L - \frac{R}{R+R_C}v_C + v_i \tag{6.72}$$

$$C\frac{dv_C}{dt} = \frac{R}{R+R_C}i_L - \frac{1}{R+R_C}v_C, \tag{6.73}$$

where $R_p = \frac{RR_C}{R+R_C}$. In a similar manner, it may be shown that the state equations during the interval $(j+d_1)T_S \leq t < (j+d_1+d_2)T_S$ are

$$L\frac{di_L}{dt} = -R_p i_L - \frac{R}{R+R_C}v_C \tag{6.74}$$

$$C\frac{dv_C}{dt} = \frac{R}{R+R_C}i_L - \frac{1}{R+R_C}v_C. \tag{6.75}$$

Finally, the state equations during the interval $(j+d_1+d_2)T_S \leq t < T_S$ are

$$L\frac{di_L}{dt} = 0 \tag{6.76}$$

$$C\frac{dv_C}{dt} = -\frac{1}{R+R_C}v_C. \tag{6.77}$$

Averaging the state equations over the three intervals, we get

$$L\frac{d\langle i_L\rangle}{dt} = \langle d_1 v_i\rangle - R_p\langle (d_1+d_2)i_L\rangle - \frac{R}{R+R_C}\langle (d_1+d_2)v_C\rangle, \tag{6.78}$$

$$C\frac{d\langle v_C\rangle}{dt} = \frac{R}{R+R_C}\langle (d_1+d_2)i_L\rangle - \frac{1}{R+R_C}\langle v_C\rangle. \tag{6.79}$$

There is a discrepancy in eqn (6.79) [6]. The average inductor current should be equal to the sum of the average capacitor charging current and the average load current. That is,

$$\begin{aligned}\langle i_L \rangle &= C\frac{d\langle v_C \rangle}{dt} + \frac{\langle v_O \rangle}{R} \\ &= \frac{\langle v_O \rangle - \langle v_C \rangle}{R_C} + \frac{\langle v_O \rangle}{R},\end{aligned}$$

which, after some manipulation, gives

$$C\frac{d\langle v_C \rangle}{dt} = \frac{R}{R+R_C}\langle i_L \rangle - \frac{1}{R+R_C}\langle v_C \rangle. \tag{6.80}$$

Comparing eqn (6.79) and (6.80), the discrepancy that is noted in the former equation is the factor $(d_1 + d_2)$ that multiplies the inductor current i_L. An effective way to remove the discrepancy is to replace i_L by $i_L/(d_1 + d_2)$ wherever the term appears in the state equations. Finally, the corrected, averaged state equations for the buck converter operating in the DCM are

$$L\frac{d\langle i_L \rangle}{dt} = \langle d_1 v_i \rangle - R_p\langle i_L \rangle - \frac{R}{R+R_C}\langle (d_1 + d_2) v_C \rangle, \tag{6.81}$$

$$C\frac{d\langle v_C \rangle}{dt} = \frac{R}{R+R_C}\langle i_L \rangle - \frac{1}{R+R_C}\langle v_C \rangle. \tag{6.82}$$

To eliminate d_2 from eqn (6.81), we note that the average value of i_L is given by

$$\langle i_L \rangle = \frac{1}{2}(d_1 + d_2)\, i_{pk} = \frac{1}{2}(d_1 + d_2)\,\frac{d_1 T_S (v_i - v_O)}{L}. \tag{6.83}$$

Therefore,

$$(d_1 + d_2) = \frac{2L\langle i_L \rangle}{d_1 T_S (v_i - v_O)}. \tag{6.84}$$

Eqn (6.84) is the duty-ratio constraint that should be substituted in eqn (6.81). Finally, we obtain the averaged state equations for the buck converter operating in the DCM:

$$L\frac{d\langle i_L \rangle}{dt} = \langle d_1 v_i \rangle - R_p\langle i_L \rangle - \frac{R}{R+R_C}\frac{2L}{T_S}\frac{\langle i_L\, v_C \rangle}{\langle d_1\,(v_i - v_O) \rangle} \tag{6.85}$$

$$C\frac{d\langle v_C \rangle}{dt} = \frac{R}{R+R_C}\langle i_L \rangle - \frac{1}{R+R_C}\langle v_C \rangle. \tag{6.86}$$

6.1.10.2 Linearization of the averaged state equations

Eqn (6.85) and (6.86) are nonlinear and they should be linearized in the neighborhood of an operating point. Putting

$$\begin{aligned}
\langle i_L\rangle &= I_L(1+\tilde{i}_L/I_L)\\
\langle v_C\rangle &= V_C(1+\tilde{v}_C/V_C)\\
\langle v_i\rangle &= V_i(1+\tilde{v}_i/V_i)\\
\langle d_1\rangle &= D_1(1+\tilde{d}_1/D_1)\\
L\frac{d\langle i_L\rangle}{dt} &= L\frac{d(I_L+\tilde{i}_L)}{dt} = L\frac{d\tilde{i}_L}{dt}\\
C\frac{d\langle v_C\rangle}{dt} &= C\frac{d(V_C+\tilde{v}_C)}{dt} = C\frac{d\tilde{v}_C}{dt}
\end{aligned}$$

and noting that the inverse of any averaged variable $\langle x\rangle$ is

$$\frac{1}{\langle x\rangle} = \frac{1}{X(1+\tilde{x}/X)} \simeq \frac{1}{X}(1-\tilde{x}/X), \tag{6.87}$$

we get after simplification,

$$\begin{aligned}
L\frac{d\tilde{i}_L}{dt} &= \left(D_1Vi - R_pI_L - \frac{R}{(R+R_C)}\frac{2LI_LV_C}{D_1T_S(V_i-V_0)}\right)\\
&+\left(V_i+\frac{R}{R+R_C}\frac{2LI_LV_C}{D_1^2T_S(V_i-V_0)}\right)\tilde{d}_1 - \left(R_p+\frac{R}{R+R_C}\frac{2LV_C}{D_1T_S(V_i-V_0)}\right)\tilde{i}_L\\
&-\frac{R}{R+R_C}\frac{2LI_L}{D_1T_S(V_i-V_0)}\tilde{v}_C - \frac{R}{R+R_C}\frac{2LI_LV_C}{D_1T_S(V_i-V_0)^2}\tilde{v}_0 \qquad (6.88)\\
C\frac{d\tilde{v}_C}{dt} &= \frac{RI_L-V_C}{R+R_C}+\frac{R}{R+R_C}\tilde{i}_L - \frac{1}{R+R_C}\tilde{v}_C. \qquad (6.89)
\end{aligned}$$

The third equation that provides the link between $\tilde{v}_C$ and $\tilde{v}_0$ is

$$C\frac{d\tilde{v}_C}{dt} = \frac{\langle v_0\rangle - \langle v_C\rangle}{R_C} = \frac{V_0-V_C}{R_C}+\frac{\tilde{v}_0-\tilde{v}_C}{R_C}. \tag{6.90}$$

Equating the constant terms on the RHS of eqn (6.88)–(6.90) to zero,

$$\begin{aligned}
D_1V_i &= R_pI_L + \frac{R}{R+R_C}\frac{2LI_LV_C}{D_1T_S(V_i-V_0)} &(6.91)\\
V_C &= RI_L &(6.92)\\
V_0 &= V_C. &(6.93)
\end{aligned}$$

The direct component of the output voltage V_0 is equal to the direct component of the capacitor voltage V_C and the direct component of the load current I_0 is equal to the direct component of the inductor current I_L.

Finally, the state equations for the buck converter operating in DCM are:

$$\begin{aligned} L\frac{d\tilde{i}_L}{dt} &= \left(V_i + \frac{R}{R+R_C}\frac{2LI_LV_C}{D_1^2T_S(V_i-V_0)}\right)\tilde{d}_1 - \left(R_p + \frac{R}{R+R_C}\frac{2LV_C}{D_1T_S(V_i-V_0)}\right)\tilde{i}_L \\ &\quad -\frac{R}{R+R_C}\frac{2LI_L}{D_1T_S(V_i-V_0)}\tilde{v}_C - \frac{R}{R+R_C}\frac{2LI_LV_C}{D_1T_S(V_i-V_0)^2}\tilde{v}_0 \qquad (6.94) \\ C\frac{d\tilde{v}_C}{dt} &= \frac{R}{R+R_C}\tilde{i}_L - \frac{1}{R+R_C}\tilde{v}_C \qquad (6.95) \end{aligned}$$

with the link between $\tilde{v}_C$ and $\tilde{v}_0$ given by

$$C\frac{d\tilde{v}_C}{dt} = \frac{\tilde{v}_0 - \tilde{v}_C}{R_C}. \qquad (6.96)$$

After applying the Laplace transform to eqn (6.94)–(6.96), transfer functions such as the control to output transfer function $\tilde{v}_0(s)/\tilde{d}_1(s)$ may be obtained. However, the equations are complex and, therefore, it is worthwhile to consider a simplification.

6.1.10.3 Transfer function of the lossless buck converter operating in DCM

The eqn (6.94)–(6.96) are very much simplified if the filter capacitor is considered lossless, i.e., the ESR of the capacitor is negligible. Putting $R_C \simeq 0$ in eqn (6.91), we get

$$\frac{D_1^2}{(2L/RT_S)} = \frac{(V_0/V_i)^2}{1 - V_0/V_i}. \qquad (6.97)$$

Further, putting $M = \frac{V_0}{V_i}$ and $K = \frac{2L}{RT_S}$ results in

$$M = \frac{2}{1 + \sqrt{1 + \frac{4K}{D_1^2}}}, \qquad (6.98)$$

which is the result obtained in eqn (6.71). (Note that eqn (6.92) and (6.93) are still valid.)

Using eqn (6.91)–(6.93) and noting that $R_C \simeq 0$, the state equations for the lossless buck converter operating in DCM are

$$L\frac{d\tilde{i}_L}{dt} = 2V_i\tilde{d}_1 - \frac{D_1 R}{M}\tilde{i}_L - \frac{D_1}{M(1-M)}\tilde{v}_C \tag{6.99}$$

$$C\frac{d\tilde{v}_C}{dt} = \tilde{i}_L - \frac{1}{R}\tilde{v}_C. \tag{6.100}$$

Note that $\tilde{v}_0 = \tilde{v}_C$ because the capacitor has been considered to be lossless.

Applying the Laplace transform to eqn (6.99) and (6.100), we get after some simplification,

$$\frac{\tilde{v}_C(s)}{\tilde{d}_1(s)} = \frac{2V_i/LC}{s^2 + \left(\frac{D_1}{M}\frac{R}{L} + \frac{1}{RC}\right)s + \frac{D_1}{M}\frac{(2-M)}{(1-M)}\frac{1}{LC}}, \tag{6.101}$$

which is the control to output transfer function. Since the converter is operating in DCM, the load is comparatively low. Besides, L is of the order of a μH and C could be a few hundred μF. The conclusion is that $\frac{D_1}{M}\frac{R}{L} \gg \frac{1}{RC}$. Therefore, the location of the poles of the transfer function will be practically unaffected if $\frac{1}{RC}$ is replaced by $\frac{(2-M)}{(1-M)}\frac{1}{RC}$. This modification allows the factorization of the denominator. Thus,

$$\begin{aligned}\frac{\tilde{v}_C(s)}{\tilde{d}_1(s)} &= \frac{2V_i/LC}{s^2 + \left(\frac{D_1}{M}\frac{R}{L} + \frac{(2-M)}{(1-M)}\frac{1}{RC}\right)s + \frac{D_1}{M}\frac{(2-M)}{(1-M)}\frac{1}{LC}} \\ &= \frac{2V_i/LC}{\left(s + \frac{D_1}{M}\frac{R}{L}\right)\left(s + \frac{(2-M)}{(1-M)}\frac{1}{RC}\right)}.\end{aligned} \tag{6.102}$$

A feedback control system may be designed to stabilize the lossless converter operating in DCM, once the control to output transfer function has been derived.

6.2 The Non-isolated Boost Converter

The second basic switching power stage topology is the boost or step-up converter. The boost converter is used when the required output voltage is higher than, and of the same polarity as, the input voltage. The input current of the boost converter is the same as the inductor current, and, therefore, it is non-pulsating. On the other hand, the diode current, which is the sum of the capacitive current and the load current, is a pulsating current. The properties of the non-isolated boost converter such as the steady-state voltage conversion ratio, the input and output current waveforms, output voltage ripple, and the converter models are studied in this section.

6.2.1 Steady-state analysis

Steady-state implies that the input voltage, load current, and duty ratio are fixed and not time-varying [1]. Figure 6.14 shows the power stage of the lossless boost converter. The inductor L and the capacitor C form the output filter, even though they are connected only when the diode conducts. The load connected to the power stage is represented by R. The switch SW is repeatedly turned on for $t_{\mathrm{ON}} = D * T_{\mathrm{S}}$ seconds,where D is the duty ratio and $T_{\mathrm{S}} = 1/f_{\mathrm{S}}$ is the switching period. The power stage may operate in the continuous conduction mode (CCM) or in the discontinuous conduction mode (DCM). For most of this section, the boost converter is considered to be functioning in CCM, though its model when operating in DCM will be studied later.

Figure 6.15 shows the inductor voltage and current waveforms. The input voltage is the direct voltage V_i and the output voltage is the direct voltage V_{O}. The voltage applied to the inductor is V_i when the switch is turned on and $V_i - V_{\mathrm{O}}$ when the switch is turned off.

The average inductor voltage over a switching period is zero. Therefore,

$$DT_{\mathrm{S}}V_i - (1 - D)T_{\mathrm{S}}(V_{\mathrm{O}} - V_i) = 0. \tag{6.103}$$

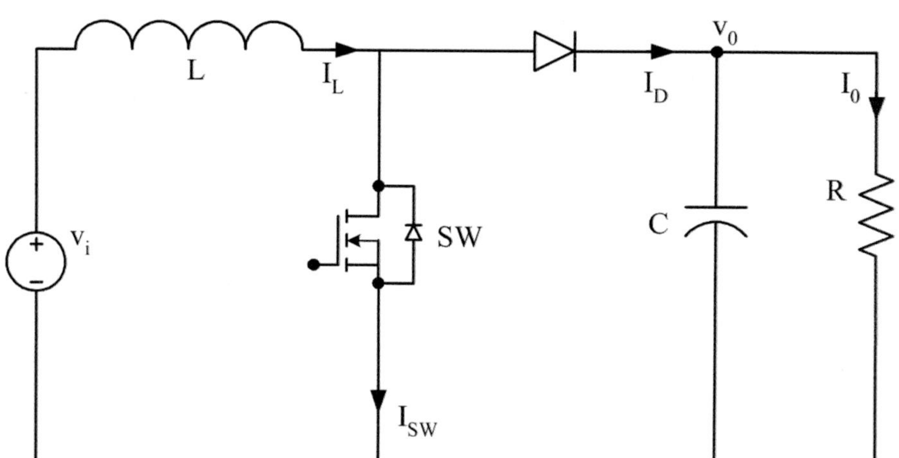

Figure 6.14 Basic boost converter: power stage.

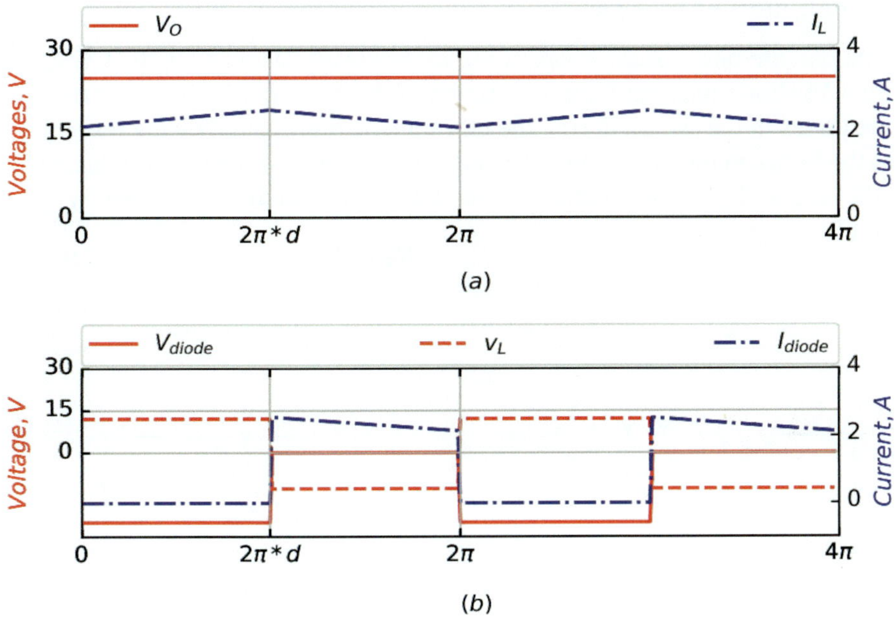

Figure 6.15 Boost converter: current and voltage waveforms.

Therefore,

$$V_0 = V_i/(1 - D) = V_i/\acute{D}. \tag{6.104}$$

The average switch current is

$$I_{SW} = I_L D T_S / T_S = D\, I_L. \tag{6.105}$$

Similarly, the average diode current is $I_D = (1 - D)I_L$. But the average diode current is given by $I_D = I_C + I_0$, where I_C, the average capacitor current, is zero. Therefore, the average load current is $I_0 = (1 - D)I_L$.

The load current is supplied by the output capacitor during the ON state of the power stage. Consequently, the peak-to-peak ripple in the output voltage is given by

$$\begin{aligned} \Delta V_0 &= \frac{1}{C}\int_0^{DT_S} I_0\, dt \\ &= \frac{I_0 D T_S}{C} \\ &= \frac{D V_0}{f_S R C}. \end{aligned} \tag{6.106}$$

6.2.2 Inclusion of losses

Assume that the power stage is operating in the steady-state and in the CCM. Figure 6.16 shows the circuit representation of the converter in the two states. Let the losses in the switch during the ON state be represented by the resistor R_{SW}, the losses in the inductor by R_L, and let the diode's voltage drop during the ON state be V_{DE}. During the ON state, the state variables are given by

$$v_L = V_i - (R_L + R_{SW})I_L \tag{6.107}$$

$$i_C = -\frac{V_O}{R}. \tag{6.108}$$

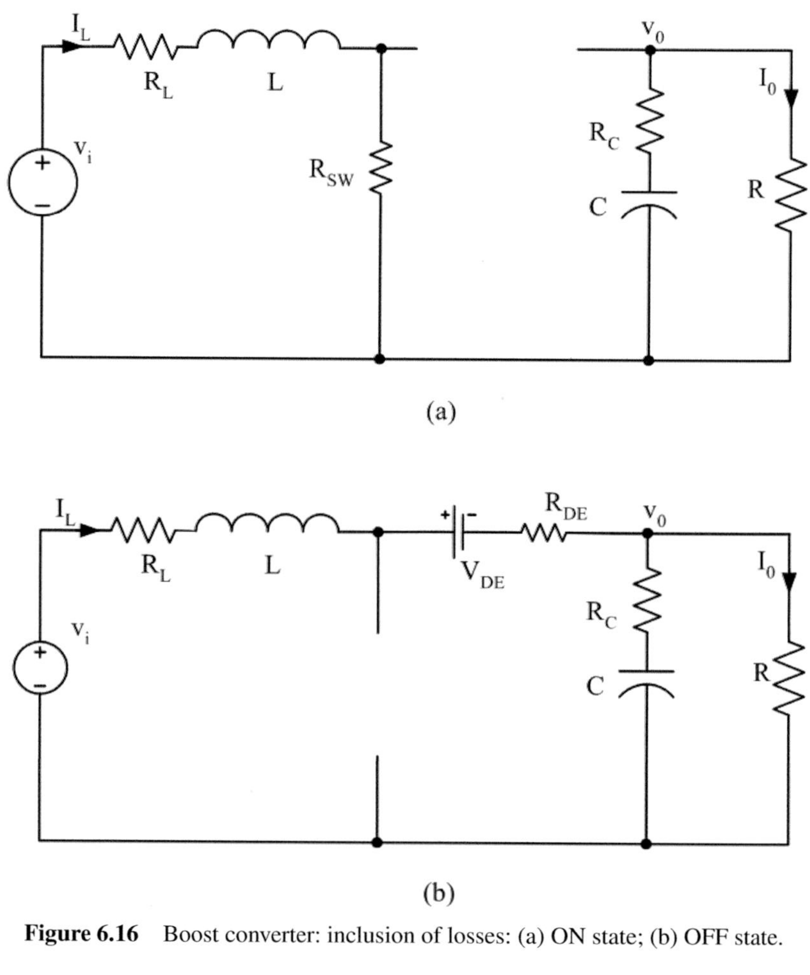

Figure 6.16 Boost converter: inclusion of losses: (a) ON state; (b) OFF state.

The corresponding equations during the OFF state are

$$v_L = V_i - V_0 - (R_L + R_{DE})I_L - V_{DE} \quad (6.109)$$

$$i_C = I_L - \frac{V_0}{R}. \quad (6.110)$$

The averages of the variables over a switching period are

$$\langle v_L \rangle = D * \{V_i - (R_L + R_{SW})I_L\} + (1 - D) * \{V_i - V_0 - (R_L + R_{DE})I_L - V_{DE}\} = 0 \quad (6.111)$$

$$\langle i_C \rangle = (1 - D)I_L - \frac{V_0}{R} = 0. \quad (6.112)$$

Simplifying eqn (6.111) and (6.112) and eliminating I_L, we get

$$V_0 = \frac{V_i - (1 - D)V_{DE}}{(1 - D) + \frac{R_L + DR_{SW} + (1-D)R_{DE}}{(1-D)R}}. \quad (6.113)$$

6.2.3 Selection of power stage components [7]

6.2.3.1 Inductors

Ferrite core inductors are the usual choice for the power stage. Current flowing through the inductor causes power dissipation due to the inductor's resistance. Power is also dissipated in the inductor's core due to the alternating component of the voltage applied across it. The maximum ratings of the inductor, usually given in its data sheet, should not be exceeded.

The inductor's current rises during the ON state and the peak-to-peak ripple is given by

$$L \frac{\Delta I_L}{DT_S} = V_i. \quad (6.114)$$

Similarly, the inductor's current falls during the OFF state and the peak-to-peak ripple is

$$L \frac{\Delta I_L}{(1 - D)T_S} = V_0 - V_i. \quad (6.115)$$

That is,

$$L = \frac{V_i(V_0 - V_i)}{V_0} * \frac{1}{f_S \Delta I_L}. \quad (6.116)$$

ΔI_L is usually 0.2–0.4 of its average value. It is usual to choose an inductance that is at least twice the value given by eqn (6.116).

6.2.3.2 Output capacitor

The function of the output capacitor is to maintain a constant output voltage by limiting the ripple in the output voltage. The ripple in the output voltage is due to (a) the discharge of the capacitor during the ON state, and (b) the voltage drop caused by the flow of the inductive ripple current in the capacitor's ESR. Usually, the ripple caused by the discharge of the capacitor is very small and may be ignored. The voltage ripple caused by the capacitor's ESR is

$$\begin{aligned}\Delta V_0 &= R_C * (I_L + \Delta I_L/2)\\ &= R_C * \left(\frac{I_0}{(1-D)} + \frac{DV_i}{2Lf_S}\right)\\ &= R_C * \left(\frac{I_0}{(1-D)} + \frac{D(1-D)V_0}{2Lf_S}\right).\end{aligned}$$

Therefore,

$$\frac{\Delta V_0}{V_0} = \frac{R_C}{(1-D)R}\left\{1 + \frac{D(1-D)^2 R}{2Lf_S}\right\}. \qquad (6.117)$$

Given that $R = 24$, $D = 0.5$, $L = 22$ μH, $f_S = 700$ kHz, R_C should be less than 218.6 mΩ if we require that $\frac{\Delta V_0}{V_0} < 0.02$. The ripple current flowing through the capacitor's ESR causes power dissipation, and consequently, temperature rise internal to the capacitor that could seriously impair the capacitor's life. As previously reviewed, aluminum electrolytic capacitors are less expensive, but their ESR is higher than that of tantalum capacitors. Besides the low ESR that is stable over the temperature range, tantalum capacitors have high ripple current capability and high capacitance per unit volume.

6.2.3.3 Power switch

The N-channel MOSFET is the usual choice for the switch in the boost converter's power stage. It conducts the inductor current when switched on and blocks the output voltage when turned off. The turn-on and turn-off times of the switch should be as small as possible to avoid excessive power dissipation during the switch transitions.

The duty ratio of the power stage is

$$D = 1 - \frac{\eta V_{i(\min)}}{V_0}, \qquad (6.118)$$

where $V_{i(\min)}$ is the minimum input voltage and η is the efficiency of the converter (which is usually estimated at 0.8). The maximum switch current is

$$I_{\mathrm{SW,max}} = \frac{I_{0,\max}}{1-D} + \frac{\Delta I_L}{2}. \tag{6.119}$$

This is the peak withstand current for the switch, the inductor, and the diode. The power dissipated by the MOSFET is given by

$$P_{\mathrm{SW}} = D\left(\frac{I_0}{1-D}\right)^2 R_{\mathrm{SW,ON}} + \frac{1}{2}V_0\left(\frac{I_0}{1-D}\right)(t_r+t_f)f_{\mathrm{S}} + Q_G V_{GS} f_{\mathrm{S}}, \tag{6.120}$$

where t_r, t_f are the MOSFET's turn-on and turn-off times, and Q_G is the MOSFET's gate-to-source capacitance charge. The power dissipation will raise the junction temperature to $T_{J,\mathrm{SW}}$ given by

$$T_{J,\mathrm{SW}} = T_A + R_{JA,\mathrm{SW}} * P_{\mathrm{SW}}, \tag{6.121}$$

where T_A is the ambient or heat-sink temperature, and $R_{JA,\mathrm{SW}}$ is the junction to heat-sink thermal resistance. The junction temperature should not exceed the value specified in the MOSFET's data sheet.

6.2.3.4 Diode

The freewheeling diode is usually a Schottky diode. Its breakdown voltage should be greater than the maximum output voltage of the power stage, with some margin added to withstand transients. The forward voltage drop is primarily responsible for the power loss in the diode, i.e.,

$$P_D = V_{DE} * I_0, \tag{6.122}$$

where V_{DE} is the forward voltage drop. The diode's junction temperature can be estimated as

$$T_{J,DE} = T_A + R_{JA,DE} * P_D. \tag{6.123}$$

6.2.4 Simulation of the boost converter

The power stage of the boost converter is shown in Figure 6.17. The transient simulation is carried out over $ncycle = 1000$ periods of the switching frequency with $n = 160$ time-steps per period. Given that the duty ratio is D, the power switch remains closed during the first $nc = round(D * n)$ time-steps of each period and it remains open for the rest of the period. The

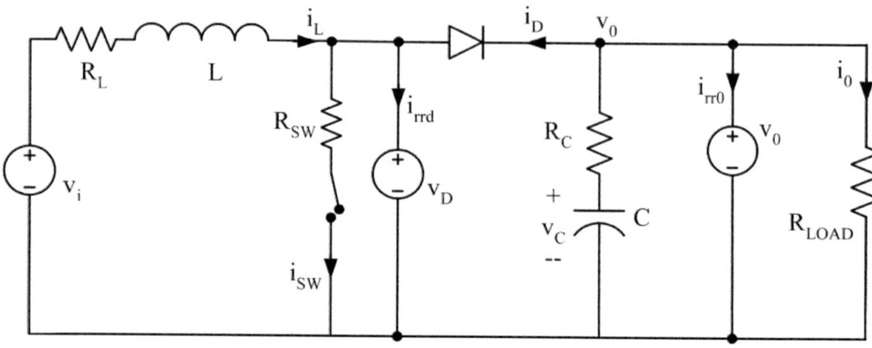

$R_{SW} = 4\text{m}\Omega$, $L = 20\mu\text{H}$, $v_i = 12\text{V}$
$R_L = 2\text{m}\Omega$, $C = 1480\mu\text{F}$, $fs = 200\text{kHz}$
$R_C = 8\text{m}\Omega$, $R_{LOAD} = 2.5\Omega$, $d = 0.3333$

Figure 6.17 Boost converter for simulation.

conductance of the switch is high ($gsw = 100$ S) when the switch is closed and it is very small ($gsw = 10^{-7}$ S) when the switch is open.

The transient equivalent circuit of the converter is shown in Figure 6.18 and the probe voltages are v_D and v_0. The past-history current sources for the inductor and the capacitor are pil and $gcp * pvc$, respectively. Applying Kirchhoff's current law at nodes 1 and 2, we get

$$grl(v_1 - v_i) + gls(v_1 - v_D) + pil = 0 \tag{6.124}$$

$$gcp * v_C + grc(v_C - v_0) = gcp * pvc. \tag{6.125}$$

The expressions for v_1 and v_C are obtained from eqn (6.124) and (6.125) as

$$v_1 = (grl * v_i + gls * v_D - pil)/(grl + gls) \tag{6.126}$$

$$v_C = (grc * v_0 + gcp * pvc)/(grc + gcp). \tag{6.127}$$

The error currents i_{rrd} and i_{rr0} flowing into the probe voltages are calculated as shown below:

$$isw = gsw * v_D \tag{6.128}$$

$$ils = grl(v_i - v_1) \tag{6.129}$$

$$i_D, g_D = fdiode(v_0 - v_D) \tag{6.130}$$

$$icp = grc(v_0 - v_C) \tag{6.131}$$

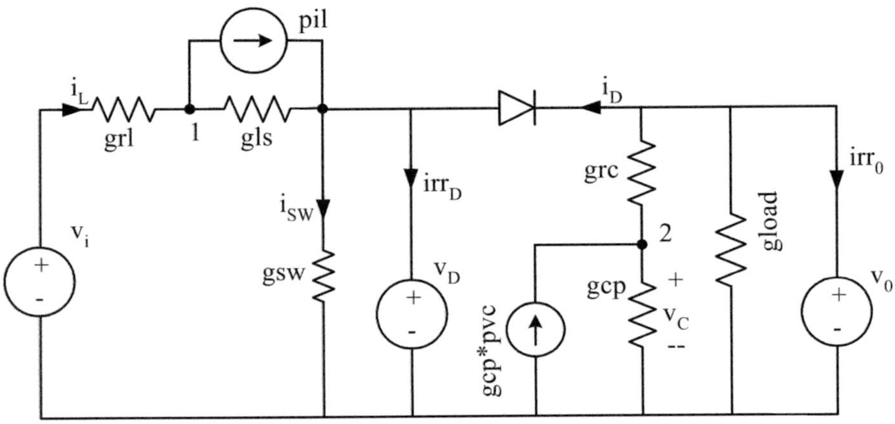

Figure 6.18 Boost converter: transient equivalent circuit.

$$iload = gload * v_0 \tag{6.132}$$
$$i_{rrd} = ils - isw + i_D \tag{6.133}$$
$$i_{rr0} = -(i_D + icp + iload). \tag{6.134}$$

The small-signal equivalent circuit is shown in Figure 6.19 and the equations for computing the corrections are

$$gser = grl * gls/(grl + gls)$$
$$gpar = gload + grc * gcp/(grc + gcp)$$

$$\begin{bmatrix} g_D + gsw + gser & -g_D \\ -g_D & g_D + gpar \end{bmatrix} \begin{bmatrix} \delta v_D \\ \delta v_0 \end{bmatrix} = \begin{bmatrix} irr_d \\ irr_0 \end{bmatrix}. \tag{6.135}$$

Figure 6.20, plotted by Program 6.20, shows the results of the simulation. Whereas Figure 6.20 (a) shows the waveforms of the inductor current and the output voltage for the entire simulation period, Figure 6.20 (b) shows the waveforms for the last two periods.

6.2.5 Boost converter model in CCM

The boost converter passes through two states during each switching period. During the jth period, the ON state stretches from $(j-1)T_S$ to $(j-1+d)T_S$ with the power switch closed. The switch is turned off from $(j-1+d)T_S$ to jT_S. We proceed to develop the state equations for the two intervals and obtain the averaged equations, using the nomenclature of Figure 6.15 [8].

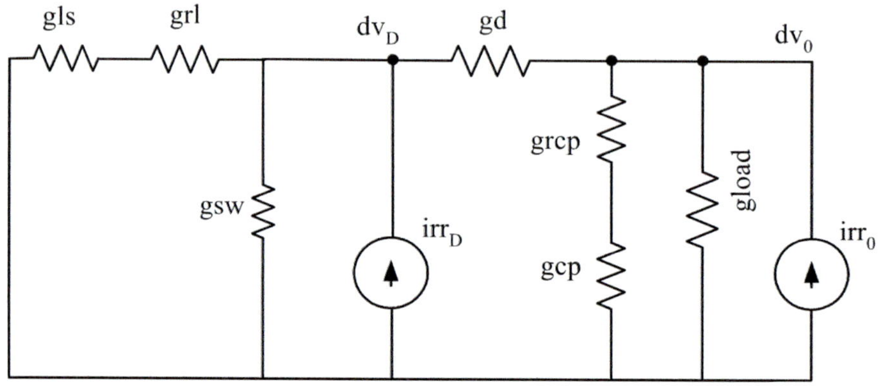

Figure 6.19 Boost converter: small-signal equivalent circuit.

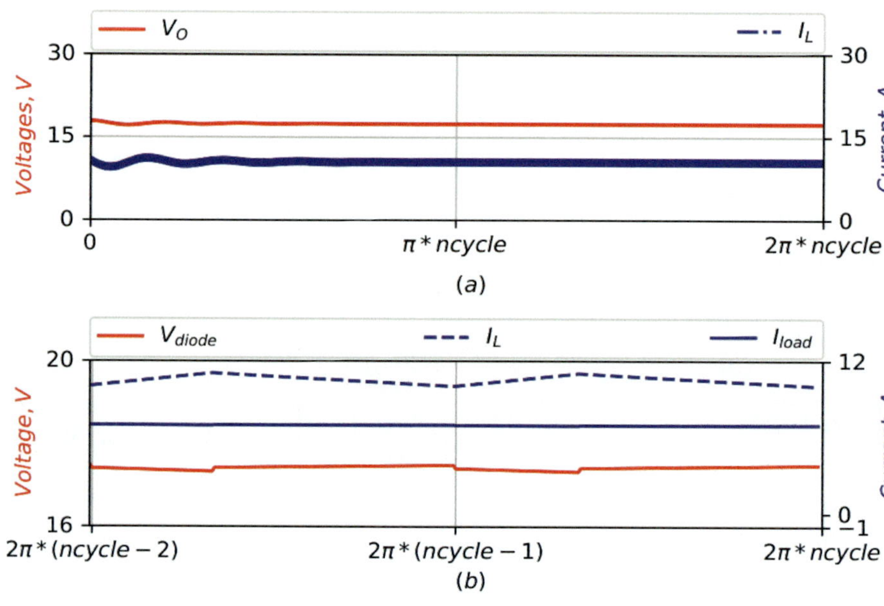

Figure 6.20 Boost converter: response waveforms.

During the ON state, the state equations are

$$L\frac{di_L}{dt} = v_i \tag{6.136}$$

$$C\frac{dv_C}{dt} = -\frac{1}{R + R_C} * v_C \tag{6.137}$$

$$v_0 = \frac{R}{R+R_C} * v_C. \tag{6.138}$$

The state equations during the OFF state are

$$L\frac{di_L}{dt} = v_i - \left(R_p * i_L + \frac{R}{R+R_C} * v_C\right) \tag{6.139}$$

$$C\frac{dv_C}{dt} = -\frac{1}{R+R_C} * v_C + R_p * i_L \tag{6.140}$$

$$v_0 = \frac{R}{R+R_C} * v_C + R_p * i_L, \tag{6.141}$$

where $R_p = R * R_C/(R+R_C)$. Putting $\acute{d} = (1-d)$ and applying state-space averaging, we get

$$L\frac{d\langle i_L\rangle}{dt} = \langle v_i\rangle - \langle\acute{d}\rangle\left(R_p * \langle i_L\rangle + \frac{R}{R+R_C} * \langle v_C\rangle\right) \tag{6.142}$$

$$C\frac{d\langle v_C\rangle}{dt} = -\frac{1}{R+R_C} * \langle v_C\rangle + R_p * \langle\acute{d}\rangle\langle i_L\rangle \tag{6.143}$$

$$\langle v_0\rangle = \frac{R}{R+R_C} * \langle v_C\rangle + R_p * \langle\acute{d}\rangle\langle i_L\rangle. \tag{6.144}$$

To linearize eqn (6.142)–(6.144), we put $\langle i_L\rangle = I_L + \tilde{i}_L$, $\langle v_C\rangle = V_C + \tilde{v}_C$, $\langle d\rangle = D + \tilde{d}$, and $\langle\acute{d}\rangle = \acute{D} - \tilde{d}$ and obtain

$$L\frac{d\tilde{i}_L}{dt} = \left\{V_i - \acute{D}\left(R_pI_L + \frac{R}{R+R_C}V_C\right)\right\} + \tilde{v}_i - \acute{D}R_p\tilde{i}_L - \acute{D}\frac{R}{R+R_C}\tilde{v}_C + \left(R_pI_L + \frac{RV_C}{R+R_C}\right)\tilde{d} \tag{6.145}$$

$$C\frac{d\tilde{v}_C}{dt} = \frac{\acute{D}RI_L - V_C}{R+R_C} + \frac{\acute{D}R}{R+R_C}i_L - \frac{1}{R+R_C}\tilde{v} - \frac{R\,I_L}{R+R_C}\tilde{d} \tag{6.146}$$

$$\tilde{v}_0 = \left(\acute{D}R_pI_L + \frac{R}{R+R_C}V_C - V_0\right) + \acute{D}R_p\tilde{i}_L + \frac{R}{R+R_C}\tilde{v}_C - R_pI_L\tilde{d}. \tag{6.147}$$

Taking into account that the variables on the LHS of eqn (6.145)–(6.147) are periodic, time-varying quantities, the constant terms on the RHS are set to zero to obtain the following DC equations:

$$V_i = V_0\left(1 - D * \frac{R}{R+R_C}\right) \tag{6.148}$$

$$V_C = V_0 \tag{6.149}$$

$$I_0 = \acute{D} * I_L. \tag{6.150}$$

Finally, the state equations for small variations in the neighborhood of the operating point are

$$L\frac{d\tilde{i}_L}{dt} = \tilde{v}_i - \acute{D}R_p\tilde{i}_L - \acute{D}\frac{R}{R+R_C}\tilde{v}_C + \frac{V_i}{\acute{D}}\tilde{d} \tag{6.151}$$

$$C\frac{d\tilde{v}_C}{dt} = \acute{D}\frac{R}{R+R_C}\tilde{i}_L - \frac{1}{R+R_C}\tilde{v}_C - \frac{RI_L}{R+R_C}\tilde{d} \tag{6.152}$$

$$\tilde{v}_0 = \acute{D}R_p\tilde{i}_L + \frac{R}{R+R_C}\tilde{v}_C - R_pI_L\tilde{d}. \tag{6.153}$$

Assuming that the variations in the input voltage are negligible, i.e., $\tilde{v}_i \simeq 0$, and applying the Laplace transform to eqn (6.151)–(6.153), we get

$$sL\tilde{i}(s) = -\acute{D}R_p\tilde{i}(s) - \acute{D}\frac{R}{R+R_C}\tilde{v}_C(s) + \frac{V_i}{\acute{D}}\tilde{d}(s) \tag{6.154}$$

$$sCR_C\tilde{v}_C(s) = \acute{D}R_p\tilde{i}(s) - \frac{R_C}{R+R_C}\tilde{v}_C(s) - R_pI_L\tilde{d}(s) \tag{6.155}$$

$$\tilde{v}_0(s) = \acute{D}R_p\tilde{i}(s) + \frac{R}{R+R_C}\tilde{v}_C(s) - R_pI_L\tilde{d}(s). \tag{6.156}$$

Considering that $R_C \ll R$ and eliminating $\tilde{i}_L(s)$ from eqn (6.154)–(6.156), we get

$$\frac{\tilde{v}_C(s)}{\tilde{d}(s)} = \frac{V_i(1 - \frac{sL}{\acute{D}^2R})}{s^2LC + s\frac{L}{R} + \acute{D}^2}. \tag{6.157}$$

From eqn (6.155) and (6.156), we get

$$\frac{\tilde{v}_0(s)}{\tilde{v}_C(s)} = (1 + sCR_C). \tag{6.158}$$

Therefore,

$$G_{vd}(s) = \frac{\tilde{v}_0(s)}{\tilde{d}(s)} = \frac{V_i}{\acute{D}^2}\cdot\frac{(1+sCR_C)\left(1 - \frac{sL}{\acute{D}^2R}\right)}{s^2\frac{LC}{\acute{D}^2} + s\frac{L}{R\acute{D}^2} + 1}. \tag{6.159}$$

The function $G_{vd}(s)$ is the control to output transfer function.

The control to output transfer function has two poles both approximately located at the resonant frequency $\omega_{LC} = \acute{D}/\sqrt{LC}$. Note that the location of the double pole depends on $\acute{D}$ and, therefore, shifts with the operating point. The transfer function also has two zeros, one in the LHP at $\omega_{\mathrm{ESR}} = 1/CR_C$ and the other in the RHP at $\omega_{RHP} = R\acute{D}^2/L$. The LHP zero is due to the ESR of the capacitor. The location of the RHP zero depends on the load, the inductance, and the duty ratio. It increases the gain slope of $G_{vd}(s)$ by 20 dB/decade for $\omega > \omega_{RHP}$, but brings down the phase by $90°$. The existence of the RHP zero is due to the power switch and the diode being connected between the inductor and the capacitor of the low-pass filter.

The RHP zero does not allow the control loop to respond immediately to deviations in the output voltage. Consider a drop in the output voltage. The response of the feedback control will be to increase the duty ratio. This would cause the output voltage to increase further due to the discharge of the capacitor over a longer interval. To stabilize the power stage properly, the bandwidth of the closed loop should be limited to a frequency that is considerably less than ω_{RHP}.

6.2.6 Feedback control of the boost converter in CCM

6.2.6.1 Voltage mode control

Figure 6.21 shows the boost converter and the typical system for voltage mode control [9]. The parameters of the converter are:

$$R_{\mathrm{S}} = 2\ \mathrm{m}\Omega\ ,\ L = 20\ \mu\mathrm{H}$$
$$R_C = 8\ \mathrm{m}\Omega\ ,\ C = 1480\ \mu\mathrm{F}$$

$$\begin{aligned} R &= 8\ \Omega, \mathrm{D} = 0.7222 \\ V_i &= 5\ \mathrm{V}, \mathrm{V_O} = 18\ \mathrm{V} \\ V_{RAMP} &= 3.4616\ \mathrm{V}, \mathrm{V_{REF}} = 2.5\ \mathrm{V} \\ R_{f1} &= 620\ \mathrm{k}\Omega, \mathrm{R_{f2}} = 100\ \mathrm{k}\Omega \\ f_{\mathrm{S}} &= 200\ \mathrm{kHz}. \end{aligned}$$

A fraction $\mu = R_{f2}/(R_{f1} + R_{f2})$ of the output voltage is compared to a reference voltage V_{REF} and the difference $v_R = -\mu\tilde{v}_{\mathrm{O}}$ is the input for the Type III compensator. The PWM pulse generator compares the output

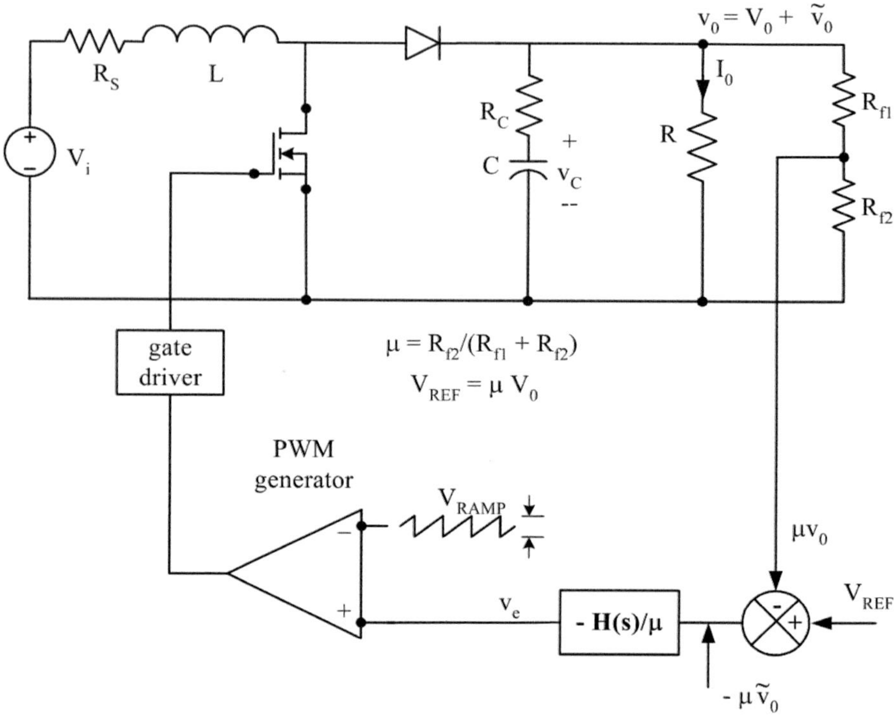

Figure 6.21 Boost converter with voltage mode control.

v_e of the compensator to a periodic, sawtooth voltage having its peak-to-peak amplitude equal to V_{RAMP} and its frequency equal to the switching frequency. (Note that $V_{\text{REF}} = D * V_{RAMP}$.)

The frequencies associated with the control to output transfer function are:

$$\begin{aligned}
\omega_{LC} &= \acute{D}/\sqrt{LC} = 1614.6 \text{ rads/sec} \\
\omega_{\text{ESR}} &= 1/(CR_C) = 84460 \text{ rads/sec} \\
\omega_{RHP} &= R\acute{D}^2/L = 23148 \text{ rads/sec} \\
\omega_{\text{S}} &= \text{switching frequency} = 200 \text{ kHz.}
\end{aligned}$$

The zero-crossing frequency ω_0 of the loop gain $G_{vd}(s)H(s)$ is chosen such that $2\omega_{LC} < \omega_0 < \omega_{RHP}/5$. The proper placement of ω_0 usually requires a large output capacitor in voltage mode control. Note that a large inductor cannot be used in the boost converter because ω_{RHP} would be pushed to a low frequency. A reasonable choice is $\omega_0 = 4000$ rads/sec.

The gain and phase of $G_{vd}(s)$ at $\omega = \omega_0$ are required for designing the compensator:

$$\begin{aligned}
|G_{vd}(j\omega_0)| &= \frac{64.8\,|1 + j(\omega_0/\omega_{\mathrm{ESR}})| * |1 - j(\omega_0/\omega_{RHP})|}{|1 - (\omega_0/\omega_{LC})^2 + j\omega_0/\omega_{RHP}|} \\
&= \frac{64.8 * 1.0011 * 1.0148}{5.1346} \\
&= 12.82 \\
\angle G_{vd}(j\omega_0) &= tan^{-1}(\omega_0/\omega_{\mathrm{ESR}}) - tan^{-1}(\omega_0/\omega_{RHP}) - \\
&\qquad tan^{-1}\{(\omega_0/\omega_{LC})/(1 - (\omega_0/\omega_{LC})^2)\} \\
&= 2.71^\circ - 9.80^\circ - 178.07^\circ \\
&= -185.2^\circ.
\end{aligned}$$

Aiming for a phase margin of 60° and a crossover frequency of 4000 rads/sec, we require a compensator $H(s)$ having

$$\begin{aligned}
|H(j\omega_0)| &= 1/12.82 \\
\angle H(j\omega_0) &= 65.2^\circ.
\end{aligned}$$

A Type III compensator has two zeros and two poles in the LHP, besides a pole at the origin. There are several ways of choosing their locations to satisfy the gain and phase constraints. A simple procedure (known as the k factor technique [12]) places both zeros at a frequency ω_z and both poles at a frequency ω_p. Thus,

$$H(s) = \frac{(1 + s/\omega_z)^2}{(s/\omega_{p0}) * (1 + s/\omega_p)^2} \tag{6.160}$$

is a good compensator for the task. The phase angle of the compensator at the crossover frequency is given by

$$\angle H(j\omega_0) = 2(tan^{-1}(\omega_0/\omega_z) - tan^{-1}(\omega_0/\omega_z)) - 90^\circ = 65.2^\circ. \tag{6.161}$$

Putting $\omega_0/\omega_z = k$ and $\omega_0/\omega_p = 1/k$, we get

$$\begin{aligned}
tan^{-1}k - tan^{-1}(1/k) &= 77.6^\circ \\
\text{But} \qquad tan^{-1}k + tan^{-1}(1/k) &= 90^\circ.
\end{aligned}$$

Therefore, $k = 9.19$. It is convenient to put $k = 10$, which will result in a slightly higher phase margin. Finally, we get $\omega_z = 400$ rads/sec and $\omega_p = 40,000$ rads/sec.

The compensator's gain constraint is met by choosing an appropriate value for ω_{p0}. We have

$$\begin{aligned}|H(j\omega_0| &= \frac{|1+j10|}{|j\omega_0/\omega_{p0}| * |1+j0.1|^2} \\ &= 1/12.82.\end{aligned}$$

Therefore, $\omega_{p0} = 3.1201$. Finally,

$$H(s) = 31,201\frac{(s+400)^2}{s(s+40,000)^2} \tag{6.162}$$

It is well known that the transfer function representation of a linear system may be converted to the state-space representation. The equivalent state-space representation of $H(s)$ is

$$\frac{d}{dt}\begin{bmatrix} x_2 \\ x_1 \\ x_0 \end{bmatrix} = \begin{bmatrix} -b_1 & -b_2 & 0 \\ 1 & 0 & 0 \\ 0 & 1 & 0 \end{bmatrix}\begin{bmatrix} x_2 \\ x_1 \\ x_0 \end{bmatrix} + \begin{bmatrix} 1 \\ 0 \\ 0 \end{bmatrix} * \tilde{v}_R \tag{6.163}$$

$$v_e = K * (x_2 + a_1 x_1 + a_2 x_0), \tag{6.164}$$

where $K = 31,201$, $a_1 = 2\omega_z$, $a_2 = \omega_z^2$, $b_1 = 2\omega_p$, and $b_2 = \omega_p^2$. The state equations are numerically integrated as

$$\frac{x_2 - px2}{\Delta t} = \tilde{v}_R - b_1 x_2 - b_2 x_1 \tag{6.165}$$

$$\frac{x_1 - px1}{\Delta t} = x_2 \tag{6.166}$$

$$\frac{x_0 - px0}{\Delta t} = x_1, \tag{6.167}$$

where $px2$, $px1$, and $px0$ are past-history values. (The simulation is usually started by putting the past-history values to zero.) Eqn (6.165)–(6.167) may be written as

$$x_2 = (px2 - b_2\Delta t + \Delta t * \tilde{v}_R)/(1 + b_1\Delta t + b_2\Delta t^2) \tag{6.168}$$

$$x_1 = px1 + \Delta t * x_2 \tag{6.169}$$

$$x_0 = px0 + \Delta t * x_1. \tag{6.170}$$

Figure 6.22 shows the Bode diagrams for $G_{vd}(j\omega)$, $H(j\omega)$ and the loop gain $H(j\omega)G_{vd}(j\omega)$ plotted by Program 6.22. The zero crossover frequency

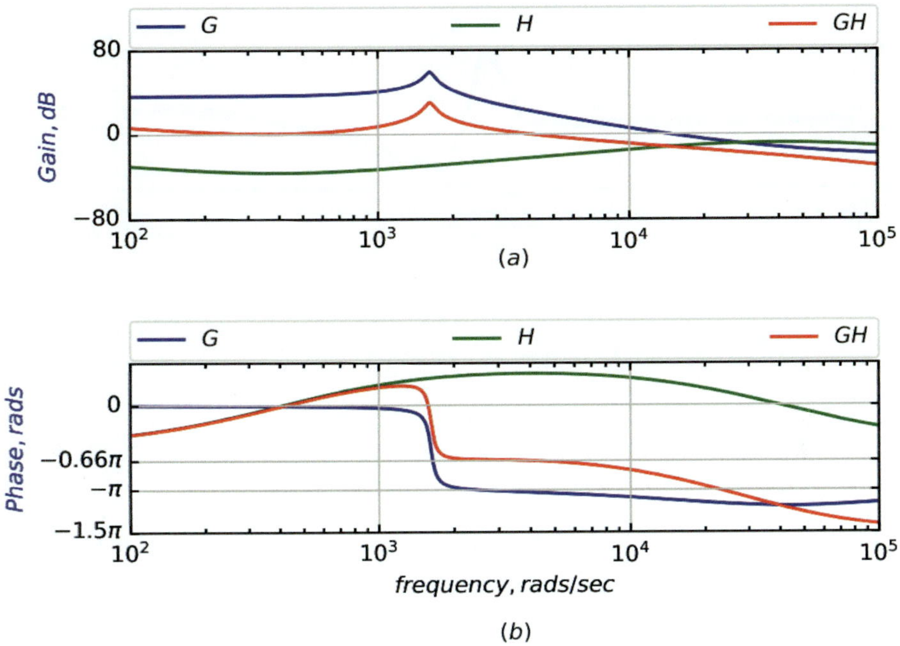

Figure 6.22 Boost converter, voltage mode control: Bode plots.

of the loop gain is 3986 rads/sec and the phase margin is $62°$. Figure 6.23 (plotted by Program 6.23) shows the output voltage and the inductor current of the boost converter, applying voltage mode control. The simulation was extended over 1000 periods of the switching frequency. Disturbances were introduced in the middle of the simulation. Between the 450th and the 500th periods, the load was decreased by 20% and the input voltage was increased by 20% of their nominal values. For the next 50 periods, the load was increased by 20% and the input voltage was decreased by 20% below their nominal values. Feedback control is able to maintain the output voltage constant.

6.2.6.2 Current mode control

Another widely used control scheme is the current mode control (CMC) comprising two nested control loops [10]. The inner loop controls the inductor current and its reference i_e is provided by the outer loop that controls the output voltage v_0. A current sensing resistor R_{SN} is connected in series with the power switch to monitor the inductor current.

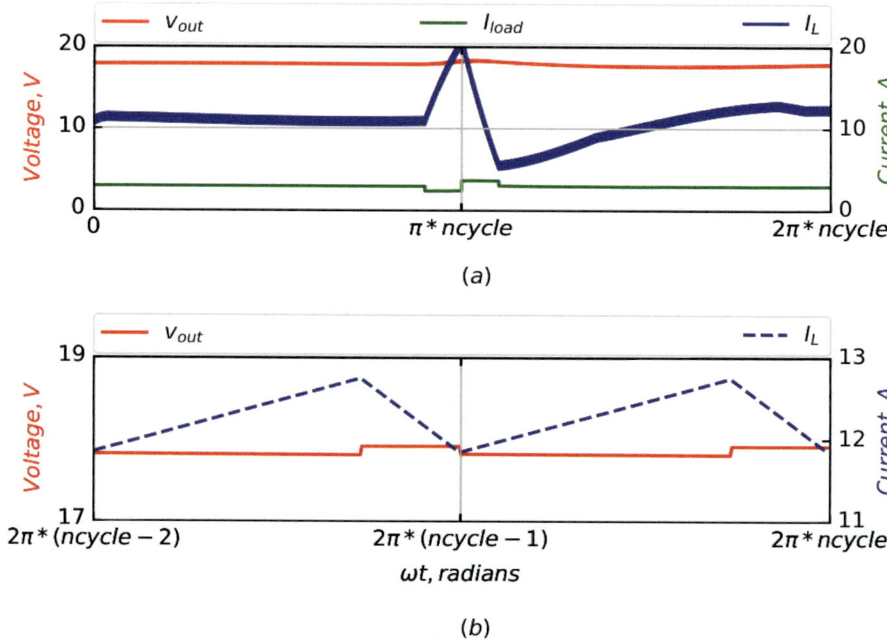

Figure 6.23 Boost converter, voltage mode control: transient response.

The block diagram of current mode control is shown in Figure 6.24. A fraction of the output voltage is compared to a reference voltage V_{REF} and the difference is the error signal $\tilde{v}_R$, which is the input of the compensator. The output of the compensator $\tilde{v}_e$ is converted to an equivalent current reference $\tilde{i}_e = \tilde{v}_e / R_{\mathrm{SN}}$. The outer loop is closed through the control current to output transfer function G_{vi}. In the inner loop, the switch current (which is equal to the inductor current during the ON state) is compared to the reference current $\tilde{i}_e$ and the error is converted to the variation $\tilde{d}$ in the duty ratio.

6.2.6.3 Modeling the current mode controlled, boost converter

At the start of each period, the switch is turned on and the switch current is compared to the control current i_e. A compensating ramp current of slope $-m_C$ is usually added to i_e [11]. Without the compensating ramp, the current loop may generate sub-harmonic oscillations (as will be demonstrated later in the simulation). Figure 6.25 shows the current waveforms during an ON state. (A simplifying assumption has been made in the inductor current waveform,

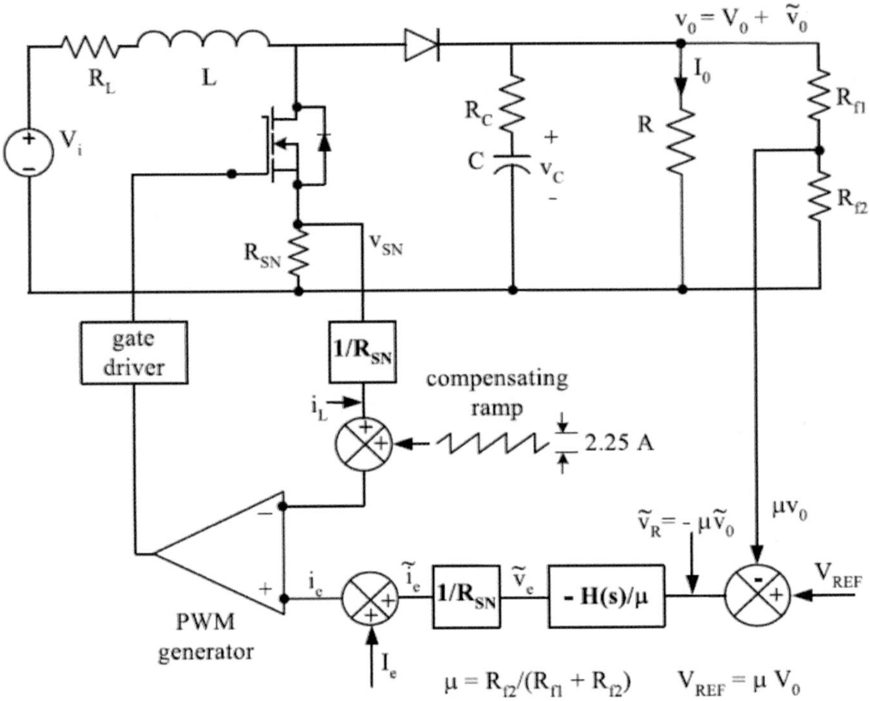

Figure 6.24 Boost converter: peak current mode control.

i.e., $i_L(0) = i_L(T_S)$. Without this assumption, the transfer functions become quite complex.)

From Figure 6.25, the averaged current is given by

$$\begin{aligned}\langle i_L\rangle &= \left((\langle i_e\rangle - m_C)T_S\langle d\rangle) - \frac{M_1}{2}T_S\langle d\rangle\right)\\ &= \langle i_e\rangle - \left(m_C + \frac{M_1}{2})T_S\langle d\rangle\right)\\ &= \langle i_e\rangle - I_M\langle d\rangle.\end{aligned} \tag{6.171}$$

Putting $i_L = I_L + \tilde{i}_L$, $i_e = I_e + \tilde{i}_e$ and $d = D + \tilde{d}$, we get

$$I_L + \tilde{i}_L = I_e + \tilde{i}_e - I_M(D + \tilde{d}). \tag{6.172}$$

The DC equation is given by

$$I_L = I_e - I_M D. \tag{6.173}$$

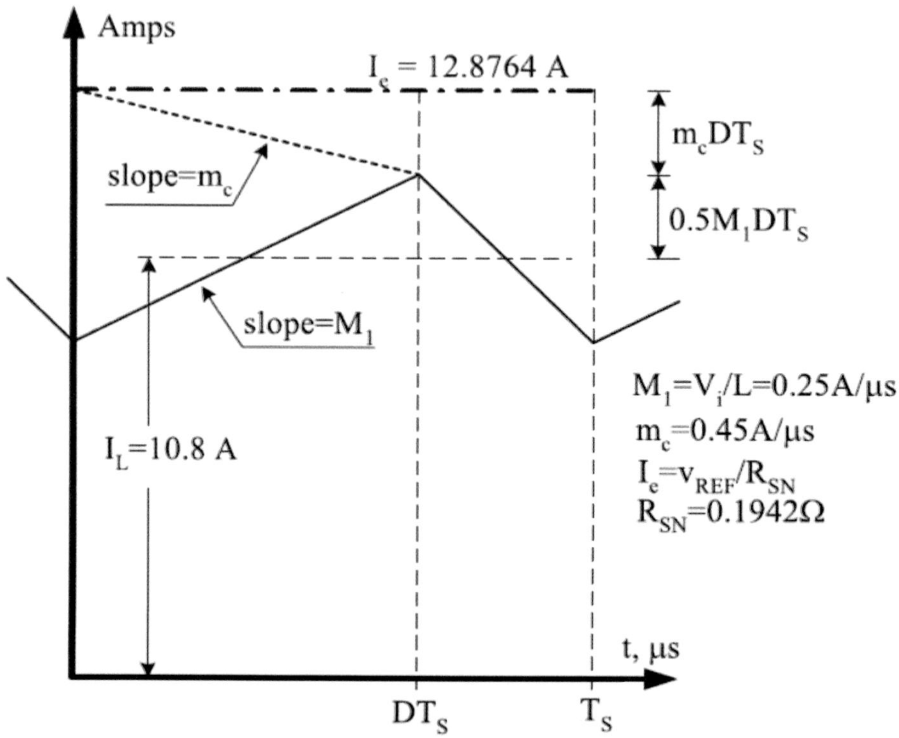

Figure 6.25 Boost converter, current mode control: ON state current waveforms.

The periodic, time-varying components satisfy the equation

$$\tilde{i}_L = \tilde{i}_e - I_M\tilde{d}. \tag{6.174}$$

The state equations and the control to output transfer function were obtained in the previous subsection and, for the sake of convenience, these equations are repeated below:

$$\begin{aligned}
(sL + \acute{D}R_p)\tilde{i}_L(s) + \acute{D}\frac{R}{R+R_C}\tilde{v}_C(s) &= \frac{V_i}{\acute{D}}\tilde{d}(s)\\
-\acute{D}\frac{R}{R+R_C}\tilde{i}_L(s) + \left(sC + \frac{1}{R+R_C}\right)\tilde{v}_C &= -\frac{RI_L}{R+R_C}\tilde{d}(s)\\
G_{vd}(s) = \frac{\tilde{v}_0(s)}{\tilde{d}(s)} &= \frac{(V_i/\acute{D}^2)(1+sCR_C)(1-sL/\acute{D}^2R)}{s^2LC/\acute{D}^2 + sL/\acute{D}^2R + 1}.
\end{aligned}$$

Eliminating $\tilde{v}_C(s)$ in the state equations, we get

$$\tilde{i}_L(s) = \frac{(V_i/\acute{D})(sC + 2/R)}{s^2LC + sL/R + \acute{D}^2}\tilde{d}(s). \tag{6.175}$$

From eqn (6.174) and (6.175), we get

$$\begin{aligned}\tilde{i}_e(s) &= \left(\frac{(V_i/\acute{D})(sC + 2/R)}{s^2LC + sL/R + \acute{D}^2} + I_M\right)\tilde{d}(s) \\ &= \frac{s^2I_MLC + s(I_ML/R + V_iC/\acute{D}) + (I_M\acute{D}^2 + 2V_i/(\acute{D}R))}{s^2LC + sL/R + \acute{D}^2}\tilde{d}(s).\end{aligned} \tag{6.176}$$

Using the control to output transfer function $G_{vd}(s)$ and eqn (6.176), we get finally

$$\begin{aligned}G_{vi}(s) &= \frac{\tilde{v}_O(s)}{\tilde{i}_e(s)} \\ &= \frac{V_i(1 + sCR_C)(1 - sL/(\acute{D}^2R))}{s^2I_MLC + s(I_ML/R + V_iC/\acute{D}) + (I_M\acute{D}^2 + 2V_i/(\acute{D}R))}.\end{aligned} \tag{6.177}$$

Consider a boost converter with the following parameters: $f_S = 200$ kHz, $V_i = 5$ V, $V_O = 18$ V, $L = 20$ µH, $C = 480$ µF, $R_C = 8$ mΩ, $R = 6$ Ω, $R_{SN} = 0.1942$ Ω, $M_1 = V_i/L = 0.25$ A/µs. The downslope of the inductor current during the OFF state is

$$M_2 = \frac{V_O - V_i}{L} = 0.65 \text{ A/µs}. \tag{6.178}$$

The downslope of the compensating ramp m_C should at least be greater than $M_2/2$. We choose $m_C = 0.45$ A/µs. Then

$$I_M = (m_C + M_1/2)T_S = 2.875 \text{ A}. \tag{6.179}$$

The output of the exterior loop compensator at the operating point is $V_e = V_{REF} = 2.5$ V. Therefore, the current reference at the operating point is $I_e = V_e/R_{SN} = 12.8764$ A. The average inductor current at the operating point is $I_L = 10.8$ A. Note that $(I_e - I_L)/I_M = 0.7222 = D$.

Substituting the values of the nominal currents and the parameter values listed above in the expression for the transfer function $G_{vi}(s)$, its poles are located in the LHP at $115\ Hz$ and 51 kHz. Since one of the poles is located far to the left of the origin, the frequency response of $G_{vi}(s)$ resembles that of a single pole plant though the RHP zero is still present.

6.2.6.4 Compensator design in current mode control

Let the compensator's transfer function be $H(s)$. Then, the loop gain is given by $H(s)G_{vi(s)}$. The zero-crossing frequency ω_0 of the loop gain is chosen to be 3800 rads/sec, which is less than $\omega_{RHP}/5$. From the Bode plots of $G_{vi}(s)$, we get

$$\begin{aligned} 20\log_{10}|G_{vi}(j\omega_0)| &= -2.0975\ dB \\ \angle G_{vi}(j\omega_0) &= -88.33^\circ. \end{aligned}$$

That is, $|G_{vi}(j\omega_0)| = 0.7855$. Specifying the phase margin of the loop gain to be 75°, we require that

$$\begin{aligned} |H(j\omega_0)| &= 1.273 \\ \angle H(j\omega_0) &= -16.67^\circ. \end{aligned}$$

Choose the Type II compensator of the form

$$H(s) = \frac{(1 + s/\omega_z)}{(s/\omega_{po})(1 + s/\omega_p)}. \tag{6.180}$$

Using the k factor technique described in the previous section, we get $\omega_{po} = 708.73$ rads/sec, $\omega_z = 556.73$ rads/sec, and $\omega_p = 25,937$ rads/sec. Therefore,

$$H(s) = \frac{v_e(s)}{v_R(s)} = 33,018\frac{(s + 556.73)}{s(s + 25937)}, \tag{6.181}$$

Figure 6.26 shows the frequency responses of $H(s)$, $G_{vi}(s)$ and the loop gain $H(s)G_{vi}(s)$. The zero-crossing frequency of the loop gain is $3780\ rad/sec$ and the phase margin is nearly 75°.

Eqn (6.181) is equivalent to the state equations

$$\frac{dx_1}{dt} = x_2 \tag{6.182}$$

$$\frac{dx_2}{dt} + \omega_p x_2 = 33,018 * \tilde{v}_R \tag{6.183}$$

$$\tilde{v}_e = x_2 + \omega_z x_1. \tag{6.184}$$

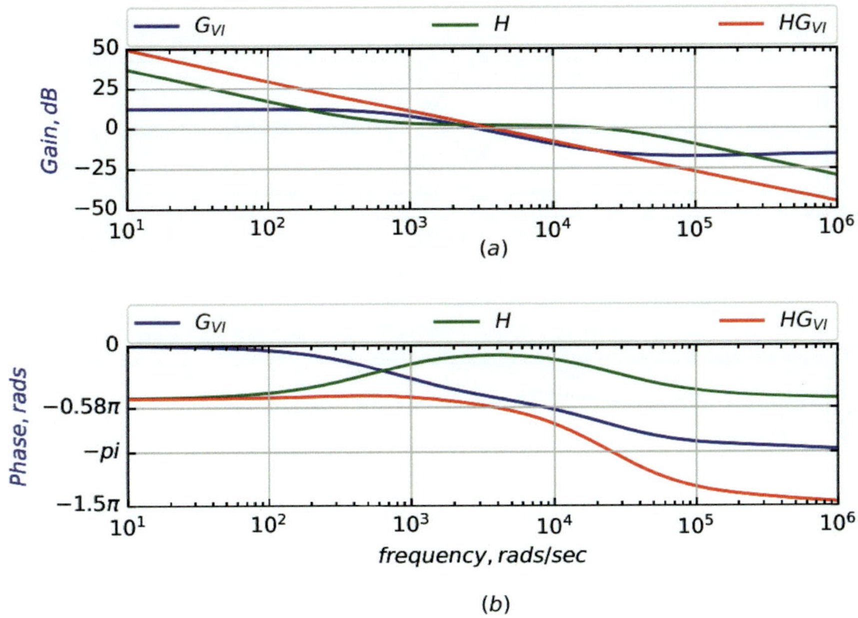

Figure 6.26 Boost converter, current mode control: Bode plots.

Eqn (6.182)–(6.184) are numerically integrated in a simulation program.

Figure 6.27 shows the responses of the boost converter with current mode control. The simulation was extended over 1000 periods with 256 time-steps per period. Disturbances were introduced over an interval of 100 periods stretching on either sides of the 500th period. Over the interval comprising the 450th period to the 500th period, the input voltage was increased by 30% above the nominal value and the load was decreased by 30%. Over the next 50 periods, the input voltage was reduced by 30% and the load was increased by 30%. The output voltage is controlled well at 18 V. Figure 6.28 shows the second harmonic oscillations in the inductor current when the compensating ramp is removed.

6.2.7 Boost converter in DCM

During the jth period, the boost converter is in the ON state during the interval $jT_S \leq t \leq (j + d_1)T_S$, in the OFF state during the interval $(j + d_1)T_S \leq t \leq (j + d_1 + d_2)T_S$, and, finally, in the IDLE state during the interval $(j + d_1 + d_2)T_S \leq t \leq (j + 1)T_S$. The switch conducts during the ON state

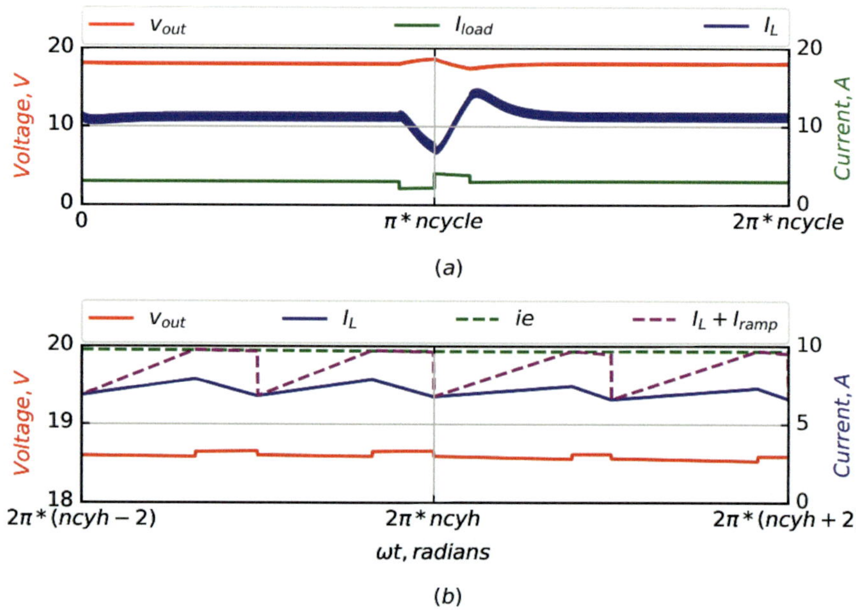

Figure 6.27 Boost converter, current mode control: transient response.

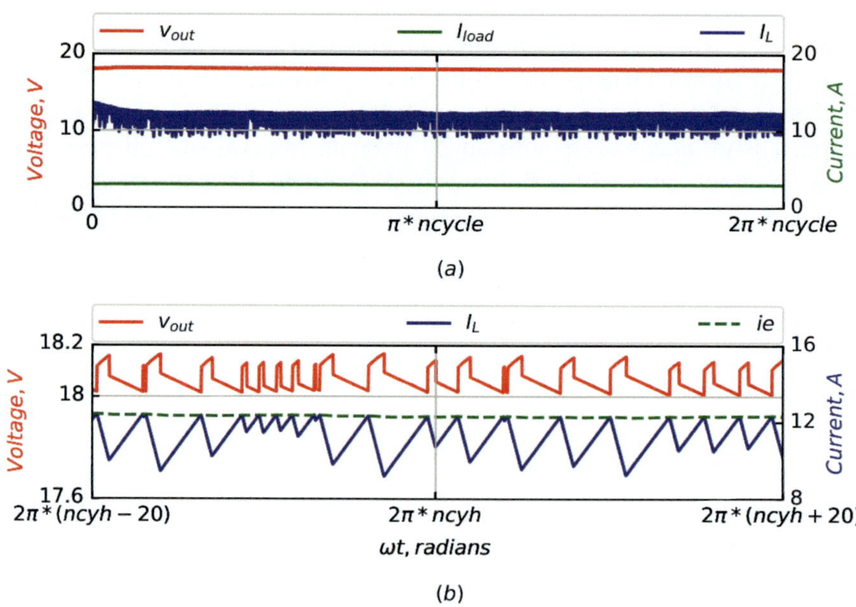

Figure 6.28 Absence of compensating ramp: second harmonic generation.

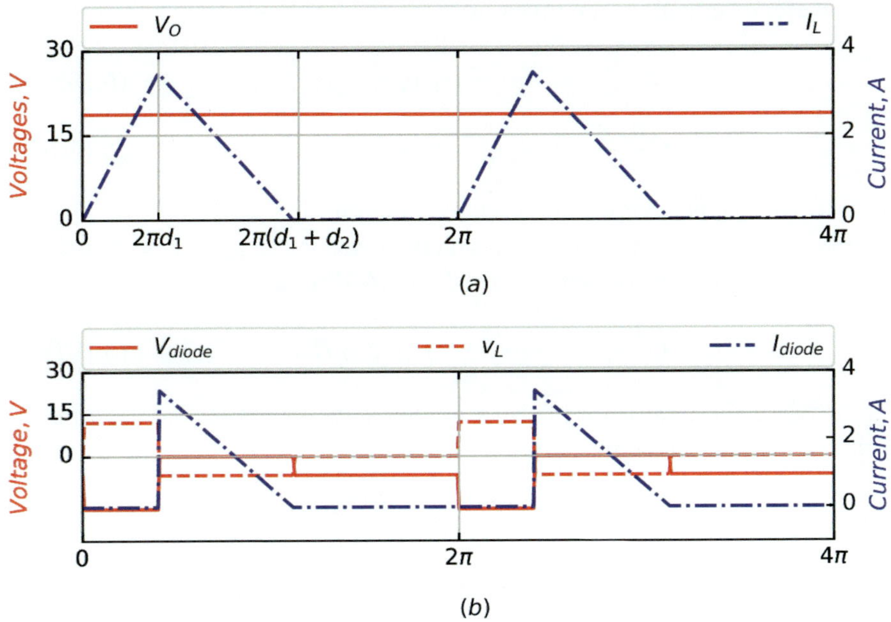

Figure 6.29 The boost converter in DCM.

but remains blocked during the OFF and IDLE states. On the other hand, the diode conducts during the OFF state but remains blocked during the ON and IDLE states. Figure 6.29 shows the inductor current waveform during a switching period.

6.2.8 Small-signal analysis in DCM

To simplify the development of the small-signal model, the ESR of the capacitor is considered to be negligible. Then, the state equations for the three intervals are

$$(i)\, L\frac{di_L}{dt} = v_i, \quad C\frac{dv_C}{dt} = -v_C/R$$

$$\begin{aligned}(ii)\, L\frac{di_L}{dt} &= v_i - v_C, \quad C\frac{dv_C}{dt} = i_L - v_C/R \\ (iii)\, L\frac{di_L}{dt} &= 0, \qquad\quad C\frac{dv_C}{dt} = -v_C/R.\end{aligned}$$

The averaged equations are given by

$$L\frac{di_L}{dt} = (d_1 + d_2)v_i - d_2 v_C \tag{6.185}$$
$$C\frac{dv_C}{dt} = d_2 i_L - v_C/R. \tag{6.186}$$

Eqn (6.186) is the charge conservation equation.

Applying the same reasoning presented in Section 1.10.1, eqn (6.186) requires correction. The corrected form of eqn (6.186) is

$$C\frac{dv_C}{dt} = \frac{d_2}{d_1 + d_2} i_L - v_C/R. \tag{6.187}$$

To arrive at the final form of the averaged state equations, we need the duty ratio constraint given by

$$d_1 + d_2 = \frac{2\langle i_L\rangle}{i_{Lpk}} = \frac{2L\langle i_L\rangle}{d_1 T_S v_i}. \tag{6.188}$$

Using the duty ratio constraint, the averaged state equations for the boost converter in DCM are

$$L\frac{d\langle i_L\rangle}{dt} = \frac{2L\langle i_L\rangle}{\langle d_1\rangle T_S}\left(1 - \frac{\langle v_C\rangle}{\langle v_i\rangle}\right) + \langle d_1\rangle\langle v_C\rangle \tag{6.189}$$
$$C\frac{d\langle v_C\rangle}{dt} = \langle i_L\rangle - \langle v_C\rangle/R - \frac{\langle d_1\rangle^2 T_S \langle v_i\rangle}{2L}. \tag{6.190}$$

Putting $\langle i_L\rangle = I_L + \tilde{i}_L$, $\langle v_C\rangle = V_C + \tilde{v}_C$, $\langle d_1\rangle = D_1 + \tilde{d}_1$, $\langle v_i\rangle = V_i + \tilde{v}_i \simeq V_i$, $M = V_C/V_i$ in eqn (6.189) and (6.190) and after linearization, we get

(i) DC equations

$$I_L = \frac{D_1^2 T_S}{2L}\frac{M}{M-1}V_i \tag{6.191}$$
$$\frac{M(M-1)}{R} = \frac{D_1^2 T_S}{2L}. \tag{6.192}$$

(ii) AC equations

$$\frac{d\tilde{i}_L}{dt} = \frac{2(1-M)}{D_1 T_S}\tilde{i}_L - \frac{2M}{D_1 T_S R}\tilde{v}_C + \frac{2MV_i}{L}\tilde{d}_1 \tag{6.193}$$
$$\frac{d\tilde{v}_C}{dt} = \frac{\tilde{i}_L}{C} - \frac{\tilde{v}_C}{RC} - \frac{D_1 T_S V_i}{LC}\tilde{d}_1. \tag{6.194}$$

Taking the Laplace transforms of eqn (6.193) and (6.194) and eliminating either $i_L(s)$ or $v_C(s)$, we get the following transfer functions:

$$G_{vd}(s) = \frac{v_C(s)}{d_1(s)} = \frac{\frac{D_1 T_S V_i}{LC}\left(-s + \frac{2}{D_1 T_S}\right)}{s^2 + s\left(\frac{1}{RC} + \frac{2(M-1)}{D_1 T_S}\right) + \frac{2(2M-1)}{D_1 T_S RC}} \quad (6.195)$$

$$G_{id}(s) = \frac{i_L(s)}{d_1(s)} = \frac{2M\frac{V_i}{L}\left(s + \frac{2}{RC}\right)}{s^2 + s\left(\frac{1}{RC} + \frac{2(M-1)}{D_1 T_S}\right) + \frac{2(2M-1)}{D_1 T_S RC}}. \quad (6.196)$$

Given the transfer functions $G_{vd}(s)$ and $G_{id}(s)$, a feedback control system can be designed to stabilize the operation of the converter at an operating point.

6.3 The Non-isolated Buck–Boost Converter

The last of the basic converter topologies is the buck–boost (or up-down) converter. The magnitude of the output voltage may be either higher than or lower than that of the input voltage. However, there is a polarity reversal in the output voltage. It is assumed that the converter is operating in CCM, unless otherwise stated. The converter cycles through two states in CCM: the ON state with the switch turned on and the diode reverse biased and the OFF state with the switch turned off and the diode forward biased.

6.3.1 Steady-state analysis [1]

Figure 6.30 shows the power stage of the lossless buck–boost converter. When the switch is turned on, the diode is reverse biased and the output capacitor supplies the load current. Energy flows from the DC voltage source into the inductor's magnetic field. During the OFF state, the inductor current supplies the load current and also charges the capacitor. Figure 6.31 shows the inductor current and voltage waveforms.

The inductor voltage is V_i during the ON state and V_0 during the OFF state. Since the average inductor voltage over a switching period is zero,

$$D\,V_i + (1-D)\,V_0 = 0. \quad (6.197)$$

Therefore,

$$\frac{V_0}{V_i} = -\frac{D}{1-D}. \quad (6.198)$$

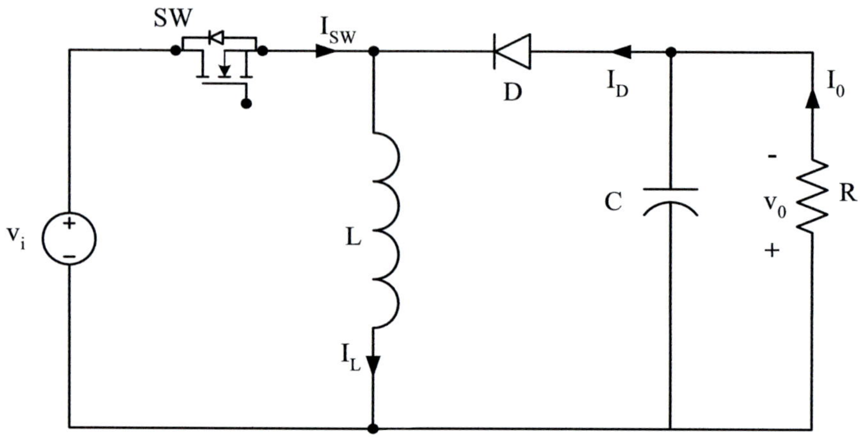

Figure 6.30 Buck–boost converter's power stage.

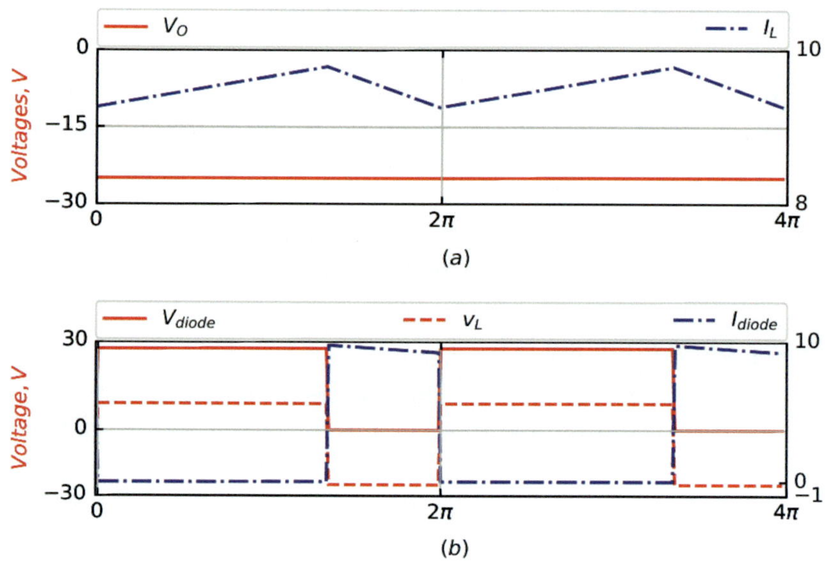

Figure 6.31 Buck–boost converter's current and voltage waveforms.

The negative sign in eqn (6.198) indicates a polarity reversal in the output voltage. From eqn (6.198), we get

$$D = \frac{V_0}{V_0 - V_i}. \tag{6.199}$$

For $V_i = 15\,V$ and $D = 0.25$, we get $V_O = -5\,V$.

The inductor current increases linearly during the ON state. The peak-to-peak ripple in the inductor current is given by

$$\Delta i_L = \frac{DT_S V_i}{L} = \frac{DV_i}{f_S L}. \tag{6.200}$$

The switch current is equal to the inductor current during the ON state, and, therefore, the average switch current is

$$I_{SW} = DI_L. \tag{6.201}$$

Similarly, the diode current is equal to the inductor current during the OFF state. Thus, the average diode current is

$$I_{DE} = (1 - D)I_L = I_O, \tag{6.202}$$

where I_O is the load current.

The capacitor current is obtained by subtracting the load current from the diode current:

$$i_C = i_{DE} - I_O. \tag{6.203}$$

The capacitor current is a pulsating current and its average value is zero. The peak-to-peak ripple in the output voltage is given by

$$\begin{aligned} \Delta V_O &= \frac{1}{C}\int_0^{DT_S} I_O\, dt \\ &= \frac{V_O}{RC} DT_S. \end{aligned}$$

That is,

$$\frac{\Delta V_O}{V_O} = \frac{D}{f_S RC}. \tag{6.204}$$

The equivalent series resistance R_C of the capacitor can contribute significantly to the ripple in the output voltage. The peak-to-peak variation in the capacitor current is

$$\Delta i_C = I_L + \frac{\Delta I_L}{2}. \tag{6.205}$$

The output voltage ripple due to the capacitor's ESR is

$$\Delta V_{O,\text{ESR}} = R_C\left(I_L + \frac{\Delta I_L}{2}\right). \tag{6.206}$$

The inductor current must remain positive in CCM. The boundary between CCM and DCM is obtained by setting $I_L = \frac{\Delta I_L}{2}$. That is,

$$\frac{I_O}{1-D} = \frac{1}{2}\frac{DV_i}{f_S L}. \tag{6.207}$$

For a given duty ratio and load, the minimum inductance for operation in CCM is

$$\begin{aligned} L_{min} &= \frac{1}{2f_S}\frac{D(1-D)V_i}{I_O} \\ &= (1-D)^2\frac{R}{2f_S}. \end{aligned} \tag{6.208}$$

The input power to the converter is $P_i = V_i I_{SW}$ and the output power is $P_O = V_O I_O$. The two powers are equal since the converter has been considered to be lossless.

6.3.2 Inclusion of losses

Assume that the power stage is operating in CCM. Figure 6.32 shows the circuit representation of the converter in the ON and OFF states. The losses are taken into account by the resistors R_{SW}, R_L, R_C, and R_{DE}. There is a voltage drop V_{DE} in the diode when it is forward biased.

During the ON state, the inductor's voltage and the capacitive current are

$$\begin{aligned} v_L &= v_i - (R_L + R_{SW})i_L \\ i_C &= -\frac{v_O}{R}. \end{aligned}$$

During the OFF state, the same variables are given by

$$\begin{aligned} v_L &= v_O - (R_L + R_{DE})i_L - V_{DE} \\ i_C &= -\frac{v_O}{R} - i_L. \end{aligned}$$

Taking the average over a switching period and neglecting the small variations in the variables, we get

$$\begin{aligned} \langle v_L \rangle &= DV_i - D(R_L + R_{SW}) + \acute{D}V_O - \acute{D}(R_L + R_{DE})I_L - \acute{D}V_{DE} \\ \langle i_C \rangle &= -\frac{V_O}{R} - \acute{D}I_L, \end{aligned}$$

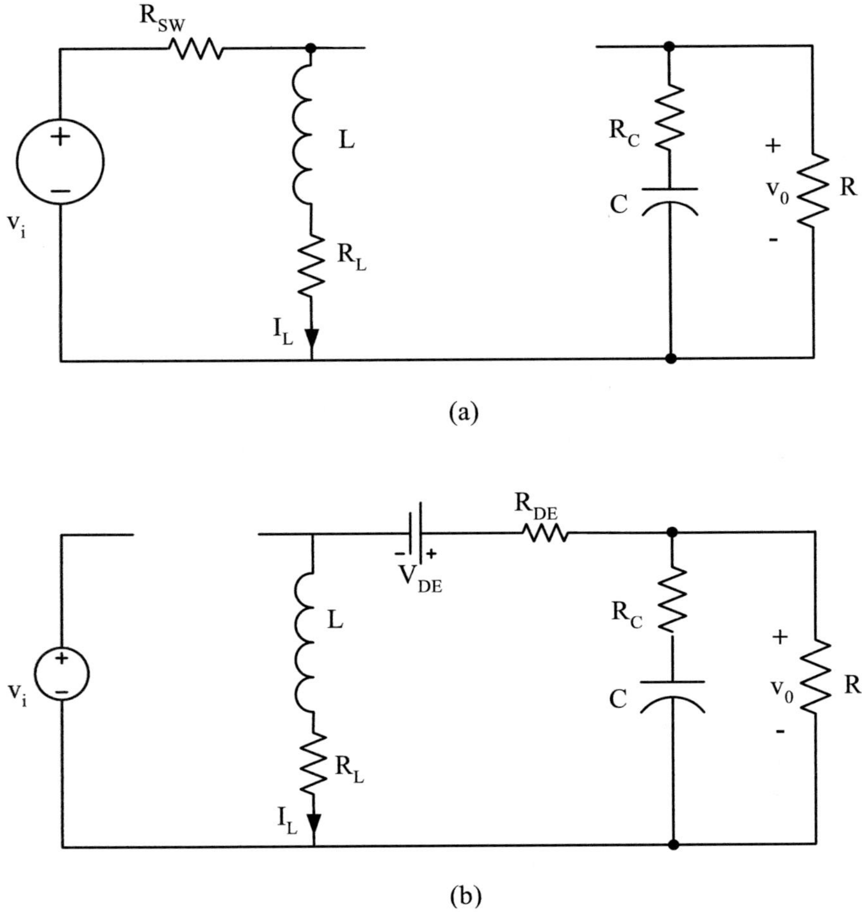

Figure 6.32 Buck–boost converter with losses: (a) ON state; (b) OFF state.

where $\acute{D} = 1 - D$. Setting $\langle v_L \rangle = 0$ and $\langle i_C \rangle = 0$, we get

$$I_L = -\frac{V_0}{\acute{D}R} \tag{6.209}$$

$$DV_i + \acute{D}(V_0 - V_{DE}) = \{D(R_L + R_{\mathrm{SW}}) + \acute{D}(R_L + R_{DE})\}I_L. \tag{6.210}$$

Substituting eqn (6.209) in eqn (6.210) and after simplification, we get

$$V_0 = \frac{-\frac{D}{\acute{D}}V_i + V_{DE}}{1 + \frac{R_L + DR_{\mathrm{SW}} + \acute{D}R_{DE}}{\acute{D}^2 R}}. \tag{6.211}$$

6.3.3 Selection of components [13]

In the discussion that follows, it is assumed that the converter is operating in CCM.

6.3.3.1 Filter capacitor

The filter capacitor of a buck–boost power stage is selected to limit the output voltage ripple to the level required by the specifications. The output voltage ripple is caused by the cyclic charging and discharging of the capacitor and also by the alternating capacitive current flowing through the ESR of the capacitor.

Considering that the converter is operating in CCM, eqn (6.204) allows the determination of the amount of capacitance required to limit the output voltage ripple due to the capacitance to the specified value of ΔV_0:

$$C \geq \frac{V_0}{\Delta V_0} \frac{D_{\max}}{f_S R}. \tag{6.212}$$

where $D_{\max}$ is the maximum duty ratio. The specified ripple in the output voltage also sets a limit to the capacitor's ESR as given by eqn (6.206):

$$\begin{aligned} R_C &\leq \frac{\Delta V_0}{I_L + \Delta I_L/2} \\ &\leq \frac{\Delta V_0}{\left(\frac{I_{0,\max}}{1-D_{\max}} + \Delta I_L/2\right)}. \end{aligned} \tag{6.213}$$

The capacitor having a limiting value given by eqn (6.213) may have a capacitance much greater than the value given by eqn (6.212). Besides the series resistance, the capacitor also has a series inductance that could cause ringing in the low megahertz range. The ringing may be controlled by selecting low ESL capacitors and by limiting the lead length.

As discussed in the previous sections, aluminum electrolytic capacitors are the least expensive, but their ESRs are high. Solid tantalum capacitors (available as surface mount devices) have low ESRs that are stable over the temperature range, high ripple-current capabilities, low series inductances, and high ratios of capacitance to volume. Organic semiconductor electrolytic capacitors (such as the Sanyo OS-CON series) are another good choice though they are supplied in lead-mounted packages.

6.3.3.2 Inductor

The usual choice for the core material is ferrite. The value of the inductance is selected such that the peak-to-peak ripple current is limited. The peak-to-peak inductor current ripple is inversely proportional to the inductance and directly proportional to the applied voltage and the duty ratio. Other important factors to be considered when selecting the inductor are the peak inductor current and the maximum switching frequency.

Current flowing through the inductor causes power dissipation in the inductor's resistance. Power is also dissipated in the core due to the alternating voltage applied to the inductor. The total power dissipation leads to temperature rise in the inductor that should not exceed the specified values.

6.3.3.3 Power switch

The power switch is required to conduct the inductive current when it is turned on and block the difference between the input and output voltages when turned off. The transition from one state to the other should take place rapidly to avoid excessive power dissipation.The best choice in terms of cost and performance is the N-channel MOSFET. P-channel MOSFETs are more expensive though driving the gate is simpler.

The power dissipated by the switch is given by

$$P_{D,\mathrm{SW}} = DR_{DS,on}\, I_{\mathrm{rms}}^2 + \frac{1}{2}(V_i - V_O)\frac{I_O}{1-D}(t_r + t_f)f_S + Q_G\, v_{GS}\, f_S, \tag{6.214}$$

where t_r, t_f are the turn-on and turn-off times of the MOSFET, and Q_G is the total gate charge. As discussed previously, the calculated power dissipation allows the estimation of the junction temperature, which should not exceed the specified value.

6.3.3.4 Diode

Fast switching, adequate breakdown voltage and current rating and low forward voltage drop are important criteria for selecting the power diode. Schottky diodes are the usual choice (as discussed in previous sections). The power dissipated by the diode is given by

$$P_{D,\mathrm{Diode}} = (1 - D)V_{DE}I_O, \tag{6.215}$$

where V_{DE} is the forward voltage drop. The junction temperature may be estimated using the calculated value of $P_{D,\mathrm{Diode}}$ and it should be lower than the value specified for the diode.

6.3.4 Simulation of the buck–boost converter

The buck–boost converter that will be simulated is shown in Figure 6.33. The transient simulation will be carried out over $ncycle = 1000$ periods with $n = 256$ time-steps per period. The transient equivalent circuit is shown in Figure 6.34.

The probe voltages are v_D and v_0 and the error currents flowing into the probe voltages are i_{rrd} and i_{rr0}. Analysis of the transient equivalent circuit and the determination of the error currents are left as exercises. The

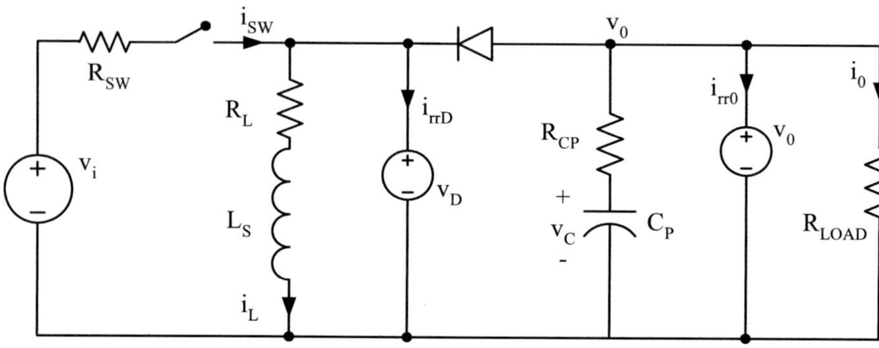

$R_{SW} = 4\text{m}\Omega$	$L_S = 22\ \mu\text{H}$	$v_i = 12\text{V}$
$R_L = 2\text{m}\Omega$	$C_P = 100\ \mu\text{F}$	$f_s = 200\text{kHz}$
$R_{CP} = 8\text{m}\Omega$	$R_{LOAD} = 8\ \Omega$	$d = 0.6667$

Figure 6.33 Buck–boost converter for simulation.

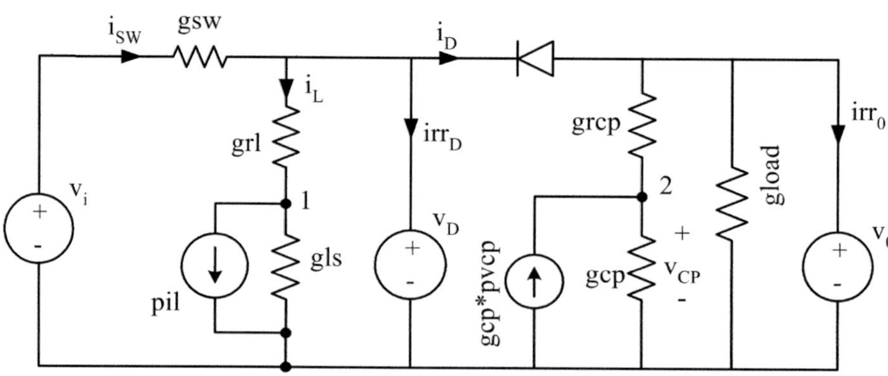

Figure 6.34 Buck–boost converter: transient equivalent circuit.

small-signal equivalent circuit is shown in Figure 6.35. Determination of the corrections δv_D and δv_0 are also left as exercises. The response waveforms, plotted by Program 6.36, are shown in Figure 6.36.

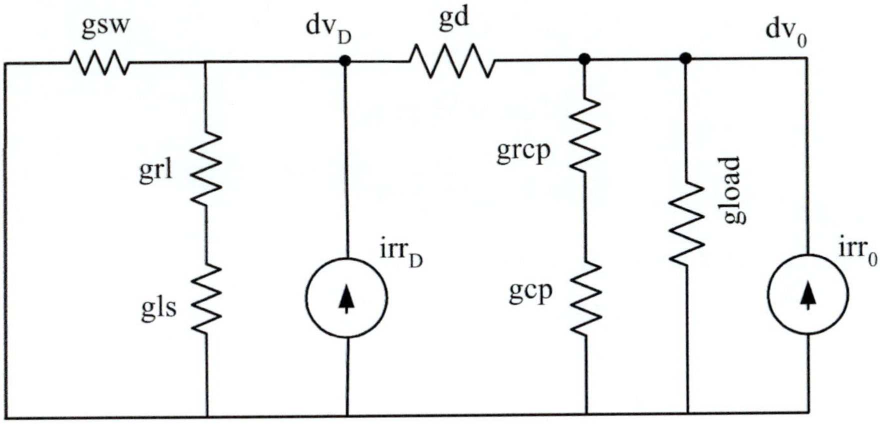

Figure 6.35 Buck–boost converter: small-signal equivalent circuit.

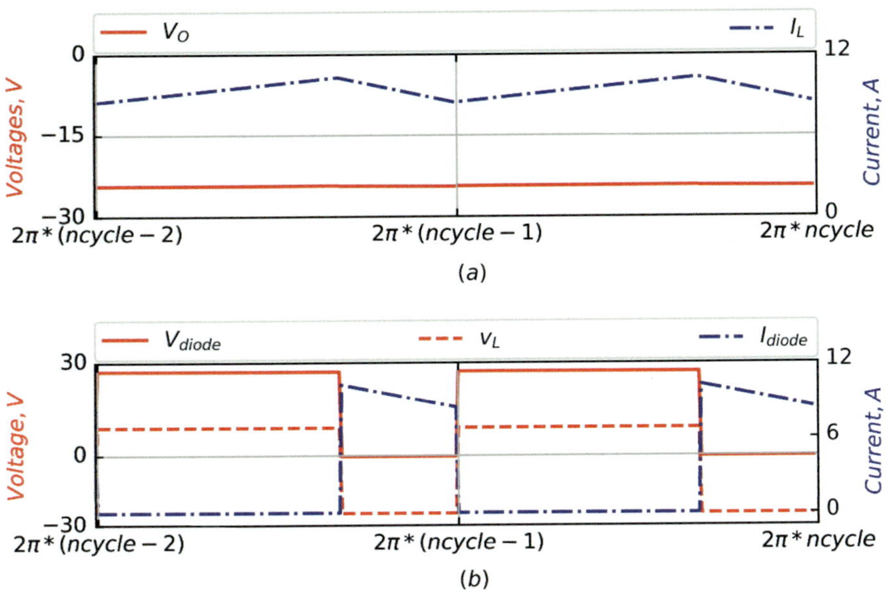

Figure 6.36 Buck–boost converter: response waveforms.

6.3.5 Buck–boost converter model in CCM

The development of the buck–boost converter model closely follows that of the boost converter described in Section 2.5. The state equations during the ON state are (Figure 6.30) [14, 15]

$$
\begin{aligned}
L\frac{di_L}{dt} &= v_i \\
C\frac{dv_C}{dt} + \frac{v_O}{R} &= 0 \\
v_O &= \frac{R}{R+R_C}v_C.
\end{aligned}
$$

The state equations during the OFF state are

$$
\begin{aligned}
L\frac{di_L}{dt} &= v_O. \\
C\frac{dv_C}{dt} + \frac{v_O}{R} + i_L &= 0 \\
v_O &= \frac{R}{R+R_C}\left(v_C - R_C i_L\right).
\end{aligned}
$$

Putting $\acute{d} = (1 - d)$ and applying state-space averaging,

$$L\frac{d\langle i_L\rangle}{dt} = \langle d\rangle\langle v_i\rangle + \langle \acute{d}\rangle\langle v_O\rangle \tag{6.216}$$

$$C\frac{d\langle v_C\rangle}{dt} + \frac{\langle v_O\rangle}{R} + \langle \acute{d}\rangle\langle i_L\rangle = 0 \tag{6.217}$$

$$\langle v_O\rangle = \frac{R}{R+R_C}\Big(\langle v_C\rangle - R_C\langle \acute{d}\rangle\langle i_L\rangle\Big). \tag{6.218}$$

Eqn (6.216)–(6.218) are linearized by putting $\langle v_i\rangle = V_i + \tilde{v}_i \simeq V_i$, $\langle i_L\rangle = I_L + \tilde{i}_L$, $\langle v_C\rangle = V_C + \tilde{v}_C$, $\langle v_O\rangle = V_O + \tilde{v}_O$, $\langle d\rangle = D + \tilde{d}$, and $\langle \acute{d}\rangle = \acute{D} - \tilde{d}$. Then

$$
\begin{aligned}
L\frac{d\tilde{i}_L}{dt} &= DV_i + \acute{D}V_O + \acute{D}\tilde{v}_O + (V_i - V_O)\tilde{d} \\
C\frac{d\tilde{v}_C}{dt} &= -\frac{1}{R+R_C}\left(V_C + \acute{D}RI_L + \tilde{v}_C + \acute{D}R\tilde{i}_L - RI_L\tilde{d}\right) \\
V_O + \tilde{v}_O &= \frac{R}{R+R_C}\left(V_C - \acute{D}R_CI_L + \tilde{v}_C - \acute{D}R_C\tilde{i}_L + R_CI_L\tilde{d}\right).
\end{aligned}
$$

The constant terms on the right-hand sides are equated to zero, giving

$$
\begin{aligned}
V_0 &= -\frac{D}{\acute{D}}V_i \\
V_C &= -\acute{D}RI_L \\
V_0 &= \frac{R}{R+R_C}\left(V_C - \acute{D}R_C I_L\right) = V_C.
\end{aligned}
$$

Finally, the state equations for small variations in the neighborhood of the operating point are

$$
\begin{aligned}
L\frac{d\tilde{i}_L}{dt} &= \acute{D}\tilde{v}_0 + (V_i - V_0)\tilde{d} && (6.219)\\
C\frac{d\tilde{v}_C}{dt} &= -\frac{1}{R+R_C}\left(\tilde{v}_C + \acute{D}R\tilde{i}_L - RI_L\tilde{d}\right) && (6.220)\\
\tilde{v}_0 &= \frac{R}{R+R_C}\left(\tilde{v}_C - \acute{D}R_C\tilde{i}_L + R_C I_L\tilde{d}\right). && (6.221)
\end{aligned}
$$

Applying the Laplace transform to eqn (6.219) and (6.220), we get

$$
\begin{aligned}
\tilde{v}_C(s) &= \frac{\frac{V_i}{\acute{D}^2}\left(s\frac{DL}{\acute{D}^2R} - 1\right)}{s^2\frac{LC}{\acute{D}^2} + s\frac{L}{\acute{D}^2R} + 1}\tilde{d}(s) && (6.222)\\
\tilde{v}_0(s) &= (1 + sCR_C)\tilde{v}_C. && (6.223)
\end{aligned}
$$

(The demonstration of eqn (6.222) is left as an exercise.) Eqn (6.222) and (6.223) lead to

$$
\frac{\tilde{v}_0(s)}{\tilde{d}(s)} = \frac{\frac{V_i}{\acute{D}^2}(1 + sCR_C)\left(s\frac{DL}{\acute{D}^2R} - 1\right)}{s^2\frac{LC}{\acute{D}^2} + s\frac{L}{\acute{D}^2R} + 1}. \qquad (6.224)
$$

The polarity reversal in the output voltage is included in eqn (6.224). Removing the polarity reversal in the output voltage, the control to output voltage transfer function is finally given by

$$
G_{vd}(s) = \frac{\tilde{v}_0(s)}{\tilde{d}(s)} = \frac{\frac{V_i}{\acute{D}^2}(1 + sCR_C)\left(1 - s\frac{DL}{\acute{D}^2R}\right)}{s^2\frac{LC}{\acute{D}^2} + s\frac{L}{\acute{D}^2R} + 1} \qquad (6.225)
$$

Eqn (6.225) is similar to the control to output transfer function of the boost converter. The transfer function $G_{vd}(s)$ has two zeros, one in the LHP at $\omega_{\mathrm{ESR}} = 1/(CR_C)$, and the other zero in the RHP at $\omega_{RHP} = \frac{\acute{D}^2R}{DL}$. There are two poles located close to the resonant frequency $\omega_{LC} = \frac{\acute{D}}{\sqrt{LC}}$.

6.3.6 Feedback control of the buck–boost converter in CCM

6.3.6.1 Voltage mode control

Figure 6.37 shows the buck–boost converter with the control system for voltage mode control [9]. The parameters of the converter are:

$$
\begin{aligned}
R_L &= 2\ \text{m}\Omega, & L &= 22\ \mu\text{H} \\
R_C &= 8\ \text{m}\Omega, & C &= 440\ \mu\text{F} \\
R &= 8\ \Omega, & D &= 0.667 \\
e_S &= 12\ \text{V}, & V_0 &= 24\ \text{V} \\
V_{RAMP} &= 3.75\ \text{V}, & V_{REF} &= 2.5\ \text{V} \\
R_{f1} &= 860\ \text{k}\Omega, & R_{f2} &= 100\ \text{k}\Omega \\
f_S &= 200\ \text{kHz}.
\end{aligned}
$$

The frequencies associated with the control to output transfer function are:

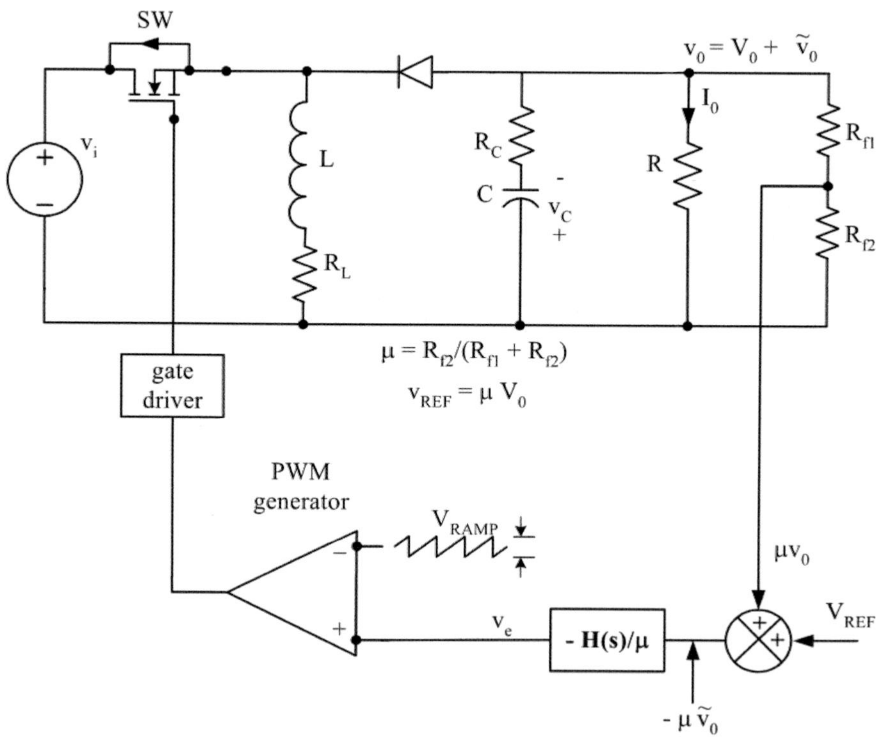

Figure 6.37 Buck–boost converter: voltage mode control.

$$\begin{aligned}
\omega_{LC} &= \acute{D}/\sqrt{LC} &&= 3388 \text{ rads/sec} \\
\omega_{\text{ESR}} &= 1/(CR_C) &&= 28,409 \text{ rads/sec} \\
\omega_{RHP} &= \acute{D}^2 R/(DL) &&= 60,606 \text{ rads/sec} \\
\omega_{\text{S}} &= \text{switching frequency} &&= 200 \text{ kHz.}
\end{aligned}$$

The zero-crossing frequency ω_0 of the loop gain $G_{vd}(s)H(s)$ is chosen such that $2\omega_{LC} < \omega_0 < \omega_{RHP}/5$. Choose $\omega_0 = 9000$ rads/sec.

The gain and phase of $G_{vd}(s)$ at $\omega = \omega_0$ are obtained as $|G_{vd}(j\omega_0)| = 18.015$ and $\angle G_{vd}(j\omega_0) = -186.34°$. Therefore, we required a compensator $H(s)$ having $|H(j\omega_0)| = 1/18.015$. Setting $\angle H(j\omega_0) = 65°$ will lead to a phase margin of about 58°. Using the k factor technique, the compensator's transfer function is obtained as

$$H(s) = 41,690\frac{(s + 985.76)^2}{s(s + 82170)^2}. \tag{6.226}$$

Figure 6.38 shows the Bode plots for $G_{vd}(j\omega)$, $H(j\omega)$ and the loop gain $G_{vd}(j\omega)H(j\omega)$. The program for plotting the Bode diagrams is Program

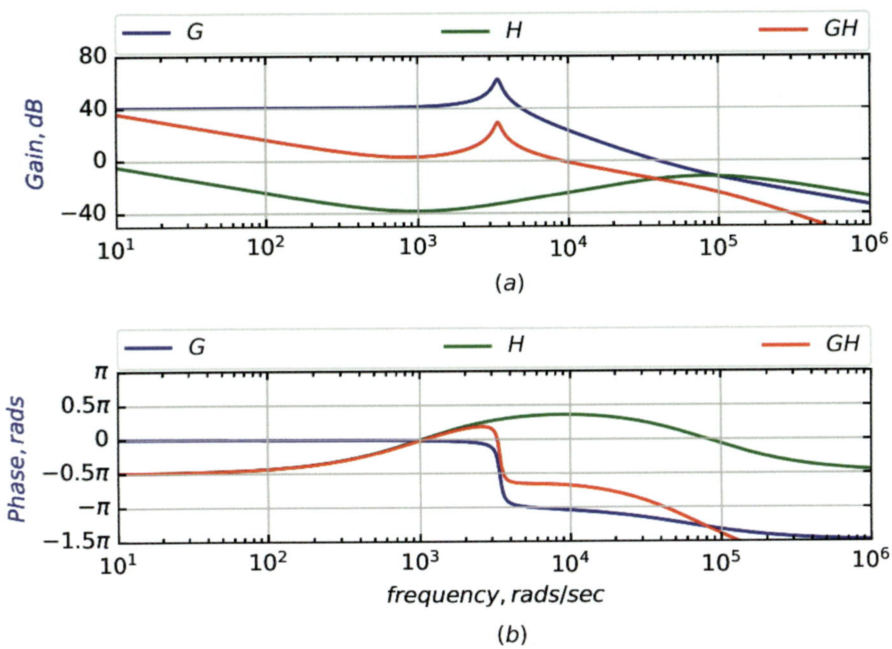

Figure 6.38 Buck–boost converter, voltage mode control: Bode plots.

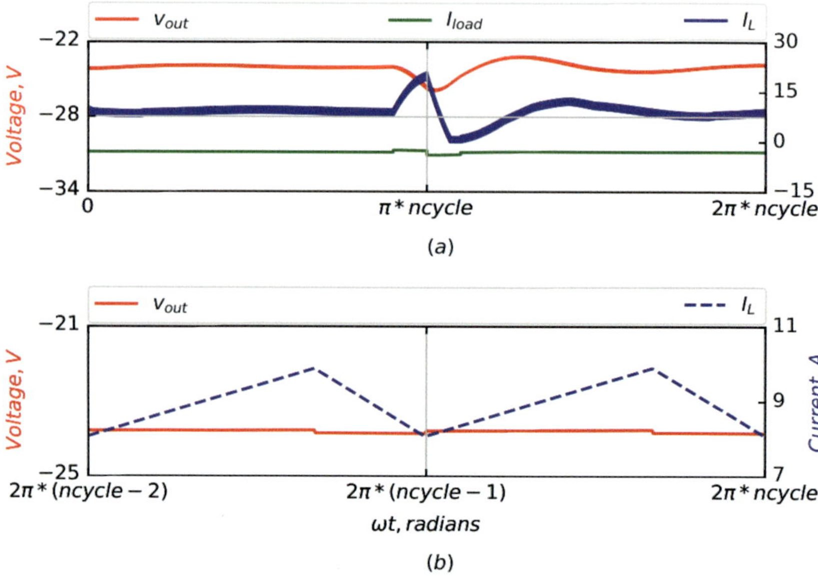

Figure 6.39 Buck–boost converter, voltage mode control: transient response.

6.38. The state-space representation of the transfer function $H(s)$ is identical to eqn (6.163) and (6.164), with the parameters being

$$K = 41,690,\ \omega_z = 985.76 \text{ rads/sec},\ \omega_\mathrm{p} = 82,170 \text{ rads/sec}$$
$$a_1 = 2\omega_z,\ a_2 = \omega_z^2,\ b_1 = 2\omega_p,\ b_2 = \omega_2^2.$$

The discrete representation of the state equations is identical to eqn (6.168)–(6.170).

Figure 6.39, plotted by Program 6.39, shows the output voltage and the inductor current of the buck–boost converter with the voltage mode control. The simulation was carried out for 1000 periods and disturbances, identical to those applied to the boost converter, were introduced in the middle of the simulation window. Feedback control is able to stabilize the output voltage, though there are considerable oscillations in the waveforms soon after the disturbances were applied.

6.3.7 Current mode control

Much of the theoretical developments in Sections 2.6.2 and 2.6.3 apply to the current mode control of the buck–boost converter. The parameters of the converter are

$$L = 22\ \mu\text{H},\ \text{C} = 440\ \mu\text{F},\ \text{R} = 8\ \Omega\text{s},\ \text{R}_\text{C} = 8\ \text{m}\Omega\text{s}$$
$$V_i = 12\ \text{V},\ \text{D} = 0.66667,\ \text{f}_\text{S} = 200\ \text{kHz}.$$

The current sensing resistor $R_{\text{SN}} = 0.201465\ \Omega$. When the switch is turned on, the inductor current slope $M_1 = V_i/L = 0.5454\ \text{A}/\mu\text{s}$. Choose the compensating current slope $m_C = 0.65\ \text{A}/\mu\text{s}$. The parameter $I_M = (m_C + M_1/2)T_\text{S} = 4.613636$ A. The current reference at the operating point $I_e = V_{\text{REF}}/R_{\text{SN}} = 12.07576$ A.

The control to output voltage transfer function is

$$G_{vd} = \frac{v_O(s)}{d(s)} = \frac{\frac{V_i}{\acute{D}^2}\left(1 - \frac{s}{\omega_{RHP}}\right)}{s^2LC + sL/R + \acute{D}^2}. \tag{6.227}$$

The control to inductor current transfer function is,

$$G_{id} = \frac{i_L(s)}{d(s)} = \frac{\frac{V_i}{\acute{D}^2}\left(sC + \frac{1+D}{R}\right)}{s^2LC + sL/R + \acute{D}^2}. \tag{6.228}$$

Demonstration of the expression for G_{id} is left as an exercise.

Now, $i_L(s) = i_e(s) - I_M d(s)$, where $i_e(s)$ is the alternating component of the current reference for the inner current loop. Using this relation, we get

$$\frac{i_e(s)}{d(s)} = \frac{s^2 I_M LC + s\left(\frac{V_i C}{\acute{D}} + \frac{LI_M}{R}\right) + \left(\frac{1+D}{\acute{D}}\frac{V_i}{R} + \acute{D}^2 I_M\right)}{s^2LC + sL/R + \acute{D}^2}. \tag{6.229}$$

Finally, the reference current to output voltage transfer function is given by

$$\begin{aligned} G_{vi}(s) &= v_O(s)/i_e(s) \\ &= \frac{V_i\,(1 - s/\omega_{RHP})}{s^2 I_M LC + s\left(\frac{V_i C}{\acute{D}} + \frac{LI_M}{R}\right) + \left(\frac{1+D}{\acute{D}}\frac{V_i}{R} + \acute{D}^2 I_M\right)}. \end{aligned} \tag{6.230}$$

Substituting the parameter values in eqn (6.230), we obtain

$$G_{vi}(s) = \frac{12\,(1 - s/60606)}{(0.4466e - 07)\,s^2 + (0.1585e - 01)\,s + 8.012626}. \tag{6.231}$$

For $s = j\omega_0 = j9000$, $G_{vi} = 0.085\angle{-96.68^\circ}$.

Specifying the phase margin of the loop gain to be $70°$, we require that $H(j\omega_0) = 1/0.085\angle{-13.32°}$. Putting

$$H(s) = \frac{1}{R_{SN}} \frac{(1 + s/\omega_z)}{(s/\omega_{po})(1 + s/\omega_p)} \tag{6.232}$$

and using the k factor technique, we get

$$H(s) = 12,363.4 \frac{(1 + s/1050.87)}{s(1 + s/77077.8)}. \tag{6.233}$$

The demonstration of eqn (6.233) is left as an exercise.

Figure 6.40, plotted by Program 6.40, shows the Bode diagrams of $G_{vi}(s)$, $H(s)$ and the loop gain $H(s)G_{vi}(s)$. The zero-crossing frequency of the loop gain is $8998.7\,\text{rads/sec}$ and the phase margin is approximately $70°$. Figure 6.41, plotted by Program 6.41, shows the output voltage and the inductor current variations during a simulation window of 1000 periods.

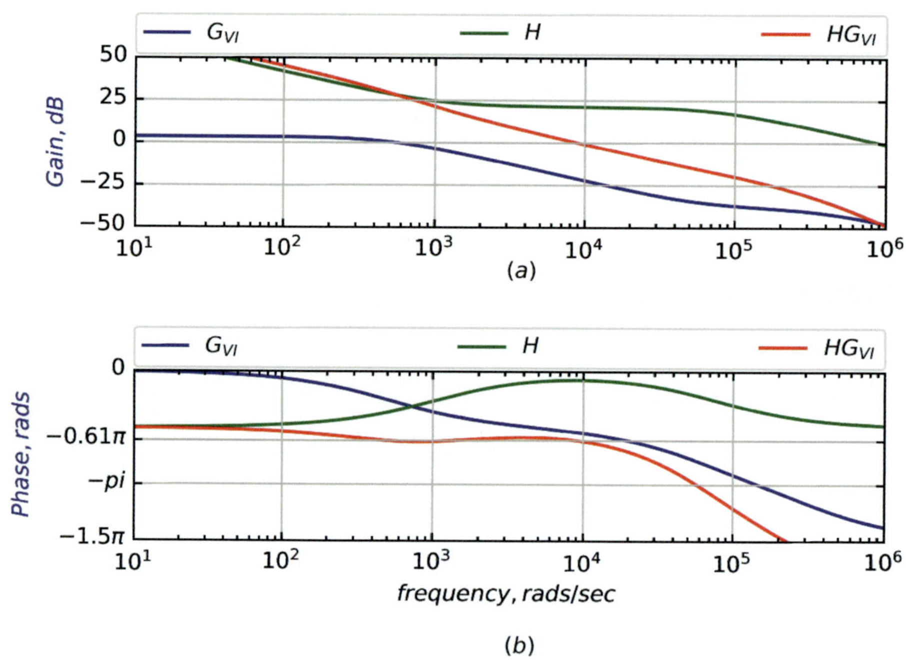

Figure 6.40 Buck–boost converter, current mode control: Bode plots.

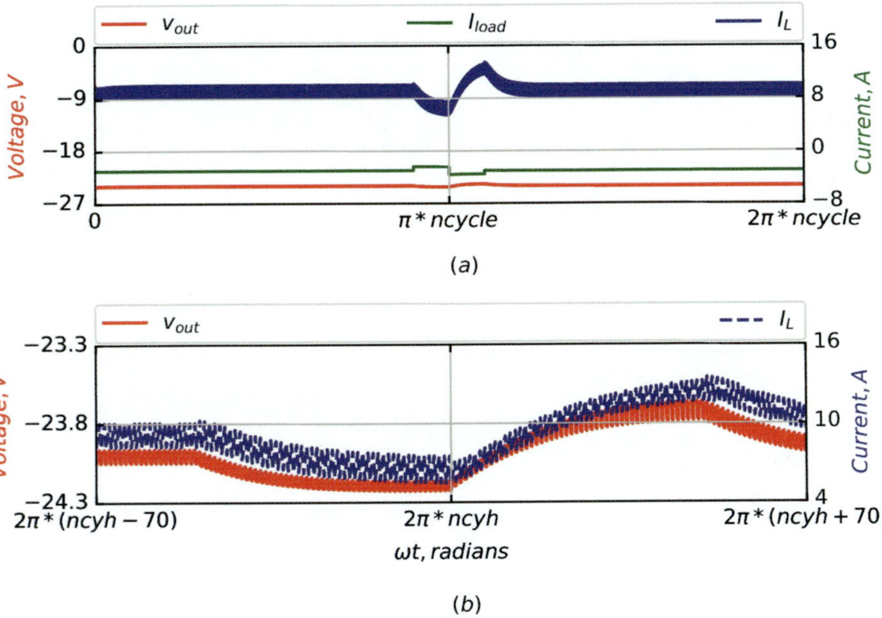

Figure 6.41 Buck–boost converter, current mode control: transient response.

Figure 6.41 (b) is a zoom of the central part of Figure 6.41 (a). Disturbances, identical to those applied to the boost converter, were introduced in the middle of the simulation window. Feedback control maintains the output voltage constant.

6.3.8 Buck–boost converter in DCM

For a given load, duty ratio, and switching frequency, there is a minimum inductance required for operation in CCM (see eqn (6.208)). When operating in DCM, the buck–boost converter passes through (i) the ON state during $jT_S \leq t \leq (j + d_1)T_S$; (ii) the OFF state during $(j + d_1)T_S \leq t \leq (j + d_1 + d_2)T_S$; (iii) the IDLE state during $(j + d_1 + d_2)T_S \leq t \leq (j + 1)T_S$. As in the case of the boost converter, the switch conducts during the ON state but remains blocked during the OFF and IDLE states. Similarly, the diode conducts during the OFF state but is reverse biased during the ON and IDLE states. Figure 6.42 shows the inductor current waveform when the buck–boost converter is operating in DCM.

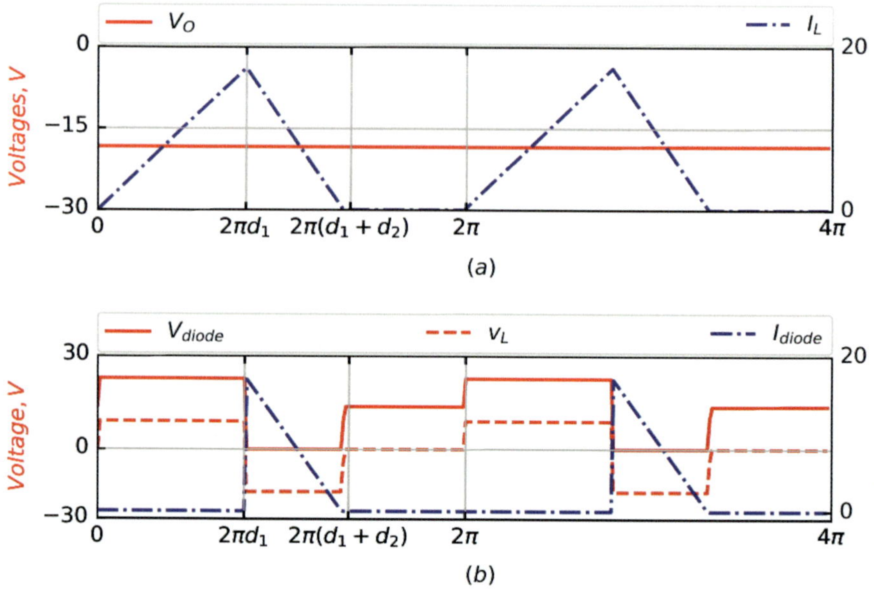

Figure 6.42 The buck–boost converter in DCM.

6.3.9 Small-signal analysis of the buck–boost converter operating in DCM

The ESR of the capacitor is considered to be negligible in order to simplify the development of the small-signal model. The state equations for the three intervals are:

$$\begin{aligned}
&\text{(i) } L\frac{di_L}{dt} = v_i, \quad C\frac{dv_C}{dt} = -v_C/R \\
&\text{(ii) } L\frac{di_L}{dt} = v_C, \quad C\frac{dv_C}{dt} = -v_C/R - i_L \\
&\text{(iii) } L\frac{di_L}{dt} = 0, \quad C\frac{dv_C}{dt} = -v_C/R.
\end{aligned}$$

The averaged equations are:

$$L\frac{d\langle i_L\rangle}{dt} = \langle d_1\rangle\langle v_i\rangle + d_2\langle v_C\rangle \tag{6.234}$$

$$C\frac{d\langle v_C\rangle}{dt} = -\frac{\langle v_C\rangle}{R} - d_2\langle i_L\rangle. \tag{6.235}$$

The corrected form of eqn (6.235) is

$$C\frac{d\langle v_C\rangle}{dt} = -\frac{\langle v_C\rangle}{R} - \frac{d_2}{d_1+d_2}\langle i_L\rangle. \tag{6.236}$$

The duty ratio constraint is

$$(d_1+d_2) = \frac{2L}{T_S V_i}\frac{\langle i_L\rangle}{\langle d_1\rangle}. \tag{6.237}$$

Put $\langle i_L\rangle = I_L + \tilde{i}_L$, $\langle v_C\rangle = V_C + \tilde{v}_C$, $\langle d_1\rangle = D_1 + \tilde{d}_1$. Further, we consider that the input voltage $\langle v_i\rangle = V_i$.

Substituting the averaged variables in eqn (6.234) and (6.236) and using the usual technique of linearization, we get the following sets of DC and AC equations:

DC equations:

$$V_i - V_C + \frac{2LI_L}{D_1^2 T_S}\frac{V_C}{V_i} = 0 \tag{6.238}$$

$$I_L + \frac{V_C}{R} - \frac{D_1^2 T_S}{2L}V_i = 0. \tag{6.239}$$

AC equations:

$$L\frac{d\tilde{i}_L}{dt} - \frac{2LV_C}{D_1 T_S V_i}\tilde{i}_L - \left(\frac{2LI_L}{D_1 T_S V_i} - D_1\right)\tilde{v}_C = \left(V_i - V_C - \frac{2LI_L}{D_1^2 T_S}\frac{V_C}{V_i}\right)\tilde{d}_1 \tag{6.240}$$

$$C\frac{d\tilde{v}_C}{dt} + \frac{\tilde{v}_C}{R} + \tilde{i}_L = \frac{D_1 T_S V_i}{L}\tilde{d}_1. \tag{6.241}$$

From the DC equations, the voltage gain in DCM is obtained as

$$\frac{V_C}{V_i} = D_1\sqrt{\frac{RT_S}{2L}}. \tag{6.242}$$

Note that the output voltage V_C is a negative voltage due to the polarity inversion in the output. Applying the Laplace transform to eqn (6.240) and (6.241) and putting $M = -\left(\frac{V_C}{V_i}\right)$, we obtain

$$\left(sL + \frac{2ML}{D_1 T_S}\right)\tilde{i}_L(s) - \frac{2ML}{D_1 T_S R}\tilde{v}_C(s) = 2(V_i - V_C)\,\tilde{d}_1(s) \tag{6.243}$$

$$\tilde{i}_L(s) + \left(sC + \frac{1}{R}\right)\tilde{v}_C(s) = \frac{D_1 T_S V_i}{L}\,\tilde{d}_1(s). \tag{6.244}$$

Eliminating $\tilde{i}_L(s)$ from eqn (6.243) and (6.244), we get

$$\begin{aligned} G_{vd}(s) &= \frac{\tilde{v}_C(s)}{\tilde{d}_1(s)} \\ &= \frac{\frac{2V_i}{LC}\left(s\frac{D_1 T_S}{2} - 1\right)}{s^2 + s\left(\frac{1}{RC} + \frac{2M}{D_1 T_S}\right) + \frac{2}{RC}\frac{2M}{D_1 T_S}} \\ &\simeq \frac{\frac{2V_i}{LC}\left(s\frac{D_1 T_S}{2} - 1\right)}{s^2 + s\left(\frac{2}{RC} + \frac{2M}{D_1 T_S}\right) + \frac{2}{RC}\frac{2M}{D_1 T_S}} \\ &= \frac{\frac{2V_i}{LC}\left(s\frac{D_1 T_S}{2} - 1\right)}{\left(s + \frac{2}{RC}\right)\left(s + \frac{2M}{D_1 T_S}\right)}. \end{aligned} \tag{6.245}$$

The approximation used in eqn (6.245) is valid since $\frac{1}{RC} \ll \frac{2M}{D_1 T_S}$. The buck–boost converter in DCM essentially behaves like a single-pole system.

6.4 Computational Programs

6.4.1 Program 6.07

```
#
def fdiode(vd):
    """ Diode characteristic.
        Returns diode current and its conductance"""
    #
    glow = 0.1e-06
    ghigh = 0.1e+03
    Vr = -0.3
    #
    if vd >= Vr:
        id = glow*(vd - Vr)
        gd = glow
    else:
        id = ghigh*(vd - Vr)
        gd = ghigh
    #
    return id, gd
#
```

```
#
def ch6fig07_solver(es, gs, gls, gc, gcp, gload, nc, ncycle, n):
    """Assumes fdiode is a function.
        Returns the lists vout, cinduct"""
    #
    import numpy as np
    #
    vd = 1.8
    v0 = 1.8
    pil = 11.2
    pvc = 1.8
    #
    erlim = 0.1e-07
    #
    N = n*ncycle
    gser = gs*gls/(gs + gls)
    gpar = gc*gcp/(gc + gcp)
    #
    vout = []
    vcap = []
    cinduct = []
    #
    for k in range(ncycle):
        #
        for j in range(n):
            #
            if j <= nc:
                gsw = 0.1e+03
            else:
                gsw = 0.1e-07
            #
            for its in range(1000):
                #
                v1 = (gs*vd + gls*v0 - pil)/(gs + gls)
                vc = (gc*v0 + gcp*pvc)/(gc + gcp)
                isw = gsw*(es - vd)
                il = gs*(vd - v1)
                id, gd = fdiode(vd)
                irrd = isw - id - il
                icap = gc*(v0 - vc)
                iload = gload*v0
                irr0 = il - icap - iload
                #
                err = np.sqrt(irrd**2 + irr0**2)
                #
                if err < erlim:
                    break
                else:
                    gmat = np.array([[(gsw + gd + gser), -gser],
                                     [-gser, (gload + gpar + gser)]])
                    bcol = np.array([irrd, irr0])
                    xdv = np.linalg.solve(gmat, bcol)
                    #
```

```
                dvd = xdv[0]
                dv0 = xdv[1]
                vd += dvd
                v0 += dv0
        #
        if its >= 999:
            print('time-step= ', (k*n + j), 'solver does not converge')
            break
        #
        pil = il
        pvc = vc
        #
        vout.append(v0)
        vcap.append(vc)
        cinduct.append(il)
        #
    #
#
print('final value of v0= ', np.mean(vout[N-2*n:N-1]))
print('mean value of Il = ', np.mean(cinduct[N-2*n:N-1]))
print('mean value of Vc = ', np.mean(vcap[N-2*n:N-1]))
#
return vout, vcap, cinduct
#
#
def ch6fig07(es, freq, rl, ls, rc, cp, rload, d, ncycle, n):
    """ Main program, plots results
        Buck converter"""
    #
    #calling command:
    #ch6fig07(12.0, 625000.0, 0.08, 0.68e-06, 0.015, 0.49e-03, 0.15,     \
    #                                                    0.22, 200, 160)
    #
    from math import pi
    import numpy as np
    import matplotlib.pyplot as plt
    from mpl_toolkits.axes_grid1 import host_subplot
    #
    nc = int(np.rint(d*n))
    #
    delt = 1.0/(freq*n)
    dteta = 2.0*pi/n
    gs = 1.0/rl
    gls = delt/ls
    gcp = cp/delt
    gc = 1.0/rc
    gload = 1.0/rload
    #
    vout, vcap, cinduct = ch6fig07_solver(es, gs, gls, gc, gcp, gload,    \
                                                  nc, ncycle, n)
    #
    N = ncycle*n
    ang = dteta*np.arange(N)
```

```
#
ax1 = host_subplot(2, 1, 1)
ax2 = ax1.twinx()
ax1.set_xlim([0, 2*pi*ncycle])
ax1.set_xticks([0, pi*ncycle, 2*pi*ncycle])
ax1.set_xticklabels(['$0$', '$\pi *ncycle$', '$2\pi *ncycle$'])
ax1.set_ylim([0, 2.0])
ax1.set_yticks([0, 0.5, 1.0, 2.0])
ax1.set_yticklabels(['$0$', '$\ $', '$1.0$', '$2.0$'])
ax1.set_ylabel('$Voltage, V$')
ax2.set_ylim(0, 20)
ax2.set_yticks([0, 5, 12, 20])
ax2.set_yticklabels(['$0$', '$5$', '$12$', '$20$'])
ax2.set_ylabel('$Current, A$')
ax1.text(pi*ncycle, -0.85, '$(a)$', size =10)
p1, = ax1.plot(ang, vout, 'r-', label = '$v_{0}$')
p2, = ax2.plot(ang, cinduct, 'g-', label = '$I_{L}$')
ax1.yaxis.get_label().set_color(p1.get_color())
ax2.yaxis.get_label().set_color(p2.get_color())
ax1.legend(bbox_to_anchor=(0., 1.02, 1., .102), loc=3, ncol=2,      \
                               mode="expand", borderaxespad=0.)
ax1.grid(True)
plt.subplots_adjust(hspace=0.85)
#
ax1 = host_subplot(2, 1, 2)
ax2 = ax1.twinx()
ax1.set_xlim([2*pi*(ncycle-2), 2*pi*ncycle])
ax1.set_xticks([2*pi*(ncycle-2), 2*pi*(ncycle-1), 2*pi*ncycle])
ax1.set_xticklabels(['$2\pi *(ncycle-2)$', '$2\pi *(ncycle-1)$',     \
                                                  '$2\pi *ncycle$'])
ax1.set_xlabel('$\omega t, radians$')
ax1.set_ylim([1.5, 1.6])
ax1.set_yticks([1.5, 1.55, 1.6])
ax1.set_yticklabels(['$1.5$', '$\ $', '$1.6$'])
ax1.set_ylabel('$Voltage, V$')
ax2.set_ylim(0, 18)
ax2.set_yticks([0, 9, 18])
ax2.set_yticklabels(['$0$', '$9$', '$18$'])
ax2.set_ylabel('$Current, A$')
ax1.text(2*pi*(ncycle-1), 1.435, '$(b)$', size =10)
p1, = ax1.plot(ang[N-2*n:N-1], vcap[N-2*n:N-1], 'r--',       \
                                   label = '$V_{C}$')
p2, = ax1.plot(ang[N-2*n:N-1], vout[N-2*n:N-1], 'r',       \
                                   label = '$V_{0}$')
p3, = ax2.plot(ang[N-2*n:N-1], cinduct[N-2*n:N-1], 'b--',    \
                                   label = '$I_{L}$')
ax1.yaxis.get_label().set_color(p1.get_color())
ax2.yaxis.get_label().set_color(p3.get_color())
ax1.legend(bbox_to_anchor=(0., 1.02, 1., .102), loc=3, ncol=3,  \
          mode="expand", borderaxespad=0.)
ax1.grid(True)
#
plt.show()
```

```
    #
    #plt.draw
    #plt.savefig('Fig607.ps', dpi=400, bbox_inches='tight')
    #
```

6.4.2 Program 6.10

```
#
def ch6fig10_fresp(es, vramp, ls, rcp, cp, rload, r1, rf1, c1, c2):
    """ Main program, plots results"""
    #
    #calling command:
    #ch6fig10_fresp(12.0, 1.8, 0.68e-06, 0.015, 0.49e-03, 0.15, 3204.0,  \
    #                                     1200.0, 0.76e-08, 0.16e-9)
    #
    from math import pi
    import numpy as np
    from scipy import signal
    import matplotlib.pyplot as plt
    from mpl_toolkits.axes_grid1 import host_subplot
    #
    ag1 = 1.0/(rcp*cp)
    bg1 = (rcp/ls) + 1.0/(rload*cp)
    bg2 = 1.0/(ls*cp)
    Ggain = (es/vramp)*(rcp/ls)
    wg = np.logspace(3, 7, num=1000)
    #
    ah1 = 1.0/(r1*c1)
    bh1 = 1.0/(r1*c2)
    Hgain = 1.0/(rf1*c2)
    #
    numG = [Ggain, Ggain*ag1]
    denG = [1, bg1, bg2]
    tfG = signal.lti(numG, denG)
    numH = [Hgain, Hgain*ah1]
    denH = [1, bh1, 0]
    tfH = signal.lti(numH, denH)
    wg, magG, phaseG = signal.bode(tfG, wg)
    wg, magH, phaseH = signal.bode(tfH, wg)
    #
    agh1 = ag1 + ah1
    agh2 = ag1*ah1
    bgh1 = bh1+bg1
    bgh2 = bg1*bh1 + bg2
    bgh3 = bg2*bh1
    GHgain = Ggain*Hgain
    numGH = [GHgain, GHgain*agh1, GHgain*agh2]
    denGH = [1, bgh1, bgh2, bgh3, 0]
    tfGH = signal.lti(numGH, denGH)
    wg, magGH, phaseGH = signal.bode(tfGH, wg)
    #
    minGH = min(abs(magGH))
```

```
for jwg in range(len(wg)):
    if abs(magGH[jwg]) == minGH:
        wgmin = wg[jwg]
        phmin = phaseGH[jwg]
#
cfreq = wgmin/(2000*pi)
print('crossover frequency = ', cfreq, 'kHz')
print('phase margin = ', 180.0 + phmin,'degrees')
#
ax1 = host_subplot(2, 1, 1)
ax1.set_xlim([1e+4, 1e+7])
ax1.set_xticks([1e+4, 1e+5, 1e+6, 1e+7])
ax1.set_xticklabels(['$10^{4}$', '$10^{5}$', '$10^{6}$', '$10^{7}$'])
ax1.set_ylim([-50, 50])
ax1.set_yticks([-50, 0, 50])
ax1.set_yticklabels(['$-50$', '$0$', '$50$'])
ax1.set_ylabel('$Gain, dB$')
ax1.text(3e+5, -80, '$(a)$', size =10)
p1, = ax1.semilogx(wg, magG, 'b-', label = '$G$')
p2, = ax1.semilogx(wg, magH, 'g-', label = '$H$')
p3, = ax1.semilogx(wg, magGH, 'r-', label = '$GH$')
ax1.yaxis.get_label().set_color(p1.get_color())
ax1.legend(bbox_to_anchor=(0., 1.02, 1., .102), loc=3, ncol=3,        \
                                mode="expand", borderaxespad=0.)
ax1.grid(True)
plt.subplots_adjust(hspace=0.85)
#
ax1 = host_subplot(2, 1, 2)
ax1.set_xlim([1e+4, 1e+7])
ax1.set_xticks([1e+4, 1e+5, 1e+6, 1e+7])
ax1.set_xticklabels(['$10^{4}$', '$10^{5}$', '$10^{6}$', '$10^{7}$'])
ax1.set_xlabel('$frequency, rads/sec$')
ax1.set_ylim([-180, 0])
ax1.set_yticks([-180, -90, 0])
ax1.set_yticklabels(['$-\pi$', '$-0.5\pi$', '$0$'])
ax1.set_ylabel('$Phase, rads$')
ax1.text(3e+5, -300, '$(b)$', size =10)
p1, = ax1.semilogx(wg, phaseG, 'b-', label = '$G$')
p2, = ax1.semilogx(wg, phaseH, 'g-', label = '$H$')
p3, = ax1.semilogx(wg, phaseGH, 'r-', label = '$GH$')
ax1.yaxis.get_label().set_color(p1.get_color())
ax1.legend(bbox_to_anchor=(0., 1.02, 1., .102), loc=3, ncol=3,        \
                                mode="expand", borderaxespad=0.)
ax1.grid(True)
#
plt.show()
#
#plt.draw()
#plt.savefig('Fig610.ps', dpi=400, bbox_inches='tight')
#
```

6.4.3 Program 6.12

```
#
def fdiode(vd):
    """ Diode characteristic.
        Returns diode current and its conductance"""
    #
    glow = 0.1e-06
    ghigh = 0.1e+03
    Vr = -0.3
    #
    if vd >= Vr:
        id = glow*(vd - Vr)
        gd = glow
    else:
        id = ghigh*(vd - Vr)
        gd = ghigh
    #
    return id, gd
#
def ch6fig12_solver(es0, gs, gls, gc, gcp, gload0, g1, gf1,          \
                    gf2, gc1, gc2, vref, vramp, d, ncycle, n):
    """Assumes fdiode is a function.
        Returns the lists vout, cinduct"""
    #
    import numpy as np
    #
    vd = 1.7987
    v0 = 1.8
    #
    pil = 12.8
    pvc = 1.8
    pvc1 = 0.2
    pvc2 = 0.17
    #
    erlim = 0.1e-07
    #
    nhcy = int(ncycle/2.0)
    gser = gs*gls/(gs + gls)
    gpar = gc*gcp/(gc + gcp)
    #
    vout = []
    cinduct = []
    #
    es = es0
    gload = gload0
    #
    for k in range(ncycle):
        #
        if k >= nhcy:
            #
            gload = 0.7*gload0     # case 1: 30% decrease in the load
            #
```

```
#
nc = int(np.rint(d*n))
vesum = 0.0
#
for j in range(n):
    #
    if j <= nc:
        gsw = 0.1e+03
    else:
        gsw = 0.1e-07
    #
    for its in range(1000):
        #
        v1 = (gs*vd + gls*v0 - pil)/(gs + gls)
        vc = (gc*v0 + gcp*pvc)/(gc + gcp)
        isw = gsw*(es - vd)
        il = gs*(vd - v1)
        id, gd = fdiode(vd)
        irrd = isw - id - il
        icap = gc*(v0 - vc)
        iload = gload*v0
        irr0 = il - icap - iload
        #
        err = np.sqrt(irrd**2 + irr0**2)
        #
        if err < erlim:
            break
        else:
            gmat = np.array([[(gsw + gd + gser), -gser],
                              [-gser, (gload + gpar + gser)]])
            bcol = np.array([irrd, irr0])
            xdv = np.linalg.solve(gmat, bcol)
            #
            dvd = xdv[0]
            dv0 = xdv[1]
            vd += dvd
            v0 += dv0
    #
    if its >= 999:
        print('time-step= ', (k*n + j), 'solver does not converge')
        break
    #
    pil = il
    pvc = vc
    #
    vout.append(v0)
    cinduct.append(il)
    #
    b1 = (gf1 + gf2 + gc2 + g1)*vref - gf1*v0 - gc2*pvc2
    b2 = g1*vref + gc1*pvc1
    gcntl = np.array([[(g1 + gc2), g1],
                      [g1, (g1 + gc1)]])
    bcnt = np.array([b1, b2])
```

```
            xvcnt = np.linalg.solve(gcntl, bcnt)
            ve = xvcnt[0]
            vc1 = xvcnt[1]
            vc2 = vref - ve
            pvc1 = vc1
            pvc2 = vc2
            #
            vesum += ve
            #
        #
        Ve_mean = vesum/n
        d = Ve_mean/vramp
        #
    #
    return vout, cinduct
#
#
def ch6fig12(es, freq, rl, ls, rc, cp, rload, r1,                    \
             rf1, rf2, c1, c2, vref, vramp, d, ncycle, n):
    """ Main program, plots results
        Buck converter with feedback control
        Frequency responses"""
    #
    #calling command:
    #ch6fig12(12.0, 625000.0, 0.08, 0.68e-06, 0.015, 0.49e-03, 0.15,      \
    #   3204.0, 1200.0, 764.0, 0.76e-08, 0.16e-9, 0.7, 1.8, 0.15, 120, 160)
    #
    from math import pi
    import numpy as np
    import matplotlib.pyplot as plt
    from mpl_toolkits.axes_grid1 import host_subplot
    #
    delt = 1.0/(freq*n)
    dteta = 2.0*pi/n
    gs = 1.0/rl
    gls = delt/ls
    gcp = cp/delt
    gc = 1.0/rc
    gload = 1.0/rload
    #
    gf1 = 1.0/rf1
    gf2 = 1.0/rf2
    g1 = 1.0/r1
    gc1 = c1/delt
    gc2 = c2/delt
    #
    vout, cinduct = ch6fig12_solver(es, gs, gls, gc, gcp, gload, g1,     \
                        gf1, gf2, gc1, gc2, vref, vramp, d, ncycle, n)
    #
    N = ncycle*n
    Nh = int(N/2)
    ang = dteta*np.arange(N)
    #
```

```
print(vout[Nh-1], vout[Nh], vout[Nh+1])
#
ax1 = host_subplot(2, 1, 1)
ax2 = ax1.twinx()
ax1.set_xlim([0, 2*pi*ncycle])
ax1.set_xticks([0, pi*ncycle, 2*pi*ncycle])
ax1.set_xticklabels(['$0$', '$\pi *ncycle$', '$2\pi *ncycle$'])
ax1.set_ylim([1.2, 2.0])
ax1.set_yticks([1.2, 1.6, 2.0])
ax1.set_yticklabels(['$1.2$', '$1.6$', '$2.0$'])
ax1.set_ylabel('$Voltage, V$')
ax2.set_ylim(0, 30)
ax2.set_yticks([0, 15, 30])
ax2.set_yticklabels(['$0$', '$15$', '$30$'])
ax2.set_ylabel('$Current, A$')
ax1.text(pi*ncycle, 0.85, '$(a)$', size =10)
p1, = ax1.plot(ang, vout, 'r-', label = '$v_{O}$')
p2, = ax2.plot(ang, cinduct, 'g-', label = '$I_{L}$')
ax1.yaxis.get_label().set_color(p1.get_color())
ax2.yaxis.get_label().set_color(p2.get_color())
ax1.legend(bbox_to_anchor=(0., 1.02, 1., .102), loc=3, ncol=2,     \
                                mode="expand", borderaxespad=0.)
ax1.grid(True)
plt.subplots_adjust(hspace=0.88)
#
ax1 = host_subplot(2, 1, 2)
ax2 = ax1.twinx()
ax1.set_xlim([pi*(ncycle-4), pi*(ncycle+4)])
ax1.set_xticks([pi*(ncycle-4), pi*ncycle, pi*(ncycle+4)])
ax1.set_xticklabels(['$\pi *(ncycle-4)$', '$\pi *ncycle$',         \
                                     '$\pi *(ncycle+4)$'])
ax1.set_xlabel('$\omega t, radians$')
ax1.set_ylim([1.2, 2.0])
ax1.set_yticks([1.2, 1.6, 2.0])
ax1.set_yticklabels(['$1.2$', '$1.6$', '$2.0$'])
ax1.set_ylabel('$Voltage, V$')
ax2.set_ylim(0, 30)
ax2.set_yticks([0, 15, 30])
ax2.set_yticklabels(['$0$', '$15$', '$30$'])
ax2.set_ylabel('$Current, A$')
ax1.text(pi*ncycle, 0.7, '$(b)$', size =10)
p1, = ax1.plot(ang[Nh-2*n:Nh+2*n-1], vout[Nh-2*n:Nh+2*n-1], 'r-',  \
                                         label = '$v_{O}$')
p2, = ax2.plot(ang[Nh-2*n:Nh+2*n-1], cinduct[Nh-2*n:Nh+2*n-1], 'g-', \
                                         label = '$I_{L}$')
ax1.yaxis.get_label().set_color(p1.get_color())
ax2.yaxis.get_label().set_color(p2.get_color())
ax1.legend(bbox_to_anchor=(0., 1.02, 1., .102), loc=3, ncol=2,       \
                                mode="expand", borderaxespad=0.)
ax1.grid(True)
#
plt.show()
#
```

```
    #plt.draw()
    #plt.savefig('Fig612.ps', dpi=400, bbox_inches='tight')
    #
```

6.4.4 Program 6.20

```
#
def fdiode(vd):
    """ Diode characteristic.
        Returns diode current and its conductance"""
    #
    glow = 0.1e-06
    ghigh = 0.1e+03
    Vr = -0.5
    #
    if vd >= Vr:
        id = glow*(vd - Vr)
        gd = glow
    else:
        id = ghigh*(vd - Vr)
        gd = ghigh
    #
    return id, gd
#
#
def ch6fig20_solver(es, grl, gls, grcp, gcp, gload, nc, ncycle, n):
    """Assumes fdiode is a function.
        Returns the lists vout, cinduct"""
    #
    import numpy as np
    #
    vd = 18.39
    v0 = 17.99
    pils = 10.19
    pvcp = 17.93
    #
    erlim = 0.1e-07
    #
    gser = grl*gls/(grl + gls)
    gpar = gload + grcp*gcp/(grcp + gcp)
    #
    vout = []
    vinduct = []
    vdiode = []
    cinduct = []
    cload = []
    cdiode = []
    #
    for k in range(ncycle):
        #
        for j in range(n):
            #
```

```
        if j <= nc:
            gsw = 0.25e+03
        else:
            gsw = 0.1e-07
        #
        for its in range(1000):
            #
            v1 = (grl*es + gls*vd - pils)/(grl + gls)
            vcp = (grcp*v0 + gcp*pvcp)/(grcp + gcp)
            #
            isw = gsw* vd
            id, gd = fdiode(v0 - vd)
            ils = grl*(es - v1)
            icp = grcp*(v0 - vcp)
            iload = gload*v0
            #
            irrd = ils - isw + id
            irr0 = -(id + icp + iload)
            #
            err = np.sqrt(irrd**2 + irr0**2)
            #
            if err < erlim:
                break
            else:
                g1 = gsw + gser
                gmat = np.array([[(g1 + gd), -gd],
                                  [-gd, (gd + gpar)]])
                bcol = np.array([irrd, irr0])
                xdv = np.linalg.solve(gmat, bcol)
                #
                dvd = xdv[0]
                dv0 = xdv[1]
                vd += dvd
                v0 += dv0
        #
        if its >= 999:
            print('time-step= ', (k*n + j), 'solver does not converge')
            break
        #
        pils = ils
        pvcp = vcp
        #
        vout.append(v0)
        vinduct.append(vd)
        vdiode.append(vd - v0)
        cinduct.append(ils)
        cload.append(iload)
        cdiode.append(-id)
        #
    #
#
return vout, cinduct, cload, vinduct, vdiode, cdiode
#
```

```
#
def ch6fig20(es, freq, rl, ls, rcp, cp, rload, d, ncycle, n):
    """ Main program, plots results
        Boost converter: Response waveforms"""
    #
    #calling command:
    #ch6fig20(12.0, 200000.0, 0.002, 0.2e-04, 0.008, 0.148e-03, 2.5,    \
    #                                             0.3333, 1000, 160)
    #
    from math import pi
    import numpy as np
    import matplotlib.pyplot as plt
    from mpl_toolkits.axes_grid1 import host_subplot
    #
    nc = int(np.rint(d*n))
    #
    delt = 1.0/(freq*n)
    dteta = 2.0*pi/n
    grl = 1.0/rl
    gls = delt/ls
    gcp = cp/delt
    grcp = 1.0/rcp
    gload = 1.0/rload
    #
    vout, cinduct, cload, vinduct, vdiode, cdiode =  ch6fig20_solver     \
                          (es, grl, gls, grcp, gcp, gload, nc, ncycle, n)
    #
    N = ncycle*n
    ang = dteta*np.arange(N)
    m0 = 2*(ncycle - 2)
    m1 = 2*(ncycle - 1)
    m2 = 2*ncycle
    #
    ax1 = host_subplot(2, 1, 1)
    ax2 = ax1.twinx()
    ax1.set_xlim([0, m2*pi])
    ax1.set_xticks([0, ncycle*pi, m2*pi])
    ax1.set_xticklabels(['$0$', '$\pi *ncycle$', '$2\pi *ncycle$'])
    ax1.set_ylim([0, 30])
    ax1.set_yticks([0, 15, 30])
    ax1.set_yticklabels(['$0$', '$15$', '$30$'])
    ax1.set_ylabel('$Voltages, V$')
    ax1.text(ncycle*pi, -13.0, '$(a)$', size =10)
    ax2.set_ylim(0, 30)
    ax2.set_yticks([0, 15, 30])
    ax2.set_yticklabels(['$0$', '$15$', '$30$'])
    ax2.set_ylabel('$Current, A$')
    p1, = ax1.plot(ang, vout, 'r-', label = '$V_{O}$')
    p2, = ax2.plot(ang, cinduct, 'b-.', label = '$I_{L}$')
    ax1.yaxis.get_label().set_color(p1.get_color())
    ax2.yaxis.get_label().set_color(p2.get_color())
    ax1.legend(bbox_to_anchor=(0., 1.02, 1., .102), loc=3, ncol=3,       \
                                  mode="expand", borderaxespad=0.)
```

```
ax1.grid(True)
plt.subplots_adjust(hspace = 0.85)
#
#
ax1 = host_subplot(2, 1, 2)
ax2 = ax1.twinx()
ax1.set_xlim([m0*pi-dteta, m2*pi])
ax1.set_xticks([m0*pi, m1*pi, m2*pi])
ax1.set_xticklabels(['$2\pi *(ncycle-2)$', '$2\pi *(ncycle-1)$',      \
                                        '$2\pi *ncycle$'])
ax1.set_ylim([16, 20.0])
ax1.set_yticks([16, 20])
ax1.set_yticklabels(['$16$', '$20$'])
ax1.set_ylabel('$Voltage, V$')
ax1.text(m1*pi, 14.5, '$(b)$', size =10)
p1, = ax1.plot(ang[N-2*n-1:N-1], vout[N-2*n-1:N-1], 'r-',              \
                                   label = '$V_{diode}$')
ax2.set_ylim(-1.0, 12.0)
ax2.set_yticks([-1, 0, 12])
ax2.set_yticklabels(['$-1$', '$0$', '$12$'])
ax2.set_ylabel('$Current, A$')
p2, = ax2.plot(ang[N-2*n-1:N-1], cinduct[N-2*n-1:N-1], 'b--',          \
                                   label = '$I_{L}$')
p3, = ax2.plot(ang[N-2*n-1:N-1], cload[N-2*n-1:N-1], 'b',              \
                                   label = '$I_{load}$')
ax1.yaxis.get_label().set_color(p1.get_color())
ax2.yaxis.get_label().set_color(p3.get_color())
ax1.legend(bbox_to_anchor=(0., 1.02, 1., .102), loc=3, ncol=3,         \
                               mode="expand", borderaxespad=0.)
ax1.grid(True)
#
plt.show()
#
#plt.draw()
#plt.savefig('Fig620.ps', dpi=400, bbox_inches='tight')
#
```

6.4.5 Program 6.22

```
def ch6fig22_fresp(Ggain, Hgain, wz, wp, cg1, cg2, bg1, bg2):
    """ Main program, plots results
        Transient response of boost converter
        Voltage mode control"""
    #
    #calling command:
    #ch6fig22_fresp(0.0864, 31201, 400, 40000, 84460,                  \
    #                          23148, 112.61, 2606773)
    #
    import numpy as np
    from scipy import signal
    import matplotlib.pyplot as plt
    from mpl_toolkits.axes_grid1 import host_subplot
```

```
#
ag1 = cg1 - cg2
ag2 = cg1*cg2
numG = [-Ggain, -Ggain*ag1, Ggain*ag2]
denG = [1, bg1, bg2]
#
ah1 = 2*wz
ah2 = wz*wz
bh1 = 2*wp
bh2 = wp*wp
numH = [Hgain, Hgain*ah1, Hgain*ah2]
denH = [1, bh1, bh2, 0]
#
GHgain = Ggain*Hgain
agh1 = ag1 + ah1
agh2 = (ag2 - ah1*ag1 - ah2)
agh3 = (ah1*ag2 - ag1*ah2)
agh4 = ag2*ah2
bgh1 = bh1+bg1
bgh2 = bh2 + bg1*bh1 + bg2
bgh3 = bg2*bh1 + bg1*bh2
bgh4 = bg2*bh2
numGH = [-GHgain, -GHgain*agh1, GHgain*agh2, GHgain*agh3,
        GHgain*agh4]
denGH = [1, bgh1, bgh2, bgh3, bgh4, 0]
#
tfG = signal.lti(numG, denG)
tfH = signal.lti(numH, denH)
tfGH = signal.lti(numGH, denGH)
#
wg = np.logspace(2, 5, num=1000)
wg, magG, phaseG = signal.bode(tfG, wg)
wg, magH, phaseH = signal.bode(tfH, wg)
wg, magGH, phaseGH = signal.bode(tfGH, wg)
#
magGH_min = min(abs(magGH))
for j in range(len(wg)):
    if abs(magGH[j]) == magGH_min:
        wg_min = wg[j]
        phaseGH_min = phaseGH[j]
print('wg_min = ', wg_min, 'magGH_min = ', magGH_min)
print('phase_margin = ', 180.0 + phaseGH_min)
#
ax1 = host_subplot(2, 1, 1)
ax1.set_xlim([1e+2, 1e+5])
ax1.set_xticks([1e+2, 1e+3, 1e+4, 1e+5])
ax1.set_xticklabels(['$10^{2}$', '$10^{3}$', '$10^{4}$', '$10^{5}$'])
ax1.set_ylim([-80, 80])
ax1.set_yticks([-80, 0, 80])
ax1.set_yticklabels(['$-80$', '$0$', '$80$'])
ax1.set_ylabel('$Gain, dB$')
ax1.text(3e+3, -125, '$(a)$', size =10)
p1, = ax1.semilogx(wg, magG, 'b-', label = '$G$')
```

```
    p2, = ax1.semilogx(wg, magH, 'g-', label = '$H$')
    p3, = ax1.semilogx(wg, magGH, 'r-', label = '$GH$')
    ax1.yaxis.get_label().set_color(p1.get_color())
    ax1.legend(bbox_to_anchor=(0., 1.02, 1., .102), loc=3, ncol=3,        \
                               mode="expand", borderaxespad=0.)
    ax1.grid(True)
    plt.subplots_adjust(hspace=0.85)
    #
    ax1 = host_subplot(2, 1, 2)
    ax1.set_xlim([1e+2, 1e+5])
    ax1.set_xticks([1e+2, 1e+3, 1e+4, 1e+5])
    ax1.set_xticklabels(['$10^{2}$', '$10^{3}$', '$10^{4}$', '$10^{5}$'])
    ax1.set_ylim([-270, 90])
    ax1.set_yticks([-270, -180, -118, 0])
    ax1.set_yticklabels(['$-1.5\pi$', '$-\pi$', '$-0.66\pi$', '$0$'])
    ax1.set_xlabel('$frequency, rads/sec$')
    ax1.set_ylabel('$Phase, rads$')
    ax1.text(3e+3, -500, '$(b)$', size =10)
    p1, = ax1.semilogx(wg, phaseG, 'b-', label = '$G$')
    p2, = ax1.semilogx(wg, phaseH, 'g-', label = '$H$')
    p3, = ax1.semilogx(wg, phaseGH, 'r-', label = '$GH$')
    ax1.yaxis.get_label().set_color(p1.get_color())
    ax1.legend(bbox_to_anchor=(0., 1.02, 1., .102), loc=3, ncol=3,        \
                               mode="expand", borderaxespad=0.)
    ax1.grid(True)
    #
    plt.show()
    #
    #plt.draw
    #plt.savefig('Fig622.ps', dpi=400, bbox_inches='tight')
    #
```

6.4.6 Program 6.23

```
#
def fdiode(vd):
    """ Diode characteristic.
        Returns diode current and its conductance"""
    #
    glow = 0.1e-06
    ghigh = 0.1e+03
    Vr = -0.3
    #
    if vd >= Vr:
        id = glow*(vd - Vr)
        gd = glow
    else:
        id = ghigh*(vd - Vr)
        gd = ghigh
    #
    return id, gd
#
```

```
def ch6fig23_solver(es0, grl, gls, grcp, gcp, gload0, K, a1, a2, c0,     \
                           c1, c2, c3, vref, vramp, v0_dc, ncycle, n):
    """Assumes fdiode is a function.
        Returns the lists vout, cinduct, cout"""
    #
    import numpy as np
    #
    vd = 18.35365
    v0 = 17.94839
    pils = 10.526
    pvcp = 17.8881
    #
    px0 = 0.0
    px1 = 0.0
    px2 = 0.0
    #
    erlim = 0.1e-07
    ghigh = 0.25e+03
    glow = 0.1e-07
    v0_dc = 18.0
    #
    vout = []
    cinduct = []
    cout = []
    #
    es = es0
    gload = gload0
    gser = grl*gls/(grl + gls)
    gpar = grcp*gcp/(grcp + gcp)
    #
    for k in range(ncycle):
        #
        gsw = ghigh
        ntrig = 1
        #
        if (k >= int(0.45*ncycle)) and (k < int(0.5*ncycle)):
            gload = 0.8*gload0
            es = 1.2*es0
        elif (k >= int(0.5*ncycle)) and (k < int(0.55*ncycle)):
            gload = 1.2*gload0
            es = 0.8*es0
        else:
            gload = gload0
            es = es0
        #
        for j in range(n):
            #
            for its in range(1000):
                #
                v1s = (grl*es + gls*vd - pils)/(grl + gls)
                vcp = (grcp*v0 + gcp*pvcp)/(grcp + gcp)
                #
                isw = gsw*(es - vd)
```

```
        #
        isw = gsw* vd
        id, gd = fdiode(v0 - vd)
        ils = gls*(v1s - vd) + pils
        icp = grcp*(v0 - vcp)
        iload = gload*v0
        #
        irrd = ils - isw + id
        irr0 = -(id + icp + iload)
        #
        err = np.sqrt(irrd**2 + irr0**2)
        #
        if err < erlim:
            break
        else:
            g1 = gsw + gser
            gmat = np.array([[(g1 + gd), -gd],
                             [-gd, (gd + gpar)]])
            bcol = np.array([irrd, irr0])
            xdv = np.linalg.solve(gmat, bcol)
            #
            dvd = xdv[0]
            dv0 = xdv[1]
            vd += dvd
            v0 += dv0
    #
    if its >= 999:
        print('time-step= ', (k*n + j), 'solver does not converge')
        break
    #
    pils = ils
    pvcp = vcp
    #
    vout.append(v0)
    cinduct.append(ils)
    cout.append(gload*v0)
    #
    vrr = v0_dc - v0
    x2 = c1*vrr + c0*px2 + c2*px1
    x1 = px1 +c3*x2
    x0 = px0 + c3*x1
    dve = K*(x2 + a1*x1 +a2*x0)
    #
    px0 = x0
    px1 = x1
    px2 = x2
    #
    ve = vref + dve
    if (ve <= (j*vramp/n)) and (ntrig == 1):
        gsw = glow
        ntrig = 0
    #
#
```

```
    #
    return vout, cinduct, cout
    #
#
def ch6fig23(es, freq, rl, ls, rcp, cp, rload, K, z, p, vref, vramp,     \
                                                v0_dc, ncycle, n):
    """ Main program, plots results
        Boost converter with voltage mode control
        Transient eesponse"""
    #
    #calling command:
    #ch6fig23(5.0, 200000.0, 0.002, 0.20e-04, 0.008, 0.148e-02,          \
    #        6.0, 31201.0, 400.0, 40000.0, 2.5, 3.4616, 18.0, 1000, 256)
    #
    from math import pi
    import numpy as np
    import matplotlib.pyplot as plt
    from mpl_toolkits.axes_grid1 import host_subplot
    #
    delt = 1.0/(freq*n)
    dteta = 2.0*pi/n
    grl = 1.0/rl
    gls = delt/ls
    gcp = cp/delt
    grcp = 1.0/rcp
    gload = 1.0/rload
    #
    a1 = 2*z
    a2 = z**2
    b1 = 2*p
    b2 = p**2
    c0 = 1.0/(1.0 + b1*delt + b2*delt**2)
    c1 = c0*delt
    c2 = -b2*c1
    c3 = delt
    #
    vout, cinduct, cout = ch6fig23_solver(es, grl, gls, grcp, gcp,      \
        gload, K, a1, a2, c0, c1, c2, c3, vref, vramp, v0_dc, ncycle, n)
    #
    N = ncycle*n
    ang = dteta*np.arange(N)
    Nh = int(N/2)
    ncyh = int(ncycle/2)
    #
    ax1 = host_subplot(2, 1, 1)
    ax2 = ax1.twinx()
    ax1.set_xlim([0, 2*pi*ncycle])
    ax1.set_xticks([0, pi*ncycle, 2*pi*ncycle])
    ax1.set_xticklabels(['$0$', '$\pi *ncycle$', '$2\pi *ncycle$'])
    ax1.set_ylim([0, 20])
    ax1.set_yticks([0, 10, 20])
    ax1.set_yticklabels(['$0$', '$10$', '$20$'])
    ax1.set_ylabel('$Voltage, V$')
```

```
ax2.set_ylim(0, 20)
ax2.set_yticks([0, 10, 20])
ax2.set_yticklabels(['$0$', '$10$', '$20$'])
ax2.set_ylabel('$Current, A$')
ax1.text(pi*ncycle, -9, '$(a)$', size =10)
p1, = ax1.plot(ang[(Nh-ncyh*n):(Nh+ncyh*n-1)],                        \
       vout[(Nh-ncyh*n):(Nh+ncyh*n-1)], 'r-', label = '$v_{out}$')
p2, = ax2.plot(ang[(Nh-ncyh*n):(Nh+ncyh*n-1)],                        \
       cout[(Nh-ncyh*n):(Nh+ncyh*n-1)], 'g-', label = '$I_{load}$')
p3, = ax2.plot(ang[(Nh-ncyh*n):(Nh+ncyh*n-1)],                        \
       cinduct[(Nh-ncyh*n):(Nh+ncyh*n-1)], 'b-', label = '$I_{L}$')
ax1.yaxis.get_label().set_color(p1.get_color())
ax2.yaxis.get_label().set_color(p2.get_color())
ax1.legend(bbox_to_anchor=(0., 1.02, 1., .102), loc=3, ncol=4,      \
                              mode="expand", borderaxespad=0.)
ax1.grid(True)
plt.subplots_adjust(hspace=0.9)
#
ax1 = host_subplot(2, 1, 2)
ax2 = ax1.twinx()
ax1.set_xlim([2*pi*(ncycle-2), 2*pi*ncycle])
ax1.set_xticks([2*pi*(ncycle-2), 2*pi*(ncycle-1), 2*pi*ncycle])
ax1.set_xticklabels(['$2\pi *(ncycle-2)$', '$2\pi *(ncycle-1)$',    \
                                             '$2\pi *ncycle$'])
ax1.set_xlabel('$\omega t, radians$')
ax1.set_ylim([17, 19])
ax1.set_yticks([17, 19])
ax1.set_yticklabels(['$17$', '$19$'])
ax1.set_ylabel('$Voltage, V$')
ax2.set_ylim(11, 13)
ax2.set_yticks([11, 12, 13])
ax2.set_yticklabels(['$11$', '$12$', '$13$'])
ax2.set_ylabel('$Current, A$')
ax1.text(2*pi*(ncycle-1), 15.6, '$(b)$', size =10)
p1, = ax1.plot(ang[N-2*n:N-1], vout[N-2*n:N-1], 'r-',                \
                                label = '$v_{out}$')
p2, = ax2.plot(ang[N-2*n:N-1], cinduct[N-2*n:N-1], 'b--',            \
                                label = '$I_{L}$')
ax1.yaxis.get_label().set_color(p1.get_color())
ax2.yaxis.get_label().set_color(p2.get_color())
ax1.legend(bbox_to_anchor=(0., 1.02, 1., .102), loc=3, ncol=2,      \
                              mode="expand", borderaxespad=0.)
ax1.grid(True)
#
plt.show()
#
#plt.draw()
#plt.savefig('Fig623.ps', dpi=400, bbox_inches='tight')
#
```

6.4.7 Program 6.36

```
#
def fdiode(vd):
    """ Diode characteristic.
        Returns diode current and its conductance"""
    #
    glow = 0.1e-06
    ghigh = 0.1e+03
    Vr = -0.5
    #
    if vd >= Vr:
        id = glow*(vd - Vr)
        gd = glow
    else:
        id = ghigh*(vd - Vr)
        gd = ghigh
    #
    return id, gd
#
#
def ch6fig36_solver(es, grl, gls, grcp, gcp, gload, nc, ncycle, n):
    """Assumes fdiode is a function.
        Returns the lists vout, cinduct"""
    #
    import numpy as np
    #
    vd = 0.0
    v0 = -24.0
    pils = 9.1164357
    pvcp = -24.533384
    #
    erlim = 0.1e-07
    #
    gser = grl*gls/(grl + gls)
    gpar = gload + grcp*gcp/(grcp + gcp)
    #
    vout = []
    vinduct = []
    vdiode = []
    cinduct = []
    cload = []
    cdiode = []
    #
    for k in range(ncycle):
        #
        for j in range(n):
            #
            if j <= nc:
                gsw = 0.25e+03
            else:
                gsw = 0.1e-07
            #
```

```
            for its in range(1000):
                #
                v1 = (grl*vd - pils)/(grl + gls)
                vcp = (grcp*v0 + gcp*pvcp)/(grcp + gcp)
                #
                isw = gsw* (es - vd)
                id, gd = fdiode(vd - v0)
                ils = grl*(vd - v1)
                icp = grcp*(v0 - vcp)
                iload = gload*v0
                #
                irrd = isw - id - ils
                irr0 = id - icp - iload
                #
                err = np.sqrt(irrd**2 + irr0**2)
                #
                if err < erlim:
                    break
                else:
                    g1 = gsw + gser
                    gmat = np.array([[(g1 + gd), -gd],
                                      [-gd, (gd + gpar)]])
                    bcol = np.array([irrd, irr0])
                    xdv = np.linalg.solve(gmat, bcol)
                    #
                    dvd = xdv[0]
                    dv0 = xdv[1]
                    vd += dvd
                    v0 += dv0
            #
            if its >= 999:
                print('time-step= ', (k*n + j), 'solver does not converge')
                break
            #
            pils = ils
            pvcp = vcp
            #
            vout.append(v0)
            vinduct.append(vd)
            vdiode.append(vd - v0)
            cinduct.append(ils)
            cload.append(iload)
            cdiode.append(-id)
            #
        #
    #
    return vout, cinduct, cload, vinduct, vdiode, cdiode
#
#
def ch6fig36(es, freq, rl, ls, rcp, cp, rload, d, ncycle, n):
    """ Main program, plots results
        Buck-boost converter
        Response waveforms"""
```

```
#
#calling command:
#ch6fig36(12.0, 200000.0, 0.002, 0.22e-04, 0.008, 0.1e-03,          \
#                               8.0, 0.66667, 1000, 160)
#
from math import pi
import numpy as np
import matplotlib.pyplot as plt
from mpl_toolkits.axes_grid1 import host_subplot
#
nc = int(np.rint(d*n))
#
delt = 1.0/(freq*n)
dteta = 2.0*pi/n
grl = 1.0/rl
gls = delt/ls
gcp = cp/delt
grcp = 1.0/rcp
gload = 1.0/rload
#
vout, cinduct, cload, vinduct, vdiode, cdiode =                         \
          ch6fig36_solver(es, grl, gls, grcp, gcp, gload, nc, ncycle, n)
#
N = ncycle*n
ang = dteta*np.arange(N)
m0 = 2*(ncycle - 2)
m1 = 2*(ncycle - 1)
m2 = 2*ncycle
#
ax1 = host_subplot(2, 1, 1)
ax2 = ax1.twinx()
ax1.set_xlim([m0*pi-dteta, m2*pi])
ax1.set_xticks([m0*pi, m1*pi, m2*pi])
ax1.set_xticklabels(['$2\pi *(ncycle-2)$', '$2\pi *(ncycle-1)$',     \
                                          '$2\pi *ncycle$'])
ax1.set_ylim([-30, 0])
ax1.set_yticks([-30, -15, 0])
ax1.set_yticklabels(['$-30$', '$-15$', '$0$'])
ax1.set_ylabel('$Voltages, V$')
ax1.text(m1*pi, -42, '$(a)$', size =10)
p1, = ax1.plot(ang[N-2*n-1:N-1], vout[N-2*n-1:N-1], 'r-',           \
                                     label = '$V_{O}$')
ax2.set_ylim(0, 12)
ax2.set_yticks([0, 12])
ax2.set_yticklabels(['$0$', '$12$'])
ax2.set_ylabel('$Current, A$')
p2, = ax2.plot(ang[N-2*n-1:N-1], cinduct[N-2*n-1:N-1], 'b-.',       \
                                     label = '$I_{L}$')
ax1.yaxis.get_label().set_color(p1.get_color())
ax2.yaxis.get_label().set_color(p2.get_color())
ax1.legend(bbox_to_anchor=(0., 1.02, 1., .102), loc=3, ncol=3,       \
                          mode="expand", borderaxespad=0.)
ax1.grid(True)
```

```
    plt.subplots_adjust(hspace = 0.9)
    #
    ax1 = host_subplot(2, 1, 2)
    ax2 = ax1.twinx()
    ax1.set_xlim([m0*pi-dteta, m2*pi])
    ax1.set_xticks([m0*pi, m1*pi, m2*pi])
    ax1.set_xticklabels(['$2\pi *(ncycle-2)$',                          \
                '$2\pi *(ncycle-1)$', '$2\pi *ncycle$'])
    ax1.set_ylim([-30.0, 40.0])
    ax1.set_yticks([-30, 0, 40])
    ax1.set_yticklabels(['$-30$', '$0$', '$30$'])
    ax1.set_ylabel('$Voltage, V$')
    ax1.text(m1*pi, -57.0, '$(b)$', size =10)
    ax2.set_ylim(-1.0, 12.0)
    ax2.set_yticks([0, 6, 12])
    ax2.set_yticklabels(['$0$', '$6$', '$12$'])
    ax2.set_ylabel('$Current, A$')
    p1, = ax1.plot(ang[N-2*n-1:N-1], vdiode[N-2*n-1:N-1], 'r-',          \
                                  label = '$V_{diode}$')
    p2, = ax1.plot(ang[N-2*n-1:N-1], vinduct[N-2*n-1:N-1], 'r--',        \
                                  label = '$v_{L}$')
    p3, = ax2.plot(ang[N-2*n-1:N-1], cdiode[N-2*n-1:N-1], 'b-.',         \
                                  label = '$I_{diode}$')
    ax1.yaxis.get_label().set_color(p1.get_color())
    ax2.yaxis.get_label().set_color(p3.get_color())
    ax1.legend(bbox_to_anchor=(0., 1.02, 1., .102), loc=3, ncol=3,   \
                                  mode="expand", borderaxespad=0.)
    ax1.grid(True)
    #
    plt.show()
    #
    #plt.draw()
    #plt.savefig('Fig636.ps', dpi=400, bbox_inches='tight')
    #
```

6.4.8 Program 6.38

```
#
def ch6fig38_fresp(Ls, Cp, R, Vi, D, Hgain, wz, wp):
    """ Main program, plots results
        Buck-boost converter
        Frequency responses"""
    #
    #calling command:
    #ch6fig38_fresp(0.22e-04, 0.44e-03, 8.0, 12.0, 0.66667,           \
    #                             41690, 985.76, 82170)
    #
    from math import pi
    import numpy as np
    from scipy import signal
    import matplotlib.pyplot as plt
    from mpl_toolkits.axes_grid1 import host_subplot
```

```
#
Ds = 1.0 - D
#
Ggain = -Vi
yrhp = D*Ls/(Ds*Ds*R)
bg1 = Ls*Cp
bg2 = Ls/R
bg3 = Ds*Ds
numG = [Ggain*yrhp, -Ggain]
denG = [bg1, bg2, bg3]
#
numH = [Hgain, Hgain*2.0*wz, Hgain*wz*wz]
denH = [1, 2.0*wp, wp*wp, 0]
#
tfG = signal.lti(numG, denG)
tfH = signal.lti(numH, denH)
#
wg = np.logspace(1, 6, num=100000)
wg, magG, phaseG = signal.bode(tfG, wg)
wg, magH, phaseH = signal.bode(tfH, wg)
magGH = magG + magH
phaseGH = phaseG + phaseH
#
minG = min(abs(magGH))
#
print('minG = ', minG)
#
for jwg in range(len(wg)):
    if abs(magGH[jwg]) == minG:
        wgmin = wg[jwg]
        phmin = phaseGH[jwg]
#
print('crossover frequency = ', wgmin/(2.0*pi),                    \
       'phase margin = ', 180.0 + phmin,'degrees')
#
ax1 = host_subplot(2, 1, 1)
ax1.set_xlim([1e+1, 1e+6])
ax1.set_xticks([1e+1, 1e+2, 1e+3, 1e+4, 1e+5, 1e+6])
ax1.set_xticklabels(['$10^{1}$', '$10^{2}$', '$10^{3}$',            \
                              '$10^{4}$', '$10^{5}$', '$10^{6}$'])
ax1.set_ylim([-50, 80])
ax1.set_yticks([-40, 0, 40, 80])
ax1.set_yticklabels(['$-40$', '$0$', '$40$', '$80$'])
ax1.set_ylabel('$Gain, dB$')
ax1.text(3.16e+3, -100, '$(a)$', size =10)
p1, = ax1.semilogx(wg, magG, 'b-', label = '$G$')
p2, = ax1.semilogx(wg, magH, 'g-', label = '$H$')
p3, = ax1.semilogx(wg, magGH, 'r-', label = '$GH$')
ax1.yaxis.get_label().set_color(p1.get_color())
ax1.legend(bbox_to_anchor=(0., 1.02, 1., .102), loc=3, ncol=3,      \
                          mode="expand", borderaxespad=0.)
ax1.grid(True)
plt.subplots_adjust(hspace=0.85)
```

```
    #
    ax1 = host_subplot(2, 1, 2)
    ax1.set_xlim([1e+1, 1e+6])
    ax1.set_xticks([1e+1, 1e+2, 1e+3, 1e+4, 1e+5, 1e+6])
    ax1.set_xticklabels(['$10^{1}$', '$10^{2}$', '$10^{3}$',              \
                         '$10^{4}$', '$10^{5}$', '$10^{6}$'])
    ax1.set_ylim([-270, 180])
    ax1.set_yticks([-270, -180, -90, 0, 90, 180])
    ax1.set_yticklabels(['$-1.5\pi$', '$-\pi$', '$-0.5\pi$',              \
                                 '$0$', '$0.5\pi$', '$\pi$'])
    ax1.set_xlabel('$frequency, rads/sec$')
    ax1.set_ylabel('$Phase, rads$')
    ax1.text(3.16e+3, -540, '$(b)$', size =10)
    p1, = ax1.semilogx(wg, phaseG, 'b-', label = '$G$')
    p2, = ax1.semilogx(wg, phaseH, 'g-', label = '$H$')
    p3, = ax1.semilogx(wg, phaseGH, 'r-', label = '$GH$')
    ax1.yaxis.get_label().set_color(p1.get_color())
    ax1.legend(bbox_to_anchor=(0., 1.02, 1., .102), loc=3, ncol=3,          \
                                         mode="expand", borderaxespad=0.)
    ax1.grid(True)
    #
    plt.show()
    #
    #plt.draw()
    #plt.savefig('Fig638.ps', dpi=400, bbox_inches='tight')
    #
```

6.4.9 Program 6.39

```
#
def fdiode(vd):
    """ Diode characteristic.
        Returns diode current and its conductance"""
    #
    glow = 0.1e-06
    ghigh = 0.1e+03
    Vr = -0.3
    #
    if vd >= Vr:
        id = glow*(vd - Vr)
        gd = glow
    else:
        id = ghigh*(vd - Vr)
        gd = ghigh
    #
    return id, gd
#
def ch6fig39_solver(es0, grl, gls, grcp, gcp, gload0, K, a1, a2,          \
                              c0, c1, c2, c3, vref, vramp, v0_dc, ncycle, n):
    """Assumes fdiode is a function.
        Returns the lists vout, cinduct, cout"""
    #
```

```
import numpy as np
#
vd = -24.623394
v0 = -24.239877
pils = 8.351717
pvcp = -24.19730
#
px0 = 0.0
px1 = 0.0
px2 = 0.0
#
erlim = 0.1e-07
ghigh = 0.25e+03
glow = 0.1e-07
#
vout = []
cinduct = []
cout = []
#
es = es0
gload = gload0
gser = grl*gls/(grl + gls)
gpar = grcp*gcp/(grcp + gcp)
#
for k in range(ncycle):
    #
    gsw = ghigh
    ntrig = 1
    #
    if (k >= int(0.45*ncycle)) and (k < int(0.5*ncycle)):
        gload = 0.8*gload0
        es = 1.2*es0
    elif (k >= int(0.5*ncycle)) and (k < int(0.55*ncycle)):
        gload = 1.2*gload0
        es = 0.8*es0
    else:
        gload = gload0
        es = es0
    #
    for j in range(n):
        #
        for its in range(1000):
            #
            v1 = (grl*vd - pils)/(grl + gls)
            vcp = (grcp*v0 + gcp*pvcp)/(grcp + gcp)
            #
            isw = gsw* (es - vd)
            id, gd = fdiode(vd - v0)
            ils = grl*(vd - v1)
            icp = grcp*(v0 - vcp)
            iload = gload*v0
            #
            irrd = isw - id - ils
```

```
            irr0 = id - icp - iload
            #
            err = np.sqrt(irrd**2 + irr0**2)
            #
            if err < erlim:
                break
            else:
                g1 = gsw + gser
                gmat = np.array([[(g1 + gd), -gd],
                                  [-gd, (gd + gpar)]])
                bcol = np.array([irrd, irr0])
                xdv = np.linalg.solve(gmat, bcol)
                #
                dvd = xdv[0]
                dv0 = xdv[1]
                vd += dvd
                v0 += dv0
        #
        if its >= 999:
            print('time-step= ', (k*n + j), 'solver does not converge')
            break
        #
        pils = ils
        pvcp = vcp
        #
        vout.append(v0)
        cinduct.append(ils)
        cout.append(gload*v0)
        #
        vrr = v0_dc + v0
        x2 = c1*vrr + c0*px2 + c2*px1
        x1 = px1 +c3*x2
        x0 = px0 + c3*x1
        dve = K*(x2 + a1*x1 +a2*x0)
        #
        px0 = x0
        px1 = x1
        px2 = x2
        #
        ve = vref + dve
        if (ve <= (j*vramp/n)) and (ntrig == 1):
            gsw = glow
            ntrig = 0
        #
    #
#
return vout, cinduct, cout
#
#
def ch6fig39(es, freq, rl, ls, rcp, cp, rload, K, z, p, vref,              \
                                                  vramp, v0_dc, ncycle, n):
    """ Main program, plots results
        Buck-boost converter: Voltage mode control
```

```
    Transient response"""
#
#calling command:
#ch6fig39(12.0, 200000.0, 0.002, 0.22e-04, 0.008, 0.44e-03,          \
#        8.0, 41690.0, 985.76, 82170.0, 2.5, 3.75, 24.0, 1000, 256)
#
from math import pi
import numpy as np
import matplotlib.pyplot as plt
from mpl_toolkits.axes_grid1 import host_subplot
#
delt = 1.0/(freq*n)
dteta = 2.0*pi/n
grl = 1.0/rl
gls = delt/ls
gcp = cp/delt
grcp = 1.0/rcp
gload = 1.0/rload
#
a1 = 2*z
a2 = z**2
b1 = 2*p
b2 = p**2
c0 = 1.0/(1.0 + b1*delt + b2*delt**2)
c1 = c0*delt
c2 = -b2*c1
c3 = delt
#
vout, cinduct, cout = ch6fig39_solver(es, grl, gls, grcp, gcp,        \
    gload, K, a1, a2, c0, c1, c2, c3, vref, vramp, v0_dc, ncycle, n)
#
N = ncycle*n
ang = dteta*np.arange(N)
Nh = int(N/2)
ncyh = int(ncycle/2)
#
ax1 = host_subplot(2, 1, 1)
ax2 = ax1.twinx()
ax1.set_xlim([0, 2*pi*ncycle])
ax1.set_xticks([0, pi*ncycle, 2*pi*ncycle])
ax1.set_xticklabels(['$0$', '$\pi *ncycle$', '$2\pi *ncycle$'])
ax1.set_ylim([-34, -22])
ax1.set_yticks([-34, -28, -22])
ax1.set_yticklabels(['$-34$', '$-28$', '$-22$'])
ax1.set_ylabel('$Voltage, V$')
ax2.set_ylim(-15, 30)
ax2.set_yticks([-15, 0, 15, 30])
ax2.set_yticklabels(['$-15$', '$0$', '$15$', '$30$'])
ax2.set_ylabel('$Current, A$')
ax1.text(pi*ncycle, -39, '$(a)$', size =10)
p1, = ax1.plot(ang[(Nh-ncyh*n):(Nh+ncyh*n-1)],                        \
      vout[(Nh-ncyh*n):(Nh+ncyh*n-1)], 'r-', label = '$v_{out}$')
p2, = ax2.plot(ang[(Nh-ncyh*n):(Nh+ncyh*n-1)],                        \
```

```
        cout[(Nh-ncyh*n):(Nh+ncyh*n-1)], 'g-', label = '$I_{load}$')
p3, = ax2.plot(ang[(Nh-ncyh*n):(Nh+ncyh*n-1)],                    \
        cinduct[(Nh-ncyh*n):(Nh+ncyh*n-1)], 'b-', label = '$I_{L}$')
ax1.yaxis.get_label().set_color(p1.get_color())

ax2.yaxis.get_label().set_color(p2.get_color())
ax1.legend(bbox_to_anchor=(0., 1.02, 1., .102), loc=3, ncol=3,    \
                        mode="expand", borderaxespad=0.)
ax1.grid(True)
plt.subplots_adjust(hspace=0.9)
#
ax1 = host_subplot(2, 1, 2)
ax2 = ax1.twinx()
ax1.set_xlim([2*pi*(ncycle-2), 2*pi*ncycle])
ax1.set_xticks([2*pi*(ncycle-2), 2*pi*(ncycle-1), 2*pi*ncycle])
ax1.set_xticklabels(['$2\pi *(ncycle-2)$', '$2\pi *(ncycle-1)$',  \
                                '$2\pi *ncycle$'])
ax1.set_xlabel('$\omega t, radians$')
ax1.set_ylim([-25, -21])
ax1.set_yticks([-25, -21])
ax1.set_yticklabels(['$-25$', '$-21$'])
ax1.set_ylabel('$Voltage, V$')
ax2.set_ylim(7, 11)
ax2.set_yticks([7, 9, 11])
ax2.set_yticklabels(['$7$', '$9$', '$11$'])
ax2.set_ylabel('$Current, A$')
ax1.text(2*pi*(ncycle-1), -27.5, '$(b)$', size =10)
p1, = ax1.plot(ang[N-2*n:N-1], vout[N-2*n:N-1], 'r-',             \
                        label = '$v_{out}$')
p2, = ax2.plot(ang[N-2*n:N-1], cinduct[N-2*n:N-1], 'b--',         \
                        label = '$I_{L}$')
ax1.yaxis.get_label().set_color(p1.get_color())
ax2.yaxis.get_label().set_color(p2.get_color())
ax1.legend(bbox_to_anchor=(0., 1.02, 1., .102), loc=3, ncol=2,    \
                        mode="expand", borderaxespad=0.)
ax1.grid(True)
#
plt.show()
#
#plt.draw()
#plt.savefig('Fig639.ps', dpi=400, bbox_inches='tight')
#
```

6.4.10 Program 6.40

```
#
def ch6fig40_fresp(Vi, Im, ls, R, cp, rsn, Ds, Hgain, ah1, bh1):
    """ Main program, plots results
        Buck-boost converter: Current mode control
        Frequency responses"""
    #
    #calling command:
```

```
#ch6fig40_fresp(12.0, 4.613636, 0.22e-04, 8.0, 0.44e-03, 0.201465,   \
#                            0.333333, 12363.4, 1050.886, 77077.8)
#
import numpy as np
from scipy import signal
import matplotlib.pyplot as plt
from mpl_toolkits.axes_grid1 import host_subplot
#
D = 1.0 - Ds
#
Ggain = Vi
wRHP = (Ds*Ds*R)/(D*ls)
ag1 = -1.0/wRHP
ag0 = 1.0
#
bg2 = Im*ls*cp
bg1 = Vi*cp/Ds + Im*ls/R
bg0 = Im*(Ds**2) + Vi*(1.0 + D)/(R*Ds)
#
numG = [Ggain*ag1, Ggain*ag0]
denG = [bg2, bg1, bg0]
#
numH = [Hgain/ah1, Hgain]
denH = [1/bh1, 1, 0]
#
tfG = signal.lti(numG, denG)
tfH = signal.lti(numH, denH)
#
wg = np.logspace(1, 6, num=100000)
wg, magG, phaseG = signal.bode(tfG, wg)
wg, magH, phaseH = signal.bode(tfH, wg)
#
magGH = magG + magH
phaseGH = phaseG + phaseH
#
magGH_min = min(abs(magGH))
for j in range(len(wg)):
    if abs(magGH[j]) == magGH_min:
        wg_min = wg[j]
        phaseGH_min = phaseGH[j]
print('wg_min = ', wg_min, 'magGH_min = ', magGH_min,                   \
               'phase_margin = ', 180.0 + phaseGH_min)
#
ax1 = host_subplot(2, 1, 1)
ax1.set_xlim([1e+1, 1e+6])
ax1.set_xticks([1e+1, 1e+2, 1e+3, 1e+4, 1e+5, 1e+6])
ax1.set_xticklabels(['$10^{1}$', '$10^{2}$', '$10^{3}$',                \
                     '$10^{4}$', '$10^{5}$', '$10^{6}$'])
ax1.set_ylim([-50, 50])
ax1.set_yticks([-50, -25, 0, 25, 50])
ax1.set_yticklabels(['$-50$', '$-25$', '$0$', '$25$', '$50$'])
ax1.set_ylabel('$Gain, dB$')
ax1.text(3e+3, -80, '$(a)$', size =10)
```

```
    p1, = ax1.semilogx(wg, magG, 'b-', label = '$G_{VI}$')
    p2, = ax1.semilogx(wg, magH, 'g-', label = '$H$')
    p3, = ax1.semilogx(wg, magGH, 'r-', label = '$HG_{VI}$')
    ax1.axis['left'].label.set_color(p1.get_color())
    ax1.legend(bbox_to_anchor=(0., 1.02, 1., .102), loc=3, ncol=3,         \
                                mode="expand", borderaxespad=0.)
    ax1.grid(True)
    plt.subplots_adjust(hspace=0.85)
    #
    ax1 = host_subplot(2, 1, 2)
    ax1.set_xlim([1e+1, 1e+6])
    ax1.set_xticks([1e+1, 1e+2, 1e+3, 1e+4, 1e+5, 1e+6])
    ax1.set_xticklabels(['$10^{1}$', '$10^{2}$', '$10^{3}$',                 \
                         '$10^{4}$', '$10^{5}$', '$10^{6}$'])
    ax1.set_ylim([-270, 0])
    ax1.set_yticks([-270, -180, -110, 0])
    ax1.set_yticklabels(['$-1.5\pi$', '$-pi$', '$-0.61\pi$', '$0$'])
    ax1.set_xlabel('$frequency, rads/sec$')
    ax1.set_ylabel('$Phase, rads$')
    ax1.text(3e+3, -450, '$(b)$', size =10)
    p1, = ax1.semilogx(wg, phaseG, 'b-', label = '$G_{VI}$')
    p2, = ax1.semilogx(wg, phaseH, 'g-', label = '$H$')
    p3, = ax1.semilogx(wg, phaseGH, 'r-', label = '$HG_{VI}$')
    ax1.yaxis.get_label().set_color(p1.get_color())
    ax1.legend(bbox_to_anchor=(0., 1.02, 1., .102), loc=3, ncol=3,         \
                                mode="expand", borderaxespad=0.)
    ax1.grid(True)
    #
    plt.show()
    #
    #plt.draw()
    #plt.savefig('Fig640.ps', dpi=400, bbox_inches='tight')
    #
```

6.4.11 Program 6.41

```
#
def fdiode(vd):
    """ Diode characteristic.
        Returns diode current and its conductance"""
    #
    glow = 0.1e-06
    ghigh = 0.1e+03
    Vr = -0.3
    #
    if vd >= Vr:
        id = glow*(vd - Vr)
        gd = glow
    else:
        id = ghigh*(vd - Vr)
        gd = ghigh
    #
    return id, gd
#
def ch6fig41_solver(es0, v0_dc, grl, gls, grcp, gcp, gload0,           \
```

```
                 gsn, c0, c1, c2, c3, vref, Im, d, ncycle, n):
    """Assumes fdiode is a function.
       Returns the lists vout, cinduct"""
    #
    import numpy as np
    #
    vd = -24.44084946
    v0 = -24.058252
    pils = 8.2597747
    pvcp = -24.01623178
    #
    px1 = 0.74746
    px2 = -9.9858742
    #
    erlim = 0.1e-07
    ghigh = 0.25e+03
    glow = 0.1e-07
    Ie = gsn*vref
    #
    vout = []
    cinduct = []
    cout = []
    #
    es = es0
    gload = gload0
    gser = grl*gls/(grl + gls)
    gpar = grcp*gcp/(grcp + gcp)
    #
    for k in range(ncycle):
        #
        ntrig = 1
        gsw = ghigh
        #
        if (k >= int(0.45*ncycle)) and (k < int(0.5*ncycle)):
            gload = 0.8*gload0
            es = 1.2*es0
        elif (k >= int(0.5*ncycle)) and (k < int(0.55*ncycle)):
            gload = 1.2*gload0
            es = 0.8*es0
        else:
            gload = gload0
            es = es0
        #
        for j in range(n):
            #
            for its in range(1000):
                #
                v1 = (grl*vd - pils)/(grl + gls)
                vcp = (grcp*v0 + gcp*pvcp)/(grcp + gcp)
                #
                isw = gsw* (es - vd)
                id, gd = fdiode(vd - v0)
                ils = grl*(vd - v1)
```

```
        icp = grcp*(v0 - vcp)
        iload = gload*v0
        #
        irrd = isw - id - ils
        irr0 = id - icp - iload
        #
        err = np.sqrt(irrd**2 + irr0**2)
        #
        if err < erlim:
            break
        else:
            g1 = gsw + gser
            gmat = np.array([[(g1 + gd), -gd],
                             [-gd, (gd + gpar)]])
            bcol = np.array([irrd, irr0])
            xdv = np.linalg.solve(gmat, bcol)
            #
            dvd = xdv[0]
            dv0 = xdv[1]
            vd += dvd
            v0 += dv0
    #
    if its >= 999:
        print('time-step= ', (k*n + j), 'solver does not converge')
        break
    #
    pils = ils
    pvcp = vcp
    #
    vout.append(v0)
    cinduct.append(ils)
    cout.append(gload*v0)
    #
    vrr = v0 - v0_dc
    x2 = (px2 + c1*vrr)/(1.0 + c0)
    x1 = px1 +c2*x2
    die = x1 + c3*x2
    #
    px1 = x1
    px2 = x2
    #
    ie = Ie + die
    #
#
#           without the compensating ramp
#
#            if (ie <= ils) and (ntrig == 1):
#                gsw = glow
#                ntrig = 0
    #
#
#           with the compensating ramp
#
```

```
            if (ie <= (ils + j*Im/n)) and (ntrig == 1):
                gsw = glow
                ntrig = 0
            #
        #
    #
    return vout, cinduct, cout
    #
#
def ch6fig41(es, v0_dc, freq, rl, ls, rcp, cp, rload, rsn,                    \
                        K, wz, wp, vref, Im, d, ncycle, n):
    """ Main program, plots results
        Buck-boost converter: Current mode control
        Transient response"""
    #
    #calling command:
    #ch6fig41(12.0, -24.0, 200000.0, 0.002, 0.22e-04, 0.008,                  \
    #           0.44e-03, 8.0, 0.20702635, 12363.4, 1050.886,                 \
    #             77077.8, 2.5, 4.61363625, 666667, 1000, 256)
    #
    from math import pi
    import numpy as np
    import matplotlib.pyplot as plt
    from mpl_toolkits.axes_grid1 import host_subplot
    #
    delt = 1.0/(freq*n)
    dteta = 2.0*pi/n
    grl = 1.0/rl
    gls = delt/ls
    gcp = cp/delt
    grcp = 1.0/rcp
    gload = 1.0/rload
    gsn = 1.0/rsn
    #
    c0 = delt*wp
    c1 = K*c0
    c2 = delt
    c3 = 1.0/wz
    #
    vout, cinduct, cout = ch6fig41_solver(es, v0_dc, grl,                     \
                              gls, grcp, gcp, gload, gsn, c0,                 \
                              c1, c2, c3, vref, Im, d, ncycle, n)
    #
    N = ncycle*n
    ang = dteta*np.arange(N)
    Nh = int(N/2)
    ncyh = int(ncycle/2)
    Nhb = ncyh*n
    #
    ax1 = host_subplot(2, 1, 1)
    ax2 = ax1.twinx()
    ax1.set_xlim([0, 2*pi*ncycle])
    ax1.set_xticks([0, pi*ncycle, 2*pi*ncycle])
```

```
ax1.set_xticklabels(['$0$', '$\pi *ncycle$', '$2\pi *ncycle$'])
ax1.set_ylim([-27, 0])
ax1.set_yticks([-27, -18, -9, 0])
ax1.set_yticklabels(['$-27$', '$-18$', '$-9$', '$0$'])
ax1.set_ylabel('$Voltage, V$')
ax1.text(pi*ncycle, -39, '$(a)$', size =10)
ax2.set_ylim(-8, 16)
ax2.set_yticks([-8, 0, 8, 16])
ax2.set_yticklabels(['$-8$', '$0$', '$8$', '$16$'])
ax2.set_ylabel('$Current, A$')
p1, = ax1.plot(ang[(Nh-ncyh*n):(Nh+ncyh*n-1)],                    \
       vout[(Nh-ncyh*n):(Nh+ncyh*n-1)], 'r-', label = '$v_{out}$')
p2, = ax2.plot(ang[(Nh-ncyh*n):(Nh+ncyh*n-1)],                    \
       cout[(Nh-ncyh*n):(Nh+ncyh*n-1)], 'g-', label = '$I_{load}$')
p3, = ax2.plot(ang[(Nh-ncyh*n):(Nh+ncyh*n-1)],                    \
       cinduct[(Nh-ncyh*n):(Nh+ncyh*n-1)], 'b-', label = '$I_{L}$')
ax1.yaxis.get_label().set_color(p1.get_color())
ax2.yaxis.get_label().set_color(p2.get_color())
ax1.legend(bbox_to_anchor=(0., 1.02, 1., .102), loc=3,            \
              ncol=3, mode="expand", borderaxespad=0.)
ax1.grid(True)
plt.subplots_adjust(hspace=0.9)
#
ax1 = host_subplot(2, 1, 2)
ax2 = ax1.twinx()
ax1.set_xlim([ang[Nhb-70*n], ang[Nhb+70*n]])
ax1.set_xticks([2*pi*(ncyh-70), 2*pi*ncyh, 2*pi*(ncyh + 70)])
ax1.set_xticklabels(['$2\pi *(ncyh-70)$', '$2\pi *ncyh$',        \
                     '$2\pi *(ncyh+70)$'])
ax1.set_xlabel('$\omega t, radians$')
ax1.set_ylim([-24.3, -23.3])
ax1.set_yticks([-24.3, -23.8, -23.3])
ax1.set_yticklabels(['$-24.3$', '$-23.8$', '$-23.3$'])
ax1.set_ylabel('$Voltage, V$')
ax1.text(2*pi*ncyh, -25.0, '$(b)$', size =10)
ax2.set_ylim(4, 16)
ax2.set_yticks([4, 10, 16])
ax2.set_yticklabels(['$4$', '$10$', '$16$'])
ax2.set_ylabel('$Current, A$')
p1, = ax1.plot(ang[Nhb-120*n:Nhb+120*n],                          \
       vout[Nhb-120*n:Nhb+120*n], 'r-', label = '$v_{out}$')
p2, = ax2.plot(ang[Nhb-120*n:Nhb+120*n],                          \
       cinduct[Nhb-120*n:Nhb+120*n], 'b--', label = '$I_{L}$')
ax1.yaxis.get_label().set_color(p1.get_color())
ax2.yaxis.get_label().set_color(p2.get_color())
ax1.legend(bbox_to_anchor=(0., 1.02, 1., .102), loc=3,            \
              ncol=2, mode="expand", borderaxespad=0.)
ax1.grid(True)
#
plt.show()
#

#plt.draw()
#plt.savefig('Fig641.ps', dpi=400, bbox_inches='tight')
#
```

References

[1] N. Mohan, T. M. Undeland, W. P. Robbins, *Power Electronics: Converters, Applications, and Design*, 2^{nd}Ed., John Wiley & Sons Inc., USA, 1995.

[2] B. Hauke, *Basic calculations of a buck converter's power stage*, Texas Instruments Application Report SLVA477, August 2015.

[3] G. Maimone, *Selecting L and C components in the power stage of MC34700 switching regulators*, International Rectifier Application Note AN4067, December 2010.

[4] B. Lynch, K. Hesse, *Under the hood of low-voltage DC/DC converters*, SLUP206, 2002 Texas Instruments Power Supply Design Seminar.

[5] A. H. Rahimi, P. Parto, P. Asadi, *Compensator design procedure for buck converter with voltage-mode error amplifier*, International Rectifier Application Note AN-1162.

[6] J. Sun, D. M. Mitchell, M. F. Greuel, P. T. Krein, R. M. Bass, *Averaged modeling of PWM converters operating in discontinuous conduction mode*, IEEE Trans. on Power Electronics, vol.16, no.4, July 2001, pp. 482-492.

[7] B. Hauke, *Basic calculations of a boost converter's power stage*, Texas Instruments Application Report SLVA372, July 2010.

[8] E. Rogers, *Understanding boost power stages in switch-mode power supplies*, Texas Instruments Application Report SLVA061, March 1999.

[9] S. W. Lee, *Practical feedback loop analysis for voltage-mode boost converter*, Texas Instruments Application Report SLVA633, Jan. 2014.

[10] S. W. Lee, *Practical feedback loop analysis for current-mode boost converter*, Texas Instruments Application Report SLVA636, March 2014.

[11] AN-1994, *Modeling and design of current mode control boost converters*, Texas Instruments Application Report SNVA408, April 2013.

[12] H. D. Venable, *The k-factor: A new mathematical tool for stability analysis and synthesis*, Proc. of Powercon 10, CA, March 22-24, 1983.

[13] I. Weiss, *Basic calculations of an inverting buck-boost power stage*, Texas Instruments Application Report SLVA721, August 2017.

[14] V. Choudhary, T. Hegarty, D. Pace, *Under the hood of a non-inverting buck-boost converter*, Texas Instruments Power Supply Seminar, Sept. 2016.

[15] E. Rogers, *Understanding buck-boost power stages in switch mode power supplies*, Texas Instruments Application Report SLVA059, Nov. 2002.

7

Isolated DC–DC Converters

Converters that have their input connected directly to the electric utility system are known as off-line converters. Regulations usually require electric isolation between the converter input and output in off-line applications. Even if the converter input is not connected to the utility system, including an isolation transformer may result in an optimized converter design when a large step-up or step-down voltage conversion ratio is required. The power processed by the switch and diodes is lower in an isolated converter. There is the further advantage of obtaining multiple DC outputs by adding secondary windings and converter circuits. By ensuring that the transformer operates at the switching frequency (and not at the utility system frequency), the size, weight, and cost of the isolation transformer may be reduced.

The simplified equivalent circuit of the isolation transformer is shown in Figure 7.1. The equivalent circuit assumes that the transformer is lossless and that its windings are perfectly coupled. The magnetizing inductance L_M, referred to the primary, is shown on the input side.

The magnetizing current is given by

$$L_M \frac{di_M}{dt} = v_P. \tag{7.1}$$

Integrating eqn (7.1) over the switching period T_S,

$$i_M(T_S) - i_M(0) = \frac{1}{L_M} \int_0^{T_S} v_P \, dt. \tag{7.2}$$

The direct component of the voltage applied to the magnetizing inductance is zero in the steady-state. Therefore,

$$\frac{1}{L_M} \int_0^{T_S} v_P dt = 0. \tag{7.3}$$

In the steady-state, the magnetizing current (and, consequently, the magnetic flux) should be reset to the initial value at the start of each switching

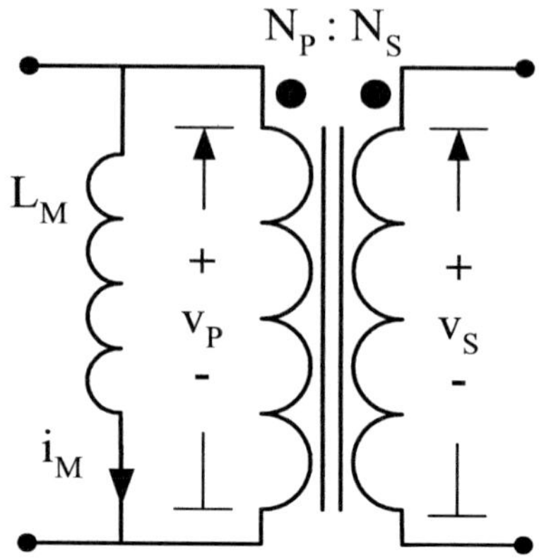

Figure 7.1 Equivalent circuit of isolation transformer.

period. This is an additional constraint in the operation of isolated DC–DC converters.

Each of the basic converter topologies studied in the previous chapter has several isolated versions. This chapter is restricted to the study of commonly used, isolated DC–DC converters.

7.1 Single-ended Forward Converter

The converter is single-ended because power flows through the transformer for only one polarity of the primary winding voltage [1, 2, 4–6]. Power flows to the load only when the switch is turned on and for this reason, the converter is also known as a forward converter. The single-ended forward converter is widely used for output power less than 200 watts and input voltage in the range 60 V–200 V.

One version of the single-ended forward converter is shown in Figure 7.2. Beyond the diode D_1, it is identical to the buck converter with the output filter comprising L_S, C_P and the free wheeling diode being D_2. The voltage source e_S is connected to the buck converter by means of a three-winding transformer. The tertiary winding is for the purpose of resetting the magnetic flux to zero at the start of each switching period. The diode D_3 controls the power flow out of the tertiary winding. In the discussion that follows, it is assumed that the converter operates in the continuous conduction mode, i.e., $i_L \geq 0$. Figure 7.3 shows the typical waveforms of the voltages and currents.

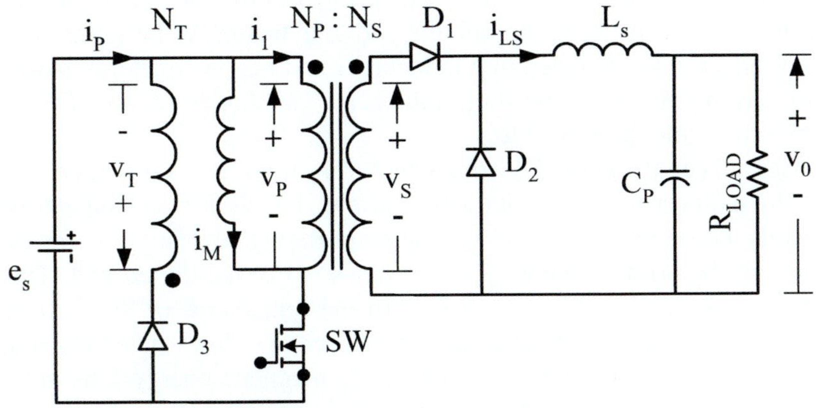

Figure 7.2 Single-ended forward converter, with reset winding.

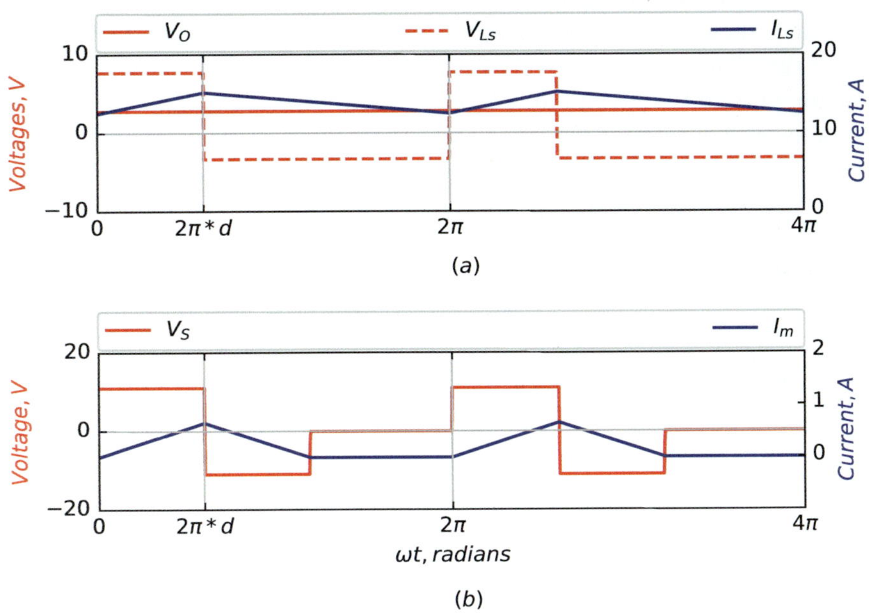

Figure 7.3 Single-ended forward converter waveforms.

During each period of the switching frequency, the converter passes through the ON state with the power switch SW turned on and the OFF state with the switch turned off. The duration of the ON state is dT_S, where d is the duty ratio. The diode D_1 is forward biased during the ON state and it enables the flow of current to the load through the output inductor. At the same time, the current flows in the magnetizing inductance and magnetic energy stored in the core sets up the magnetic flux in the core. The diodes D_2 and D_3 are reverse biased during the ON state.

The free wheeling diode D_2 conducts during the entire OFF state and provides the path for the output inductor current. The OFF state consists of two intervals. During the first interval, the magnetizing current i_M is forced to circulate in the primary winding and it flows out of the dotted-end. This causes a current $i_T = \frac{N_P}{N_T} i_M$ to flow into the dotted-end of the tertiary winding and the diode D_3 to turn on. The current i_T flows into the input voltage source and it returns to the source, the magnetic energy remaining in the core. The voltage across the tertiary when D_3 conducts is $v_T = -e_S$ and the primary winding voltage is $v_P = -\frac{N_P}{N_T} e_S$. The negative voltage appearing across the primary winding discharges the magnetizing inductance and brings the current i_M to zero. The diode D_3 prevents i_T from going negative, and, therefore, the magnetizing current falls linearly to zero and remains at zero. The second interval of the OFF state is the duration when the magnetizing current is zero and the diode D_3 is reverse biased.

The equations during the ON state are

$$
\begin{aligned}
v_P &= e_S, \; v_S = \frac{N_S}{N_P} e_S, \; v_T = \frac{N_T}{N_P} e_S \\
i_P &= i_M + \frac{N_S}{N_P} i_{LS} \\
L_S \frac{di_{LS}}{dt} &= \frac{N_S}{N_P} e_S - v_0 \\
L_M \frac{di_M}{dt} &= e_S.
\end{aligned}
$$

The equations during the OFF state are

First interval :

$$
\begin{aligned}
v_P &= -\frac{N_P}{N_S} e_S, \; v_S = -\frac{N_S}{N_T} e_S, \; v_T = -e_S \\
L_M \frac{di_M}{dt} &= -\frac{N_P}{N_T} e_S \\
L_S \frac{di_{LS}}{dt} &= -v_0,
\end{aligned}
$$

Second interval :

$$
\begin{aligned}
v_P &= 0,\ v_S = 0,\ v_T = 0 \\
L_M \frac{di_M}{dt} &= 0 \\
L_S \frac{di_{LS}}{dt} &= -v_O.
\end{aligned}
$$

Let $d_2 T_S$ be the interval over which the diode D_3 is forward biased, where $f_S = 1/T_S$ is the switching frequency. Note that $d_2 T_S \leq (1-d)T_S$. The direct component of the magnetizing inductor's voltage is zero in the steady-state. Therefore,

$$
d\,e_S + d_2 \left(-\frac{N_P}{N_T} e_S \right) = 0. \tag{7.4}
$$

That is,

$$
d_2 = \frac{N_T}{N_P} d \leq (1-d). \tag{7.5}
$$

Or,

$$
d \leq \frac{1}{1 + \frac{N_T}{N_P}}. \tag{7.6}
$$

In most applications, the tertiary winding has the same number of turns as the primary winding and $d < 0.5$.

In the steady-state, the direct component of the voltage across the output filter inductor is also zero. Therefore,

$$
d \left(\frac{N_S}{N_P} e_S - v_O \right) + (1-d)(-v_O) = 0. \tag{7.7}
$$

That is

$$
v_O = d \frac{N_S}{N_P} e_S. \tag{7.8}
$$

7.1.1 Simulation of the forward converter

Figure 7.4 shows the forward converter that will be simulated. The transient equivalent circuit is shown in Figure 7.5. Three probe voltages have been applied in the circuit. The past history current sources are $pilm$, $pils$, and $gcp * pvc$.

The equations obtained from the transient equivalent circuit are

$$
vd1 = vd2 - v1/N_{\mathrm{tr}}
$$

$$
\begin{aligned}
vd3 &= es + v1 \\
id1,\, gd1 &= fdiode(vd1) \\
id2,\, gd2 &= fdiode(vd2) \\
id3,\, gd3 &= fdiode(vd3) \\
i1 &= id3 - id1/N_{\mathrm{tr}} \\
ilm &= glm\, v1 + pilm \\
isw &= gsw(es - v1) \\
iload &= gload\, v0.
\end{aligned}
$$

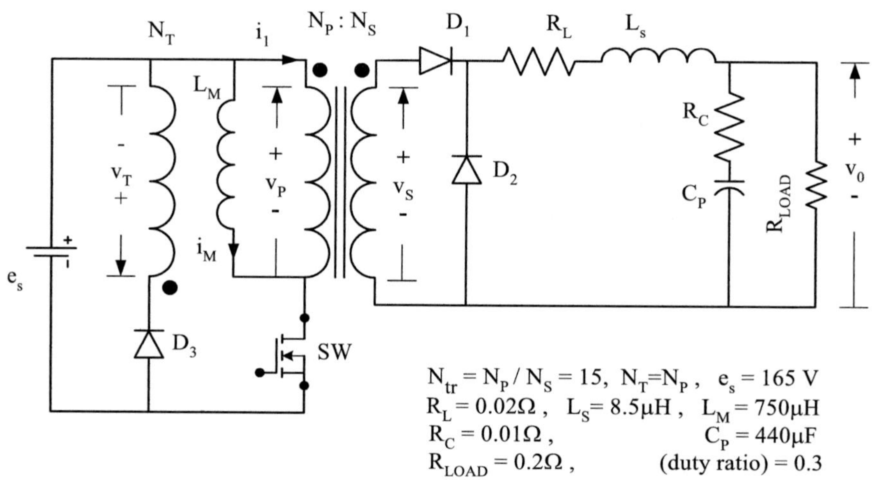

Figure 7.4 Forward converter for simulation, with reset winding.

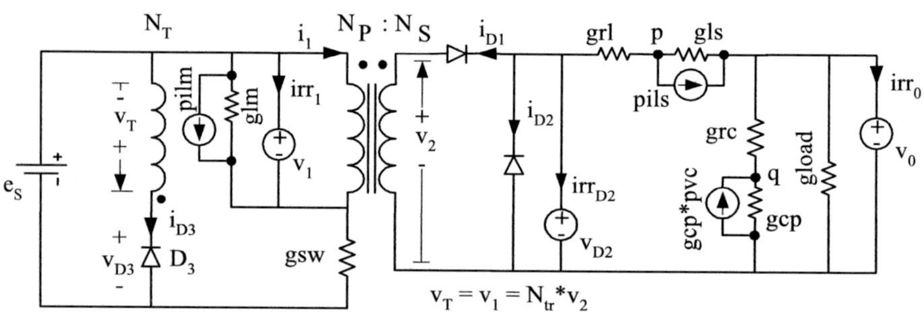

Figure 7.5 Transient equivalent circuit of the forward converter.

Applying KCL at nodes p and q, we get

$$\begin{aligned} grl(vp - vd2) + gls(vp - v0) + pils &= 0 \\ grc(vc - v0) + gcp(vc - pvc) &= 0. \end{aligned}$$

Therefore, the node voltages and the output filter currents are given by

$$\begin{aligned} vp &= (grl\,vd2 + gls\,v0 - pils)/(grl + gls) \\ vc &= (grc\,v0 + gcp\,pvc)/(grc + gcp) \\ ils &= grl(vd2 - vp) \\ icp &= grc(v0 - vc). \end{aligned}$$

Finally, the error currents are

$$\begin{aligned} irr1 &= isw - i1 - ilm \\ irrD2 &= -(id1 + id2 + ils) \\ irr0 &= ils - icp - iload. \end{aligned}$$

The small–signal equivalent circuit for correcting the probe voltages is shown in Figure 7.6. From this circuit, we obtain the corrections to be added to the probe voltages as

$$\begin{bmatrix} g_{11} & -gd1/N_{\mathrm{tr}} & 0 \\ -gd1/N_{\mathrm{tr}} & g_{22} & -gser \\ 0 & -gser & g_{33} \end{bmatrix} \begin{bmatrix} \delta v1 \\ \delta vd2 \\ \delta v0 \end{bmatrix} = \begin{bmatrix} irr1 \\ irrD2 \\ irro, \end{bmatrix}, \qquad (7.9)$$

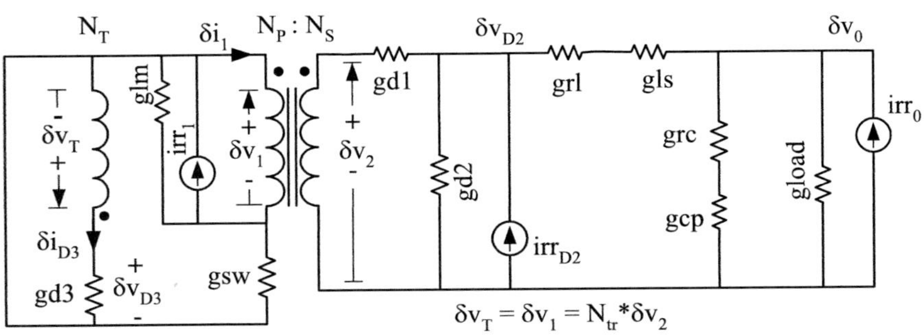

Figure 7.6 Small signal equivalent of the forward converter.

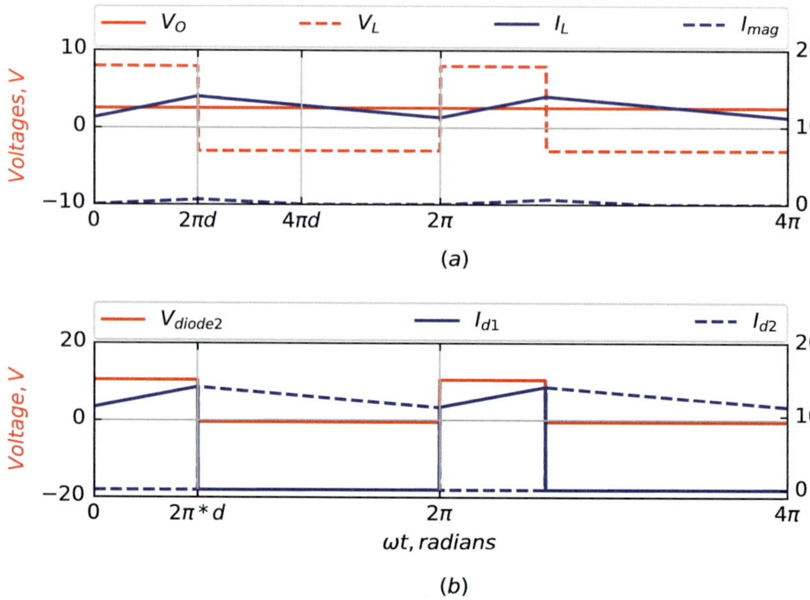

Figure 7.7 Forward converter with reset winding: response waveforms.

where

$$
\begin{aligned}
gser &= (grl\, gls)/(grl + gls) \\
gpar &= (grc\, gcp)/(grc + gcp) \\
g_{11} &= gsw + glm + gd3 + gd1/N_{\text{tr}}^2 \\
g_{22} &= gd1 + gd2 + gser \\
g_{33} &= gser + gpar + gload.
\end{aligned}
$$

Figure 7.7, plotted by Program 7.07, shows the results of the simulation.

7.1.2 Effects of leakage inductances [2]

7.1.2.1 Leakage inductance between primary and secondary

Let the leakage inductance between the primary and secondary windings be L_K, seen from the secondary side. The filter inductor's current flows through the diode D_1 during the ON state. In the absence of the leakage inductance, the inductor current is transferred instantaneously from the diode D_1 to the diode D_2 when the switch is turned off. Similarly, the inductor current is

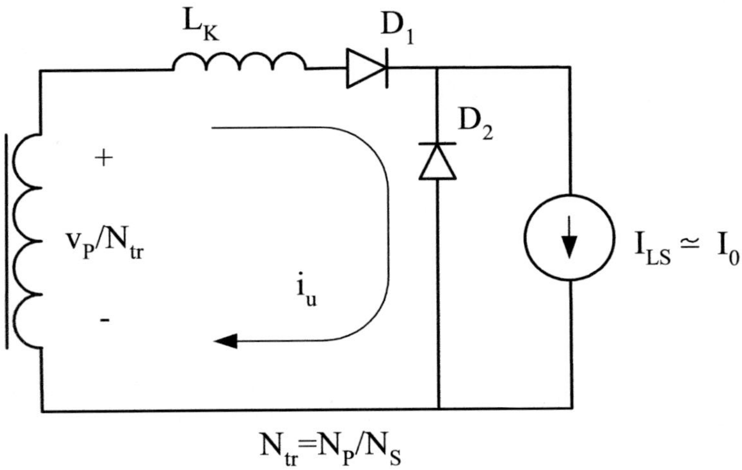

Figure 7.8 Commutation of the inductor current.

transferred instantaneously from the diode D_2 to D_1 when the switch is turned on at the start of the ON state. Instantaneous transfer of the inductor current from one diode to the other cannot take place in the presence of the leakage inductance L_K. Instead, a transient current i_u is initiated in the loop shown in Figure 7.8 and the inductor current is transferred during a finite commutation interval τ_u. Both diodes are forward biased during the commutation interval.

Consider that the switch is turned on at the start of the $(k+1)$th period and let the commutation interval last from $t = k\,T_S$ to $t = k\,T_S + \tau_u$. Applying KVL to the loop, we get

$$L_K \frac{di_u}{dt} = v_P/N_{\text{tr}}. \tag{7.10}$$

Observing the boundary conditions $i_L = i_u = 0$ at $t = k\,T_S$ and $i_L = I_0$ at $t = k\,T_S + \tau_u$ and integrating over the commutation interval,

$$L_K(I_0 - 0) = \tau_u \frac{v_P}{N_{\text{tr}}}. \tag{7.11}$$

Or,

$$\tau_u = \frac{L_K\, I_0}{(v_P/N_{\text{tr}})}. \tag{7.12}$$

The voltage across the diode D_2 is

$$\begin{aligned} v_{d2} &= v_P/N_{\text{tr}} \text{ for } \tau_u \leq t \leq dT_S \\ &= 0 \text{ outside the interval.} \end{aligned}$$

The output voltage is the average of v_{d2} over the switching period. That is,

$$\begin{aligned} v_0 &= \frac{1}{T_S}\int_{\tau_u}^{dT_S} (v_P/N_{\text{tr}})dt \\ &= d\frac{v_1}{N_{\text{tr}}} - f_S L_K I_0. \end{aligned}$$

The transformer's leakage inductance is responsible for the drop in the output voltage.

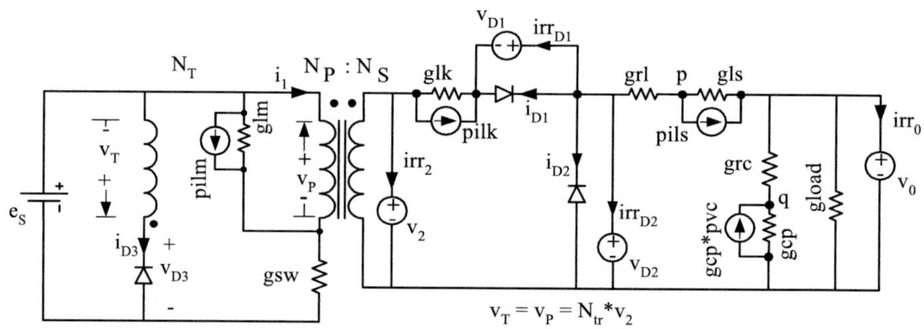

Figure 7.9 Equivalent circuit of forward converter: leakage inductance L_K.

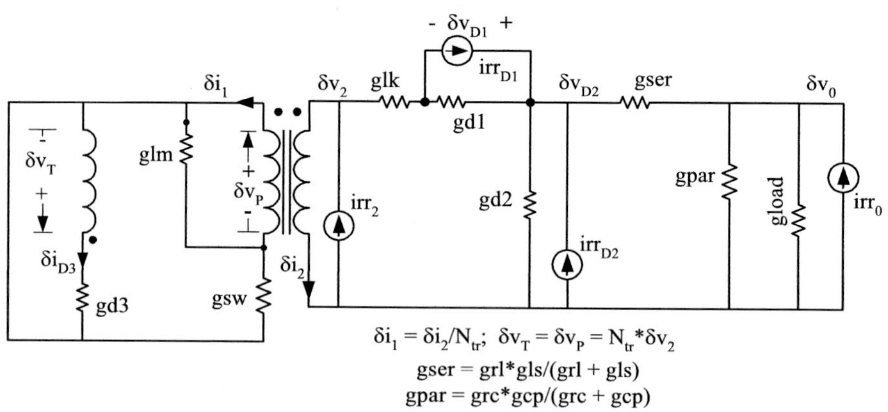

Figure 7.10 Forward converter with leakage inductance L_K: small–signal equivalent circuit.

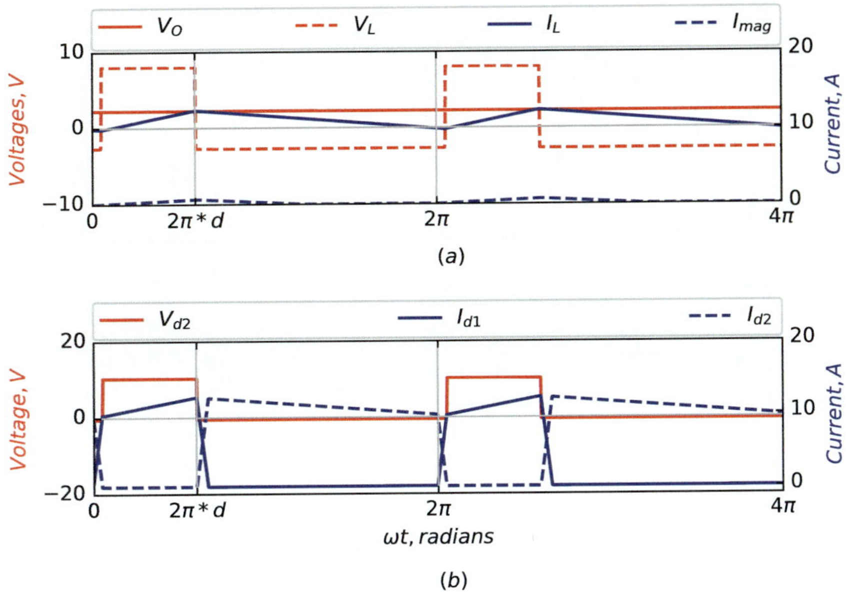

Figure 7.11 Response waveforms of the forward converter including the effects of L_K.

Consider the forward converter shown in Figure 7.4. Let the leakage inductance between the primary winding and the secondary winding be 0.3 μH seen from the secondary side. The transient equivalent circuit and the small–signal equivalent circuit are shown in Figures 7.9 and 7.10, respectively.

Note that in Figure 7.9, an additional probe voltage v_{D1} has been applied across the diode D_1. The development of Figures 7.9 and 7.10, the equations for calculating the error currents, and the corrections that should be added to the probe voltages are all left as exercises. The effects of the leakage inductance on the voltage and current waveforms of the forward converter are shown in Figure 7.11, which has been plotted by Program 7.11.

7.1.2.2 Leakage inductance between the primary and the tertiary

The leakage inductance L_{PT} between the primary and the tertiary is shown in Figure 7.12. Assume that the primary and tertiary windings have the same number of turns and let $L_{PT} = 5$ μH.

When the power switch is turned off, the magnetizing current is diverted into the dot-less end of the primary winding. The reflected current in the

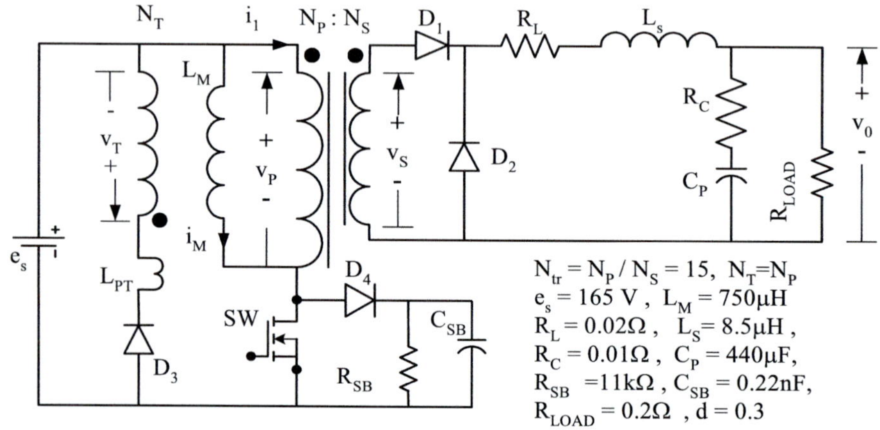

Figure 7.12 Forward converter with leakage inductance L_{PT}.

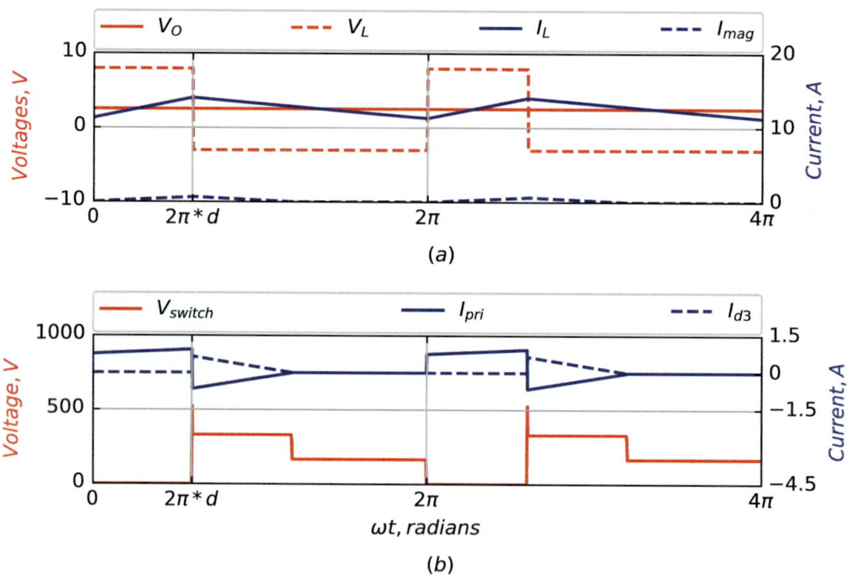

Figure 7.13 Forward converter with leakage inductance L_{PT}, without snubber.

tertiary winding flows into the dotted end. Thus, a steeply rising current begins to flow into the leakage inductance generating a voltage spike v_{PT} across it. The tertiary winding's voltage is then given by $v_T = e_S + v_{\mathrm{PT}} + v_{D3}$

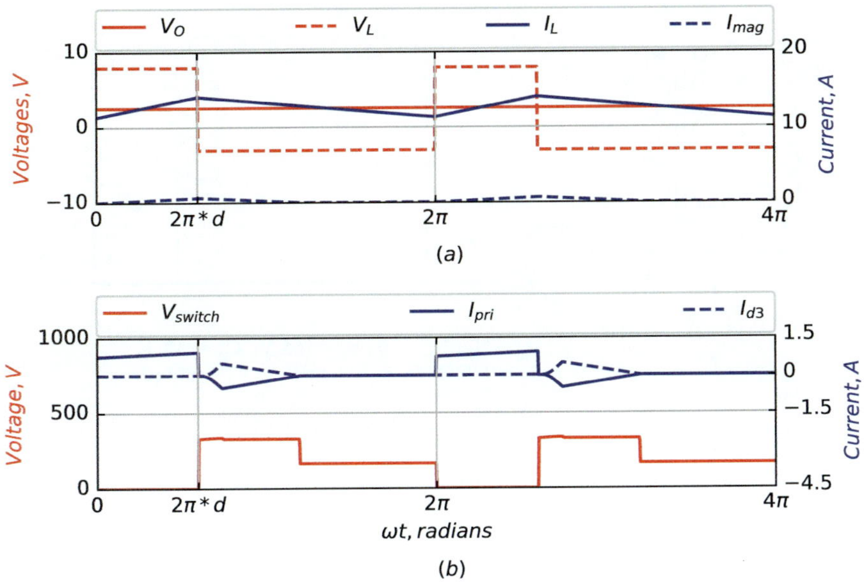

Figure 7.14 Forward converter with leakage inductance $L_{\rm PT}$, with snubber.

where v_{D3} is the voltage drop across diode D_3. The voltage across the switch when it is turned off is

$$\begin{aligned} v_{\rm SW} &= e_S + v_P \\ &= e_S + v_T \\ &= 2\,e_S + v_{PT} + v_{D3} \\ &\simeq (2.5 \text{ to } 3)\,e_S. \end{aligned}$$

This is essentially equal to twice the input voltage with a spike $v_{\rm PT}$ added to it.

The voltage and current waveforms obtained from a simulation of the converter without the RCD snubber is shown in Figure 7.13, which has been plotted by Program 7.13. It is usual to protect the switch with an RCD snubber (D_4, R_{SB}, C_{SB}). Figure 7.14 shows the converter waveforms (plotted by Program 7.14), when the switch is provided with the snubber.

7.1.3 Forward converter using resonant reset

Figure 7.15 shows the forward converter that uses the resonant reset technique [7]. The isolating transformer is a compact two-winding transformer, which is

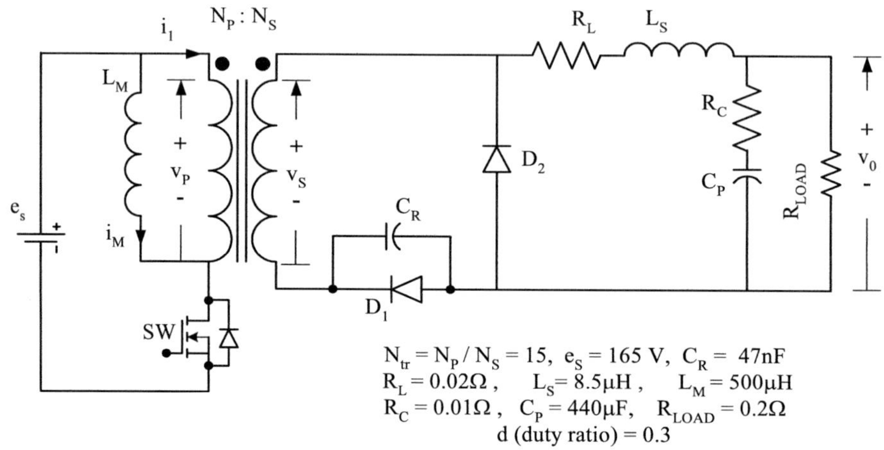

Figure 7.15 Forward converter with resonant reset.

less expensive than the three-winding transformer used in forward converters using the auxiliary reset procedure. An external capacitor C_R is connected in parallel with the rectifying diode D_1. Other capacitors appear in parallel with C_R when the power switch is turned off. These are the switch capacitance C_{SW}, the transformer's primary winding capacitance C_{TP}, and the diode's capacitance C_{D1}.

The magnetizing current rises linearly during the ON state, from i_{Mmin} at the start of the ON state to i_{Mmax} at the end of the ON state. That is,

$$i_{Mmax} = i_{Mmin} + \frac{dT_S}{L_M} e_S. \tag{7.13}$$

In the steady-state, the magnetizing current rises symmetrically from a negative value $-I_M$ to the positive value I_M. That is,

$$I_M = -I_M + \frac{dT_S}{L_M} e_S. \tag{7.14}$$

Or,

$$I_M = \frac{dT_S}{2L_S} e_S. \tag{7.15}$$

In the OFF state, the free wheeling diode D_2 conducts and effectively places the capacitor C_R in parallel with the magnetizing inductance which

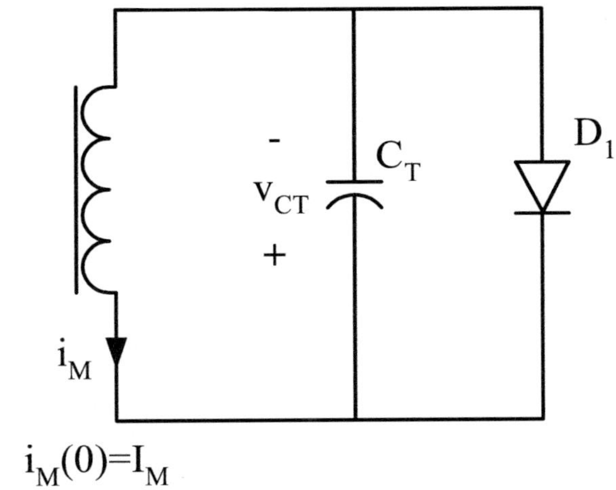

Figure 7.16 Resonant circuit during the OFF state.

is carrying an initial current of I_M amps. The capacitance that appears in parallel with L_M is

$$C_T = \left(\frac{N_S}{N_P}\right)^2 (C_R + C_{D1}) + C_{\mathrm{SW}} + C_{\mathrm{TP}}. \tag{7.16}$$

It is usual that $C_R \gg C_{D1}, C_{\mathrm{SW}}, C_{\mathrm{TP}}$, so that $C_T \simeq \left(\frac{N_S}{N_P}\right)^2 C_R$. In short, a resonant circuit is formed at the instant the power switch is turned off (Figure 7.16).

The magnetizing current that circulates in the resonant circuit and the capacitor voltage are given by

$$i_M = I_M \cos \omega_0 t \tag{7.17}$$
$$v_{CT} = I_M \sqrt{L_M/C_T} \sin \omega_0 t, \tag{7.18}$$

where $\omega_0 = 1/\sqrt{L_M C_T}$. After $t_R = \pi\sqrt{L_M C_T}$ seconds, the magnetizing current swings to $-I_M$. The capacitor voltage follows a sinusoidal variation starting from $v_{CT} = 0$, reaches a peak value $v_{CTmax} = I_M \sqrt{L_M/C_T}$, and discharges back to zero at precisely the instant that the magnetizing current is reset to $-I_M$ amps. Beyond that instant, the diode D_1 clamps the capacitor voltage to zero and the magnetizing current remains at $-I_M$ until the start of the next switching period. The proper functioning of the resonant reset technique requires that $t_R \leq (1-d)T_S$. Therefore,

$$C_T \leq \left(\frac{1-d}{\pi f_S}\right)^2 \frac{1}{L_M} \tag{7.19}$$

$$C_R \leq \left(\frac{N_P}{N_S}\right)^2 C_T. \tag{7.20}$$

The simulation of the forward converter of Figure 7.15 is left as an exercise. The transient equivalent circuit and the small–signal equivalent circuit for the converter are shown in Figure 7.17 and Figure 7.18 respectively.

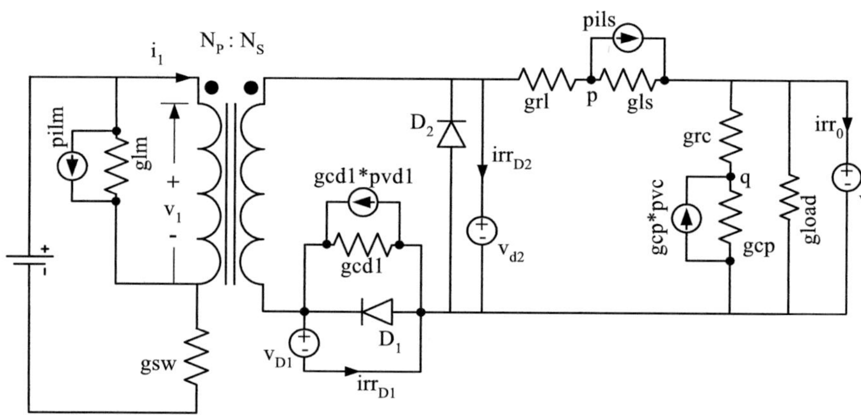

Figure 7.17 Forward converter, resonant reset, transient equivalent circuit.

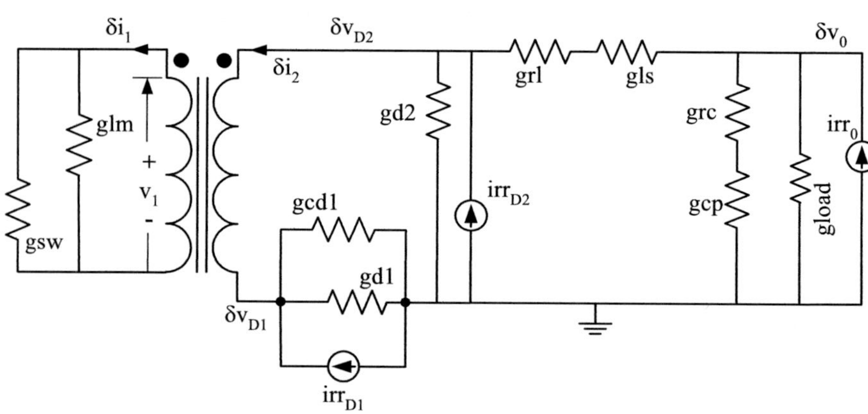

Figure 7.18 Forward converter with resonant reset, small–signal equivalent.

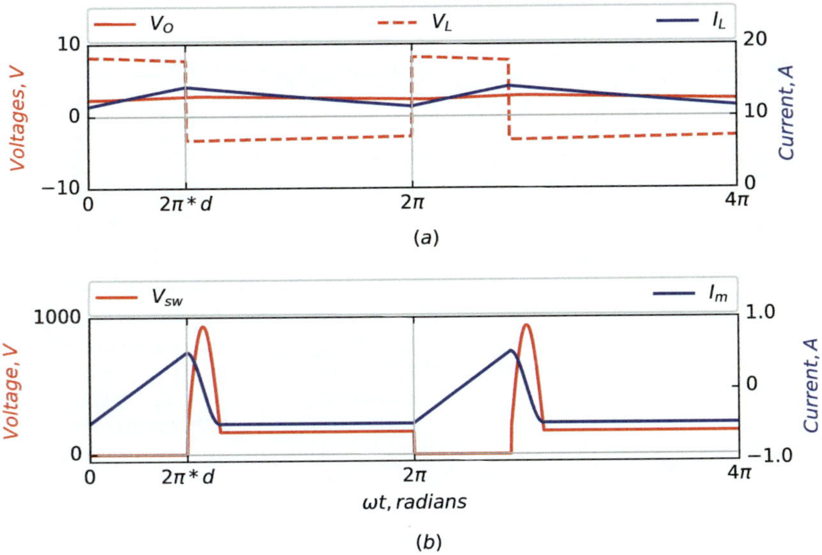

Figure 7.19 Forward converter with resonant reset: response waveforms.

The probe voltages that are suggested for the simulation are v_{d1}, v_{d2}, and v_0. Figure 7.19, plotted by Program 7.19, shows the converter's current and voltage waveforms.

7.2 Isolated Single-ended Hybrid-bridge Converter

The isolated hybrid-bridge converter shown in Figure 7.20 is built with two switches that turn on and off simultaneously [1, 2]. When the switches are turned on at the start of a switching period, a positive voltage e_S is applied to the primary winding of the transformer causing D_1 to conduct and D_2 to be reverse biased. Power flows to the load and the magnetizing current rises linearly from zero. The diodes D_3 and D_4 are reverse biased. When the switches are turned off, the magnetizing current forces D_3 and D_4 to conduct and a negative voltage $-e_S$ is applied to the primary winding. The magnetizing current falls linearly to zero due to the negative voltage applied to the magnetizing inductance. At the same time, the magnetizing current flows through D_3 and D_4 and returns the stored magnetic energy to the source. The transformer core is reset, with the reset time being equal to the ON time.

The voltages appearing across the switches when they are turned off are never more than a diode drop greater than the input voltage. Energy

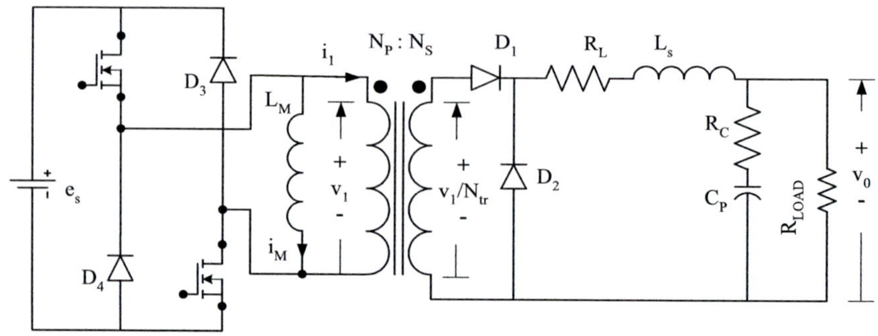

$N_{tr} = N_P / N_S = 15$, $e_s = 165$ V
$R_L = 0.02\Omega$, $L_S = 8.5\mu H$, $L_M = 750\mu H$
$R_C = 0.01\Omega$, $C_P = 440\mu F$
$R_{LOAD} = 0.2\Omega$, d (duty ratio) = 0.3

Figure 7.20 Isolated hybrid-bridge converter.

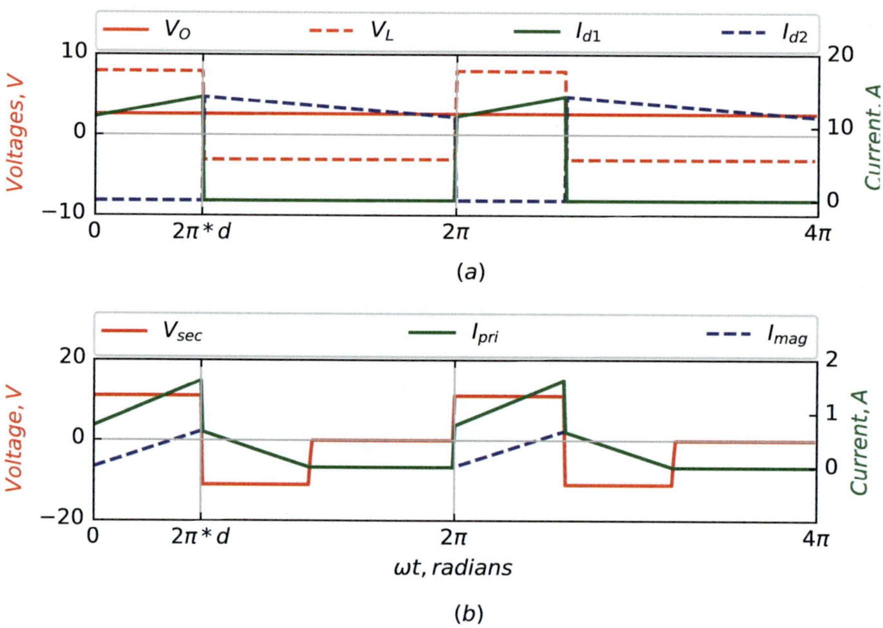

Figure 7.21 Isolated hybrid-bridge converter: response waveforms.

stored in the leakage inductance during the ON time is discharged into the source when the switches are turned off. However, the leakage inductance leads to a commutation interval during which the diodes D_1 and D_2 conduct simultaneously. The duty ratio is limited to less than 0.5. Another inconvenience is that there are two switches at different potentials that need to be controlled.

Figure 7.21 shows the voltage and current waveforms of the converter of Figure 7.20. The waveforms have been plotted by Program 7.21.

7.3 The Double-ended, Full-bridge Forward Converter

The double-ended, full-bridge forward converter is shown in Figure 7.22 [1, 2]. On the front end of the converter, there is a full-bridge formed by the bi-directional switches S_1 to S_4. The switches are triggered in such a manner that the voltage v_1 at the output of the bridge is a series of alternating pulses. A full-wave, center-tapped rectifier is shown on the secondary side. (A full-bridge rectifier would also function well.) Beyond the secondary side rectifier, the output LC filter and load complete the converter circuit. The voltage v_2 applied to the output filter and load is a rectified form of v_1, i.e., it consists of a series of pulses of positive polarity. There are several primary side switching sequences that can generate the waveform v_1. The two commonly used sequences are shown in Table 7.1.

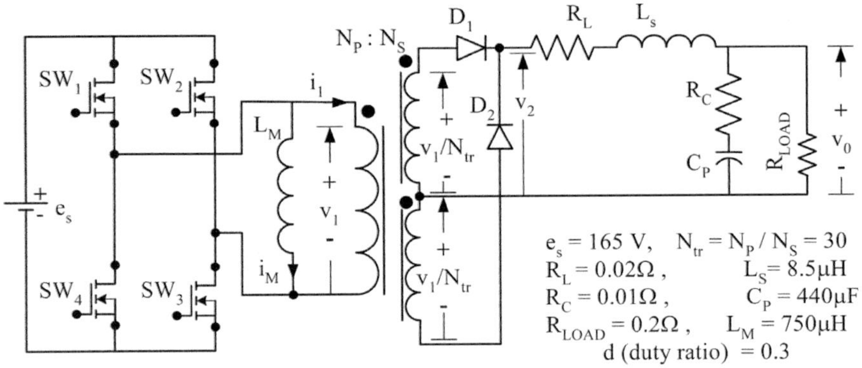

Figure 7.22 Full-bridge forward converter: circuit diagram.

Table 7.1 Switching sequences.

Interval	Sequence A				Sequence B			
	S_1	S_2	S_3	S_4	S_1	S_2	S_3	S_4
$T_1 : 0 \leq t < dT_S/2$	On	Off	On	Off	On	Off	On	Off
$T_2 : dT_S/2 \leq t < T_S/2$	Off	Off	Off	Off	Off	Off	On	On
$T_3 : T_S/2 \leq t < (1+d)T_S/2$	Off	On	Off	On	Off	On	Off	On
$T_4 : (1+d)T_S/2 \leq t < T_S$	Off	Off	Off	Off	On	On	Off	Off

The switch positions are identical for both sequences during intervals T_1 and T_3. In the steady-state, the magnetizing current increases linearly during T_1 from $-I_M$ to $+I_M$; the diode D_1 is forward biased and D_2 is reverse biased. Similarly, the magnetizing current decreases linearly during interval T_3 from $+I_M$ to $-I_M$. The diode D_2 is forward biased, whereas the diode D_1 is reverse biased. Power flows to the load during the two intervals T_1 and T_3. The switch positions differ for the two sequences during the intervals T_2 and T_4. All the switches are turned off during both these intervals in sequence A. In sequence B, a short-circuit is created at the output of the bridge by turning on S_3 and S_4 during T_2 and by turning on S_1 and S_2 during T_4.

Consider the interval T_2 of sequence A. Since all the switches are turned off, the magnetizing current is forced to flow into the primary winding. It flows into the dot-less terminal of the winding and out of the dotted terminal. The reflected magnetizing current $N_{\text{tr}} I_M$ on the secondary side circulates in the loop formed by the diode D_1, the diode D_2, and the two sections of the secondary winding. It flows in the forward direction in D_2, and in the reverse direction in D_1. At the same time, the output inductor current i_{Lmax} flows into the parallel combination of the output capacitor, and the load and returns to the center-tap of the secondary winding where it divides and flows in the forward directions of the diodes D_1 and D_2. Thus, the total current flowing in D_1 is $(\frac{1}{2}i_{Lmax} - N_{\text{tr}} I_M)$ and that in D_2 is $(\frac{1}{2}i_{Lmax} + N_{\text{tr}} I_M)$. The conclusion is that sequence A may be used only if $\frac{1}{2}i_{Lmax} > N_{\text{tr}} I_M$. The magnetizing current remains constant at I_M amps due to the short-circuit applied on the primary winding by the simultaneous conduction of D_1 and D_2. Similar reasoning shows that during the interval T_4, the magnetizing current is constant at $-I_M$, and the currents flowing in D_1 and D_2 are $(\frac{1}{2}i_{Lmax}+N_{\text{tr}} I_M)$ and $(\frac{1}{2}i_{Lmax}-N_{\text{tr}} I_M)$, respectively. The main advantage of switching sequence A is that the primary switches carry unipolar currents. On the other hand, the transformer's leakage inductance will generate voltage spikes when the primary switches are turned off.

Consider intervals T_2 and T_4 of switching sequence B. Closure of the switches S_3 and S_4 during the interval T_2 (or S_1 and S_2 during the interval T_4) creates a short-circuit at the output terminals of the full-bridge. The output inductor current will, again, split at the center-tap of the secondary winding and flow in the forward directions of the diodes D_1 and D_2. The short-circuit that results across the secondary winding will be reflected to the primary side. If the characteristics of the diodes are identical and those of the primary switches are also identical, then the magnetizing current will circulate in the short-circuit created by the full-bridge. Otherwise, the magnetizing

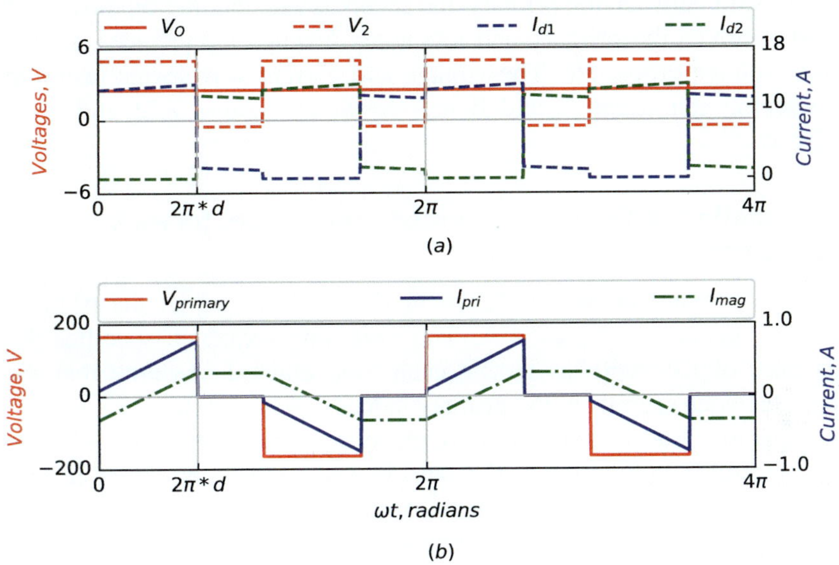

Figure 7.23 Full-bridge forward converter: sequence A: response waveforms.

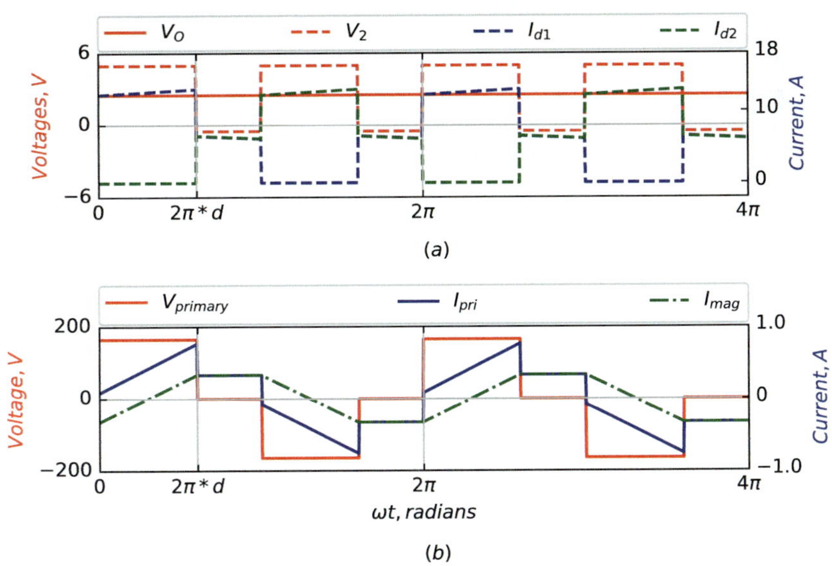

Figure 7.24 Full-bridge forward converter: sequence B, response waveforms.

current will split between the primary and secondary depending on the circuit parameters such as the primary and secondary leakage inductances, switch and diode voltage drops, etc. The voltage and current waveforms resulting from switching sequences A and B are shown in Figures 7.23 and 7.24, respectively.

7.3.1 Simulation of the double-ended, full-bridge, forward converter

Consider the forward converter shown in Figure 7.22 and let the primary switching sequence be sequence A. The switching sequence specifies the conductances of the switches during each time interval. Assume that the leakage inductance seen from the primary side is $L_K = 2\ \mu\text{H}$.

It is usual to connect a small non-polarized capacitor C_b in series with the primary winding. The function of the capacitor is to block any direct current bias that may be present in the primary current. Such a bias may be present if the average value of the voltage applied to the primary winding is not exactly zero. Direct current bias in the primary current will cause magnetic flux imbalance and the transformer core will drift up or down the hysteresis loop into saturation. The blocking capacitor should not cause more than 10% droop in the voltage pulses applied to the primary. Assuming a maximum duty ratio of 0.8,

$$I_P = \frac{(P_0/\eta)}{0.8\, V_{\text{DCmin}}}, \tag{7.21}$$

where, P_0 = output power $\simeq$ 66 watts, and η = efficiency of the converter $\simeq$ 0.8. Putting

$$C_b = \frac{I_P\, \Delta t}{\Delta V}, \tag{7.22}$$

where

$$\Delta t = \frac{0.8}{2\, f_S}, \qquad f_S = 100,000$$

$$\begin{aligned} V_{\text{DCmin}} &= 0.85\, V_{\text{DC}}, \quad V_{\text{DC}} = 140\ V \\ \Delta V &= 0.1\, V_{\text{DCmin}}. \end{aligned}$$

For the converter of Figure 7.22, a good value for C_b is 0.3 µF.

The transient equivalent circuit for the converter is shown in Figure 7.25. The probe voltages are v_1, v_{d1}, and v_0. The equations for calculating the

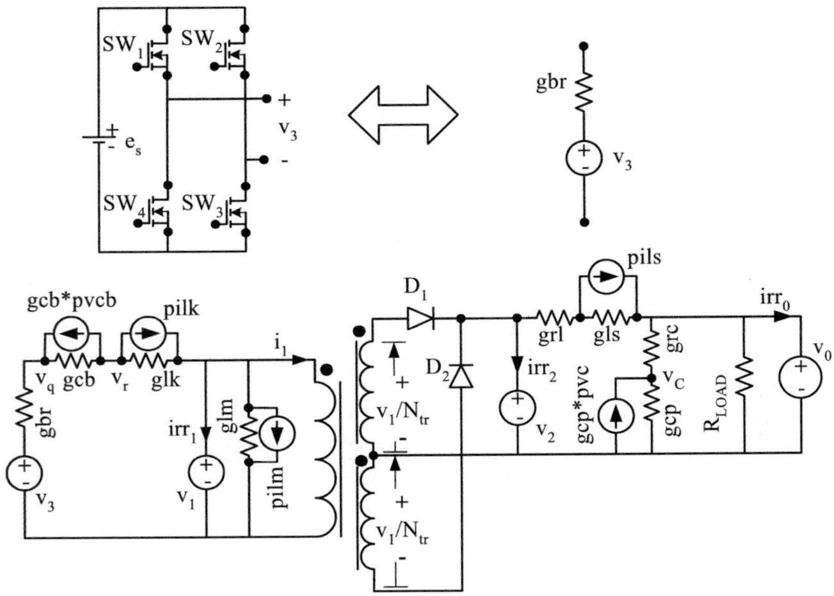

Figure 7.25 Full-bridge forward converter: transient equivalent circuit.

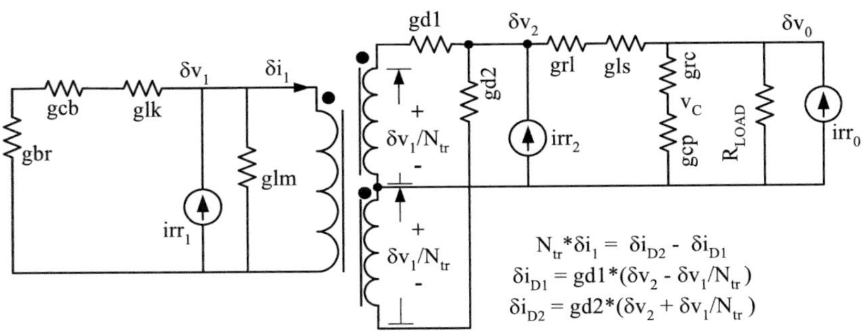

Figure 7.26 Full-bridge forward converter: small–signal equivalent circuit.

error currents are given in Program 7.27. The small–signal equivalent circuit for calculating the corrections to be added to the probe voltages is shown in Figure 7.26. Figure 7.27 shows the response waveforms plotted by Program 7.27.

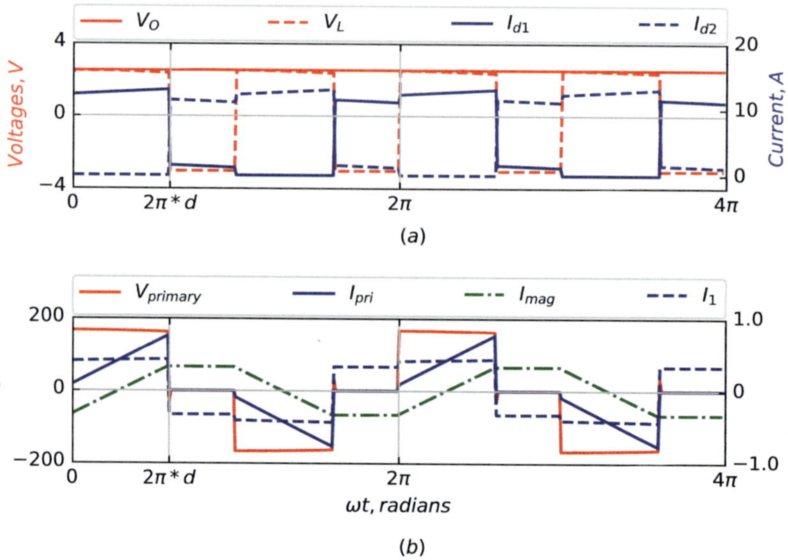

Figure 7.27 Full-bridge forward converter: response waveforms.

7.4 The Push–Pull Converter

Figure 7.28 shows the push–pull converter [2, 3, 8, 9], which is also known as the isolated, double-ended, half-bridge converter. There are two switches on the primary side and a full-wave, center-tapped rectifier on the secondary side. (A full-bridge rectifier could be used instead.) Beyond the rectifier on the secondary side, the circuit is identical to the latter half of the buck converter. The transformer consists of center-tapped primary and secondary windings. The lower half-section of the primary winding is referred to as winding P_1 and the upper half-section as winding P_2. Similarly, the half-sections of the secondary winding are referred to as windings S_1 and S_2. The magnetizing inductance is placed in parallel with winding P_1. (It can be placed across any section of the primary or secondary winding, or even across the entire primary or secondary winding.) The low-pass output filter is formed by the inductor L_S and the capacitor C_P. The equivalent series resistances of L_S and C_P are R_L and R_C. The load at the output is represented by the resistor R_{LOAD}.

The switches SW_1 and SW_2 are either unidirectional or bidirectional. Table 7.2 shows the switching sequence. Unidirectional switches are not provided with the anti-parallel diodes and are used when $N_{\mathrm{tr}}\, i_{\mathrm{Mmax}} < i_{\mathrm{Lmin}}$. That is, the magnetizing inductance should be sufficiently high to permit the

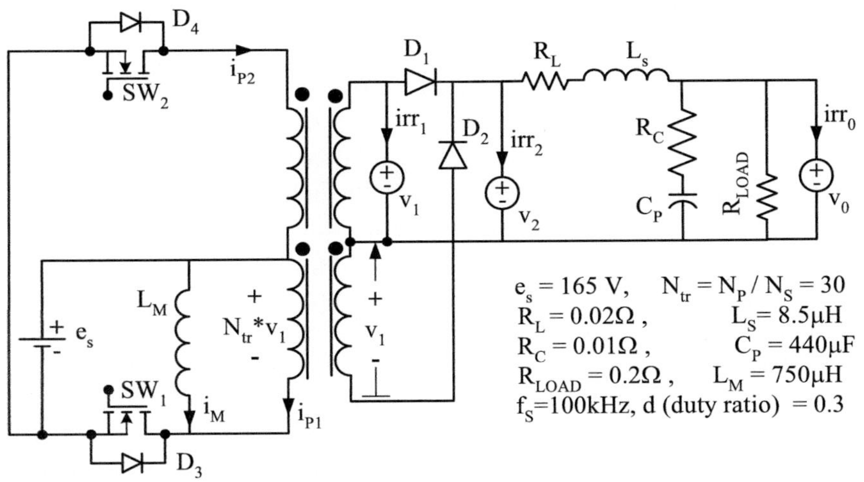

Figure 7.28 The push–pull converter.

use of unidirectional switches. On the other hand, bidirectional switches with anti-parallel diodes may be used in the push–pull converter even when the magnetizing inductance is low.

Table 7.2 Switching sequence.

Interval	Switches	
	SW_1	SW_2
$T_1 : 0 \leq t < dT_S/2$	On	Off
$T_2 : dT_S/2 \leq t < T_S/2$	Off	Off
$T_3 : T_S/2 \leq t < (1+d)T_S/2$	Off	On
$T_4 : (1+d)T_S/2 \leq t < T_S$	Off	Off

7.4.1 Low magnetizing current, high output load current

A low magnetizing current implies a high magnetizing inductance. Suppose that the inequality $N_{\text{tr}} I_{\text{Mmax}} < I_{\text{Lmin}}$ holds. (Typical values for the converter of Figure 7.28 are $L_M = 750$ μH, $R_{\text{LOAD}} = 0.2$ Ω.) Assume that the converter is operating in the steady-state and ignore leakage inductances. The source voltage e_S is applied to the winding P_1 during the interval T_1 with the dotted end of the winding being positive. The voltage appearing across each section of the secondary winding is e_S/N_{tr}. The output inductor current rises from its minimum value I_{Lmin} to its maximum value I_{Lmax}. The current flowing in the winding P_1 is $i_{P1} = I_L/N_{\text{tr}}$. Power flows to the load and the output capacitor gains charge. The magnetizing current rises from its minimum $-I_M$ to its maximum $+I_M$.

Both switches SW_1 and SW_2 are turned off during the interval T_2. The magnetizing current is forced to flow out of the dotted end of the winding P_1. The reflected current on the secondary side circulates in the loop formed by the two sections of the secondary winding and the diodes D_1 and D_2. It flows in the forward direction of D_2, but in the reverse direction of D_1. (Note that the magnetizing current flows through the winding P_1 on the primary side but the reflected current flows through the entire secondary winding.) At the same time, the output inductor current splits equally at the center-tap of the secondary winding and each of the diodes will carry one half of that current. Thus, the current flowing in the diode D_1 is $(i_L - N_{\text{tr}}I_M)/2$ and that in diode D_2 is $(i_L + N_{\text{tr}}I_M)/2$. The magnetizing current remains constant at I_M throughout the interval T_2 due to the short-circuit applied to the secondary winding by the conducting diodes.

During the interval T_3, the source voltage e_S is applied to the winding P_2 with reversed polarity and a negative voltage is induced in the other windings. The magnetizing current that continues to flow in the winding P_1 decreases linearly from $+I_M$ to $-I_M$. On the secondary side, the diode D_2

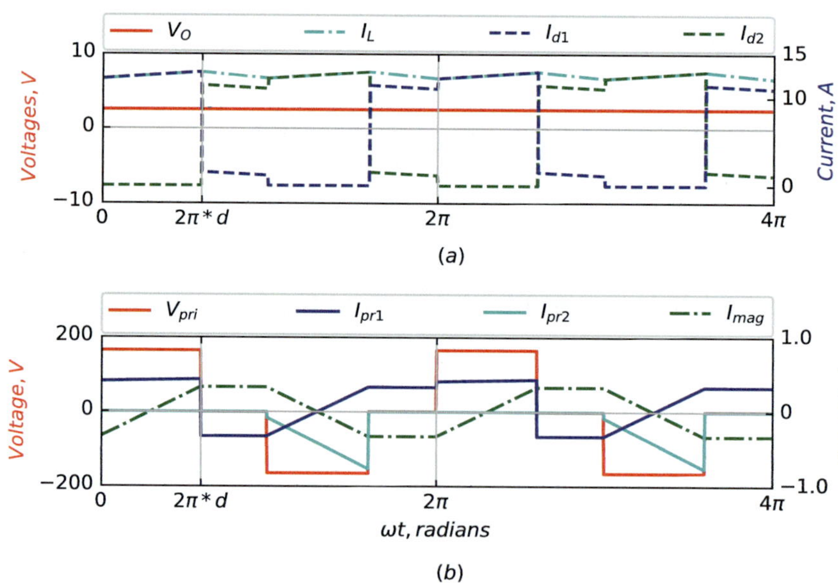

Figure 7.29 Push-pull converter, high L_M: response waveforms.

is forward biased, though the diode D_1 is reverse biased. The output inductor current flows through D_2 and increases linearly as in interval T_1 and power is supplied to the load.

Both switches are turned off and both diodes conduct during interval T_4. The current flowing in diode D_1 is $(i_L + N_{\text{tr}} I_M)/2$ and that flowing in D_2 is $(i_L - N_{\text{tr}} I_M)/2$. The magnetizing current remains constant at $-I_M$ amps. Figure 7.29, plotted by Program 7.29, shows the voltage and current waveforms for the push–pull converter.

7.4.2 High magnetizing current, low output load current

Typical values for the converter of Figure 7.28 are $L_M = 300$ µH and $R_{\text{LOAD}} = 0.4\ \Omega$. Consider that the push–pull is in the steady-state and let leakage inductances be ignored. Switch SW_1 is turned on during the interval T_1 and the diode D_1 is forward biased. The voltage applied to the winding P_1 is the source voltage e_S and the voltages reflected in the other windings are $v_{P2} = e_S$, $v_{S1} = v_{S2} = e_S/N_{\text{tr}}$. The magnetizing current i_M rises linearly from $-I_{\text{min}}$ to $+I_{\text{max}}$. The current in the winding P_1 is $i_{P1} = i_L/N_{\text{tr}}$ (which is nearly constant). Observe that $i_{P2} = 0$ and $i_{S2} = 0$.

During the interval T_2, the switches SW_1 and SW_2 are both turned off and the diode D_1 is reverse biased. The magnetizing current is forced to flow into the winding P_1 so that $i_{P1} = -i_M$. Similarly, the output inductor current i_L is forced to flow through the diode D_2. Thus, $i_{S1} = 0$ and $i_{S2} = i_L$. Equating the algebraic sum of the ampere-turns of the windings to zero, we get

$$\begin{aligned} i_{P2} &= -i_{P1} - i_{S2}/N_{\text{tr}} \\ &= i_M - i_L/N_{\text{tr}}. \end{aligned}$$

The current i_{P2} flows into the source through the diode D_4. Since D_4 is conducting, the voltage appearing across winding P_2 (and reflected to winding P_1) is $-e_S$. The magnetizing current decreases linearly due to the negative voltage appearing across the winding P_1. When the decreasing magnetizing current $i_M = i_L/N_{\text{tr}}$, the current $i_{P2} = 0$ and the diode D_4 ceases to conduct. The voltage across the winding P_2 (and all other windings) collapses to zero and remains at zero for the rest of the interval T_2. The magnetizing current is constant at $+I_{\text{min}}$.

The reasoning presented for the intervals T_1 and T_2 may be extended to intervals T_3 and T_4. During the interval T_3, the switch SW_2 is turned ON and the switch SW_1 remains in the OFF state. A negative voltage $-e_S$ is applied to winding P_2 and the same negative voltage is reflected on the winding

P_1. The reflected voltages on the half-sections of the secondary winding are $v_{S1} = v_{S2} = -e_S/N_{\mathrm{tr}}$. The magnetizing current decreases linearly from $+I_{\max}$ to $-I_{\min}$ and it flows into the winding P_1 so that $i_{P1} = -i_M$. Setting the algebraic sum of the ampere-turns of the windings to zero, we get $i_{P2} = i_M - i_L/N_{\mathrm{tr}}$.

Finally, the switches SW_1 and SW_2 are both turned off during the interval T_4. Assume that the diode D_1 provides the path for the output inductor current and that D_2 is reverse biased during T_4. (The assumption will be justified below.) The current i_L is reflected in the winding P_1 as i_L/N_{tr} and it enters the winding P_1 at its dotted end. At the start of interval T_4, the magnetizing current is $-I_{\max}$ and it flows out of the dotted end of the winding P_1. Since $|-I_{\max}| > i_L/N_{\mathrm{tr}}$, the difference $(I_{\max} - i_L/N_{\mathrm{tr}})$ flows in the loop formed by the input voltage source, the diode D_3, and the winding P_1. The source voltage is now applied across the magnetizing inductance and the winding P_1.

The voltages of the dotted ends of the winding sections are positive with respect to the dot-less ends so that the diode D_1 is forward biased and D_2 is reverse biased. The magnetizing current increases linearly from $-I_{\max}$ due to the voltage applied to the winding P_1. When the magnetizing current increases to $(-i_L/N_{\mathrm{tr}})$, the diode D_3 ceases to conduct and the

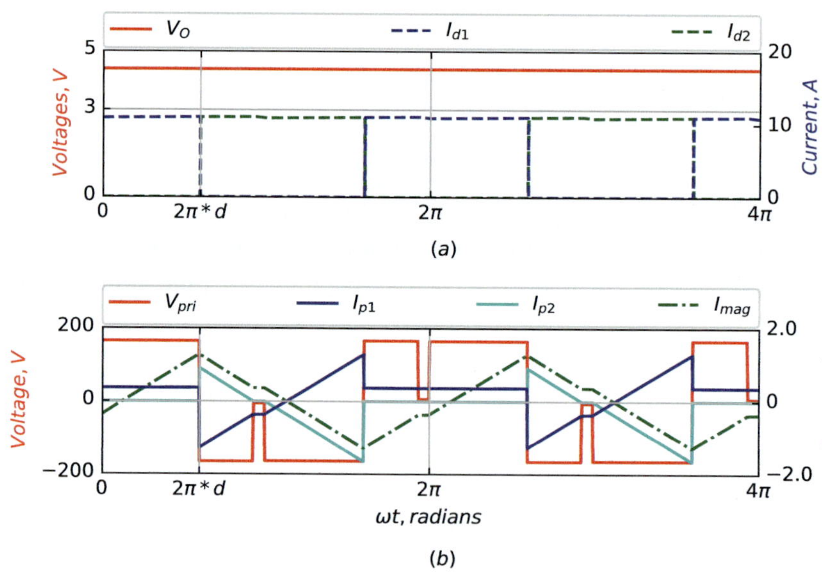

Figure 7.30 Push-pull converter, low L_M: response waveforms.

voltage applied to the winding P_1 collapses to zero. The magnetizing current remains constant at $(-i_L/N_{\rm tr})$ for the rest of the interval T_4. Figure 7.30 plotted by Program 7.30 shows the voltage and current waveforms when the magnetizing inductance is low and the output load current is low as well.

7.5 The Flyback Converter

The flyback converter is the isolated form of the up/down converter [1–3, 10]. Figure 7.31 shows the simplified form of the flyback converter. The transformer has been considered to be ideal with no magnetic leakage.

When the switch SW is turned on, the diode on the secondary side is reverse biased. Energy flows into the magnetizing inductance and the magnetizing current increases linearly from $I_{\rm min}$ to $I_{\rm max}$. When the switch is turned off, the magnetizing current is forced to flow into the dotted end of the primary winding. The reflected current $N_{\rm tr} i_M$ flows out of the dotted end of the secondary winding and power is supplied to the load.

The voltage applied to the magnetizing inductance is e_S during the ON state and $-N_{\rm tr} V_0$ during the OFF state. Equating the average magnetizing inductor voltage over a switching period to zero, we obtain

$$\frac{1}{T_S}\left(d T_S\, e_S \;-\; (1-d) T_S\, N_{tr} V_0\right) \;=\; 0. \tag{7.23}$$

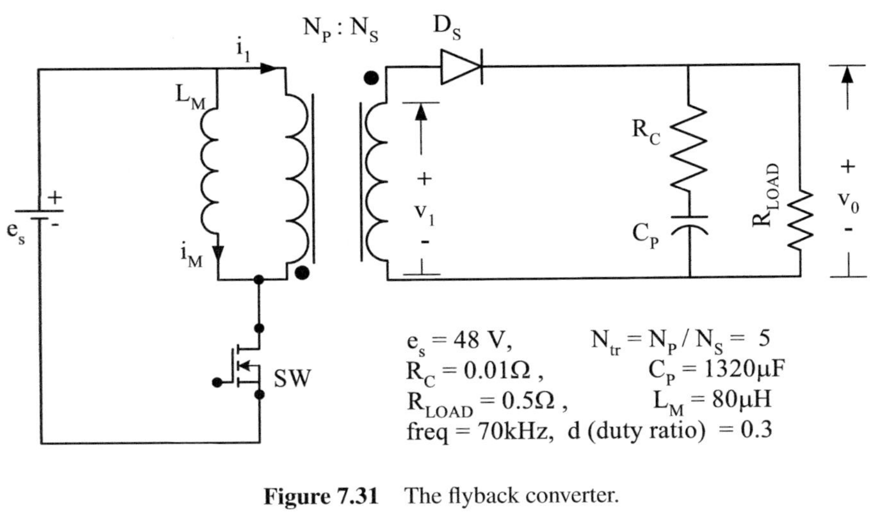

Figure 7.31 The flyback converter.

That is,

$$V_0 = \frac{1}{N_{tr}}\left(\frac{d}{1-d}\right)e_S. \tag{7.24}$$

Since a negative voltage $-N_{\text{tr}}V_0$ appears across the magnetizing inductance during the OFF state, the magnetizing current decreases linearly from $I_{\max}$ to $I_{\min}$. The transformer used in the flyback converter isolates the input voltage source from the load and also stores energy during the ON state. Its magnetizing inductance is designed to have a specified value in order to store energy. (See Chapter 2, Design Examples.) Primary side leakage inductance will also store energy that remains trapped when the power switch is turned off and that will generate a voltage spike across the power switch. It is essential to protect the power switch with an RCD snubber.

7.5.1 Simulation of the flyback converter

Figure 7.32 shows the flyback converter that will be simulated. The isolation transformer has a turns ratio of 5, magnetizing inductance of 80 μH, and primary side leakage inductance of 3 μH. An RCD snubber (C_K, R_K, D_P) has been provided to protect the power switch against turn-off voltage spikes. The secondary side circuit is similar to that of the up/down converter. The output capacitor $C_P = 1320\,\mu$ F has an ESR of $0.01\ \Omega$.

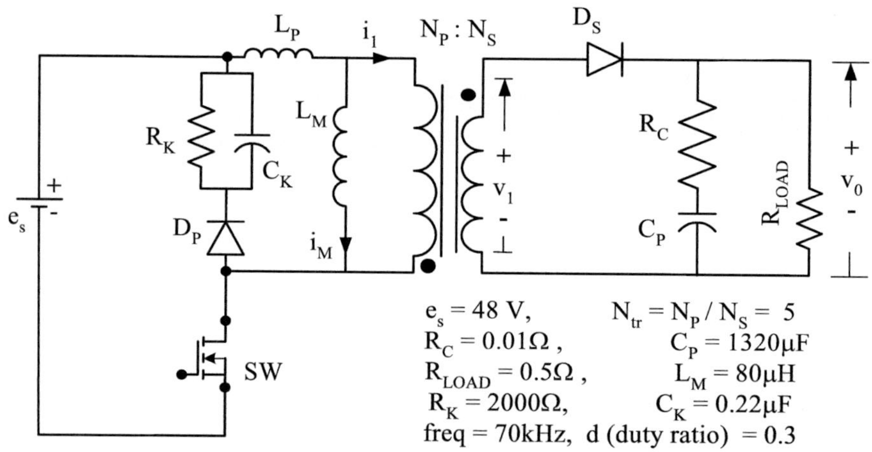

Figure 7.32 The flyback converter with RCD snubber.

The snubber is inactive during the ON state since the diode D_P is reverse biased. During the OFF state, the primary side of the converter is practically disconnected from the input voltage source. Therefore, the transient equivalent circuit of the converter during the ON state is different from that valid during the OFF state. Figure 7.33 shows the two equivalent circuits and the past-history current sources are updated seamlessly while passing from one state to the other. The secondary side remains the same for both equivalent circuits. The probe voltages are vsw, v_1, and v_0 for the ON state and vdp, v_1, and v_0 for the OFF state. The equivalent circuits are analyzed and the error currents irr_w, irr_1, irr_0 for the ON state and irr_{dp}, irr_1, irr_0 for the OFF state are computed using the equations that follow:

ON state:

$$\begin{aligned} vlp &= es + N_{\text{tr}}v1 - vsw \\ ilp &= glp\, vlp + pilp \\ isw,\ gw1 &= fswitch(vsw,\ gsw1) \\ irrw &= ilp - isw. \end{aligned}$$

OFF state:

$$idp,\ gdp = fdiode(vdp)$$

$$ick = grk\, vck + gck\,(vck - pvck) \tag{7.25}$$

$$ilp = glp\, vlp + pilp \tag{7.26}$$

$$ick = -\,ilp \tag{7.27}$$

$$vlp = vck + N_{\text{tr}}v1 + vdp \tag{7.28}$$

$$vck = (gck\, pvck - pilp - glp(N_{\text{tr}}v1 + vdp))\,/(grk + gck + glp$$

(from eqn (7.25)–(7.28))

$$irrdp = (grk + gck)\, vck - gck\, pvck - idp.$$

The secondary side equations are the same for the ON and OFF states:

$$\begin{aligned} i1 &= ilm + ilp \\ ids,\ gds &= fdiode(v0 - v1) \\ vc &= (grc\, v0 + gcp\, pvc)/(grc + gcp) \\ icp &= grc\,(v0 - vc) \\ iload &= gload\, v0 \\ irr1 &= ids - N_{\text{tr}}i1 \\ irr0 &= -(ids + icp + iload). \end{aligned}$$

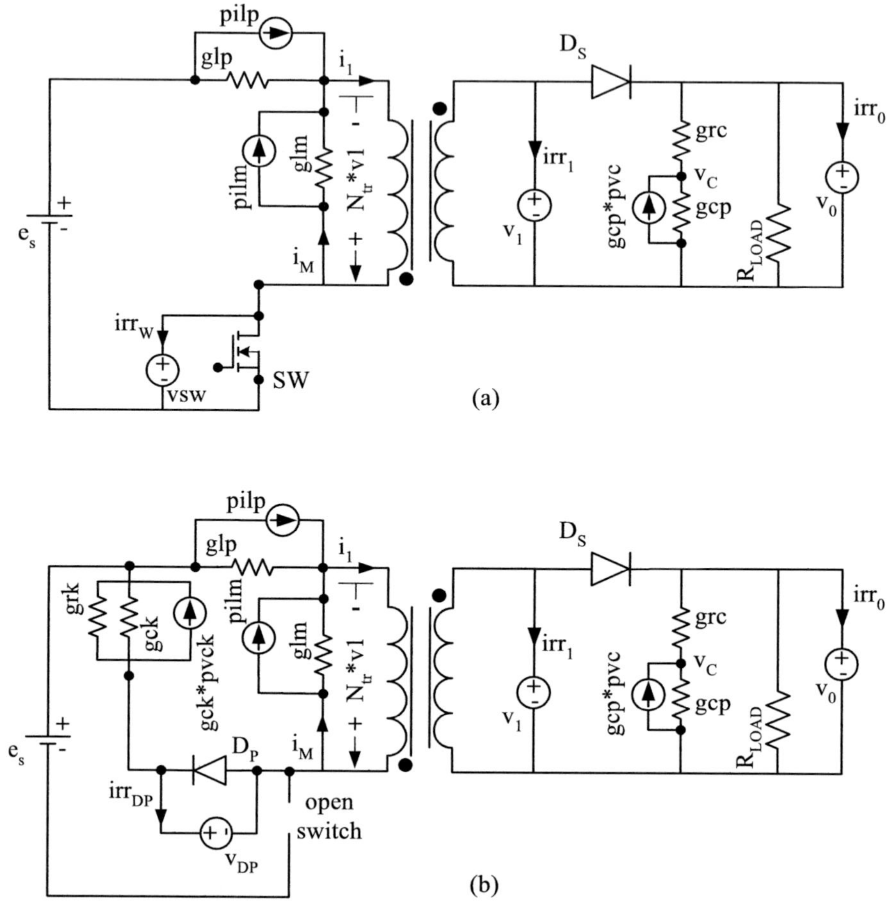

Figure 7.33 Flyback converter: transient equivalent circuit.

Figure 7.34 shows the small–signal equivalent circuits for calculating the corrections to be applied to the probe voltages. During the ON state, the corrections are given by

$$\begin{bmatrix} (glp + gw1) & -N_{\text{tr}}glp & 0 \\ -N_{\text{tr}}glp & gds + N_{\text{tr}}^2(glm + glp) & -gds \\ 0 & -gds & gds + gpar + gload \end{bmatrix} \begin{bmatrix} \delta vsw \\ \delta v1 \\ \delta v0 \end{bmatrix}$$

$$= \begin{bmatrix} irrw \\ irr1 \\ irro. \end{bmatrix}. \tag{7.29}$$

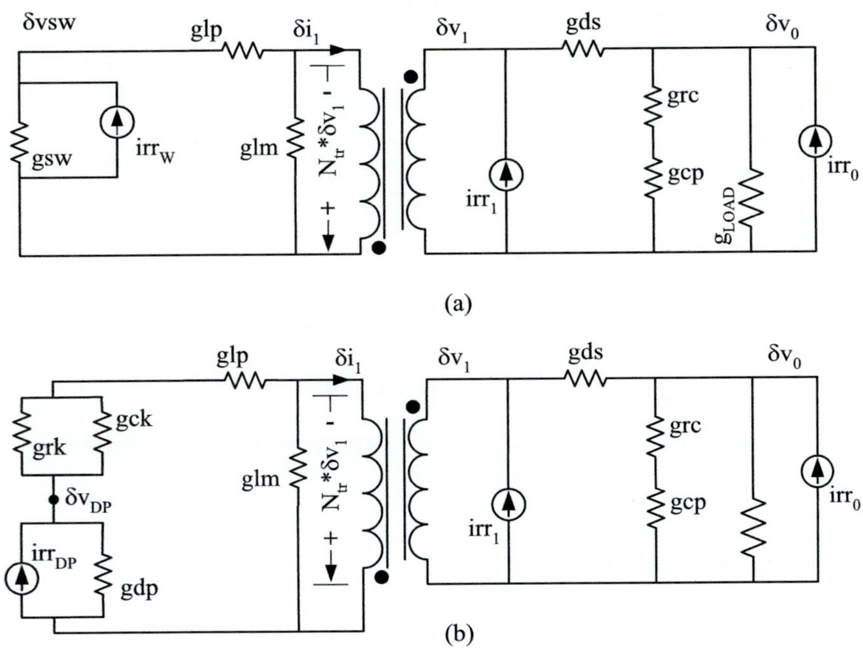

Figure 7.34 Flyback converter: small–signal equivalent circuit.

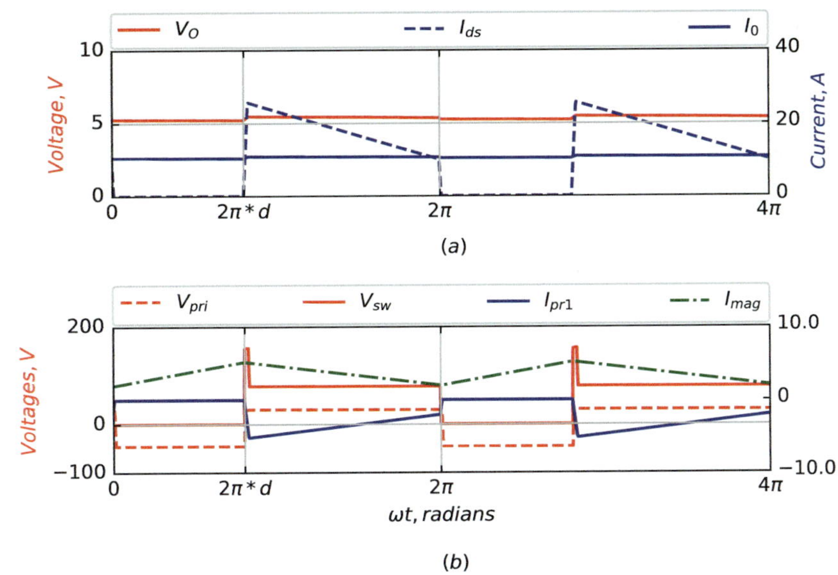

Figure 7.35 Flyback converter: response waveforms.

The corresponding equations for the OFF state are

$$\begin{bmatrix} gdp + gkp & N_{\mathrm{tr}}gkp & 0 \\ N_{\mathrm{tr}}gkp & gds + N_{\mathrm{tr}}^2 glm & -gds \\ 0 & -gds & gds + gpar + gload \end{bmatrix} \begin{bmatrix} \delta vdp \\ \delta v1 \\ \delta v0 \end{bmatrix}$$

$$= \begin{bmatrix} irrdp \\ irr1 \\ irro. \end{bmatrix} \quad (7.30)$$

In the above equations,

$$\begin{aligned} gpar &= grc\,gcp/(grc + gcp) \\ gkp &= glp(grk + gck)/(glp + grk + gck). \end{aligned}$$

Figure 7.35 shows the voltage and current waveforms plotted by Program 7.35.

7.6 Computational Programs

7.6.1 Program 7.07

```
#
def fdiode(vd):
    """ Diode characteristic.
        Returns diode current and its conductance"""
    #
    glow = 0.1e-07
    ghigh = 0.4e+03
    Vr = -0.5
    #
    if vd >= Vr:
        id = glow*(vd - Vr)
        gd = glow
    else:
        id = ghigh*(vd - Vr)
        gd = ghigh
    #
    return id, gd
#
#
def ch7fig07_solver(es, Ntr, glm, grl, gls, grc, gcp, gload, nc, ncycle, n):
    """ Assumes fdiode is a function.
        Returns the lists vout, cinduct"""
    #
    import numpy as np
    #
    N = n*ncycle
    v1 = 160.0
```

```
vd2 = 4.3
v0 = 3.3
pilm = 0
pils = 7.5
pvc = 3.0
erlim = 0.1e-04
Ntrsq = Ntr**2
gser = grl*gls/(grl + gls)
gpar = grc*gcp/(grc + gcp)
#
vout = []
vinduct = []
vdiode2 = []
cmag = []
cinduct = []
cdiode1 = []
cdiode2 = []
#
for k in range(ncycle):
    #
    for j in range(n):
        #
        if j <= nc:
            gsw = 250.0
        else:
            gsw = 0.1e-07
        #
        for its in range(1000):
            #
            v2 = v1/Ntr
            vd1 = vd2 - v2
            vd3 = es + v1
            #
            id1, gd1 = fdiode(vd1)
            id2, gd2 = fdiode(vd2)
            id3, gd3 = fdiode(vd3)
            #
            isw = gsw*(es - v1)
            ilm = glm*v1 + pilm
            i1 = id3 - id1/Ntr
            #
            vp = (grl*vd2 + gls*v0 - pils)/(grl + gls)
            vq = (grc*v0 + gcp*pvc)/(grc + gcp)
            #
            ils = grl*(vd2 - vp)
            icp = grc*(v0 - vq)
            iload = gload*v0
            #
            irr1 = isw - i1 - ilm
            irrd2 = -(id1 + id2 + ils)
            irr0 = ils - icp - iload
            #
            err = np.sqrt(irr1**2 + irrd2**2 + irr0**2)
```

```
                #
                if err < erlim:
                    break
                else:
                    gmat = np.array([[(gsw+glm+gd3 + gd1/Ntrsq), -gd1/Ntr, 0],
                                      [-gd1/Ntr, (gd1+gd2+gser), -gser],
                                      [0, -gser, (gser+gpar+gload)]])
                    bcol = np.array([irr1, irrd2, irr0])
                    xdv = np.linalg.solve(gmat, bcol)
                    #
                    dv1 = xdv[0]
                    dvd2 = xdv[1]
                    dv0 = xdv[2]
                    v1 += dv1
                    vd2 += dvd2
                    v0 += dv0
            #
            if its >= 999:
                print('time-step= ', (k*n + j), 'j = ', j, 'k = ',     \
                      k, 'solver does not converge')
                break
            #
            pilm = ilm
            pils = ils
            pvc = vq
            #
            cinduct.append(ils)
            cmag.append(ilm)
            cdiode1.append(-id1)
            cdiode2.append(-id2)
            vout.append(v0)
            vinduct.append(vd2 - v0)
            vdiode2.append(vd2)
            #
        #
    #
    print(' v0= ', np.mean(vout[N-2*n:N-1]))
    #
    return vout, vinduct, vdiode2, cinduct, cmag, cdiode1, cdiode2
#
#
def ch7fig07(es, freq, Ntr, lm, rl, ls, rc, cp, rload, d, ncycle, n):
    """ Main program, plots results
        Forward converter, tertiary reset winding,
        Prog.7.07, generates Figure~7.07"""
    #
    #calling command:
    #ch7fig07(165.0, 100000.0, 15.0, 0.75e-03, 0.02, 0.85e-05,          \
    #                        0.01, 0.44e-03, 0.2, 0.3, 200, 600)
    #
    from math import pi
    import numpy as np
    import matplotlib.pyplot as plt
```

```
from mpl_toolkits.axes_grid1 import host_subplot
#
N = ncycle*n
nc = int(np.rint(d*n))
#
delt = 1.0/(freq*n)
dteta = 2.0*pi/n
glm = delt/lm
grl = 1.0/rl
gls = delt/ls
grc = 1.0/rc
gcp = cp/delt
gload = 1.0/rload
#
vout, vinduct, vdiode2, cinduct, cmag, cdiode1, cdiode2 =          \
                ch7fig07_solver(es, Ntr, glm, grl, gls, grc,       \
                                     gcp, gload, nc, ncycle, n)
#
m0 = 2*(ncycle - 2)
m1 = 2*(ncycle - 1)
m2 = 2*ncycle
ang = dteta*np.arange(N)
#
ax1 = host_subplot(2, 1, 1)
ax2 = ax1.twinx()
ax1.set_xlim([m0*pi, m2*pi])
ax1.set_xticks([m0*pi, (m0+2*d)*pi, (m0+4*d)*pi, m1*pi, m2*pi])
ax1.set_xticklabels(['$0$', '$2\pi d$', '$4\pi d$',                \
                               '$2\pi$', '$4\pi$'])
ax1.set_ylim([-10.0, 10.0])
ax1.set_yticks([-10.0, 0, 10.0])
ax1.set_yticklabels(['$-10$', '$0$', '$10$'])
ax1.set_ylabel('$Voltages, V$')
ax1.text(m1*pi, -18.0, '$(a)$', size =10)
ax2.set_ylim(0, 20)
ax2.set_yticks([0, 10, 20])
ax2.set_yticklabels(['$0$', '$10$', '$20$'])
ax2.set_ylabel('$Current, A$')
p1, = ax1.plot(ang[N-2*n:N-1], vout[N-2*n:N-1], 'r',               \
                                  label = '$V_{O}$')
p2, = ax1.plot(ang[N-2*n:N-1], vinduct[N-2*n:N-1], 'r--',          \
                                  label ='$V_{L}$')
p3, = ax2.plot(ang[N-2*n:N-1], cinduct[N-2*n:N-1], 'b-',           \
                                   label = '$I_{L}$')
p4, = ax2.plot(ang[N-2*n:N-1], cmag[N-2*n:N-1], 'b--',             \
                                   label = '$I_{mag}$')
ax1.yaxis.get_label().set_color(p1.get_color())
ax2.yaxis.get_label().set_color(p3.get_color())
ax1.legend(bbox_to_anchor=(0., 1.02, 1., .102), loc=3, ncol=4,     \
                            mode="expand", borderaxespad=0.)
ax1.grid(True)
plt.subplots_adjust(hspace = 0.9)
#
```

```
    ax1 = host_subplot(2, 1, 2)
    ax2 = ax1.twinx()
    ax1.set_xlim([m0*pi, m2*pi])
    ax1.set_xticks([m0*pi, (m0+2*d)*pi, m1*pi, m2*pi])
    ax1.set_xticklabels(['$0$', '$2\pi *d$', '$2\pi$', '$4\pi$'])
    ax1.set_xlabel('$\omega t, radians$')
    ax1.set_ylim([-20.0, 20.0])
    ax1.set_yticks([-20.0, 0, 20.0])
    ax1.set_yticklabels(['$-20$', '$0$', '$20$'])
    ax1.set_ylabel('$Voltage, V$')
    ax1.text(m1*pi, -45, '$(b)$', size =10)
    ax2.set_ylim(-1.0, 20.0)
    ax2.set_yticks([0, 10, 20])
    ax2.set_yticklabels(['$0$', '$10$', '$20$'])
    ax2.set_ylabel('$Current, A$')
    p1, = ax1.plot(ang[N-2*n:N-1], vdiode2[N-2*n:N-1], 'r',            \
                                          label = '$V_{diode2}$')
    p2, = ax2.plot(ang[N-2*n:N-1], cdiode1[N-2*n:N-1], 'b',            \
                                          label = '$I_{d1}$')
    p3, = ax2.plot(ang[N-2*n:N-1], cdiode2[N-2*n:N-1], 'b--',          \
                                          label = '$I_{d2}$')
    ax1.yaxis.get_label().set_color(p1.get_color())
    ax2.yaxis.get_label().set_color(p2.get_color())
    ax1.legend(bbox_to_anchor=(0., 1.02, 1., .102), loc=3, ncol=3,   \
                              mode="expand", borderaxespad=0.)
    ax1.grid(True)
    #
    plt.show()
    #
    #plt.draw()
    #plt.savefig('Fig707.ps', dpi=400, bbox_inches='tight')
    #
```

7.6.2 Program 7.11

```
#
def fdiode(vd):
    """ Diode characteristic.
        Returns diode current and its conductance"""
    #
    glow = 0.1e-07
    ghigh = 0.4e+03
    Vr = -0.5
    #
    if vd >= Vr:
        id = glow*(vd - Vr)
        gd = glow
    else:
        id = ghigh*(vd - Vr)
        gd = ghigh
    #
    return id, gd
```

```
#
#
def ch7fig11_solver(es, Ntr, glk, glm, grl, gls, grc, gcp, gload,          \
                                                            nc, ncycle, n):
    """Assumes fdiode is a function.
       Returns the lists vout, cinduct"""
    #
    import numpy as np
    #
    N = n*ncycle
    v2 = 5.0
    vd1 = 0.5
    vd2 = 4.3
    v0 = 3.3
    pilk = 7.3
    pilm = 0.0
    pils = 7.5
    pvc = 3.0
    erlim = 0.1e-04
    Ntrsq = Ntr**2
    gser = grl*gls/(grl + gls)
    gpar = grc*gcp/(grc + gcp)
    #
    vout = []
    vinduct = []
    vdiode2 = []
    cmag = []
    cinduct = []
    cdiode1 = []
    cdiode2 =[]
    #
    for k in range(ncycle):
        #
        for j in range(n):
            #
            if j <= nc:
                gsw = 250.0
            else:
                gsw = 0.1e-07
            #
            for its in range(5000):
                #
                vd3 = es + Ntr*v2
                #
                id1, gd1 = fdiode(vd1)
                id2, gd2 = fdiode(vd2)
                id3, gd3 = fdiode(vd3)
                #
                isw = gsw*(es - Ntr*v2)
                ilm = Ntr*glm*v2 + pilm
                ilk = glk*(v2 - vd2 + vd1) + pilk
                i1 = isw - ilm
                i2 = Ntr*(i1 - id3)
```

```
        #
        vp = (grl*vd2 + gls*v0 - pils)/(grl + gls)
        vc = (grc*v0 + gcp*pvc)/(grc + gcp)
        #
        #
        ils = grl*(vd2 - vp)
        icp = grc*(v0 - vc)
        iload = gload*v0
        #
        irr2 = i2 - ilk
        irrd1 = -(id1 + ilk)
        irrd2 = -(id2 + ils - ilk)
        irr0 = ils - icp - iload
        #
        err = np.sqrt(irr2**2 + irrd1**2 + irrd2**2 + irr0**2)
        #
        if err < erlim:
            break
        else:
            gmat = np.array([[(Ntrsq*(gsw+glm+gd3)+glk),        \
                                                  glk, -glk, 0],
                              [glk, (gd1+glk), -glk, 0],
                              [0, gd1, (gd2+gser), -gser],
                              [0, 0, -gser, (gser+gpar+gload)]])
            bcol = np.array([irr2, irrd1, irrd1 + irrd2, irr0])
            xdv = np.linalg.solve(gmat, bcol)
            #
            dv2 = xdv[0]
            dvd1 = xdv[1]
            dvd2 = xdv[2]
            dv0 = xdv[3]
            v2 += dv2
            vd1 += dvd1
            vd2 += dvd2
            v0 += dv0
    #
    if its >= 999:
        print('time-step= ', (k*n + j), 'j = ', j, 'k = ', k,  \
                                  'solver does not converge')
        break
    #
    pilm = ilm
    pilk = ilk
    pils = ils
    pvc = vc
    #
    cinduct.append(ils)
    cdiode1.append(-id1)
    cdiode2.append(-id2)
    cmag.append(ilm)
    vout.append(v0)
    vinduct.append(vd2 - v0)
    vdiode2.append(vd2)
```

```
            #
        #
    #
    print(' v0= ', np.mean(vout[N-2*n:N-1]))
    #
    return vout, vinduct, vdiode2, cinduct, cmag, cdiode1, cdiode2
#
#
def ch7fig11(es, freq, Ntr, lk, lm, rl, ls, rc, cp, rload, d, ncycle, n):
    """ Main program, plots results
        Forward converter with leakage between primary and
        secondary windings"""
    #
    #calling command:
    #ch7fig11(165.0, 100000.0, 15.0, 0.3e-06, 0.75e-03,               \
    #                             0.02, 0.85e-05, 0.01,                \
    #                             0.44e-03, 0.2, 0.3, 200, 600)
    #
    from math import pi
    import numpy as np
    import matplotlib.pyplot as plt
    from mpl_toolkits.axes_grid1 import host_subplot
    #
    N = ncycle*n
    nc = int(np.rint(d*n))
    #
    delt = 1.0/(freq*n)
    dteta = 2.0*pi/n
    glk = delt/lk
    glm = delt/lm
    grl = 1.0/rl
    gls = delt/ls
    grc = 1.0/rc
    gcp = cp/delt
    gload = 1.0/rload
    #
    vout, vinduct, vdiode2, cinduct, cmag, cdiode1, cdiode2 =          \
                ch7fig11_solver(es, Ntr, glk, glm, grl,                \
                    gls, grc, gcp, gload, nc, ncycle, n)
    #
    m0 = 2*(ncycle - 2)
    m1 = 2*(ncycle - 1)
    m2 = 2*ncycle
    ang = dteta*np.arange(N)
    #
    ax1 = host_subplot(2, 1, 1)
    ax2 = ax1.twinx()
    ax1.set_xlim([m0*pi, m2*pi])
    ax1.set_xticks([m0*pi, (m0+2*d)*pi, m1*pi, m2*pi])
    ax1.set_xticklabels(['$0$', '$2\pi *d$', '$2\pi$', '$4\pi$'])
    ax1.set_ylim([-10.0, 10.0])
    ax1.set_yticks([-10.0, 0, 10.0])
    ax1.set_yticklabels(['$-10$', '$0$', '$10$'])
```

```
ax1.set_ylabel('$Voltages, V$')
ax1.text(m1*pi, -18.0, '$(a)$', size =10)
p1, = ax1.plot(ang[N-2*n:N-1], vout[N-2*n:N-1], 'r',                \
                                 label = '$V_{O}$')
p2, = ax1.plot(ang[N-2*n:N-1], vinduct[N-2*n:N-1], 'r--',           \
                                 label = '$V_{L}$')
ax2.set_ylim(0, 20)
ax2.set_yticks([0, 10, 20])
ax2.set_yticklabels(['$0$', '$10$', '$20$'])
ax2.set_ylabel('$Current, A$')
p3, = ax2.plot(ang[N-2*n:N-1], cinduct[N-2*n:N-1], 'b',             \
                                 label = '$I_{L}$')
p4, = ax2.plot(ang[N-2*n:N-1], cmag[N-2*n:N-1], 'b--',              \
                                 label = '$I_{mag}$')
ax1.yaxis.get_label().set_color(p1.get_color())
ax2.yaxis.get_label().set_color(p3.get_color())
ax1.legend(bbox_to_anchor=(0., 1.02, 1., .102), loc=3, ncol=4,      \
                             mode="expand", borderaxespad=0.)
ax1.grid(True)
plt.subplots_adjust(hspace = 0.9)
#
ax1 = host_subplot(2, 1, 2)
ax2 = ax1.twinx()
ax1.set_xlim([m0*pi, m2*pi])
ax1.set_xticks([m0*pi, (m0+2*d)*pi, m1*pi, m2*pi])
ax1.set_xticklabels(['$0$', '$2\pi *d$', '$2\pi$', '$4\pi$'])
ax1.set_xlabel('$\omega t, radians$')
ax1.set_ylim([-20.0, 20.0])
ax1.set_yticks([-20.0, 0, 20.0])
ax1.set_yticklabels(['$-20$', '$0$', '$20$'])
ax1.set_ylabel('$Voltage, V$')
ax1.text(m1*pi, -45, '$(b)$', size =10)
ax2.set_ylim(-1.0, 20.0)
ax2.set_yticks([0, 10, 20])
ax2.set_yticklabels(['$0$', '$10$', '$20$'])
ax2.set_ylabel('$Current, A$')
p1, = ax1.plot(ang[N-2*n:N-1], vdiode2[N-2*n:N-1], 'r',             \
                                  label = '$V_{d2}$')
p2, = ax2.plot(ang[N-2*n:N-1], cdiode1[N-2*n:N-1], 'b',             \
                                  label = '$I_{d1}$')
p3, = ax2.plot(ang[N-2*n:N-1], cdiode2[N-2*n:N-1], 'b--',           \
                                  label='$I_{d2}$')
ax1.yaxis.get_label().set_color(p1.get_color())
ax2.yaxis.get_label().set_color(p2.get_color())
ax1.legend(bbox_to_anchor=(0., 1.02, 1., .102), loc=3, ncol=3,     \
                             mode="expand", borderaxespad=0.)
ax1.grid(True)
#
plt.show()
#
#plt.draw()
#plt.savefig('Fig711.ps', dpi=400, bbox_inches='tight')
#
```

7.6.3 Program 7.13

```
#
def fdiode(vd):
    """ Diode characteristic.
        Returns diode current and its conductance"""
    #
    glow = 0.1e-07
    ghigh = 0.4e+03
    Vr = -0.5
    #
    if vd >= Vr:
        id = glow*(vd - Vr)
        gd = glow
    else:
        id = ghigh*(vd - Vr)
        gd = ghigh
    #
    return id, gd
#
def ch7fig13_solver(es, Ntr, glt, glm, grl, gls, grc,                   \
                            gcp, gload, nc, ncycle, n):
    """Assumes fdiode is a function.
       Returns the lists vout, cinduct
       """
    #
    import numpy as np
    #
    N = n*ncycle
    v1 = 160.0
    vd2 = 4.3
    vd3 = 300.0
    v0 = 3.3
    pilt = 0
    pilm = 0
    pils = 7.5
    pvc = 3.0
    erlim = 0.1e-04
    Ntrsq = Ntr**2
    gser = grl*gls/(grl + gls)
    gpar = grc*gcp/(grc + gcp)
    #
    vout = []
    vinduct = []
    vdiode2 = []
    vswitch = []
    cprimary = []
    cinduct = []
    cdiode3 = []
    cmag =[]
    #
    for k in range(ncycle):
        #
```

```
for j in range(n):
    #
    if j <= nc:
        gsw = 250.0
    else:
        gsw = 0.1e-07
    #
    for its in range(1000):
        #
        vd1 = vd2 - v1/Ntr
        #
        id1, gd1 = fdiode(vd1)
        id2, gd2 = fdiode(vd2)
        id3, gd3 = fdiode(vd3)
        #
        ilt = glt*(es + v1 - vd3) + pilt
        isw = gsw*(es - v1)
        ilm = glm*v1 + pilm
        i1 = ilt - id1/Ntr
        #
        vp = (grl*vd2 + gls*v0 - pils)/(grl + gls)
        vc = (grc*v0 + gcp*pvc)/(grc + gcp)
        #
        #
        ils = grl*(vd2 - vp)
        icp = grc*(v0 - vc)
        iload = gload*v0
        #
        irr1 = isw -ilm - i1
        irrd3 = ilt - id3
        irrd2 = -(id1 + id2 + ils)
        irr0 = ils - icp - iload
        #
        err = np.sqrt(irr1**2 + irrd2**2 + irrd3**2 + irr0**2)
        if err < erlim:
            break
        else:
            #
            gmat = np.array([ [(gsw+glm+glt+gd1/Ntrsq), -glt,     \
                                                    -gd1/Ntr, 0],
                              [-glt, (glt+gd3), 0, 0],
                              [-gd1/Ntr, 0, (gd1+gd2+gser), -gser],
                              [0, 0, -gser, (gser+gpar+gload)]])
            #
            bcol = np.array([irr1, irrd3, irrd2, irr0])
            xdv = np.linalg.solve(gmat, bcol)
            #
            dv1 = xdv[0]
            dvd3 = xdv[1]
            dvd2 = xdv[2]
            dv0 = xdv[3]
            v1 += dv1
            vd3 += dvd3
```

```
                vd2 += dvd2
                v0 += dv0
            #
            if its >= 999:
                print('time-step= ', (k*n + j), 'j = ', j, 'k = ',      \
                                      k, 'solver does not converge')
                break
            #
            pilt = ilt
            pilm = ilm
            pils = ils
            pvc = vc
            #
            cmag.append(ilm)
            cinduct.append(ils)
            cprimary.append(i1)
            cdiode3.append(-id3)
            vout.append(v0)
            vinduct.append(vd2 - v0)
            vdiode2.append(vd2)
            vswitch.append(es - v1)
            #
        #
    #
    print(' v0= ', np.mean(vout[N-2*n:N-1]))
    print('vswitch_max = ', max(vswitch))
    #
    return vout, vinduct, vdiode2, vswitch, cinduct,                    \
                            cprimary, cdiode3, cmag
#
#
def ch7fig13(es, freq, Ntr, lpt, lm, rl, ls, rc, cp, rload,             \
                                            d, ncycle, n):
    """ Main program, plots results
        Forward converter; leakage inductance
        between primary and tertiary windings.
        No RCD snubber placed across the power switch"""
    #
    #calling command:
    #ch7fig13(165.0, 100000.0, 15.0, 0.5e-05, 0.75e-03,                 \
    #         0.02, 0.85e-05, 0.01, 0.44e-03, 0.2, 0.3, 200, 600)
    #
    from math import pi
    import numpy as np
    import matplotlib.pyplot as plt
    from mpl_toolkits.axes_grid1 import host_subplot
    #
    N = ncycle*n
    nc = int(np.rint(d*n))
    #
    delt = 1.0/(freq*n)
    dteta = 2.0*pi/n
    glt = delt/lpt
```

```
glm = delt/lm
grl = 1.0/rl
gls = delt/ls
grc = 1.0/rc
gcp = cp/delt
gload = 1.0/rload
#
vout, vinduct, vdiode2, vswitch, cinduct, cprimary, cdiode3, cmag = \
                ch7fig13_solver(es, Ntr, glt, glm, grl, gls,        \
                            grc, gcp, gload, nc, ncycle, n)
#
m0 = 2*(ncycle - 2)
m1 = 2*(ncycle - 1)
m2 = 2*ncycle
ang = dteta*np.arange(N)
#
ax1 = host_subplot(2, 1, 1)
ax2 = ax1.twinx()
ax1.set_xlim([m0*pi, m2*pi])
ax1.set_xticks([m0*pi, (m0+2*d)*pi, m1*pi, m2*pi])
ax1.set_xticklabels(['$0$', '$2\pi *d$', '$2\pi$', '$4\pi$'])
ax1.set_ylim([-10.0, 10.0])
ax1.set_yticks([-10.0, 0, 10.0])
ax1.set_yticklabels(['$-10$', '$0$', '$10$'])
ax1.set_ylabel('$Voltages, V$')
ax1.text(m1*pi, -18.0, '$(a)$', size =10)
ax2.set_ylim(0, 20)
ax2.set_yticks([0, 10, 20])
ax2.set_yticklabels(['$0$', '$10$', '$20$'])
ax2.set_ylabel('$Current, A$')
p1, = ax1.plot(ang[N-2*n:N-1], vout[N-2*n:N-1], 'r',                \
                                label = '$V_{O}$')
p2, = ax1.plot(ang[N-2*n:N-1], vinduct[N-2*n:N-1], 'r--',           \
                                label='$V_{L}$')
p3, = ax2.plot(ang[N-2*n:N-1], cinduct[N-2*n:N-1], 'b',             \
                                label = '$I_{L}$')
p4, = ax2.plot(ang[N-2*n:N-1], cmag[N-2*n:N-1], 'b--',              \
                                label = '$I_{mag}$')
ax1.yaxis.get_label().set_color(p1.get_color())
ax2.yaxis.get_label().set_color(p3.get_color())
ax1.legend(bbox_to_anchor=(0., 1.02, 1., .102), loc=3, ncol=4,      \
                          mode="expand", borderaxespad=0.)
ax1.grid(True)
plt.subplots_adjust(hspace = 0.9)
#
ax1 = host_subplot(2, 1, 2)
ax2 = ax1.twinx()
ax1.set_xlim([m0*pi, m2*pi])
ax1.set_xticks([m0*pi, (m0+2*d)*pi, m1*pi, m2*pi])
ax1.set_xticklabels(['$0$', '$2\pi *d$', '$2\pi$', '$4\pi$'])
ax1.set_xlabel('$\omega t, radians$')
ax1.set_ylim([0, 1000])
ax1.set_yticks([0, 500, 1000])
```

```
    ax1.set_yticklabels(['$0$', '$500$', '$1000$'])
    ax1.set_ylabel('$Voltage, V$')
    ax1.text(m1*pi, -590, '$(b)$', size =10)
    p1, = ax1.plot(ang[N-2*n:N-1], vswitch[N-2*n:N-1], 'r',               \
                                    label = '$V_{switch}$')
    ax2.set_ylim(-4.5, 1.5)
    ax2.set_yticks([-4.5,-1.5, 0, 1.5])
    ax2.set_yticklabels(['$-4.5$','$-1.5$', '$0$', '$1.5$'])
    ax2.set_ylabel('$Current, A$')
    p2, = ax2.plot(ang[N-2*n:N-1], cprimary[N-2*n:N-1], 'b',              \
                                    label='$I_{pri}$')
    p3, = ax2.plot(ang[N-2*n:N-1], cdiode3[N-2*n:N-1], 'b--',             \
                                    label='$I_{d3}$')
    ax1.yaxis.get_label().set_color(p1.get_color())
    ax2.yaxis.get_label().set_color(p2.get_color())
    ax1.legend(bbox_to_anchor=(0., 1.02, 1., .102), loc=3, ncol=3,        \
                                  mode="expand", borderaxespad=0.)
    ax1.grid(True)
    #
    plt.show()
    #
    #plt.draw()
    #plt.savefig('Fig713.ps', dpi=400, bbox_inches='tight')
    #
```

7.6.4 Program 7.14

```
#
def fdiode(vd):
    """ Diode characteristic.
        Returns diode current and its conductance"""
    #
    glow = 0.1e-07
    ghigh = 0.4e+03
    Vr = -0.5
    #
    if vd >= Vr:
        id = glow*(vd - Vr)
        gd = glow
    else:
        id = ghigh*(vd - Vr)
        gd = ghigh
    #
    return id, gd
#
#
def ch7fig14_solver(es, Ntr, gsb, gcb, glt, glm, grl, gls, grc,         \
                                    gcp, gload, nc, ncycle, n):
    """Assumes fdiode is a function.
       Returns the lists vout, cinduct
       """
    #
```

```
import numpy as np
#
N = n*ncycle
v1 = 160.0
vd2 = 4.3
vd3 = 300.0
vd4 = 0.0
v0 = 3.3
pvcb = 0.0
pilt = 0
pilm = 0
pils = 7.5
pvc = 3.0
erlim = 0.1e-04
Ntrsq = Ntr**2
gser = grl*gls/(grl + gls)
gpar = grc*gcp/(grc + gcp)
#
vout = []
vinduct = []
vdiode2 = []
vswitch = []
cprimary = []
cinduct = []
cdiode3 = []
cmag =[]
#
for k in range(ncycle):
    #
    for j in range(n):
        #
        if j <= nc:
            gsw = 250.0
        else:
            gsw = 0.1e-07
        #
        for its in range(1000):
            #
            vd1 = vd2 - v1/Ntr
            #
            id1, gd1 = fdiode(vd1)
            id2, gd2 = fdiode(vd2)
            id3, gd3 = fdiode(vd3)
            id4, gd4 = fdiode(vd4)
            #
            ilt = glt*(es + v1 - vd3) + pilt
            isw = gsw*(es - v1)
            ilm = glm*v1 + pilm
            i1 = ilt - id1/Ntr
            #
            vcb = es - v1 + vd4
            #
            vp = (grl*vd2 + gls*v0 - pils)/(grl + gls)
```

```
    vc = (grc*v0 + gcp*pvc)/(grc + gcp)
    #
    #
    ils = grl*(vd2 - vp)
    icp = grc*(v0 - vc)
    iload = gload*v0
    #
    irrd4 = gcb*pvcb -id4 -(gsb+gcb)*vcb
    irr1 = isw -id4 -ilm - i1 -irrd4
    irrd3 = ilt - id3
    irrd2 = -(id1 + id2 + ils)
    irr0 = ils - icp - iload
    #
    err = np.sqrt(irr1**2 + irrd2**2 + irrd3**2 +          \
                                    irrd4**2 + irr0**2)
    if err < erlim:
        break
    else:
        #
        gmat = np.array([ [(gsw+glm+gsb+gcb+glt+gd1/Ntrsq), \
                           -(gsb+gcb), -glt, -gd1/Ntr, 0],
                          [-(gsb+gcb), (gsb+gcb+gd4), 0, 0, 0],
                          [-glt, 0, (glt+gd3), 0, 0],
                          [-gd1/Ntr, 0, 0, (gd1+gd2+gser),  \
                                                    -gser],
                          [0, 0, 0, -gser, (gser+gpar+gload)]])
        #
        bcol = np.array([irr1, irrd4, irrd3, irrd2, irr0])
        xdv = np.linalg.solve(gmat, bcol)
        #
        dv1 = xdv[0]
        dvd4 = xdv[1]
        dvd3 = xdv[2]
        dvd2 = xdv[3]
        dv0 = xdv[4]
        v1 += dv1
        vd4 += dvd4
        vd3 += dvd3
        vd2 += dvd2
        v0 += dv0
#
if its >= 999:
    print('time-step= ', (k*n + j), 'j = ', j, 'k = ',   \
                                k, 'solver does not converge')
    break
#
pilt = ilt
pilm = ilm
pils = ils
pvc = vc
pvcb = vcb
#
cmag.append(ilm)
```

```
            cinduct.append(ils)
            cprimary.append(i1)
            cdiode3.append(-id3)
            vout.append(v0)
            vinduct.append(vd2 - v0)
            vdiode2.append(vd2)
            vswitch.append(es - v1)
            #
        #
    #
    print(' v0= ', np.mean(vout[N-2*n:N-1]))
    print('vswitch_max = ', max(vswitch))
    #
    return vout, vinduct, vdiode2, vswitch, cinduct,                \
                            cprimary, cdiode3, cmag
#
#
def ch7fig14(es, freq, Ntr, rsb, csb, lpt, lm, rl, ls, rc, cp, rload,   \
                                                    d, ncycle, n):
    """ Main program, plots results
        Forward converter; leakage inductance
        between primary and tertiary windings.
        RCD snubber placed across the power switch"""
    #
    #calling command:
    #ch7fig14(165.0, 100000.0, 15.0, 11000.0, 0.33e-07,                 \
    #         0.5e-05, 0.75e-03, 0.02, 0.85e-05, 0.01,                  \
    #                     0.44e-03, 0.2, 0.3, 200, 600)
    #
    from math import pi
    import numpy as np
    import matplotlib.pyplot as plt
    from mpl_toolkits.axes_grid1 import host_subplot
    #
    N = ncycle*n
    nc = int(np.rint(d*n))
    #
    delt = 1.0/(freq*n)
    dteta = 2.0*pi/n
    gsb = 1.0/rsb
    gcb = csb/delt
    glt = delt/lpt
    glm = delt/lm
    grl = 1.0/rl
    gls = delt/ls
    grc = 1.0/rc
    gcp = cp/delt
    gload = 1.0/rload
    #
    vout, vinduct, vdiode2, vswitch, cinduct, cprimary,                 \
                cdiode3, cmag = ch7fig14_solver(es, Ntr,                \
                gsb, gcb, glt, glm, grl, gls, grc, gcp,                 \
                gload, nc, ncycle, n)
```

```
#
m0 = 2*(ncycle - 2)
m1 = 2*(ncycle - 1)
m2 = 2*ncycle
ang = dteta*np.arange(N)
#
ax1 = host_subplot(2, 1, 1)
ax2 = ax1.twinx()
ax1.set_xlim([m0*pi, m2*pi])
ax1.set_xticks([m0*pi, (m0+2*d)*pi, m1*pi, m2*pi])
ax1.set_xticklabels(['$0$', '$2\pi *d$', '$2\pi$', '$4\pi$'])
ax1.set_ylim([-10.0, 10.0])
ax1.set_yticks([-10.0, 0, 10.0])
ax1.set_yticklabels(['$-10$', '$0$', '$10$'])
ax1.set_ylabel('$Voltages, V$')
ax1.text(m1*pi, -18.0, '$(a)$', size =10)
p1, = ax1.plot(ang[N-2*n:N-1], vout[N-2*n:N-1], 'r',                \
                                    label = '$V_{O}$')
p2, = ax1.plot(ang[N-2*n:N-1], vinduct[N-2*n:N-1], 'r--',           \
                                    label='$V_{L}$')
ax2.set_ylim(0, 20)
ax2.set_yticks([0, 10, 20])
ax2.set_yticklabels(['$0$', '$10$', '$20$'])
ax2.set_ylabel('$Current, A$')
p3, = ax2.plot(ang[N-2*n:N-1], cinduct[N-2*n:N-1], 'b',             \
                                           label = '$I_{L}$')
p4, = ax2.plot(ang[N-2*n:N-1], cmag[N-2*n:N-1], 'b--',              \
                                           label = '$I_{mag}$')
ax1.yaxis.get_label().set_color(p1.get_color())
ax2.yaxis.get_label().set_color(p3.get_color())
ax1.legend(bbox_to_anchor=(0., 1.02, 1., .102), loc=3, ncol=4,      \
                                mode="expand", borderaxespad=0.)
ax1.grid(True)
plt.subplots_adjust(hspace = 0.9)
#
ax1 = host_subplot(2, 1, 2)
ax2 = ax1.twinx()
ax1.set_xlim([m0*pi, m2*pi])
ax1.set_xticks([m0*pi, (m0+2*d)*pi, m1*pi, m2*pi])
ax1.set_xticklabels(['$0$', '$2\pi *d$', '$2\pi$', '$4\pi$'])
ax1.set_xlabel('$\omega t, radians$')
ax1.set_ylim([0, 1000])
ax1.set_yticks([0, 500, 1000])
ax1.set_yticklabels(['$0$', '$500$', '$1000$'])
ax1.set_ylabel('$Voltage, V$')
ax1.text(m1*pi, -590, '$(b)$', size =10)
p1, = ax1.plot(ang[N-2*n:N-1], vswitch[N-2*n:N-1], 'r',             \
                                label = '$V_{switch}$')
ax2.set_ylim(-4.5, 1.5)
ax2.set_yticks([-4.5,-1.5, 0, 1.5])
ax2.set_yticklabels(['$-4.5$','$-1.5$', '$0$', '$1.5$'])
ax2.set_ylabel('$Current, A$')
p2, = ax2.plot(ang[N-2*n:N-1], cprimary[N-2*n:N-1], 'b',            \
```

```
                                    label='$I_{pri}$')
    p3, = ax2.plot(ang[N-2*n:N-1], cdiode3[N-2*n:N-1], 'b--',              \
                                    label='$I_{d3}$')
    ax1.yaxis.get_label().set_color(p1.get_color())
    ax2.yaxis.get_label().set_color(p2.get_color())
    ax1.legend(bbox_to_anchor=(0., 1.02, 1., .102), loc=3, ncol=3,         \
                                   mode="expand", borderaxespad=0.)
    ax1.grid(True)
    #
    plt.show()
    #
    #plt.draw()
    #plt.savefig('Fig714.ps', dpi=400, bbox_inches='tight')
    #
```

7.6.5 Program 7.19

```
#
def fdiode(vd):
    """ Diode characteristic.
        Returns diode current and its conductance"""
    #
    glow = 0.1e-07
    ghigh = 0.4e+03
    Vr = -0.5
    #
    if vd >= Vr:
        id = glow*(vd - Vr)
        gd = glow
    else:
        id = ghigh*(vd - Vr)
        gd = ghigh
    #
    return id, gd
#
#
def ch7fig19_solver(es, Ntr, glm, grl, gls, grc, gcp, gcd1,              \
                                      gload, nc, ncycle, n):
    """Assumes fdiode is a function.
        Returns the lists vout, cinduct"""
    #
    import numpy as np
    #
    N = n*ncycle
    #
    vd1 = -0.0125
    vd2 = -0.025514
    v0 = 3.099153
    pilm = -0.32545
    pils = 14.96736
    pvc = 3.099153
    pvd1 = -0.0125
```

```
#
erlim = 0.1e-04
Ntrsq = Ntr**2
gser = grl*gls/(grl + gls)
gpar = grc*gcp/(grc + gcp)
#
vout = []
vinduct = []
vsw = []
cmag = []
cinduct = []
#
for k in range(ncycle):
    #
    for j in range(n):
        #
        if j <= nc:
            gsw = 250.0
        else:
            gsw = 0.1e-07
        #
        gpri = Ntrsq*(gsw + glm)
        #
        for its in range(5000):
            #
            v1 = Ntr*(vd2 - vd1)
            isw = gsw*(es - v1)
            ilm = glm*v1 + pilm
            #
            id1, gd1 = fdiode(vd1)
            id2, gd2 = fdiode(vd2)
            #
            i1 = isw - ilm
            i2 = Ntr*i1
            #
            vp = (grl*vd2 + gls*v0 - pils)/(grl + gls)
            vq = (grc*v0 + gcp*pvc)/(grc + gcp)
            ils = grl*(vd2 - vp)
            icp = grc*(v0 - vq)
            icd1 = gcd1*(vd1 - pvd1)
            iload = gload*v0
            #
            irrd1 = -(id1 + i2 + icd1)
            irrd2 = i2 - id2 -ils
            irr0 = ils - icp - iload
            #
            err = np.sqrt(irrd1**2 + irrd2**2 + irr0**2)
            #
            if err < erlim:
                break
            else:
                gmat = np.array([[(gpri + gd1 + gcd1), -gpri, 0],
                                  [-gpri, (gpri + gd2 + gser), -gser],
```

```
                                        [0, -gser, (gser + gpar + gload)]])
                bcol = np.array([irrd1, irrd2, irr0])
                xdv = np.linalg.solve(gmat, bcol)
                #
                dvd1 = xdv[0]
                dvd2 = xdv[1]
                dv0 = xdv[2]
                vd1 += dvd1
                vd2 += dvd2
                v0 += dv0
            #
            if its >= 999:
                print('time-step= ', (k*n + j), 'j = ', j, 'k = ',     \
                                         k, 'solver does not converge')
                break
            #
            pilm = ilm
            pils = ils
            pvc = v0
            pvd1 = vd1
            #
            cinduct.append(ils)
            cmag.append(ilm)
            vout.append(v0)
            vinduct.append(vd2 - v0)
            vsw.append(es - v1)
            #
        #
    #
    print(' v0= ', np.mean(vout[N-2*n:N-1]))
    #
    return vout, vinduct, vsw, cinduct, cmag
#
#
def ch7fig19(es, freq, Ntr, lm, rl, ls, rc, cp, ct, rload, d, ncycle, n):
    """ Main program, plots results
        Resonant reset forward converter
        Prog7.19 plots Ch7~Figure~19"""
    #
    #calling command:
    #ch7fig19(165.0, 100000.0, 15.0, 0.5e-03, 0.02, 0.85e-05, 0.01,      \
    #                          0.44e-03, 0.47e-07, 0.2, 0.3, 400, 200)
    #
    from math import pi
    import numpy as np
    import matplotlib.pyplot as plt
    from mpl_toolkits.axes_grid1 import host_subplot
    #
    N = ncycle*n
    nc = int(np.rint(d*n))
    #
    delt = 1.0/(freq*n)
    dteta = 2.0*pi/n
```

```
glm = delt/lm
grl = 1.0/rl
gls = delt/ls
grc = 1.0/rc
gcp = cp/delt
gcd1 = ct/delt
gload = 1.0/rload
#
vout, vinduct, vsw, cinduct, cmag = ch7fig19_solver(es, Ntr,        \
          glm, grl, gls, grc, gcp, gcd1, gload, nc, ncycle, n)
#
m0 = 2*(ncycle - 2)
m1 = 2*(ncycle - 1)
m2 = 2*ncycle
ang = dteta*np.arange(N)
#
ax1 = host_subplot(2, 1, 1)
ax2 = ax1.twinx()
ax1.set_xlim([m0*pi, m2*pi])
ax1.set_xticks([m0*pi, (m0+2*d)*pi, m1*pi, m2*pi])
ax1.set_xticklabels(['$0$', '$2\pi *d$', '$2\pi$', '$4\pi$'])
ax1.set_ylim([-10.0, 10.0])
ax1.set_yticks([-10.0, 0, 10.0])
ax1.set_yticklabels(['$-10$', '$0$', '$10$'])
ax1.set_ylabel('$Voltages, V$')
ax1.text(m1*pi, -18.0, '$(a)$', size =10)
ax2.set_ylim(0, 20)
ax2.set_yticks([0, 10, 20])
ax2.set_yticklabels(['$0$', '$10$', '$20$'])
ax2.set_ylabel('$Current, A$')
p1, = ax1.plot(ang[N-2*n:N-1], vout[N-2*n:N-1], 'r',                \
                                label = '$V_{O}$')
p2, = ax1.plot(ang[N-2*n:N-1], vinduct[N-2*n:N-1], 'r--',           \
                                label='$V_{L}$')
p3, = ax2.plot(ang[N-2*n:N-1], cinduct[N-2*n:N-1], 'b',
            label='$I_{L}$')
ax1.yaxis.get_label().set_color(p1.get_color())
ax2.yaxis.get_label().set_color(p3.get_color())
ax1.legend(bbox_to_anchor=(0., 1.02, 1., .102), loc=3, ncol=3,      \
                          mode="expand", borderaxespad=0.)
ax1.grid(True)
plt.subplots_adjust(hspace = 0.9)
#
ax1 = host_subplot(2, 1, 2)
ax2 = ax1.twinx()
ax1.set_xlim([m0*pi, m2*pi])
ax1.set_xticks([m0*pi, (m0+2*d)*pi, m1*pi, m2*pi])
ax1.set_xticklabels(['$0$', '$2\pi *d$', '$2\pi$', '$4\pi$'])
ax1.set_xlabel('$\omega t, radians$')
ax1.set_ylim([-50.0, 1000.0])
ax1.set_yticks([0,1000.0])
ax1.set_yticklabels(['$0$', '$1000$'])
ax1.set_ylabel('$Voltage, V$')
```

```
    ax1.text(m1*pi, -700, '$(b)$', size =10)
    ax2.set_ylim(-1.0, 1.0)
    ax2.set_yticks([-1.0, 0, 1.0])
    ax2.set_yticklabels(['$-1.0$', '$0$', '$1.0$'])
    ax2.set_ylabel('$Current, A$')
    p1, = ax1.plot(ang[N-2*n:N-1], vsw[N-2*n:N-1], 'r',                  \
                                        label = '$V_{sw}$')
    p2, = ax2.plot(ang[N-2*n:N-1], cmag[N-2*n:N-1], 'b',                 \
                                        label = '$I_{m}$')
    ax1.yaxis.get_label().set_color(p1.get_color())
    ax2.yaxis.get_label().set_color(p2.get_color())
    ax1.legend(bbox_to_anchor=(0., 1.02, 1., .102), loc=3, ncol=3,       \
                                     mode="expand", borderaxespad=0.)
    ax1.grid(True)
    #
    plt.show()
    #
    #plt.draw()
    #plt.savefig('Fig719.ps', dpi=400, bbox_inches='tight')
    #
```

7.6.6 Program 7.21

```
#
def fdiode(vd):
    """ Diode characteristic.
        Returns diode current and its conductance"""
    #
    glow = 0.1e-07
    ghigh = 0.4e+03
    Vr = -0.5
    #
    if vd >= Vr:
        id = glow*(vd - Vr)
        gd = glow
    else:
        id = ghigh*(vd - Vr)
        gd = ghigh
    #
    return id, gd
#
def ch7fig21_solver(es, Ntr, glm, grl, gls, grc, gcp,                    \
                                   gload, nc, ncycle, n):
    """Assumes fdiode is a function.
       Returns the lists vout, cinduct"""
    #
    import numpy as np
    #
    N = n*ncycle
    v1 = 160.0
    vd1 = 0.5
    vd2 = 4.3
```

```
v0 = 3.3
pilm = 0.0
pils = 7.5
pvc = 3.0
erlim = 0.1e-04
Ntrsq = Ntr**2
gser = grl*gls/(grl + gls)
gpar = grc*gcp/(grc + gcp)
#
vout = []
vinduct = []
vsec = []
cmag = []
cinduct = []
cprimary = []
cdiode1 = []
cdiode2 = []
#
for k in range(ncycle):
    #
    for j in range(n):
        #
        if j <= nc:
            gsw = 250.0
        else:
            gsw = 0.1e-07
        #
        for its in range(5000):
            #
            vdp = 0.5*(es + v1)
            vsw = 0.5*(es - v1)
            v2 = v1/Ntr
            vd1 = vd2 - v2
            #
            id1, gd1 = fdiode(vd1)
            id2, gd2 = fdiode(vd2)
            idp, gdp = fdiode(vdp)
            #
            isw = gsw*vsw
            ip = isw - idp
            ilm = glm*v1 + pilm
            i2 = -id1
            i1 = i2/Ntr
            #
            vp = (grl*vd2 + gls*v0 - pils)/(grl + gls)
            vc = (grc*v0 + gcp*pvc)/(grc + gcp)
            #
            #
            ils = grl*(vd2 - vp)
            icp = grc*(v0 - vc)
            iload = gload*v0
            #
            irr1 = ip - i1 - ilm
```

```
                irrd2 = -(id1 + id2 + ils)
                irr0 = ils - icp - iload
                #
                err = np.sqrt(irr1**2 + irrd2**2 + irr0**2)
                #
                if err < erlim:
                    break
                else:
                    g11 = 0.5*(gsw + gdp) + glm + gd1/Ntrsq
                    g22 = gd1 + gd2 + gser
                    g33 = gser + gpar + gload
                    gmat = np.array([[g11, -gd1/Ntr, 0],
                                     [-gd1/Ntr, g22, -gser],
                                     [0, -gser, g33]])
                    bcol = np.array([irr1, irrd2, irr0])
                    xdv = np.linalg.solve(gmat, bcol)
                    #
                    dv1 = xdv[0]
                    dvd2 = xdv[1]
                    dv0 = xdv[2]
                    v1 += dv1
                    vd2 += dvd2
                    v0 += dv0
            #
            if its >= 999:
                print('time-step= ', (k*n + j), 'j = ', j, 'k = ',   \
                                      k, 'solver does not converge')
                break
            #
            pilm = ilm
            pils = ils
            pvc = vc
            #
            cinduct.append(ils)
            cprimary.append(ip)
            cmag.append(ilm)
            cdiode1.append(-id1)
            cdiode2.append(-id2)
            vout.append(v0)
            vinduct.append(vd2 - v0)
            vsec.append(v2)
            #
        #
    #
    print(' v0= ', np.mean(vout[N-2*n:N-1]))
    #
    return vout, vinduct, vsec, cinduct, cmag, cprimary, cdiode1, cdiode2
#
#
def ch7fig21(es, freq, Ntr, lm, rl, ls, rc, cp, rload, d, ncycle, n):
    """ Main program, plots results
        Single-ended hybrid converter.
        Leakage inductances have been ignored."""
```

```
#
#calling command:
#ch7fig21(165.0, 100000.0, 15.0, 0.75e-03, 0.02,                          \
#         0.85e-05, 0.01, 0.44e-03, 0.2, 0.3, 400, 200)
#
from math import pi
import numpy as np
import matplotlib.pyplot as plt
from mpl_toolkits.axes_grid1 import host_subplot
#
N = ncycle*n
nc = int(np.rint(d*n))
#
delt = 1.0/(freq*n)
dteta = 2.0*pi/n
glm = delt/lm
grl = 1.0/rl
gls = delt/ls
grc = 1.0/rc
gcp = cp/delt
gload = 1.0/rload
#
vout, vinduct, vsec, cinduct, cmag, cprimary, cdiode1, cdiode2 =    \
   ch7fig21_solver(es, Ntr, glm, grl, gls, grc, gcp, gload, nc,     \
                   ncycle, n)
#
m0 = 2*(ncycle - 2)
m1 = 2*(ncycle - 1)
m2 = 2*ncycle
ang = dteta*np.arange(N)
#
ax1 = host_subplot(2, 1, 1)
ax2 = ax1.twinx()
ax1.set_xlim([m0*pi, m2*pi])
ax1.set_xticks([m0*pi, (m0+2*d)*pi, m1*pi, m2*pi])
ax1.set_xticklabels(['$0$', '$2\pi *d$', '$2\pi$', '$4\pi$'])
ax1.set_ylim([-10.0, 10.0])
ax1.set_yticks([-10.0, 0, 10.0])
ax1.set_yticklabels(['$-10$', '$0$', '$10$'])
ax1.set_ylabel('$Voltages, V$')
ax1.text(m1*pi, -18.0, '$(a)$', size =10)
ax2.set_ylim(-2, 20)
ax2.set_yticks([0, 10, 20])
ax2.set_yticklabels(['$0$', '$10$', '$20$'])
ax2.set_ylabel('$Current, A$')
p1, = ax1.plot(ang[N-2*n:N-1], vout[N-2*n:N-1], 'r',                \
                                 label = '$V_{O}$')
p2, = ax1.plot(ang[N-2*n:N-1], vinduct[N-2*n:N-1], 'r--',           \
                                 label='$V_{L}$')
p3, = ax2.plot(ang[N-2*n:N-1], cdiode1[N-2*n:N-1], 'g',             \
                                 label = '$I_{d1}$')
p4, = ax2.plot(ang[N-2*n:N-1], cdiode2[N-2*n:N-1], 'b--',           \
                                 label='$I_{d2}$')
```

```
ax1.yaxis.get_label().set_color(p1.get_color())
ax2.yaxis.get_label().set_color(p3.get_color())
ax1.legend(bbox_to_anchor=(0., 1.02, 1., .102), loc=3, ncol=4,      \
                          mode="expand", borderaxespad=0.)
ax1.grid(True)
plt.subplots_adjust(hspace = 0.9)
#
ax1 = host_subplot(2, 1, 2)
ax2 = ax1.twinx()
ax1.set_xlim([m0*pi, m2*pi])
ax1.set_xticks([m0*pi, (m0+2*d)*pi, m1*pi, m2*pi])
ax1.set_xticklabels(['$0$', '$2\pi *d$', '$2\pi$', '$4\pi$'])
ax1.set_xlabel('$\omega t, radians$')
ax1.set_ylim([-20.0, 20.0])
ax1.set_yticks([-20.0, 0, 20.0])
ax1.set_yticklabels(['$-20$', '$0$', '$20$'])
ax1.set_ylabel('$Voltage, V$')
ax1.text(m1*pi, -45, '$(b)$', size =10)
ax2.set_ylim(-1.0, 2.0)
ax2.set_yticks([0, 1, 2])
ax2.set_yticklabels(['$0$', '$1$', '$2$'])
ax2.set_ylabel('$Current, A$')
p1, = ax1.plot(ang[N-2*n:N-1], vsec[N-2*n:N-1], 'r',               \
                              label = '$V_{sec}$')
p2, = ax2.plot(ang[N-2*n:N-1], cprimary[N-2*n:N-1], 'g',           \
                              label='$I_{pri}$')
p3, = ax2.plot(ang[N-2*n:N-1], cmag[N-2*n:N-1], 'b--',             \
                              label = '$I_{mag}$')
ax1.yaxis.get_label().set_color(p1.get_color())
ax2.yaxis.get_label().set_color(p2.get_color())
ax1.legend(bbox_to_anchor=(0., 1.02, 1., .102), loc=3, ncol=3,    \
                          mode="expand", borderaxespad=0.)
ax1.grid(True)
#
plt.show()
#
#plt.draw()
#plt.savefig('Fig721.ps', dpi=400, bbox_inches='tight')
#
```

7.6.7 Program 7.27

```
#
def fdiode(vd):
    """ Diode characteristic.
        Returns diode current and its conductance"""
    #
    glow = 0.1e-07
    ghigh = 0.4e+03
    Vr = -0.5
    #
    if vd >= Vr:
```

```
        id = glow*(vd - Vr)
        gd = glow
    else:
        id = ghigh*(vd - Vr)
        gd = ghigh
    #
    return id, gd
#
#
def ch7fig27_solver(es, Ntr, gcb, glk, glm, grl, gls, grc,              \
                        gcp, gload, nc, nseq, ncycle, n):
    """Assumes fdiode is a function.
       Returns the lists vout, cinduct"""
    #
    import numpy as np
    #
    nstop = 0
    N = n*ncycle
    nh = int(n/2)
    ghigh = 400.0
    glow = 0.1e-07
    erlim = 0.1e-04
    Ntrsq = Ntr**2
    gser = grl*gls/(grl + gls)
    gpar = grc*gcp/(grc + gcp)
    #
    v1 = 2.489
    v2 = -0.2529
    v0 = 2.9447
    #
    pvcb = -2.4714
    pilm = -0.3275
    pilk = 0.1508
    pils = 14.3504
    pvc = 2.9473
    #
    vout = []
    vinduct = []
    vprimary = []
    cprimary = []
    cmag = []
    cdiode1 = []
    cdiode2 = []
    cw1 = []
    #
    for k in range(ncycle):
        #
        for j in range(n):
            #
            #       SEQUENCE A
            #
            if nseq == 0:
                if j < nc:
```

```
            gsw1 = ghigh
            gsw3 = ghigh
        else:
            gsw1 = glow
            gsw3 = glow
        #
        if (j >= nh) and (j < nc + nh):
            gsw2 = ghigh
            gsw4 = ghigh
        else:
            gsw2 = glow
            gsw4 = glow
#
#
#SEQUENCE B
#
if nseq == 1:
    if (j < nc) or ((j >= nc + nh) and (j < n)):
        gsw1 = ghigh
    else:
        gsw1 = glow
    #
    if (j < n/2):
        gsw3 = ghigh
    else:
        gsw3 = glow
    #
    if (j >= nc) and (j < nc + nh):
        gsw4 = ghigh
    else:
        gsw4 = glow
    #
    if (j >= nh) and (j < n):
        gsw2 = ghigh
    else:
        gsw2 = glow
#
gbr = (gsw1 + gsw4)*(gsw2 + gsw3)/(gsw1 + gsw2 + gsw3 + gsw4)
gcbr = gcb*gbr/(gcb + gbr)
gleft = gcbr*glk/(gcbr + glk)
avr = 1.0 + glk/gcbr
#
vswa = gsw1*es/(gsw1 + gsw4)
vswb = gsw2*es/(gsw2 + gsw3)
v3 = vswa - vswb
#
for its in range(10000):
    #
    vr = ((glk*v1 - pilk)/gcbr + v3 - pvcb)/avr
    vq = (gcb*vr + gbr*v3 + gcb*pvcb)/(gbr + gcb)
    ilk = gbr*(v3 - vq)
    vcb = vq - vr
    #
```

```
        vd1 = v2 - v1/Ntr
        vd2 = v2 + v1/Ntr
        id1, gd1 = fdiode(vd1)
        id2, gd2 = fdiode(vd2)
        #
        ilm = glm*v1 + pilm
        #
        vp = (grl*v2 + gls*v0 - pils)/(grl + gls)
        vc = (grc*v0 + gcp*pvc)/(grc + gcp)
        #
        ils = grl*(v2 - vp)
        icp = grc*(v0 - vc)
        iload = gload*v0
        i1 = (id2 - id1)/Ntr
        #
        irr1 = ilk -ilm - i1
        irr2 = -(id1 + id2 + ils)
        irr0 = ils - icp - iload
        #
        err = np.sqrt(irr1**2 + irr2**2 + irr0**2)
        #
        if err < erlim:
            break
        else:
            #
            gmat = np.array([ [(glm+gleft+(gd1 + gd2)/Ntrsq),   \
                                        (gd2 - gd1)/Ntr, 0],
                              [(gd2 - gd1)/Ntr, gd1+gd2+gser,   \
                                                    -gser],
                              [0, -gser, (gser+gpar+gload)]])
            #
            bcol = np.array([irr1, irr2, irr0])
            xdv = np.linalg.solve(gmat, bcol)
            #
            dv1 = xdv[0]
            dv2 = xdv[1]
            dv0 = xdv[2]
            v1 += dv1
            v2 += dv2
            v0 += dv0
    #
    if its >= 9998:
        print('time-step= ', (k*n + j), 'j = ', j, 'k = ',       \
              k, 'solver does not converge')
        nstop = 1
        break
    #
    pilk = ilk
    pilm = ilm
    pils = ils
    pvc = vc
    pvcb = vcb
    #
```

```
            cw1.append(i1)
            cprimary.append(ilk)
            cmag.append(ilm)
            cdiode1.append(-id1)
            cdiode2.append(-id2)
            vout.append(v0)
            vinduct.append(v2 - v0)
            vprimary.append(vr)
            #
        if nstop == 1:
            break
        #
    #
    print(' v0= ', np.mean(vout[N-2*n:N-1]))
    #
    return vout, vinduct, vprimary, cw1, cprimary, cmag, cdiode1, cdiode2
#
#
def ch7fig27(es, freq, Ntr, cb, lk, lm, rl, ls, rc,                    \
                          cp, rload, d, nseq, ncycle, n):
    """ Main program, plots results
        simulation of double ended full bridge forward converter
        bidirectional switches, Sequence A
        Option exists for running sequence B (nseq=1)"""
    #
    #calling command:
    #ch7fig27(165.0, 100000.0, 30.0, 0.3e-06, 0.2e-05,                  \
    #         0.75e-03, 0.02, 0.85e-05, 0.01, 0.44e-03,                 \
    #         0.2, 0.3, 0, 100, 200)
    #
    from math import pi
    import numpy as np
    import matplotlib.pyplot as plt
    from mpl_toolkits.axes_grid1 import host_subplot
    #
    N = ncycle*n
    nc = int(np.rint(d*n))
    #
    delt = 1.0/(freq*n)
    dteta = 2.0*pi/n
    gcb = cb/delt
    glk = delt/lk
    glm = delt/lm
    grl = 1.0/rl
    gls = delt/ls
    grc = 1.0/rc
    gcp = cp/delt
    gload = 1.0/rload
    #
    vout, vinduct, vprimary, cw1, cprimary, cmag, cdiode1, cdiode2 =   \
                ch7fig27_solver(es, Ntr, gcb, glk, glm, grl, gls,      \
                                grc, gcp, gload, nc, nseq, ncycle, n)
    #
```

```
m0 = 2*(ncycle - 2)
m1 = 2*(ncycle - 1)
m2 = 2*ncycle
ang = dteta*np.arange(N)
#
ax1 = host_subplot(2, 1, 1)
ax2 = ax1.twinx()
ax1.set_xlim([m0*pi, m2*pi])
#
ax1.set_xticks([m0*pi, (m0+2*d)*pi, m1*pi, m2*pi])
ax1.set_xticklabels(['$0$', '$2\pi *d$', '$2\pi$', '$4\pi$'])
ax1.set_ylim([-4.0, 4.0])
ax1.set_yticks([-4, 0, 4.0])
ax1.set_yticklabels(['$-4$', '$0$', '$4$'])
ax1.set_ylabel('$Voltages, V$')
ax1.text(m1*pi, -7.0, '$(a)$', size =10)
ax2.set_ylim(-2, 20)
ax2.set_yticks([0, 10, 20])
ax2.set_yticklabels(['$0$', '$10$', '$20$'])
ax2.set_ylabel('$Current, A$')
p1, = ax1.plot(ang[N-2*n:N-1], vout[N-2*n:N-1], 'r',                \
                                label = '$V_{O}$')
p2, = ax1.plot(ang[N-2*n:N-1], vinduct[N-2*n:N-1],'r--',             \
                                label = '$V_{L}$')
p3, = ax2.plot(ang[N-2*n:N-1], cdiode1[N-2*n:N-1],'b',               \
                                label = '$I_{d1}$')
p4, = ax2.plot(ang[N-2*n:N-1], cdiode2[N-2*n:N-1],'b--',             \
                                label ='$I_{d2}$')
ax1.yaxis.get_label().set_color(p1.get_color())
ax2.yaxis.get_label().set_color(p3.get_color())
ax1.legend(bbox_to_anchor=(0., 1.02, 1., .102), loc=3, ncol=4,       \
                         mode="expand", borderaxespad=0.)
ax1.grid(True)
plt.subplots_adjust(hspace = 0.9)
#
ax1 = host_subplot(2, 1, 2)
ax2 = ax1.twinx()
ax1.set_xlim([m0*pi, m2*pi])
ax1.set_xticks([m0*pi, (m0+2*d)*pi, m1*pi, m2*pi])
ax1.set_xticklabels(['$0$', '$2\pi *d$', '$2\pi$', '$4\pi$'])
ax1.set_xlabel('$\omega t, radians$')
ax1.set_ylim([-200, 200])
ax1.set_yticks([-200, 0, 200])
ax1.set_yticklabels(['$-200$', '$0$', '$200$'])
ax1.set_ylabel('$Voltage, V$')
ax1.text(m1*pi, -450, '$(b)$', size =10)
ax2.set_ylim(-1.0, 1.0)
ax2.set_yticks([-1.0, 0, 1.0])
ax2.set_yticklabels(['$-1.0$', '$0$', '$1.0$'])
ax2.set_ylabel('$Current, A$')
p1, = ax1.plot(ang[N-2*n:N-1], vprimary[N-2*n:N-1], 'r',             \
                                label = '$V_{primary}$')
p2, = ax2.plot(ang[N-2*n:N-1], cprimary[N-2*n:N-1], 'b',             \
```

```
                                          label='$I_{pri}$')
    p3, = ax2.plot(ang[N-2*n:N-1], cmag[N-2*n:N-1], 'g-.',                  \
                                          label = '$I_{mag}$')
    p4, = ax2.plot(ang[N-2*n:N-1], cw1[N-2*n:N-1], 'b--',                   \
                                          label = '$I_{1}$')
    ax1.yaxis.get_label().set_color(p1.get_color())
    ax2.yaxis.get_label().set_color(p2.get_color())
    ax1.legend(bbox_to_anchor=(0., 1.02, 1., .102), loc=3, ncol=4,     \
                                    mode="expand", borderaxespad=0.)
    ax1.grid(True)
    #
    plt.show()
    #
    #plt.draw()
    #plt.savefig('Fig727.ps', dpi=400, bbox_inches='tight')
    #
```

7.6.8 Program 7.29

```
#
def fdiode(vd):
    """ Diode characteristic.
        Returns diode current and its conductance"""
    #
    glow = 0.1e-07
    ghigh = 0.4e+03
    Vr = -0.5
    #
    if vd >= Vr:
        id = glow*(vd - Vr)
        gd = glow
    else:
        id = ghigh*(vd - Vr)
        gd = ghigh
    #
    return id, gd
#
#
def fswitch(vsw, gsw):
    """uni-directional switch characteristic.
       Returns switch current and conductance."""
    #
    glow = 0.1e-07
    ghigh = 0.4e+03
    Vr = 0.5
    #
    if vsw >= Vr:
        vd = (gsw*vsw + ghigh*Vr)/(ghigh + gsw)
        isw = ghigh*(vd -Vr)
        gw = ghigh*gsw/(ghigh + gsw)
    else:
        vd = (gsw*vsw + glow*Vr)/(glow + gsw)
```

```
        isw = glow*(Vr - vd)
        gw = glow*gsw/(glow + gsw)
    #
    return isw, gw
#
#
def ch7fig29_solver(es, Ntr, glm, grl, gls, grc, gcp, gload, nc, ncycle, n):
    """ Assumes fdiode is a function.
        Returns the lists vout, cinduct"""
    #
    import numpy as np
    #
    nstop = 0
    N = n*ncycle
    nh = int(n/2)
    ghigh = 400.0
    glow = 0.1e-07
    erlim = 0.1e-04
    Ntrsq = Ntr**2
    gser = grl*gls/(grl + gls)
    gpar = grc*gcp/(grc + gcp)
    #
    v1 = 0.3682
    v2 = -0.3178
    v0 = 2.9242
    #
    pilm = -0.3273
    pils = 14.2465
    pvc = 2.9269
    #
    vout = []
    cinduct = []
    vprimary = []
    cpr1 = []
    cpr2 = []
    cmag = []
    cdiode1 = []
    cdiode2 = []
    #
    for k in range(ncycle):
        #
        for j in range(n):
            #
            if j < nc:
                gsw1 = ghigh
            else:
                gsw1 = glow
            #
            if (j >= nh) and (j < nh + nc):
                gsw2 = ghigh
            else:
                gsw2 = glow
            #
```

```
        for its in range(10000):
            #
            ilm = Ntr*glm*v1 + pilm
            vs1 = es - Ntr*v1
            vs2 = es + Ntr*v1
            is1, gs1 = fswitch(vs1, gsw1)
            i2, gs2 = fswitch(vs2, gsw2)
            i1 = is1 - ilm
            vd1 = v2 - v1
            vd2 = v2 + v1
            id1, gd1 = fdiode(vd1)
            id2, gd2 = fdiode(vd2)
            #
            ia = id2 - Ntr*(i1 - i2)
            #
            vp = (grl*v2 + gls*v0 - pils)/(grl + gls)
            vc = (grc*v0 + gcp*pvc)/(grc + gcp)
            #
            ils = grl*(v2 - vp)
            icp = grc*(v0 - vc)
            iload = gload*v0
            #
            irr1 = id1 - ia
            irr2 = -(id1 + id2 + ils)
            irr0 = ils - icp - iload
            #
            err = np.sqrt(irr1**2 + irr2**2 + irr0**2)
            #
            if err < erlim:
                break
            else:
                #
                gmat = np.array([ [(gd1 + gd2 + Ntrsq*(gs1+gs2+glm)),  \
                                                  (gd2 - gd1), 0],
                                  [(gd2 - gd1), gd1+gd2+gser, -gser],
                                  [0, -gser, (gser+gpar+gload)]])
                #
                bcol = np.array([irr1, irr2, irr0])
                xdv = np.linalg.solve(gmat, bcol)
                #
                dv1 = xdv[0]
                dv2 = xdv[1]
                dv0 = xdv[2]
                v1 += dv1
                v2 += dv2
                v0 += dv0
        #
        if its >= 9998:
            print('time-step= ', (k*n + j), 'j = ', j, 'k = ',        \
                                 k, 'solver does not converge')
            nstop = 1
            break
        #
```

```
            pilm = ilm
            pils = ils
            pvc = vc
            #
            cpr1.append(i1)
            cpr2.append(-i2)
            cmag.append(ilm)
            cdiode1.append(-id1)
            cdiode2.append(-id2)
            vout.append(v0)
            cinduct.append(ils)
            vprimary.append(Ntr*v1)
            #
        if nstop == 1:
            break
        #
    #
    print(' v0= ', np.mean(vout[N-2*n:N-1]))
    #
    return vout, cinduct, vprimary, cpr1, cpr2, cmag, cdiode1, cdiode2
#
#
def ch7fig29(es, freq, Ntr, lm, rl, ls, rc, cp, rload, d, ncycle, n):
    """ Main program, plots results
        push--pull converter, unilateral switches"""
    #calling command:
    #ch7fig29(165.0, 100000.0, 30.0, 0.75e-03, 0.02, 0.85e-05,          \
    #         0.01, 0.44e-03, 0.2, 0.3, 100, 200)
    #
    from math import pi
    import numpy as np
    import matplotlib.pyplot as plt
    from mpl_toolkits.axes_grid1 import host_subplot
    #
    N = ncycle*n
    nc = int(np.rint(d*n))
    #
    delt = 1.0/(freq*n)
    dteta = 2.0*pi/n
    glm = delt/lm
    grl = 1.0/rl
    gls = delt/ls
    grc = 1.0/rc
    gcp = cp/delt
    gload = 1.0/rload
    #
    vout, cinduct, vprimary, cpr1, cpr2, cmag, cdiode1, cdiode2 =        \
                ch7fig29_solver(es, Ntr, glm, grl, gls, grc,             \
                                gcp, gload, nc, ncycle, n)
    #
    m0 = 2*(ncycle - 2)
    m1 = 2*(ncycle - 1)
    m2 = 2*ncycle
```

```
ang = dteta*np.arange(N)
#
ax1 = host_subplot(2, 1, 1)
ax2 = ax1.twinx()
ax1.set_xlim([m0*pi, m2*pi])
ax1.set_xticks([m0*pi, (m0+2*d)*pi, m1*pi, m2*pi])
ax1.set_xticklabels(['$0$', '$2\pi *d$', '$2\pi$', '$4\pi$'])
ax1.set_ylim([-10.0, 10.0])
ax1.set_yticks([-10, 0, 10.0])
ax1.set_yticklabels(['$-10$', '$0$', '$10$'])
ax1.set_ylabel('$Voltages, V$')
ax1.text(m1*pi, -18.0, '$(a)$', size =10)
ax2.set_ylim(-2, 15)
ax2.set_yticks([0, 10, 15])
ax2.set_yticklabels(['$0$', '$10$', '$15$'])
ax2.set_ylabel('$Current, A$')
p1, = ax1.plot(ang[N-2*n:N-1], vout[N-2*n:N-1], 'r',                  \
                                   label = '$V_{O}$')
p2, = ax2.plot(ang[N-2*n:N-1], cinduct[N-2*n:N-1], 'c-.',             \
                                   label='$I_{L}$')
p3, = ax2.plot(ang[N-2*n:N-1], cdiode1[N-2*n:N-1], 'b--',             \
                                   label='$I_{d1}$')
p4, = ax2.plot(ang[N-2*n:N-1], cdiode2[N-2*n:N-1], 'g--',             \
                                   label='$I_{d2}$')
ax1.yaxis.get_label().set_color(p1.get_color())
ax2.yaxis.get_label().set_color(p3.get_color())
ax1.legend(bbox_to_anchor=(0., 1.02, 1., .102), loc=3, ncol=4,        \
                             mode="expand", borderaxespad=0.)
ax1.grid(True)
plt.subplots_adjust(hspace = 0.9)
#
ax1 = host_subplot(2, 1, 2)
ax2 = ax1.twinx()
ax1.set_xlim([m0*pi, m2*pi])
ax1.set_xticks([m0*pi, (m0+2*d)*pi, m1*pi, m2*pi])
ax1.set_xticklabels(['$0$', '$2\pi *d$', '$2\pi$', '$4\pi$'])
ax1.set_xlabel('$\omega t, radians$')
ax1.set_ylim([-200, 200])
ax1.set_yticks([-200, 0, 200])
ax1.set_yticklabels(['$-200$', '$0$', '$200$'])
ax1.set_ylabel('$Voltage, V$')
ax1.text(m1*pi, -450, '$(b)$', size =10)
ax2.set_ylim(-1.0, 1.0)
ax2.set_yticks([-1.0, 0, 1.0])
ax2.set_yticklabels(['$-1.0$', '$0$', '$1.0$'])
ax2.set_ylabel('$Current, A$')
p1, = ax1.plot(ang[N-2*n:N-1], vprimary[N-2*n:N-1], 'r',             \
                                   label = '$V_{pri}$')
p2, = ax2.plot(ang[N-2*n:N-1], cpr1[N-2*n:N-1], 'b-',                \
                                   label = '$I_{pr1}$')
p3, = ax2.plot(ang[N-2*n:N-1], cpr2[N-2*n:N-1], 'c-',                \
                                   label = '$I_{pr2}$')
p4, = ax2.plot(ang[N-2*n:N-1], cmag[N-2*n:N-1], 'g-.',               \
```

```
                                        label = '$I_{mag}$')
ax1.yaxis.get_label().set_color(p1.get_color())
ax2.yaxis.get_label().set_color(p2.get_color())
ax1.legend(bbox_to_anchor=(0., 1.02, 1., .102), loc=3, ncol=4,        \
                              mode="expand", borderaxespad=0.)
ax1.grid(True)
#
plt.show()
#
#plt.draw()
#plt.savefig('Fig729.ps', dpi=400, bbox_inches='tight')
#
```

7.6.9 Program 7.30

```
#
def fdiode(vd):
    """ Diode characteristic.
        Returns diode current and its conductance"""
    #
    glow = 0.1e-07
    ghigh = 0.4e+03
    Vr = -0.5
    #
    if vd >= Vr:
        id = glow*(vd - Vr)
        gd = glow
    else:
        id = ghigh*(vd - Vr)
        gd = ghigh
    #
    return id, gd
#
#
def fswitch(vsw, gsw):
    """Bi-directional switch characteristic.
       Returns switch current and conductance."""
    #
    glow = 0.1e-07
    ghigh = 0.4e+03
    Vr = 0.5
    #
    ida, gda = fdiode(vsw)
    #
    if vsw >= Vr:
        vd = (gsw*vsw + ghigh*Vr)/(ghigh + gsw)
        iw = ghigh*(vd -Vr)
        gw = gda + ghigh*gsw/(ghigh + gsw)
    else:
        vd = (gsw*vsw + glow*Vr)/(glow + gsw)
        iw = glow*(Vr - vd)
        gw = gda + glow*gsw/(glow + gsw)
```

```
    #
    isw = iw + ida
    #
    return isw, gw
#
#
def ch7fig30_solver(es, Ntr, glm, grl, gls, grc, gcp, gload, nc, ncycle, n):
    """Assumes fdiode is a function.
       Returns the lists vout, cinduct"""
    #
    import numpy as np
    #
    nstop = 0
    N = n*ncycle
    nh = int(n/2)
    ghigh = 400.0
    glow = 0.1e-07
    erlim = 0.1e-04
    Ntrsq = Ntr**2
    gser = grl*gls/(grl + gls)
    gpar = grc*gcp/(grc + gcp)
    #
    v1 = 0.3682
    v2 = -0.3178
    v0 = 2.9242
    #
    pilm = -0.3273
    pils = 14.2465
    pvc = 2.9269
    #
    vout = []
    cinduct = []
    vprimary = []
    cpr1 = []
    cpr2 =[]
    cmag = []
    cdiode1 = []
    cdiode2 = []
    #
    for k in range(ncycle):
        #
        for j in range(n):
            #
            if j < nc:
                gsw1 = ghigh
            else:
                gsw1 = glow
            #
            if (j >= nh) and (j < nh + nc):
                gsw2 = ghigh
            else:
                gsw2 = glow
            #
```

```
for its in range(10000):
    #
    ilm = Ntr*glm*v1 + pilm
    vs1 = es - Ntr*v1
    vs2 = es + Ntr*v1
    is1, gs1 = fswitch(vs1, gsw1)
    i2, gs2 = fswitch(vs2, gsw2)
    i1 = is1 - ilm
    #
    vd1 = v2 - v1
    vd2 = v2 + v1
    id1, gd1 = fdiode(vd1)
    id2, gd2 = fdiode(vd2)
    #
    ia = id2 - Ntr*(i1 - i2)
    #
    vp = (grl*v2 + gls*v0 - pils)/(grl + gls)
    vc = (grc*v0 + gcp*pvc)/(grc + gcp)
    #
    ils = grl*(v2 - vp)
    icp = grc*(v0 - vc)
    iload = gload*v0
    #
    irr1 = id1 - ia
    irr2 = -(id1 + id2 + ils)
    irr0 = ils - icp - iload
    #
    err = np.sqrt(irr1**2 + irr2**2 + irr0**2)
    #
    if err < erlim:
        break
    else:
        #
        gmat = np.array([ [(gd1 + gd2 + Ntrsq*(gs1+gs2+glm)),  \
                                            (gd2 - gd1), 0],
                          [(gd2 - gd1), gd1+gd2+gser, -gser],
                          [0, -gser, (gser+gpar+gload)]])
        #
        bcol = np.array([irr1, irr2, irr0])
        xdv = np.linalg.solve(gmat, bcol)
        #
        dv1 = xdv[0]
        dv2 = xdv[1]
        dv0 = xdv[2]
        v1 += dv1
        v2 += dv2
        v0 += dv0
#
if its >= 9998:
    print('time-step= ', (k*n + j), 'j = ', j, 'k = ',      \
                          k, 'solver does not converge')
    nstop = 1
    break
```

```
            #
            pilm = ilm
            pils = ils
            pvc = vc
            #
            cpr1.append(i1)
            cpr2.append(-i2)
            cmag.append(ilm)
            cdiode1.append(-id1)
            cdiode2.append(-id2)
            vout.append(v0)
            cinduct.append(ils)
            vprimary.append(Ntr*v1)
            #
        if nstop == 1:
            break
        #
    #
    print(' v0= ', np.mean(vout[N-2*n:N-1]))
    #
    return vout, cinduct, vprimary, cpr1, cpr2, cmag, cdiode1, cdiode2
#
#
def ch7fig30(es, freq, Ntr, lm, rl, ls, rc, cp, rload, d, ncycle, n):
    """ Main program, plots results;
        push--pull converter, birectional switches;
        low magnetizing inductance"""
    #
    #calling command:
    #ch7fig30(165.0, 100000.0, 30.0, 0.3e-03, 0.02, 0.85e-05,          \
    #                        0.01, 0.44e-03, 0.4, 0.3, 200, 600)
    #
    from math import pi
    import numpy as np
    import matplotlib.pyplot as plt
    from mpl_toolkits.axes_grid1 import host_subplot
    #
    N = ncycle*n
    nc = int(np.rint(d*n))
    #
    delt = 1.0/(freq*n)
    dteta = 2.0*pi/n
    glm = delt/lm
    grl = 1.0/rl
    gls = delt/ls
    grc = 1.0/rc
    gcp = cp/delt
    gload = 1.0/rload
    #
    vout, cinduct, vprimary, cpr1, cpr2, cmag, cdiode1, cdiode2 =         \
                        ch7fig30_solver(es, Ntr, glm, grl, gls,          \
                                    grc, gcp, gload, nc, ncycle, n)
    #
```

```
m0 = 2*(ncycle - 2)
m1 = 2*(ncycle - 1)
m2 = 2*ncycle
ang = dteta*np.arange(N)
#
ax1 = host_subplot(2, 1, 1)
ax2 = ax1.twinx()
ax1.set_xlim([m0*pi, m2*pi])
#
ax1.set_xticks([m0*pi, (m0+2*d)*pi, m1*pi, m2*pi])
ax1.set_xticklabels(['$0$', '$2\pi *d$', '$2\pi$', '$4\pi$'])
ax1.set_ylim([0, 5.0])
ax1.set_yticks([0, 3, 5])
ax1.set_yticklabels(['$0$', '$3$', '$5$'])
ax1.set_ylabel('$Voltages, V$')
ax1.text(m1*pi, -2.0, '$(a)$', size =10)
ax2.set_ylim(0, 20)
ax2.set_yticks([0, 10, 20])
ax2.set_yticklabels(['$0$', '$10$', '$20$'])
ax2.set_ylabel('$Current, A$')
p1, = ax1.plot(ang[N-2*n:N-1], vout[N-2*n:N-1], 'r',                    \
                                            label = '$V_{O}$')
p2, = ax2.plot(ang[N-2*n:N-1], cdiode1[N-2*n:N-1], 'b--',               \
                                            label='$I_{d1}$')
p3, = ax2.plot(ang[N-2*n:N-1], cdiode2[N-2*n:N-1], 'g--',               \
                                            label='$I_{d2}$')
ax1.yaxis.get_label().set_color(p1.get_color())
ax2.yaxis.get_label().set_color(p2.get_color())
ax1.legend(bbox_to_anchor=(0., 1.02, 1., .102), loc=3, ncol=3,          \
                                   mode="expand", borderaxespad=0.)
ax1.grid(True)
plt.subplots_adjust(hspace = 0.9)
#
ax1 = host_subplot(2, 1, 2)
ax2 = ax1.twinx()
ax1.set_xlim([m0*pi, m2*pi])
ax1.set_xticks([m0*pi, (m0+2*d)*pi, m1*pi, m2*pi])
ax1.set_xticklabels(['$0$', '$2\pi *d$', '$2\pi$', '$4\pi$'])
ax1.set_xlabel('$\omega t, radians$')
ax1.set_ylim([-200, 200])
ax1.set_yticks([-200, 0, 200])
ax1.set_yticklabels(['$-200$', '$0$', '$200$'])
ax1.set_ylabel('$Voltage, V$')
ax1.text(m1*pi, -450, '$(b)$', size =10)
ax2.set_ylim(-2.0, 2.0)
ax2.set_yticks([-2.0, 0, 2.0])
ax2.set_yticklabels(['$-2.0$', '$0$', '$2.0$'])
ax2.set_ylabel('$Current, A$')
p1, = ax1.plot(ang[N-2*n:N-1], vprimary[N-2*n:N-1], 'r',                \
                                            label='$V_{pri}$')
p2, = ax2.plot(ang[N-2*n:N-1], cpr1[N-2*n:N-1], 'b',                    \
                                            label = '$I_{p1}$')
p3, = ax2.plot(ang[N-2*n:N-1], cpr2[N-2*n:N-1], 'c',                    \
```

```
                                              label = '$I_{p2}$')
    p4, = ax2.plot(ang[N-2*n:N-1], cmag[N-2*n:N-1], 'g-.',                  \
                                              label = '$I_{mag}$')
    ax1.yaxis.get_label().set_color(p1.get_color())
    ax2.yaxis.get_label().set_color(p2.get_color())
    ax1.legend(bbox_to_anchor=(0., 1.02, 1., .102), loc=3, ncol=4,          \
                                    mode="expand", borderaxespad=0.)
    ax1.grid(True)
    #
    plt.show()
    #
    #plt.draw()
    #plt.savefig('Fig730.ps', dpi=400, bbox_inches='tight')
    #
```

7.6.10 Program 7.35

```
#
def fdiode(vd):
    """ Diode characteristic.
        Returns diode current and its conductance"""
    #
    glow = 0.1e-07
    ghigh = 0.1e+03
    Vr = -0.5
    #
    if vd >= Vr:
        id = glow*(vd - Vr)
        gd = glow
    else:
        id = ghigh*(vd - Vr)
        gd = ghigh
    #
    return id, gd
#
def fswitch(vsw, gsw):
    """Bi-directional switch characteristic.
       Returns switch current and conductance."""
    #
    glow = 0.1e-07
    ghigh = 0.1e+03
    Vr = 0.5
    #
    if vsw >= Vr:
        vd = (gsw*vsw + ghigh*Vr)/(ghigh + gsw)
        isw = ghigh*(vd -Vr)
        gw = ghigh*gsw/(ghigh + gsw)
    else:
        vd = (gsw*vsw + glow*Vr)/(glow + gsw)
        isw = glow*(Vr - vd)
        gw = glow*gsw/(glow + gsw)
    #
```

```
    return isw, gw
#
def ch7fig35_solver(es, Ntr, glm, grc, gcp, gload, glp, grk, gck,        \
                                                  nc, ncycle, n):
    """Assumes fdiode is a function.
       Returns the lists vout, cinduct"""
    #
    import numpy as np
    #
    nstop = 0
    N = n*ncycle
    ghigh = 400.0
    glow = 0.1e-07
    erlim = 0.1e-04
    Ntrsq = Ntr**2
    gpar = grc*gcp/(grc + gcp)
    gkp = glp*(grk + gck)/(grk + gck + glp)
    #
    v0 = 5.7420
    v1 = 6.2687
    vdp = 0.5
    vsw = 0.2
    #
    pilm = 0.1392
    pilp = 0.1
    pvc = 5.7498
    pvck = 45.0
    #
    vout = []
    vprimary = []
    vswitch = []
    cpr1 = []
    cmag = []
    cdiode = []
    cout = []
    #
    for k in range(ncycle):
        #
        for j in range(n):
            #
            if j < nc:
                gsw1 = ghigh
            else:
                gsw1 = glow
            #
            for its in range(100000):
                #
                if j < nc:
                    vlp = es + Ntr*v1 - vsw
                    ilp = glp*vlp + pilp
                    ilm = Ntr*glm*v1 + pilm
                    isw, gw1 = fswitch(vsw, gsw1)
                    irrw = ilp - isw
```

```
        else:
            idp, gdp = fdiode(vdp)
            vck = (gck*pvck - pilp - glp*(Ntr*v1 + vdp))/(grk+gck+glp)
            vlp = Ntr*v1 + vck + vdp
            vsw = es - vck - vdp
            ilp = glp*vlp + pilp
            ilm = Ntr*glm*v1 + pilm
            irrw = (grk + gck)*vck - gck*pvck - idp
        #
        i1 = ilp + ilm
        vc = (grc*v0 + gcp*pvc)/(grc + gcp)
        vds = v0 - v1
        ids, gds = fdiode(vds)
        icp = grc*(v0 - vc)
        iload = gload*v0
        #
        irr1 = ids - Ntr*i1
        irr0 = -(ids + icp + iload)
        #
        err = np.sqrt(irrw**2 + irr1**2 + irr0**2)
        #
        if err < erlim:
            break
        else:
            #
            if j < nc:
                gmat = np.array([ [(glp + gw1), -Ntr*glp, 0],
                                  [-Ntr*glp, (gds +                    \
                                        Ntrsq*(glp + glm)), -gds],
                                  [0, -gds, (gds + gpar + gload)] ])
                bcol = np.array([irrw, irr1, irr0])
                xdv = np.linalg.solve(gmat, bcol)
                dvsw = xdv[0]
                dv1 = xdv[1]
                dv0 = xdv[2]
                vsw += dvsw
                v1 += dv1
                v0 += dv0
            else:
                gmat = np.array([ [(gdp + gkp), Ntr*gkp, 0],
                                  [Ntr*gkp,                           \
                                   (gds + Ntrsq*(glm + gkp)), -gds],
                                  [0, -gds, (gds + gpar + gload)] ])
                bcol = np.array([irrw, irr1, irr0])
                xdv = np.linalg.solve(gmat, bcol)
                dvdp = xdv[0]
                dv1 = xdv[1]
                dv0 = xdv[2]
                vdp += dvdp
                v1 += dv1
                v0 += dv0
            #
#
```

```
            if its >= 99998:
                print('time-step= ', (k*n + j), 'j = ', j, 'k = ',        \
                                          k, 'solver does not converge')
                nstop = 1
                break
            #
            pilm = ilm
            pilp = ilp
            pvc = vc
            if j >= nc:
                pvck = vck
            #
            cpr1.append(i1)
            cmag.append(-ilm)
            cdiode.append(-ids)
            cout.append(iload)
            vout.append(v0)
            vprimary.append(Ntr*v1)
            vswitch.append(vsw)
            #
        if nstop == 1:
            break
        #
    #
    print(' v0= ', np.mean(vout[N-2*n:N-1]))
    print(' spike = ', max(vswitch))
    #
    return vout, vprimary, vswitch, cpr1, cmag, cdiode, cout
#
def ch7fig35(es, freq, Ntr, lpl, lm, rc, cp, rk, ck, rload, d, ncycle, n):
    """ Main program, plots results
        flyback converter, RCD snubber for switch"""
    #
    #calling command:
    #ch7fig35(48.0, 70000.0, 5.0, 0.3e-05, 0.8e-04, 0.01, 0.132e-02,      \
    #                            2000.0, 0.22e-06, 0.5, 0.4, 200, 600)
    #
    from math import pi
    import numpy as np
    import matplotlib.pyplot as plt
    from mpl_toolkits.axes_grid1 import host_subplot
    #
    N = ncycle*n
    nc = int(np.rint(d*n))
    #
    delt = 1.0/(freq*n)
    dteta = 2.0*pi/n
    glm = delt/lm
    grc = 1.0/rc
    gcp = cp/delt
    gload = 1.0/rload
    #
    glp = delt/lpl
```

```
grk = 1/rk
gck = ck/delt
#
vout, vprimary, vswitch, cpr1, cmag, cdiode, cout =                  \
                              ch7fig35_solver(es, Ntr, glm, grc,     \
                              gcp, gload, glp, grk, gck, nc, ncycle, n)
#
m0 = 2*(ncycle - 2)
m1 = 2*(ncycle - 1)
m2 = 2*ncycle
ang = dteta*np.arange(N)
#
ax1 = host_subplot(2, 1, 1)
ax2 = ax1.twinx()
ax1.set_xlim([m0*pi, m2*pi])
#
ax1.set_xticks([m0*pi, (m0+2*d)*pi, m1*pi, m2*pi])
ax1.set_xticklabels(['$0$', '$2\pi *d$', '$2\pi$', '$4\pi$'])
ax1.set_ylim([0, 10.0])
ax1.set_yticks([0, 5, 10])
ax1.set_yticklabels(['$0$', '$5$', '$10$'])
ax1.set_ylabel('$Voltage, V$')
ax1.text(m1*pi, -4.0, '$(a)$', size =10)
ax2.set_ylim(0, 40)
ax2.set_yticks([0, 20, 40])
ax2.set_yticklabels(['$0$', '$20$', '$40$'])
ax2.set_ylabel('$Current, A$')
p1, = ax1.plot(ang[N-2*n:N-1], vout[N-2*n:N-1], 'r',                 \
                                    label = '$V_{O}$')
p2, = ax2.plot(ang[N-2*n:N-1], cdiode[N-2*n:N-1], 'b--',             \
                                    label='$I_{ds}$')
p3, = ax2.plot(ang[N-2*n:N-1], cout[N-2*n:N-1], 'b',                 \
                                    label = '$I_{0}$')
ax1.yaxis.get_label().set_color(p1.get_color())
ax2.yaxis.get_label().set_color(p2.get_color())
ax1.legend(bbox_to_anchor=(0., 1.02, 1., .102), loc=3, ncol=3,       \
                              mode="expand", borderaxespad=0.)
ax1.grid(True)
plt.subplots_adjust(hspace = 0.9)
#
ax1 = host_subplot(2, 1, 2)
ax2 = ax1.twinx()
ax1.set_xlim([m0*pi, m2*pi])
ax1.set_xticks([m0*pi, (m0+2*d)*pi, m1*pi, m2*pi])
ax1.set_xticklabels(['$0$', '$2\pi *d$', '$2\pi$', '$4\pi$'])
ax1.set_xlabel('$\omega t, radians$')
ax1.set_ylim([-100, 200])
ax1.set_yticks([-100, 0, 200])
ax1.set_yticklabels(['$-100$', '$0$', '$200$'])
ax1.set_ylabel('$Voltages, V$')
ax1.text(m1*pi, -300, '$(b)$', size =10)
ax2.set_ylim(-10.0, 10.0)
ax2.set_yticks([-10.0, 0, 10.0])
```

```
ax2.set_yticklabels(['$-10.0$', '$0$', '$10.0$'])
ax2.set_ylabel('$Current, A$')
p1, = ax1.plot(ang[N-2*n:N-1], vprimary[N-2*n:N-1], 'r--',          \
                                  label = '$V_{pri}$')
p2, = ax1.plot(ang[N-2*n:N-1], vswitch[N-2*n:N-1], 'r',             \
                                  label = '$V_{sw}$')
p3, = ax2.plot(ang[N-2*n:N-1], cpr1[N-2*n:N-1], 'b',                \
                                  label = '$I_{pr1}$')
p4, = ax2.plot(ang[N-2*n:N-1], cmag[N-2*n:N-1], 'g-.',              \
                                  label = '$I_{mag}$')
ax1.yaxis.get_label().set_color(p1.get_color())
ax2.yaxis.get_label().set_color(p3.get_color())
ax1.legend(bbox_to_anchor=(0., 1.02, 1., .102), loc=3, ncol=4,       \
                            mode="expand", borderaxespad=0.)
ax1.grid(True)
#
plt.show()
#
#plt.draw()
#plt.savefig('Fig735.ps', dpi=400, bbox_inches='tight')
#
```

References

[1] A. I. Pressman, *Switching Power Supply Design*, 2^{nd}Ed., McGraw-Hill, USA, 1997.

[2] J. G. Kassakian, M. F. Schlecht, G. C. Verghese, *Principles of Power Electronics*, 1^{st}Ed., Pearson India, 2009.

[3] R. W. Erickson, D. Maksimovic, *Fundamentals of Power Electronics*, 2^{nd}Ed., Springer International Edition, 2009.

[4] N. Mohan, T. M. Undeland, W. P. Robbins, *Power Electronics: Converters, Applications, and Design*, 3^{rd}Ed., John Wiley & Sons Inc., USA, 2003.

[5] A. Lind, *Single transistor forward converter design*, Design Note AN2013-03, Infineon Technologies, Villach, Austria, 2013.

[6] Hang-Seok Choi, *Design guidelines for off-line forward converters using Fairchild power switch*, Application Note AN4134, Fairchild Semiconductor, 2002.

[7] S. Hariharan, *Designing single-switch, resonant-reset, forward converters*, Application Note 3983, Maxim Integrated Products, 2007.

[8] A. Kamath, *Push-pull converter simplifies isolated power supply design in HEV/EV systems*, Texas Instruments Literature SLYT790b, 2020.

[9] A. Kamath, *How to prevent transformer saturation in push–pull converters*, Texas instruments Literature SLYT813, 2021.
[10] Hang-Seok Choi, *Application guidelines for off-line flyback converters*, Application Note AN4137, FPS Application Group, Fairchild Semiconductor, 2003.

8

Resonant DC–DC Converters

There are many advantages in operating DC–DC converters at high switching frequencies. The sizes and weights of the inductors, transformers, capacitors, and the losses in these components are reduced. Converters working at high frequencies are compact and they are particularly useful to the data-processing and telecommunications industries. On the other hand, semiconductor junction capacitances, transformer leakage inductances, and rectifier reverse recovery introduce high switching losses and stresses at high switching frequencies. Efforts to reduce the switching losses at high switching frequencies have led to the development of resonant converters. However, the regulation of the output voltage of resonant converters requires frequency modulation. The resonant converter consists of (a) a switch network to generate a pulsating voltage from a direct voltage source, (b) a resonant tank two-port network comprising inductors and capacitors, and (c) a rectifier circuit to convert the pulsating output voltage of the resonant tank into a direct voltage [1]. Figure 8.1 shows the structure of the resonant converter.

8.1 Basic Resonant Tank Networks

The Two-element and three-element resonant tank networks are [2, 3]:

1. Two-element resonant tank:
   - series resonant tank (Figure 8.2(a));
   - parallel resonant tank (Figure 8.2(b));
2. Three-element resonant tank:
   - LLC resonant tank (Figure 8.3(a));
   - LLC resonant tank (Figure 8.3(b)).

Consider the converter with the two-element series resonant tank. The switch network usually consists of two series connected MOSFET power switches that are triggered in a complementary manner with a duty ratio

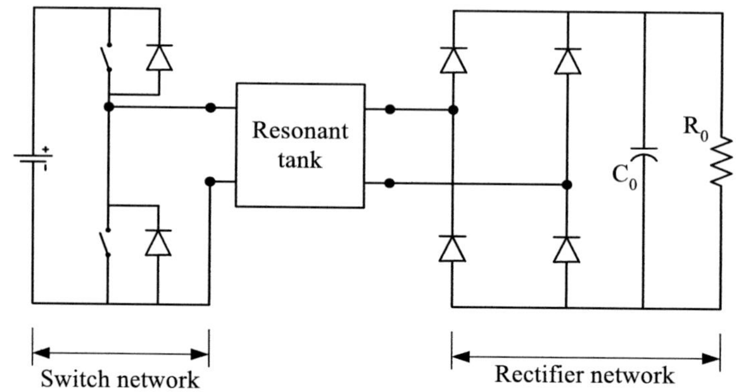

Figure 8.1 Structure of resonant converters.

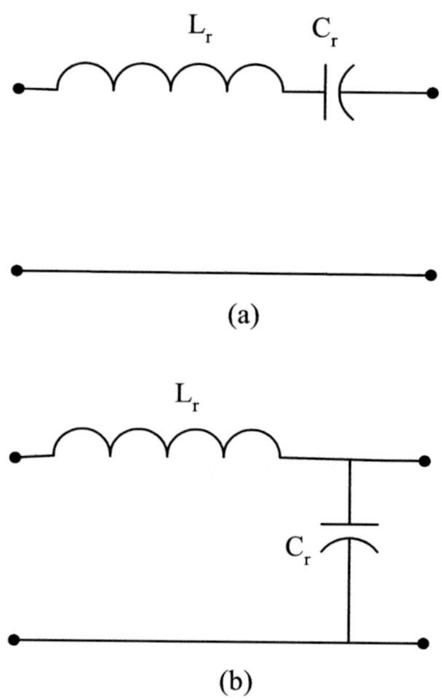

Figure 8.2 Two-element resonant tank circuits: (a) series resonant tank; (b) parallel resonant tank.

that is slightly less than 0.5. A small dead-time is necessary between the turn-off of one MOSFET and the turn-on of the other. Thus, the switch network generates a waveform that is the sum of an alternating square wave

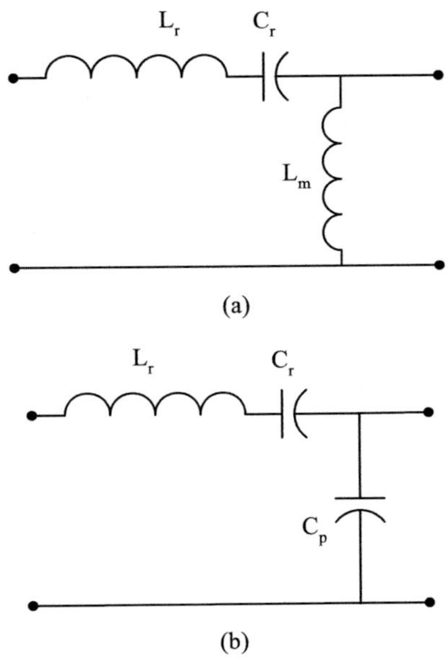

Figure 8.3 Three-element resonant tank circuits: (a) LLC tank; (b) LCC tank.

of amplitude equal to half the input direct voltage and a direct voltage of the same amplitude. The frequency of the alternating square wave is equal to the switching frequency. The direct voltage is filtered by the resonant capacitor and only the alternating square-wave voltage transmits power to the load.

A simplified calculation of the voltage gain is possible by considering only the fundamental frequency component of the alternating square-wave voltage. Let the equivalent load connected at the output of the resonant tank be represented by the resistor R. The RMS value of the fundamental frequency voltage is given by

$$\begin{aligned} E_s &= \frac{1}{\sqrt{2}}\frac{4}{\pi}\frac{e_s}{2} \\ &= \frac{\sqrt{2}}{\pi}e_s. \end{aligned}$$

The voltage gain V_0/E_s is

$$\frac{V_0}{E_s} = \frac{R}{j\omega_s L_r + 1/j\omega_s C_r + R}.$$

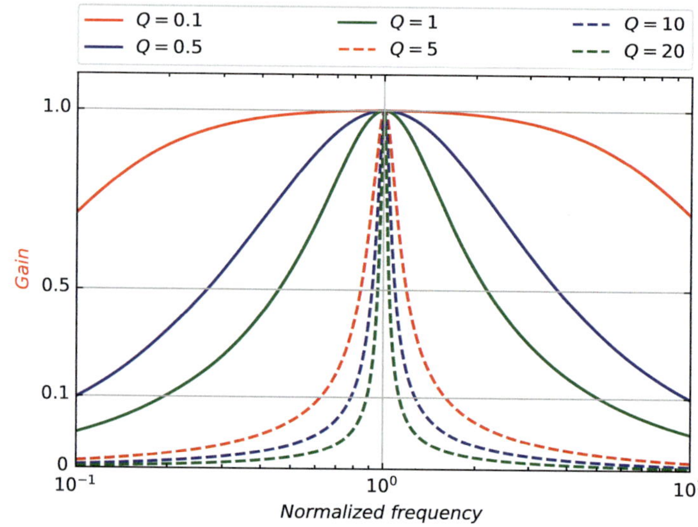

Figure 8.4 Series resonant converter: gain curves.

Putting $\omega_r = \frac{1}{\sqrt{L_r C_r}}$, $Q = \frac{\sqrt{L_r C_r}}{R}$, and $\omega_n = \frac{\omega_s}{\omega_r}$, we obtain

$$G_{\mathrm{SR}} = \frac{1}{\sqrt{1 + \frac{Q^2}{\omega_n^2}\left(\omega_n^2 - 1\right)}}. \tag{8.1}$$

Figure 8.4 (plotted by Program 8.04) shows the gain curves for several values of the quality factor Q. When the load is an open circuit, i.e., $Q = 0$, the gain is unity for all switching frequencies. Therefore, the output voltage cannot be regulated by frequency modulation at no-load or very light loads. Another disadvantage is that the output filter capacitor must carry high ripple currents.

Following a similar analysis as for the series resonant converter, the gain of the parallel resonant converter is obtained as

$$G_{\mathrm{PR}} = \frac{1}{\sqrt{(1 - \omega_n^2)^2 + Q^2\omega_n^2}}. \tag{8.2}$$

Figure 8.5 shows the gain curves for the parallel resonant converter. Around the resonant frequency, the gain is load dependent and can become very high. The gain falls above the resonant frequency and it is possible to

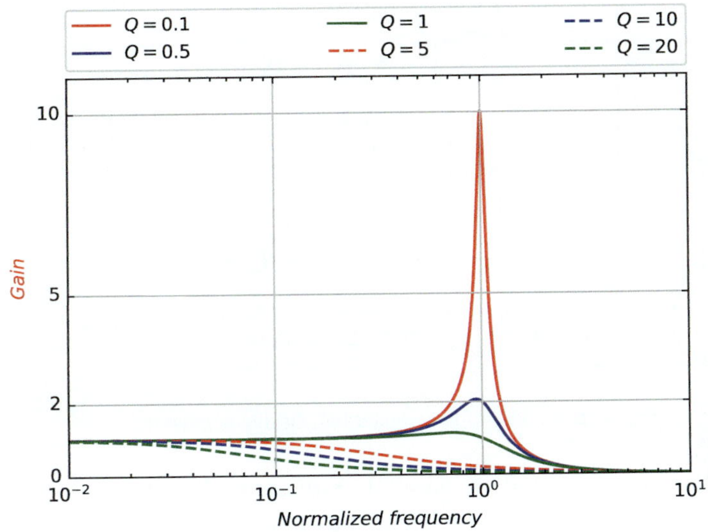

Figure 8.5 Parallel resonant converter: gain curves.

regulate the output voltage even at light loads, by operating at a frequency higher than the resonant frequency.

8.2 LLC Resonant Half-bridge Converter

The LLC converter [4–10] operates at high switching frequencies and it is possible to regulate the output voltage over wide line voltage and load variations by a relatively small variation of the switching frequency. The use of two separate inductors may be avoided by taking advantage of the leakage inductance and the magnetizing inductance of the high–frequency step-down transformer that is usually included in the circuit. Another characteristic of the LLC converter is that the switching may take place at zero-voltage or zero-current leading to low switching losses even at high switching frequencies.

Figure 8.6 shows the LLC resonant half-bridge converter circuit. It consists of a half-bridge, unipolar, square-wave voltage generator, an LLC resonant tank network, a rectifier, and the output circuit. The resonant circuit consists of the two inductors L_r, L_m and the capacitor C_r. The square–wave generator is formed by the input source voltage and the power switches S_{WA} and S_{WB}. The body diode and the drain–source capacitors of each power

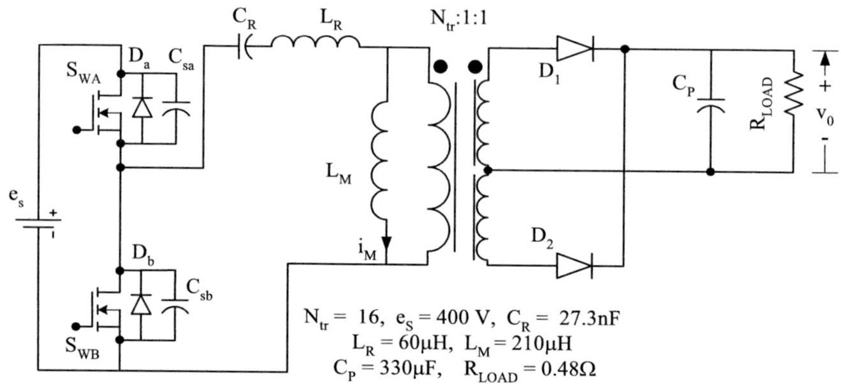

Figure 8.6 The LLC resonant half-bridge converter.

switch are shown connected in parallel. Sufficient dead-time is provided between the turn-off of one switch and turn-on of the other. A full-wave rectifier is shown on the secondary side that consists of a center-tapped winding and two diodes D_1 and D_2. The rectifier converts the alternating voltage that is applied at its input to the direct voltage that is applied to the load. The output capacitor C_P filters the output voltage.

The resonant frequency of the tank varies with the load. At no-load, the resonant frequency is given by

$$f_p = \frac{1}{2\pi\sqrt{(L_m + L_r)C_r}}. \tag{8.3}$$

When a short-circuit is applied at the output terminals, the resonant frequency is

$$f_r = \frac{1}{2\pi\sqrt{L_r C_r}}. \tag{8.4}$$

For a load in between a short-circuit and an open circuit, the resonant frequency of the tank is f_c, where $f_p < f_c < f_r$.

The switching frequency f_s of the converter is the frequency of the square–wave generated at the input terminals of the resonant tank. The switching frequency may be equal to, less than, or greater than the resonant frequency f_r. Figure 8.7 shows the square–wave voltage, the primary current, and the secondary current for (a) $f_s = f_r$, (b)$f_s < f_r$, and (c) $f_s > f_r$. The primary current is the sum of the magnetizing current and the secondary current referred to the primary. The magnetizing current flows only in the primary winding and it does not contribute to the power flow to the load.

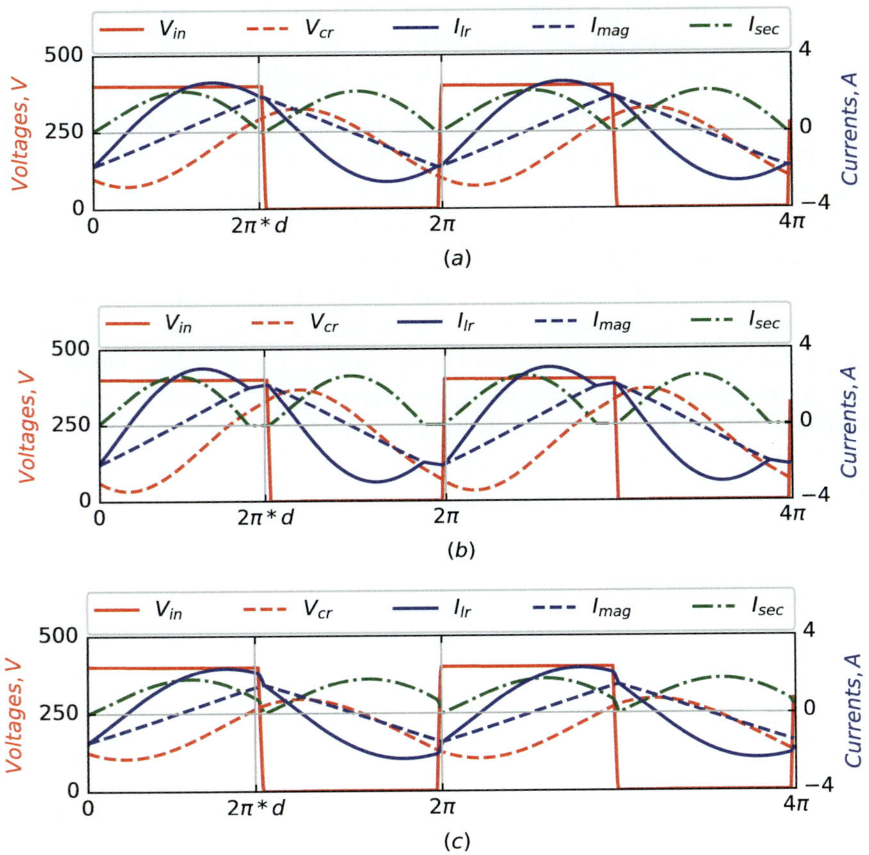

Figure 8.7 LLC converter: response waveforms. Switching frequency (a) equal to the resonant frequency f_r; (b) below f_r; (c) above f_r.

It is usual to design the LLC converter to operate at a switching frequency that is less than, but close to, the resonant frequency f_r. Each power switch is turned on for almost half of the switching period with a small dead-time inserted between successive switchings. During the dead-time, the primary current is equal to the magnetizing current. The magnetizing current is forced to flow through the drain–source capacitances of the power switches at the start of the dead-time. The drain–source capacitance of the switch that is turned off is charged, whereas the drain–source capacitance of the switch that is to be turned on is discharged so that zero-voltage switching becomes possible.

Consider the switching frequency $f_s < f_r$ (Figure 8.7(b)). In the steady-state and in each period of the switching frequency, the converter passes through eight operating modes (Figures 8.8 and 8.9)[10]:

- Mode 1. The switch S_{WA} is turned on during mode 1. This operating mode may be split into two sub-modes (mode 1a, and mode 1b). The magnetizing current rises from $-I_M$ to zero in mode 1a and it continues to rise from zero to $+I_M$ in mode 1b. Thus, the polarity of the magnetizing current is negative during mode 1a, but it becomes positive during mode 1b. The diode D_1 conducts and power flows to the load. The voltage $N_{\mathrm{tr}}V_0$ appears across the primary winding and causes the magnetizing current to increase linearly from $-I_M$ to $+I_M$. The load current reflected to the primary is a positive polarity sinusoidal pulse.
- Mode 2. The diode D_1 turns off at the end of the sinusoidal pulse of the load current reflected to the primary. The primary current becomes equal to the magnetizing current. The switch S_{WA} is turned off with zero voltage across its terminals.
- Mode 3. The dead-time starts when the switch S_{WA} is turned off. The magnetizing current is forced to split and flow through the drain–source capacitors of the power switches. The capacitor C_{sa} is charged and the capacitor C_{sb} is discharged.
- Mode 4. Both power switches remain turned off and the dead-time continues. When the capacitor C_{sb} is sufficiently discharged, the magnetizing current begins to flow through the body diode of S_{WB}. The switch S_{WB} is turned on. Note that the voltage across the switch is zero.
- Mode 5. This mode is similar to mode 1. The diode D_2 conducts and power flows to the load. The voltage across the primary winding is $-N_{\mathrm{tr}}V_0$. The negative voltage causes the magnetizing current to decrease linearly from I_M to zero (mode 5a) and continue to decrease from zero to $-I_M$ (mode 5b).
- Modes 6–8 are similar to modes 2–4.

8.2.1 Linear model of the LLC converter

The linear model of the LLC converter is shown in Figure 8.10. In this model, the unipolar square-wave input voltage is approximated by its fundamental frequency component given by

$$e_{sf}(t) = \frac{2}{\pi} e_s \sin \omega_s t, \tag{8.5}$$

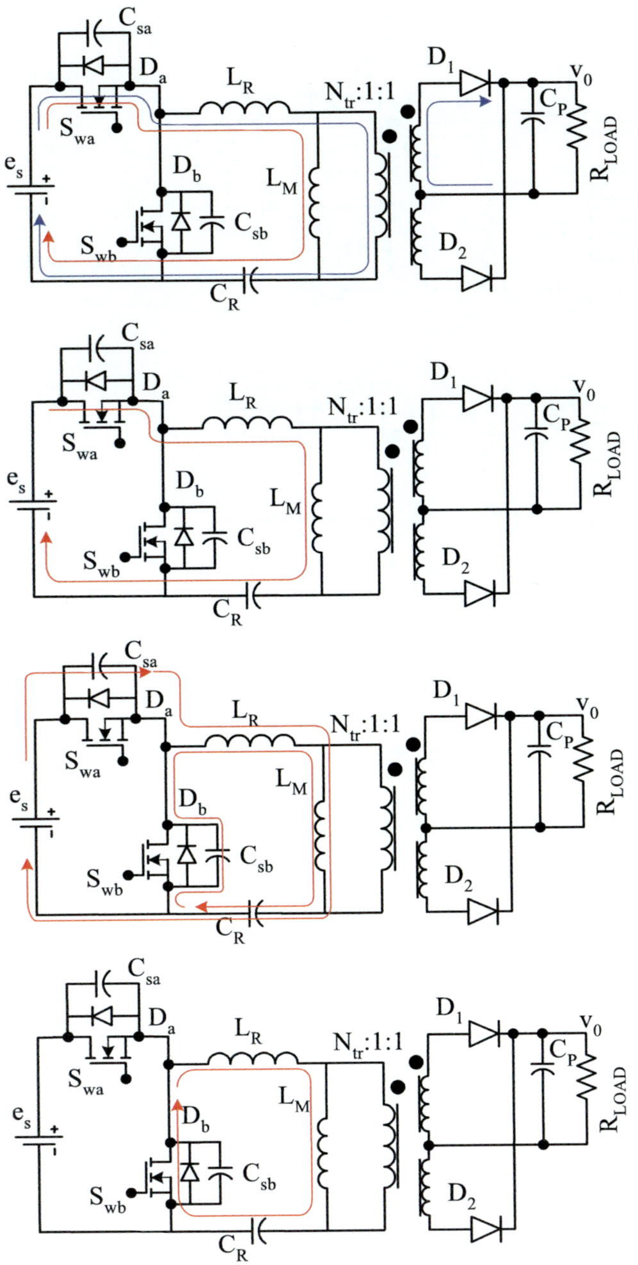

(a) Mode 1b

(b) Mode 2

(c) Mode 3

(d) Mode 4

Figure 8.8 LLC converter: operating modes 1b–4.

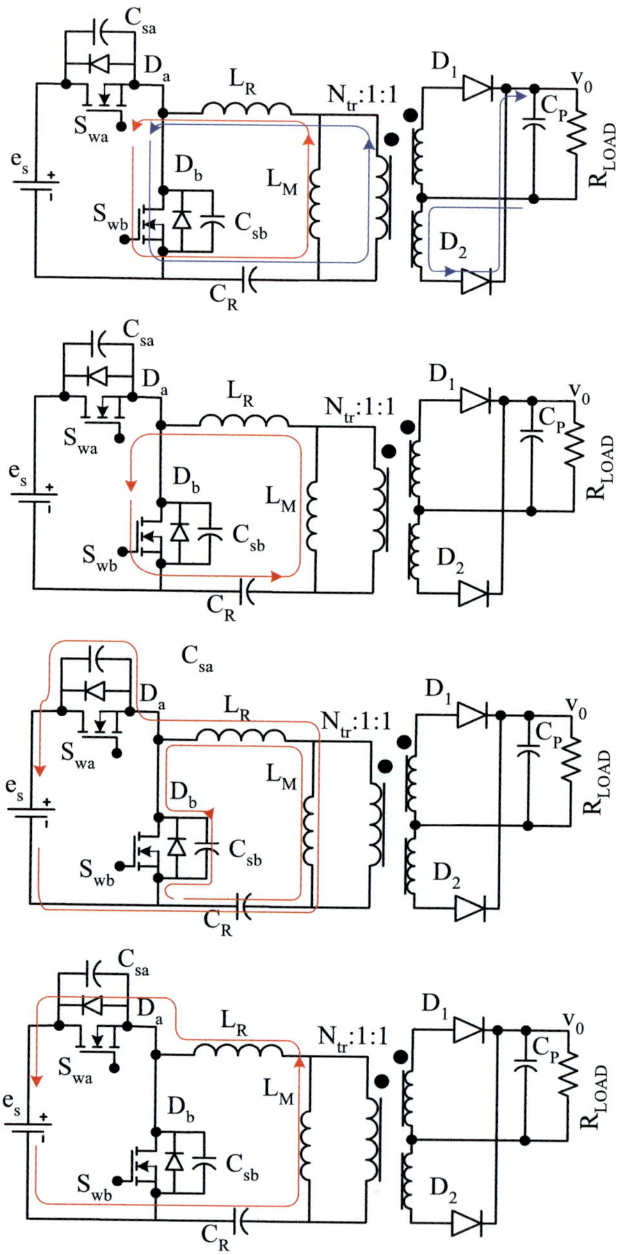

(e) Mode 5b

(f) Mode 6

(g) Mode 7

(h) Mode 8

Figure 8.9 LLC converter: operating modes 5b–8.

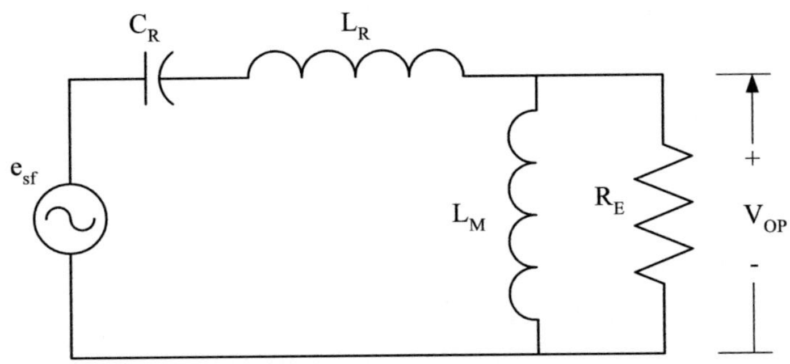

Figure 8.10 LLC converter: linear model.

where e_s is the direct voltage at the input of the converter and ω_s is the angular switching frequency in rads/sec. The RMS value of the fundamental frequency component is

$$E_{sf} = \frac{\sqrt{2}}{\pi} e_s. \tag{8.6}$$

The output voltage V_0 reflected to the primary is approximated by a square-wave whose fundamental frequency component is

$$v_{0P} = N_{\text{tr}} \frac{4}{\pi} V_0 \sin(\omega_s t - \phi_v), \tag{8.7}$$

where N_{tr} is the transformer turns ratio and ϕ_v is the phase angle between e_{sf} and v_{0P}. The RMS value of v_{0P} is

$$V_{0P} = \frac{2\sqrt{2}}{\pi} N_{\text{tr}} V_0. \tag{8.8}$$

The fundamental frequency component of the secondary current referred to the primary is

$$i_{0P} = \frac{1}{N_{\text{tr}} \frac{2}{\pi}} I_0 \sin(\omega_s t - \phi_i), \tag{8.9}$$

where ϕ_i is the phase angle between v_{0P} and i_{0P}. Note that the average value of $|i_{0P}| = \frac{I_0}{N_{\text{tr}}}$. The RMS value of i_{0P} is

$$I_{0P} = \frac{\pi}{2\sqrt{2}} \frac{I_0}{N_{\text{tr}}}. \tag{8.10}$$

Thus, the AC equivalent load is given by

$$\begin{aligned} R_e &= \frac{V_{0P}}{I_{0P}} \\ &= 8\left(\frac{N_{\text{tr}}}{\pi}\right)^2 R_{\text{LOAD}}. \end{aligned} \tag{8.11}$$

The input to output voltage gain is given by $|\frac{\tilde{V}_{0P}}{\tilde{E}_{sf}}|$, where $\tilde{V}_{0P}$ and $\tilde{E}_{sf}$ are the phasors representing the sinusoidal fundamental frequency components of the output and input voltages. From Figure 8.10, we get

$$Z_p = \frac{j\omega_s L_m R_e}{R_e + j\omega_s L_m}$$

$$G = \left|\frac{\tilde{V}_{0P}}{\tilde{E}_{sf}}\right| = \frac{Z_p}{Z_p + j\left(\omega_s L_r - \frac{1}{\omega_s C_r}\right)},$$

where G is the converter's voltage gain. Putting $Q = \sqrt{L_r/C_r}/R_e$, $\omega_r = 1/\sqrt{L_r C_r}$, $\omega_n = \omega_s/\omega_r$, and $m = (L_m + L_r)/L_r$, we get

$$G = \frac{\omega_n^2(m-1)}{\sqrt{(m\omega_n^2 - 1)^2 + Q^2\omega_n^2(m-1)^2(\omega_n^2 - 1)^2}}. \tag{8.12}$$

That is, the voltage gain $G = G(Q, m, \omega_n)$. Figure 8.11 shows the gain G as a function of ω_n for $m = 6$ and $Q = 0.1$ (light load) to $Q = 2$ (heavy load).

For each value of Q (i.e., load), the gain curve has a peak value at a normalized angular frequency less than unity and it passes through the point $G = 1$, $\omega_n = 1$. To the left of the peak, the impedance of the resonant tank is capacitive, whereas it is inductive to the right of the peak. It is usual to operate the converter in the inductive region with the switching frequency less than but close to the resonant frequency. The power switches operate with zero-voltage switching (ZVS) in this region. The output voltage may be regulated by varying the switching frequency and the converter is said to be frequency modulated.

8.2.2 Simulation of the LLC converter

The converter that will be simulated is shown in Figure 8.6 and its transient equivalent circuit is shown in Figure 8.12(a). Note the probe voltages vsb, $v1$, $v0$ inserted in this figure.

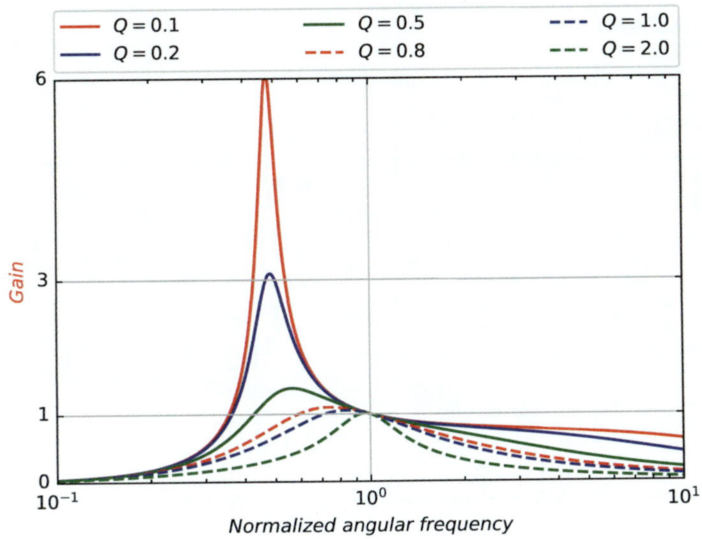

Figure 8.11 LLC converter: gain curves.

Applying Kirchhoff's laws to the transient equivalent circuit, we get the equations for the error currents:

$$\begin{aligned} irr_{sb} &= isa - isb - ilr \\ irr_1 &= ilr - ilm - i_1 \\ irr_0 &= -(id1 + id2 + ic0 + iload). \end{aligned}$$

The corrections to be added to the probe voltages are obtained from small signal equivalent circuit (Figure 8.12)(b)). Putting

$$\begin{aligned} gser &= gcr * glr/(gcr + glr) \\ gstot &= gswa + gcsa + gda + gswb + gcsb + gdb \\ g11 &= gstot + gser \\ g22 &= gser + glm + (gd1 + gd2)/N_{\mathrm{tr}}^2 \\ g33 &= gd1 + gd2 + gc0 + \mathrm{gload}, \end{aligned}$$

we get the corrections as

$$\begin{bmatrix} g11 & -gser & 0 \\ -gser & g22 & (gd2 - gd1)/N_{\mathrm{tr}} \\ 0 & (gd2 - gd1)/N_{\mathrm{tr}} & g33 \end{bmatrix} \begin{bmatrix} \delta vsb \\ \delta v1 \\ \delta v0 \end{bmatrix} = \begin{bmatrix} irr_{sb} \\ irr_1 \\ irr_0. \end{bmatrix}. \quad (8.13)$$

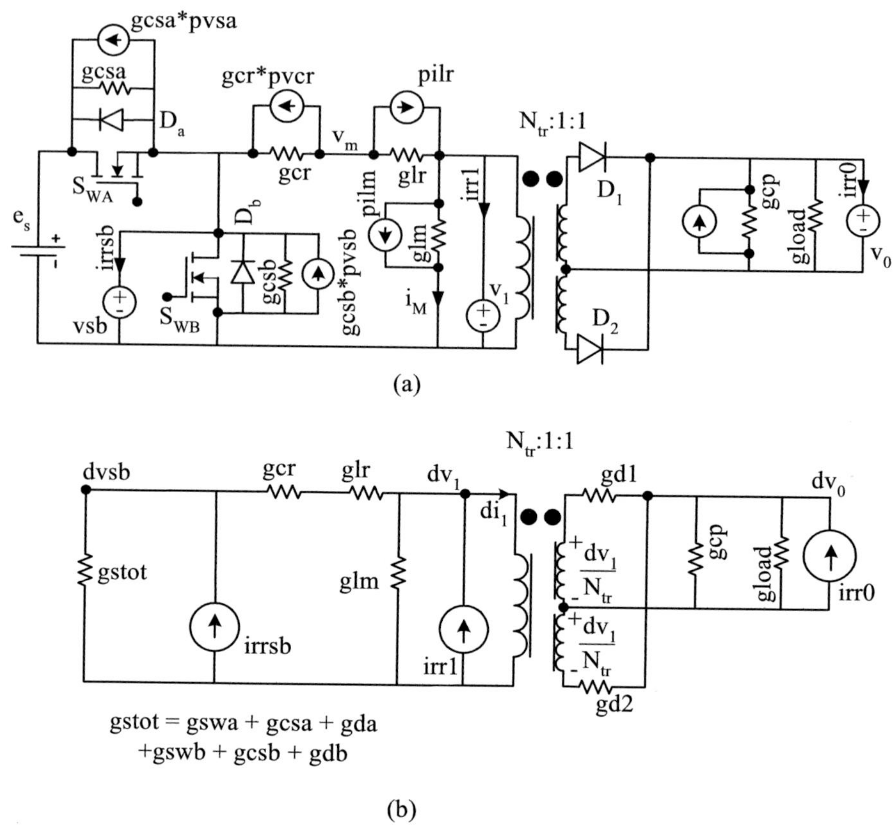

Figure 8.12 LLC converter: equivalent circuits. (a) Transient (b) Small-signal.

Figure 8.13 (plotted by Program 8.13) shows the results of the simulation. The simulation program was run for a range of switching frequencies and the voltage gain is shown plotted as a function of ω_n as well as the time-steps per period in Figure 8.14. The time-step was maintained constant at 25ns. Thus, the number of time-steps per period is inversely proportional to the switching frequency. With reference to Figure 8.14(a), the output voltage rises with ω_n, reaches a maximum, and subsequently decreases for higher switching frequencies. The decreasing part of the curve is the inductive region and it is here that the converter is operated. Figure 8.14(b) shows the variation of the output voltage as a function of the number of time-steps per period at the switching frequency.

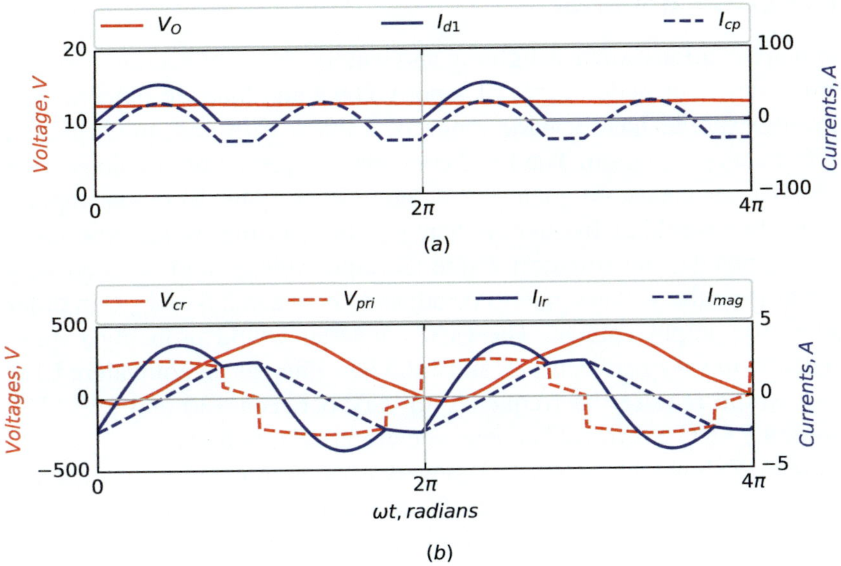

Figure 8.13 Response waveforms of the LLC converter.

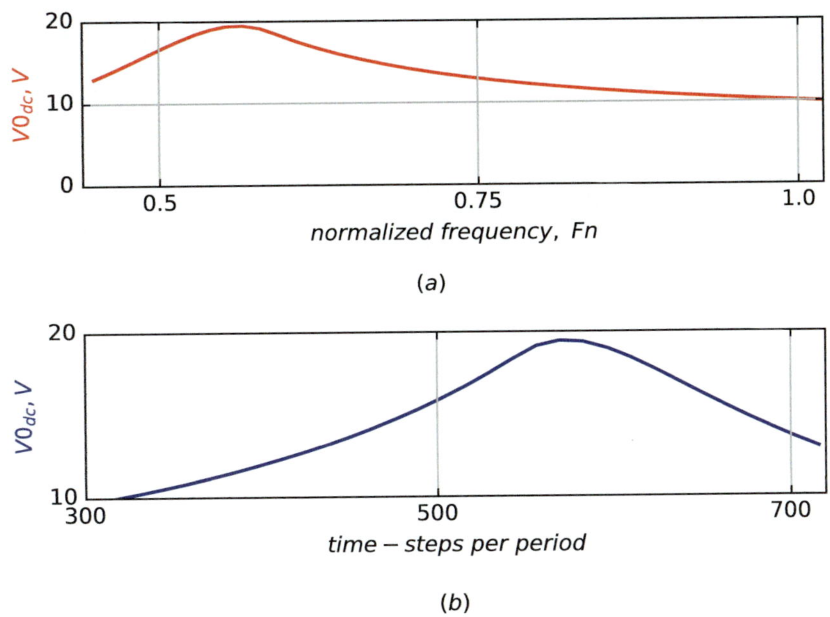

Figure 8.14 LLC converter: output voltage as a function of switching frequency.

8.3 The LCC Converter

Industrial applications, such as lighting and battery chargers, require converters that operate over wide ranges of input voltage and load. In other words, the converter should have a wide range of voltage gain and, preferably, a narrow frequency variation. The LCC converter [11] has a high voltage gain at the higher resonance frequency and the voltage gain decreases rapidly to zero as the switching frequency increases beyond the higher resonance frequency. Consider the situation where the input voltage and the load vary widely requiring the voltage gain to be adjusted between 2.5 and 1.0 in order to regulate the output voltage. The LCC converter would probably require the frequency to vary between 125 and 140 kHz, whereas an equivalent LLC converter would require the frequency to vary between 130 and 260 kHz. Therefore, the LCC converter becomes the preferred topology.

Figure 8.15 shows a typical LCC converter that is supplied from a voltage source, whose voltage varies between 400 and 450 V. The output voltage may have to vary between 100 and 200 V in order to maintain the load current constant at 1 A. Figure 8.16 shows the fundamental harmonic equivalent of the converter. The voltage gain of the converter may be shown to be

$$G = \left|\frac{\tilde{V}_{0P}}{\tilde{E}_{sf}}\right| = \frac{1}{\sqrt{\left(1 + \frac{C_p}{C_r}(1-\omega_n^2)\right)^2 + Q^2\left(\frac{1}{\omega_n} - \omega_n\right)^2}}. \tag{8.14}$$

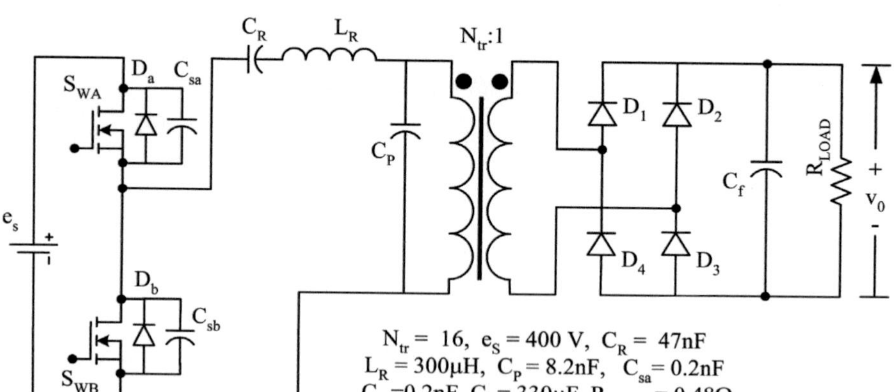

Figure 8.15 A typical LCC converter.

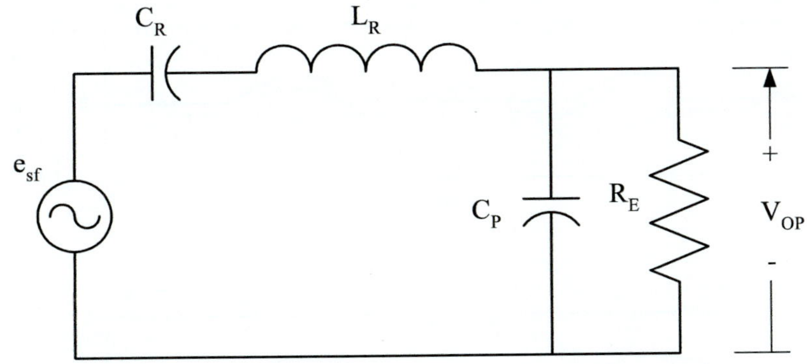

Figure 8.16 LCC converter: fundamental harmonic equivalent.

Figure 8.17 shows the LCC converter's voltage gain as a function of the normalized frequency ω_n. Figure 8.18 shows the response waveforms of the converter.

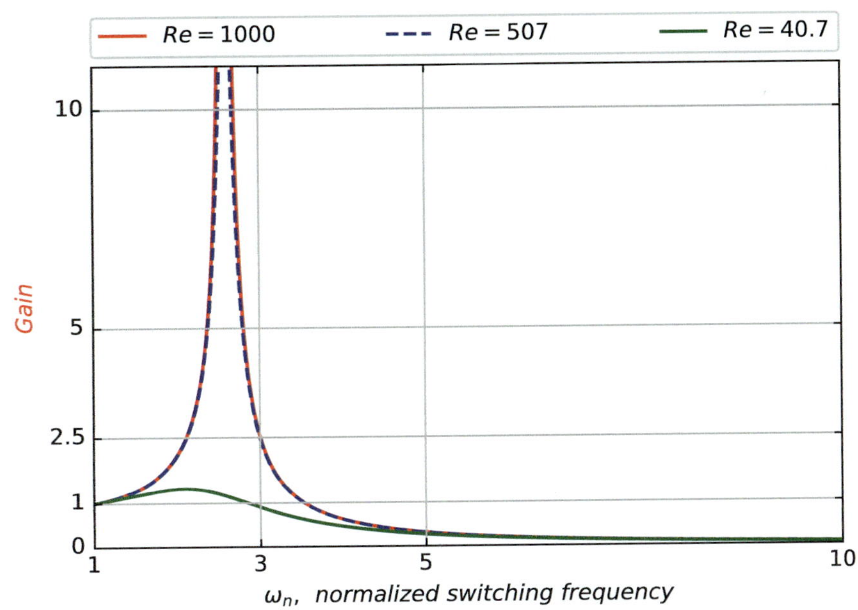

Figure 8.17 LCC converter: gain characteristics.

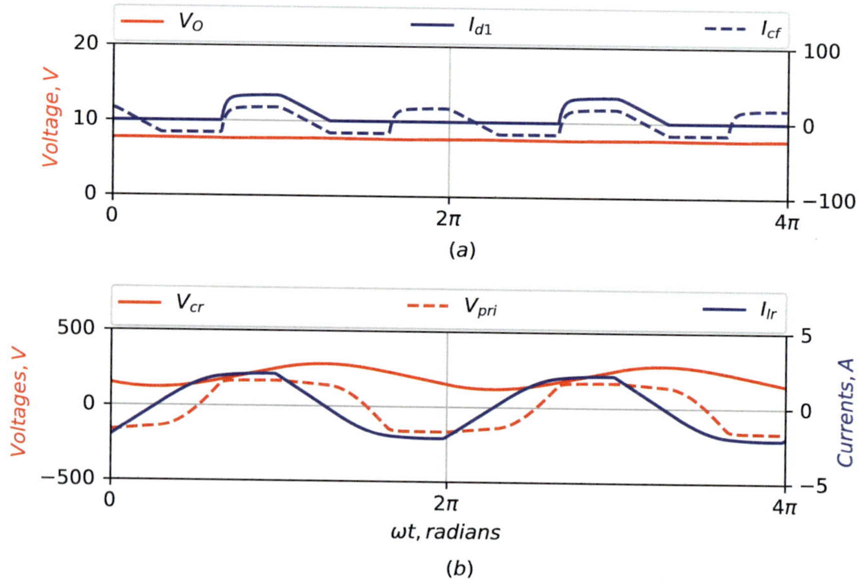

Figure 8.18 LCC converter: response waveforms.

8.4 Soft-switched, Quasi-resonant DC–DC Converters

Any of the basic DC–DC converters may be transformed into a quasi-resonant, zero-voltage switched (ZVS) converter by placing the resonant tank around the power switch [12, 14]. The resonant capacitor C_r is connected in parallel across the power switch. The resonant inductor is connected in series with the parallel combination of C_r and the power switch. (If a MOSFET is used as the power switch, the drain–source capacitance is added to the resonant capacitance.) Similarly, the zero-current switched (ZCS) converter is obtained by connecting the power switch in series with the resonant inductor L_r. The resonant capacitor C_r is connected in parallel across L_r and the power switch. Soft-switched converters were a specialized area of study and many converter topologies have been proposed. Only the ZVS, quasi-resonant buck converter is described below. Figures 8.19(a) and 8.19(b) show the quasi-resonant buck and boost converters.

8.4.1 The quasi-resonant buck converter

Consider the ZVS, quasi-resonant buck converter. The output filter consists of the inductor L_s and the capacitor C_p. The losses in the filter components are

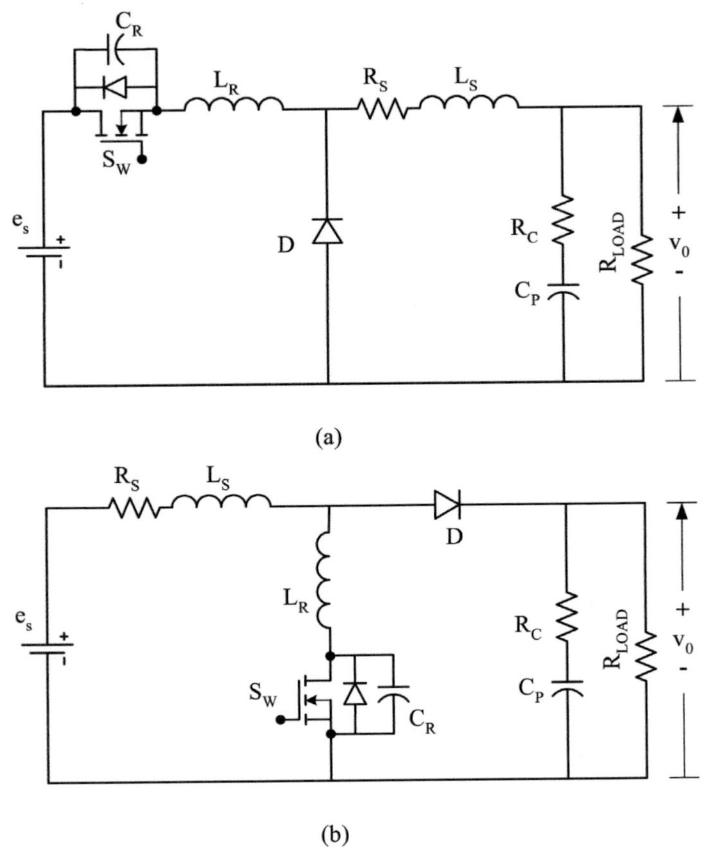

$e_S = 30$ V, $C_R = 1$nF, $L_R = 61\mu$H,
$L_S = 1.4$mH, $C_P = 17\mu$F, $R_{LOAD} = 6\ \Omega$,
$R_S = 0.6\ \Omega$, $R_C = 20$mΩ

Figure 8.19 Quasi-resonant, ZVS, DC–DC converters: (a) buck; (b) boost.

represented by the equivalent series resistances R_s and R_c. The free wheeling diode is D_0 and D_1 is connected anti-parallel with the switch. The switching frequency is $\omega_s = 2\pi f_s$ radians/sec. The power switch is turned on at $\omega_s t = 0$ and it remains closed until $\omega_s t = 2\pi d$ where d is the duty ratio. Assume that $v_{cr} = 0$, $i_{lr} = 0$ and that the free wheeling diode is forward biased at $\omega_s t = 0$. Also assume that $I_{ls} \simeq I_0$, which is the estimated load current. The parameters of the resonant tank are its angular resonant frequency $\omega_r = 1/\sqrt{L_r C_r}$ and its characteristic impedance $Z_0 = \sqrt{L_r/C_r}$. The switching frequency ω_s is significantly lower than ω_r.

Since the free wheeling diode is forward biased at $\omega_s t = 0$, the source voltage is applied across L_r causing the inductor current i_{lr} to increase linearly. At the same time, the diode's current i_{D0} decreases. When $i_{lr} = I_0$, the diode's current $i_{D0} = 0$ and the diode is reverse biased. The time elapsed from the closure of the switch at $\omega_s t = 0$ to the moment when $i_{lr} = I_0$ is $t_a = C_r * e_s / I_0$. The switch remains closed until $\omega_s t = 2\pi d$ and the resonant inductor current is constant at $i_{lr} = I_0$. At $\omega_s t = 2\pi d$, the power switch is opened.

The load current is forced to flow through the resonant capacitor C_r when the power switch is turned off. Consequently, the capacitor voltage v_{cr} increases linearly from $v_{cr=0}$ at $\omega_s t = 2\pi d$ to $v_{cr} = e_s$ at $\omega_s t = 2\pi d + \omega_s t_1$. The interval $t_1 = e_s C_r / I_0$. As soon as $v_{cr} = e_s$, the diode D_0 is forward biased and the voltage source e_s energizes the resonant tank. The initial conditions are $v_{cr} = e_s$, $i_{lr} = I_0$. The inductor current i_{lr} and the capacitor voltage v_{cr} begin to oscillate. During resonance,

$$\begin{aligned} i_{lr}(\omega_s t) &= I_0 \cos \omega_s (t - t_1) \\ v_{cr}(\omega_s t) &= e_s + \omega_s L_r I_0 \sin \omega_s (t - t_1). \end{aligned}$$

The resonant capacitor voltage v_{cr} increases in a sinusoidal manner and reaches the peak value $v_{cr,pk} = e_s + \omega_s L_r I_0$. At the same time, the resonant inductor current decreases from I_0 and it is zero when v_{cr} reaches its peak value. Subsequently, the resonant inductor current becomes negative and discharges the resonant capacitor. The resonant mode ends when v_{cr} reaches zero. Note that v_{cr} will decrease to zero only if $\omega_s L_r I_0$ is greater than e_s. The time taken to discharge the resonant capacitor to zero volts is

$$t_2 = \sqrt{L_r C_r} * \left(\pi + \sin^{-1} \left(\frac{e_s}{I_0} \sqrt{\frac{C_r}{L_r}} \right) \right). \tag{8.15}$$

The anti-parallel diode turns on when the resonant capacitor voltage $v_{cr} = 0$. Note that the output diode is also conducting and it will remain conducting as long as the resonant inductor current is less than the output current I_0. Thus, the source voltage is applied across the resonant inductor and its current will increase linearly from a negative value. The time taken by i_{lr} to reverse its polarity is $t_3 = L_r I_0 / e_s$. The power switch is closed slightly before i_{lr} reverses polarity and the closed switch clamps v_{cr} at zero. The next switching period begins.

The total duration for which the power switch is turned off is given by

$$\begin{aligned} t_{\mathrm{OFF}} &= t_1 + t_2 + 0.8 * t_3 \\ &= \frac{e_s C_r}{I_0} + \sqrt{L_r C_r} * \left(\pi + \sin^{-1}\left(\frac{e_s}{I_0}\sqrt{\frac{C_r}{L_r}}\right)\right) + 0.8\frac{L_r I_0}{e_s}. \end{aligned}$$

Suppose that the switching frequency $f_s = 250$ kHz, $L_r = 10$ µH, $C_r = 6$ nF, $e_s = 30$ V and the estimated load current is $I_0 = 2$ A. Then, $t_1 = 0.09$ µs, $t_2 = 0.8617$ µs, and $0.8t_3 = 0.533$ µs. Thus, $t_{\mathrm{OFF}} = 1.485$ µs and $t_{\mathrm{ON}} = 2.515$ µs. If there are 1000 time-steps per period, then the power switch should be closed for the first 630 time-steps. The output voltage and the duty ratio vary with the switching frequency (see Figure 8.22).

8.4.1.1 Simulation of the quasi-resonant, ZVS buck converter

The converter that will be simulated is shown in Figure 8.19(a). Figure 8.20 shows the transient and small-signal equivalent circuits. The probe voltages used in the simulation are v_0, v_1, and v_2. The error currents obtained from the analysis of the transient equivalent circuit are

$$\begin{aligned} irr_2 &= iswitch - ilr \\ irr_1 &= ilr - id0 - ils \\ irr_0 &= ils - icp. \end{aligned}$$

From Figure 8.20(b), the corrections to be added to the probe voltages are obtained as

$$\begin{bmatrix} (gswitch + glr) & -glr & 0 \\ -glr & (glr + gd0 + gser) & -gser \\ 0 & -gser & (gser + gpar) \end{bmatrix} \begin{bmatrix} \delta vsb \\ \delta v1 \\ \delta v0 \end{bmatrix} = \begin{bmatrix} irr_{sb} \\ irr_1 \\ irr_0 \end{bmatrix}, \tag{8.16}$$

where

$$\begin{aligned} gser &= grs * gls/(grs + gls) \\ gpar &= grc * gcp/(grc + gcp). \end{aligned}$$

Figure 8.21 (plotted by Program 8.21) shows the circuit variables in the steady-state during two periods at the switching frequency. The steady-state was obtained after 400 periods of transient simulation. Figure 8.22 shows the variation of the output voltage and the duty ratio with the switching frequency.

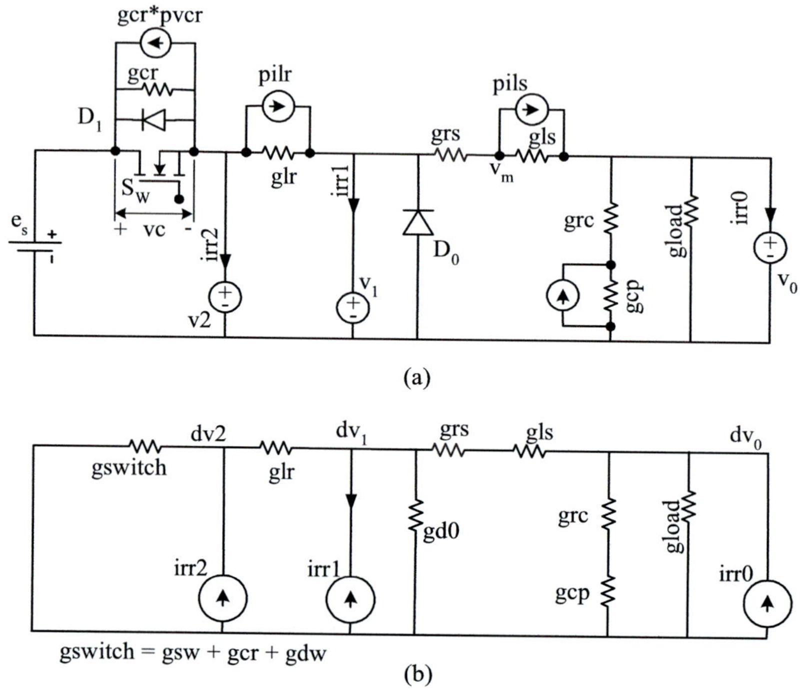

Figure 8.20 Quasi-resonant buck converter: equivalent circuits. (a) Transient (b) Small-signal.

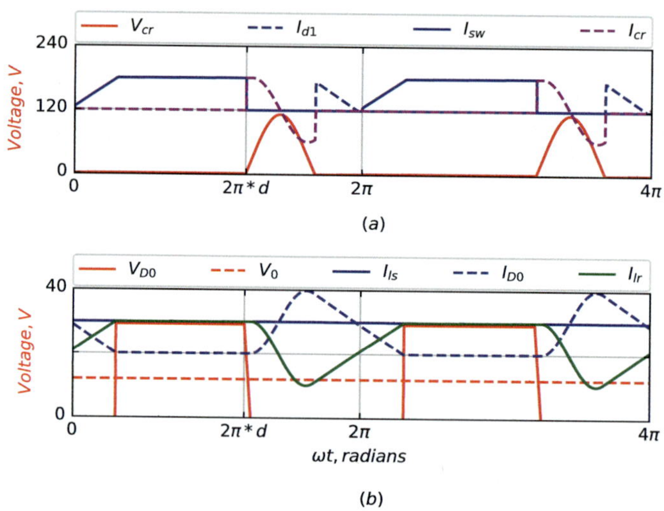

Figure 8.21 Quasi-resonant buck converter: response waveforms.

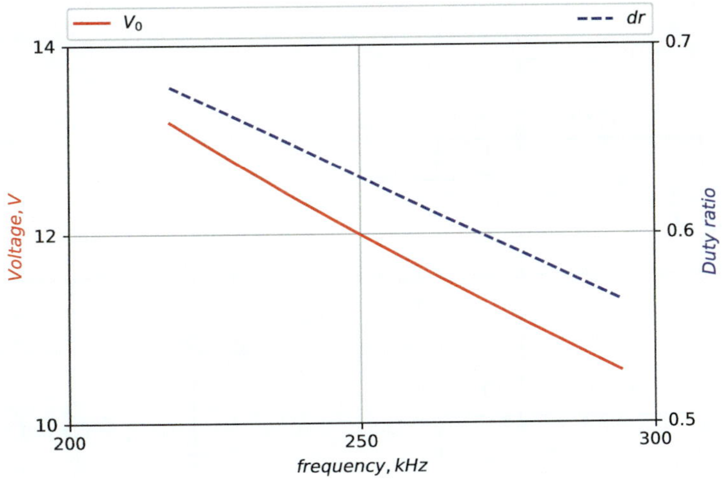

Figure 8.22 Quasi-resonant buck converter: variation of output voltage and duty ratio with f_s.

8.5 Constant Switching Frequency, Multi-resonant, ZVS Buck Converter

The multi-resonant, ZVS buck converter [13] is shown in Figure 8.23. The free-wheeling diode in the usual buck converter is replaced by a second power switch with anti-parallel diode and the shunt capacitor C_{sb}. The shunt capacitors Csa and C_{sb} need not be equal. (In the simulation that follows, they are made equal for the sake of convenience.)

The switching sequences of practical interest are the following:

Sequence	Mode 1	Mode 2	Mode 3	Mode 4
A	1-1	1-0	0-0	0-1
B	1-1	1-0	0-0	1-0
C	1-1	0-1	0-0	0-1

In the table shown above, the state of a switch is represented by 1 (closed) or 0 (open). The first digit of the pair represents the state of the switch S_{WA} and the second digit of the switch S_{WB}. The anti-parallel diode of the open switch conducts in Mode 4, so that the state of the switch becomes irrelevant. As far as the triggering of the switches are concerned, sequences A and B are

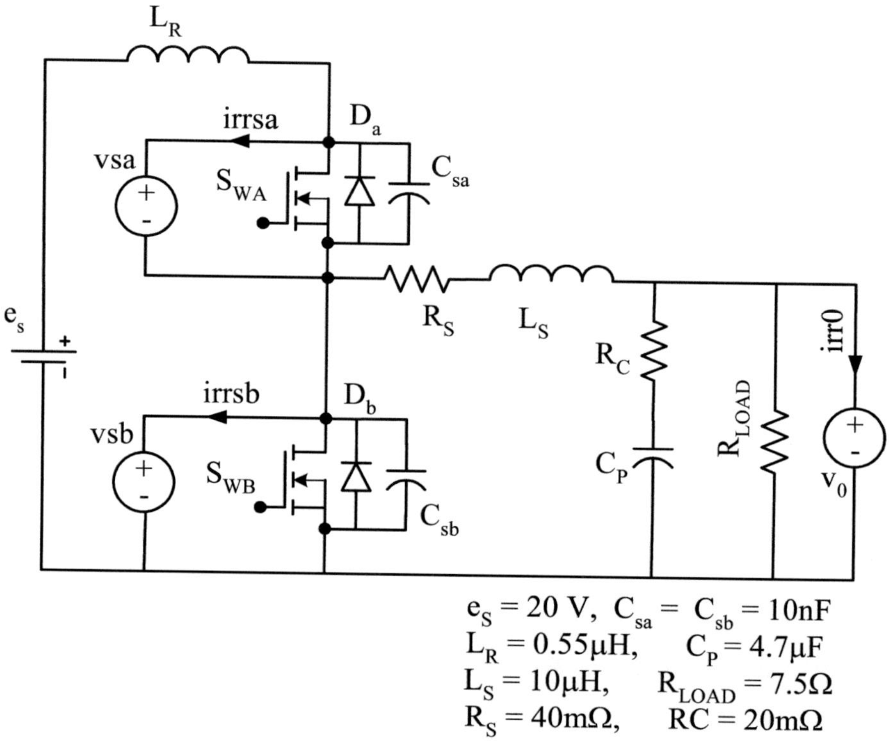

Figure 8.23 Constant frequency, multi-resonant, ZVS buck converter.

equivalent. The converter operates in sequence A from no-load to full load. It operates in sequence C at heavy loads.

Consider the operation of the converter in sequence C. Assume that both switches S_{WA} and S_{WB} are turned on at $\omega t = 0$ radians. With both switches turned on, the source voltage is applied across the inductor L_r, and therefore, the current i_r increases linearly. The converter enters Mode 2 at $\omega t = \theta_1$ radians when the switch S_{WA} is turned off. The (L_r, C_{sa}) tank resonates and the voltage of the switch S_{WA} increases from zero in a sinusoidal manner. When the switch S_{WB} is turned off at $\omega t = \theta_2$, the converter enters Mode 3, which is the multi-resonant mode involving C_{sa}, C_{sb}, and L_r. When the voltage across the switch S_{WB} drops to zero and the anti-parallel diode D_b begins to conduct, Mode 4 begins. The voltage of switch S_{WB} is clamped to zero and the tank (L_r, C_{sa}) continues to resonate so that the voltage of switch S_{WA} decreases toward zero. The switches and the diodes are switched at zero

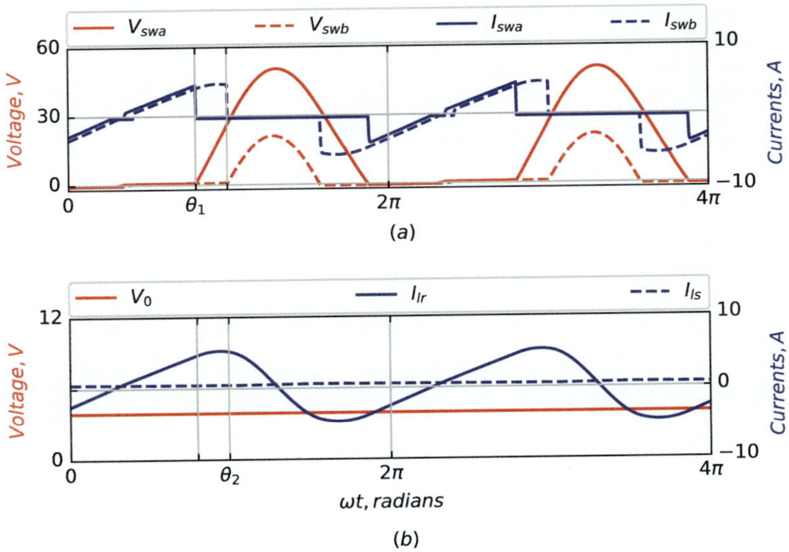

Figure 8.24 Multi-resonant converter: response waveforms.

volts so that the switching losses and the electromagnetic interferences are low. Figure 8.24 (plotted by Program 8.24) shows the response waveforms for the converter of Figure 8.23. (Development of the simulation procedure is left as an exercise. Use the probe voltages vsa, vsb, and $v0$ shown in Figure 8.23.)

8.6 Computational Programs

8.6.1 Program 8.04

```
#
def ch8fig04(Lr, Cr, Vin, V0, n):
    """ Main program, plots results"""
    #
    #calling command:
    #ch8fig04(0.6e-04, 0.273e-07, 400.0, 12.0, 10000)
    #
    from math import sqrt
    from math import pi
    import matplotlib.pyplot as plt
    from mpl_toolkits.axes_grid1 import host_subplot
    #
    wr = 1.0/sqrt(Lr*Cr)
```

```
#
wmin = 0.1*wr
wmax = 10.0*wr
wk1 = (wmax - wmin)/(wr*(n - 1))
wk2 = wmin/wr
#
Q = [0.1, 0.5, 1.0, 5.0, 10.0, 20.0]
wn = []
Mr = [[], [], [], [], [], []]
#
rs = sqrt(Lr/Cr)
G = 8.0/(pi**2)
Qs = rs/G
R0 = []
Pw = []
#
for i in range(len(Q)):
    qi = Q[i]
    rload = round(Qs/qi, 2)
    pwr = round(V0**2/rload, 2)
    R0.append(rload)
    Pw.append(pwr)
    #
    for j in range(n):
        xps = wk1*j + wk2
        d1 = (qi/xps)*(xps**2 - 1.0)
        if i == 0:
            wn.append(xps)
        gain = 1.0/sqrt(1.0 + d1**2)
        Mr[i].append(gain)
    #
#
print('load in ohms = ', R0)
print('power delivered = ', Pw)
#
ax1 = host_subplot(1, 1, 1)
ax1.set_xlim([0.1, 10.0])
ax1.set_xticks([0.1, 1.0, 10.0])
ax1.set_xticklabels(['$0.1$', '$1.0$', '$10$'])
ax1.set_xlabel('$Normalized \,\, frequency$')
ax1.set_ylim([0, 1.1])
ax1.set_yticks([0, 0.2, 0.5, 1.0])
ax1.set_yticklabels(['$0$', '$0.1$', '$0.5$', '$1.0$'])
ax1.set_ylabel('$Gain$')
p1, = ax1.semilogx(wn, Mr[0], 'r-', label = '$Q=0.1$')
p2, = ax1.semilogx(wn, Mr[1], 'b-', label = '$Q=0.5$')
p3, = ax1.semilogx(wn, Mr[2], 'g-', label = '$Q=1$')
p4, = ax1.semilogx(wn, Mr[3], 'r--', label = '$Q=5$')
p5, = ax1.semilogx(wn, Mr[4], 'b--', label = '$Q=10$')
p6, = ax1.semilogx(wn, Mr[5], 'g--', label = '$Q=20$')
ax1.yaxis.get_label().set_color(p1.get_color())
ax1.legend(bbox_to_anchor=(0., 1.02, 1., .102), loc=3,        \
               ncol=3, mode="expand", borderaxespad=0.)
```

```
    ax1.grid(True)
    #
    plt.show()
    #
    #plt.draw
    #plt.savefig('Fig804.ps', dpi=400, bbox_inches='tight')
    #
```

8.6.2 Program 8.13

```
#
def fdiode(vd):
    """ Diode characteristic.
        Returns diode current and its conductance"""
    #
    glow = 0.1e-07
    ghigh = 0.4e+02
    Vr = -0.5
    #
    if vd >= Vr:
        id = glow*(vd - Vr)
        gd = glow
    else:
        id = ghigh*(vd - Vr)
        gd = ghigh
    #
    return id, gd
#
#
def fswitch(vsw, gsw):
    """Bi-directional switch characteristic.
       Returns switch current and conductance."""
    #
    glow = 0.1e-07
    ghigh = 0.4e+02
    Vr = 0.5
    #
    if vsw >= Vr:
        vd = (gsw*vsw + ghigh*Vr)/(ghigh + gsw)
        iw = ghigh*(vd -Vr)
        gw = ghigh*gsw/(ghigh + gsw)
    else:
        vd = (gsw*vsw + glow*Vr)/(glow + gsw)
        iw = glow*(vd - Vr)
        gw = glow*gsw/(glow + gsw)
    #
    return iw, gw
#
#
def ch8_llc_fig13_solver(es, Ntr, glr, gcr, glm, gcp, gcsa, gcsb,          \
                         gload, ndt, ncycle, n):
```

```
"""Assumes fdiode is a function.
   Returns the lists vout, cinduct"""
#
import numpy as np
#
nh = int(n/2)
ghigh = 40.0
glow = 0.1e-07
erlim = 0.1e-04
Ntrsq = Ntr**2
gser = glr*gcr/(glr + gcr)
#
vsb = 200.0
v1 = 35.0
v0 = 12.0
#
pilr = -1.9816
pilm = -2.3037
pvcr = -224.2034
pvsa = -0.5495
pvsb = 400.5495
pv0 = 15.3423
#
vout = []
vac = []
vcapr = []
cindr = []
cmag = []
cdiode1 = []
cdiode2 = []
cfilt = []
#
for k in range(ncycle):
    #
    for j in range(n):
        #
        if j >= ndt and j < (nh - ndt):
            gsa = ghigh
        else:
            gsa = glow
        #
        if (j >= nh + ndt) and (j < n - ndt):
            gsb = ghigh
        else:
            gsb = glow
        #
        for its in range(1000):
            #
            vsa = es - vsb
            iswa, gswa = fswitch(vsa, gsa)
            iswb, gswb = fswitch(vsb, gsb)
            icsa = gcsa*(vsa - pvsa)
            icsb = gcsb*(vsb - pvsb)
```

```
            ida, gda = fdiode(vsa)
            idb, gdb = fdiode(vsb)
            isa = iswa + icsa + ida
            isb = iswb + icsb + idb
            #
            vm = (gcr*(vsb - pvcr) + glr*v1 - pilr)/(gcr + glr)
            ilr = glr*(vm - v1) + pilr
            ilm = glm*v1 + pilm
            #
            vd1 = 0.5*(v0 - v1/Ntr)
            vd2 = 0.5*(v0 + v1/Ntr)
            id1, gd1 = fdiode(vd1)
            id2, gd2 = fdiode(vd2)
            i1 = (id2 - id1)/Ntr
            #
            icp = gcp*(v0 - pv0)
            iload = gload*v0
            #
            irrsb = isa - isb - ilr
            irr1 = ilr - ilm - i1
            irr0 = -(id1 + id2 + icp + iload)
            #
            err = np.sqrt(irrsb**2 + irr1**2 + irr0**2)
            #
            if err < erlim:
                break
            else:
                #
                gstot = gswa + gcsa + gda + gswb + gcsb + gdb
                gmat = np.array([ [(gstot+gser), -gser, 0],
                                  [-gser, (gser+glm+(gd1+gd2)/Ntrsq),  \
                                                    (gd2-gd1)/Ntr],
                                  [0, (gd2-gd1)/Ntr,                   \
                                               (gd1+gd2+gcp+gload)] ])
                #
                bcol = np.array([irrsb, irr1, irr0])
                xdv = np.linalg.solve(gmat, bcol)
                #
                dvsb = xdv[0]
                dv1 = xdv[1]
                dv0 = xdv[2]
                vsb += dvsb
                v1 += dv1
                v0 += dv0
        #
        if its >= 998:
            print('time-step= ', (k*n + j), 'j = ', j, 'k = ',        \
                                     k, 'solver does not converge')
            break
        #
        pilr = ilr
        pilm = ilm
        pvcr = vsb - vm
```

```
            pvsb = vsb
            pvsa = vsa
            pv0 = v0
            #
            cindr.append(ilr)
            cmag.append(ilm)
            vac.append(v1)
            vcapr.append(vsb - vm)
            #
            vout.append(v0)
            cdiode1.append(-id1)
            cdiode2.append(-id2)
            cfilt.append(icp)
            #
        #
    #
    print(' output current = ', iload)
    #
    return vout, vac, vcapr, cindr, cmag, cdiode1, cdiode2, cfilt
#
#
def ch8_llc_fig13(es, delt, Ntr, lr, cr, lm, cp, csa, csb,              \
                                        rload, ndt, ncycle, n):
    """ Main program, plots results
        llc converter, birectional switches,with isolation transformer.
        Load represented by parallel RC circuit."""
    #
    #calling command:
    #ch8_llc_fig13(400.0, 0.25e-07, 16.0, 0.6e-04, 0.273e-07,           \
    #   0.21e-03, 0.33e-03, 0.2e-09, 0.2e-09, 0.48, 4, 200, 416)
    #
    from math import pi
    import numpy as np
    import matplotlib.pyplot as plt
    from mpl_toolkits.axes_grid1 import host_subplot
    #
    N = ncycle*n
    wr = 1.0/np.sqrt(lr*cr)
    omega = 2.0*pi/(n*delt)
    wn = omega/wr
    print(' wr = ', wr,' normalized switching frequency = ', wn)
    #
    glr = delt/lr
    gcr = cr/delt
    glm = delt/lm
    gcp = cp/delt
    gcsa = csa/delt
    gcsb = csb/delt
    gload = 1.0/rload
    #
    vout, vac, vcapr, cindr, cmag, cdiode1, cdiode2, cfilt =            \
           ch8_llc_fig13_solver(es, Ntr, glr, gcr, glm, gcp,            \
           gcsa, gcsb, gload, ndt, ncycle, n)
```

```
#
m0 = 2*(ncycle - 2)
m1 = 2*(ncycle - 1)
m2 = 2*ncycle
dteta = 2.0*pi/n
ang = dteta*np.arange(N)
#
plt.rcParams["figure.figsize"] = (7.5, 5.0)
#
ax1 = host_subplot(2, 1, 1)
ax2 = ax1.twinx()
ax1.set_xlim([m0*pi, m2*pi])
ax1.set_xticks([m0*pi, m1*pi, m2*pi])
ax1.set_xticklabels(['$0$', '$2\pi$', '$4\pi$'])
ax1.set_ylim([0, 20])
ax1.set_yticks([0, 10, 20])
ax1.set_yticklabels(['$0$', '$10$', '$20$'])
ax1.set_ylabel('$Voltage, V$')
ax1.text(m1*pi, -8.0, '$(a)$', size =10)
ax2.set_ylim(-100, 100)
ax2.set_yticks([-100, 0, 100])
ax2.set_yticklabels(['$-100$', '$0$', '$100$'])
ax2.set_ylabel('$Currents, $~A')
p1, = ax1.plot(ang[N-2*n:N-1], vout[N-2*n:N-1], 'r',           \
                                         label = '$V_{O}$')
p2, = ax2.plot(ang[N-2*n:N-1], cdiode1[N-2*n:N-1], 'b',        \
                                          label='$I_{d1}$')
p3, = ax2.plot(ang[N-2*n:N-1], cfilt[N-2*n:N-1], 'b--',        \
                                          label='$I_{cp}$')
ax1.yaxis.get_label().set_color(p1.get_color())
ax2.yaxis.get_label().set_color(p2.get_color())
ax1.legend(bbox_to_anchor=(0., 1.02, 1., .102), loc=3, ncol=3,   \
                        mode="expand", borderaxespad=0.)
ax1.grid(True)
plt.subplots_adjust(hspace = 0.9)
#
ax1 = host_subplot(2, 1, 2)
ax2 = ax1.twinx()
ax1.set_xlim([m0*pi, m2*pi])
ax1.set_xticks([m0*pi, m1*pi, m2*pi])
ax1.set_xticklabels(['$0$', '$2\pi$', '$4\pi$'])
ax1.set_xlabel('$\omega t, radians$')
ax1.set_ylim([-500,500])
ax1.set_yticks([-500, 0, 500])
ax1.set_yticklabels(['$-500$', '$0$', '$500$'])
ax1.set_ylabel('$Voltages, V$')
ax1.text(m1*pi, -1125, '$(b)$', size =10)
p1, = ax1.plot(ang[N-2*n:N-1], vcapr[N-2*n:N-1], 'r',          \
                                         label='$V_{cr}$')
p2, = ax1.plot(ang[N-2*n:N-1], vac[N-2*n:N-1], 'r--',          \
                                      label = '$V_{pri}$')
ax2.set_ylim(-5, 5)
ax2.set_yticks([-5, 0, 5])
```

```
    ax2.set_yticklabels(['$-5$', '$0$', '$5$'])
    ax2.set_ylabel('$Currents, $~A')
    p3, = ax2.plot(ang[N-2*n:N-1], cindr[N-2*n:N-1], 'b',        \
                                         label = '$I_{lr}$')
    p4, = ax2.plot(ang[N-2*n:N-1], cmag[N-2*n:N-1], 'b--',       \
                                         label='$I_{mag}$')
    ax1.yaxis.get_label().set_color(p1.get_color())
    ax2.yaxis.get_label().set_color(p3.get_color())
    ax1.legend(bbox_to_anchor=(0., 1.02, 1., .102), loc=3, ncol=4,  \
                          mode="expand", borderaxespad=0.)
    ax1.grid(True)
    #
    plt.show()
    #
    #plt.draw()
    #plt.savefig('Fig813.ps', dpi=400, bbox_inches='tight',          \
    #                                         transparent=True)
    #
```

8.6.3 Program 8.21

```
#
def fdiode(vd):
    """ Diode characteristic.
        Returns diode current and its conductance"""
    #
    glow = 0.1e-07
    ghigh = 0.4e+02
    Vr = -0.5
    #
    if vd >= Vr:
        id = glow*(vd - Vr)
        gd = glow
    else:
        id = ghigh*(vd - Vr)
        gd = ghigh
    #
    return id, gd
#
#
def fswitch(vsw, gsw):
    """Bi-directional switch characteristic.
       Returns switch current and conductance."""
    #
    glow = 0.1e-07
    ghigh = 0.4e+02
    Vr = 0.5
    #
    if vsw >= Vr:
        vd = (gsw*vsw + ghigh*Vr)/(ghigh + gsw)
        iw = ghigh*(vd -Vr)
        gw = ghigh*gsw/(ghigh + gsw)
```

```
    else:
        vd = (gsw*vsw + glow*Vr)/(glow + gsw)
        iw = glow*(vd - Vr)
        gw = glow*gsw/(glow + gsw)
    #
    return iw, gw
#
#
def ch8_qrbuck_fig21_solver(es, glr, gcr, grs, gls, grc,      \
                                gcp, gload, nc, ncycle, n):
    """Assumes fdiode is a function.
       Returns the lists vout, cinduct"""
    #
    import numpy as np
    #
    ghigh = 40.0
    glow = 0.1e-07
    erlim = 0.1e-04
    #
    gser = grs*gls/(grs + gls)
    gpar = gload + grc*gcp/(grc + gcp)
    #
    v2 = 30.0
    v1 = 0.0
    v0 = 12.0
    #
    pils = 2.0
    pilr = 0.0
    pvcr = 0.0
    pvcp = 12.0
    #
    vout = []
    vcapr = []
    vdiode = []
    cindr = []
    cindls = []
    cswitch = []
    cd1 = []
    ccapr = []
    cdiode = []
    #
    for k in range(ncycle):
        #
        for j in range(n):
            #
            if j <= nc:
                gsa = ghigh
            else:
                gsa = glow
            #
            for its in range(1000):
                #
                vcr = es - v2
```

```
            isw, gsw = fswitch(vcr, gsa)
            id1, gd1 = fdiode(vcr)
            icr = gcr*(vcr - pvcr)
            iswitch = isw + id1 + icr
            gswitch = gsw + gd1 + gcr
            #
            vm = (grs*v1 + gls*v0 - pils)/(grs + gls)
            vcp = (grc*v0 + gcp*pvcp)/(grc + gcp)
            #
            vlr = v2 - v1
            ils = grs*(v1 - vm)
            ilr = glr*vlr+ pilr
            id0, gd0 = fdiode(v1)
            icp = grc*(v0 - vcp)
            iload = gload*v0
            #
            irr2 = iswitch - ilr
            irr1 = ilr - id0 - ils
            irr0 = ils - icp - iload
            #
            err = np.sqrt(irr2**2 + irr1**2 + irr0**2)
            #
            if err < erlim:
                break
            else:
                #
                gmat = np.array([ [(gswitch + glr), -glr, 0],
                                  [-glr, (glr + gser + gd0), -gser],
                                  [0, -gser, (gpar + gser)] ])
                #
                bcol = np.array([irr2, irr1, irr0])
                xdv = np.linalg.solve(gmat, bcol)
                #
                dv2 = xdv[0]
                dv1 = xdv[1]
                dv0 = xdv[2]
                v2 += dv2
                v1 += dv1
                v0 += dv0
        #
        if its >= 998:
            print('time-step= ', (k*n + j), 'j = ', j, 'k = ',      \
                                    k, 'solver does not converge')
            break
        #
        pilr = ilr
        pils = ils
        pvcr = vcr
        pvcp = vcp
        #
        cindr.append(ilr)
        cindls.append(ils)
        cswitch.append(isw)
```

```
            cd1.append(-id1)
            cdiode.append(-id0)
            ccapr.append(icr)
            vout.append(v0)
            vcapr.append(vcr)
            vdiode.append(v1)
            #
        #
    #
    print(' pilr = ', ilr, ' pils = ', ils)
    print(' pvcr = ', vcr, ' pvcp = ', vcp)
    print(' v2 = ', v2, ' v1 = ', v1, ' v0 = ', v0)
    #
    return vout, vcapr, vdiode, cindr, cindls, cswitch,           \
                                            cd1, cdiode, ccapr
#
#
def ch8_qrbuck_fig21(es, delt, cr, lr, rs, ls, rc, cp, rload, nc,   \
                                                    ncycle, n):
    """ Main program, plots results
        quasi-resonant zvs buck converter.
        Similar to converter presented by Batarseh."""
    #
    #calling command:
    #ch8_qrbuck_fig21(30.0, 0.4e-08, 0.6e-08, 0.1e-04,                \
    #                 0.6, 0.8e-03, 0.02, 0.17e-04, 6.0, 600, 400, 1000)
    #
    from math import pi
    import numpy as np
    import matplotlib.pyplot as plt
    from mpl_toolkits.axes_grid1 import host_subplot
    #
    N = ncycle*n
    #
    dteta = 2.0*pi/n
    glr = delt/lr
    gcr = cr/delt
    grs = 1.0/rs
    gls = delt/ls
    grc = 1.0/rc
    gcp = cp/delt
    gload = 1.0/rload
    #
    vout, vcapr, vdiode, cindr, cindls, cswitch, cd1, cdiode, ccapr =     \
         ch8_qrbuck_fig21_solver(es, glr, gcr, grs, gls, grc, gcp, gload, \
                              nc, ncycle, n)
    #
    m0 = 2*(ncycle - 3)
    m1 = 2*(ncycle - 2)
    m2 = 2*(ncycle - 1)
    ang = dteta*np.arange(N)
    d = float(nc/n)
    #
```

```
v0 = np.mean(vout[N-2*n:N-1])
print('mean value of output voltage = ', v0)
#
ax1 = host_subplot(2, 1, 1)
ax2 = ax1.twinx()
ax1.set_xlim([m0*pi, m2*pi])
ax1.set_xticks([m0*pi, (m0+2*d)*pi, m1*pi, m2*pi])
ax1.set_xticklabels(['$0$', '$2\pi *d$', '$2\pi$', '$4\pi$'])
ax1.set_ylim([-1, 240])
ax1.set_yticks([0, 120, 240])
ax1.set_yticklabels(['$0$', '$120$', '$240$'])
ax1.set_ylabel('$Voltage, V$')
ax1.text(m1*pi, -104, '$(a)$', size =10)
ax2.set_ylim(-4.0, 4.0)
ax2.set_yticks([-4, 0, 4])
ax2.set_yticklabels([ '$-4$', '$0$', '$4$'])
ax2.set_ylabel('$Currents, $~A')
p1, = ax1.plot(ang[N-3*n:N-n-1], vcapr[N-3*n:N-n-1], 'r',                \
                                        label = '$V_{cr}$')
p2, = ax2.plot(ang[N-3*n:N-n-1], cd1[N-3*n:N-n-1], 'b--',                \
               label='$I_{d1}$')
p3, = ax2.plot(ang[N-3*n:N-n-1], cswitch[N-3*n:N-n-1], 'b',              \
               label='$I_{sw}$')
p4, = ax2.plot(ang[N-3*n:N-n-1], ccapr[N-3*n:N-n-1], 'm--',              \
               label='$I_{cr}$')
ax1.yaxis.get_label().set_color(p1.get_color())
ax2.yaxis.get_label().set_color(p3.get_color())
ax1.legend(bbox_to_anchor=(0., 1.02, 1., .102), loc=3, ncol=4,           \
                              mode="expand", borderaxespad=0.)
ax1.grid(True)
plt.subplots_adjust(hspace = 0.9)
#
ax1 = host_subplot(2, 1, 2)
ax2 = ax1.twinx()
ax1.set_xlim([m0*pi, m2*pi])
ax1.set_xticks([m0*pi, (m0+2*d)*pi, m1*pi, m2*pi])
ax1.set_xticklabels(['$0$', '$2\pi *d$', '$2\pi$', '$4\pi$'])
ax1.set_xlabel('$\omega t, radians$')
ax1.set_ylim([0, 40])
ax1.set_yticks([0, 20, 40])
ax1.set_yticklabels(['$0$', '$\ $', '$40$'])
ax1.set_ylabel('$Voltage, V$')
ax1.text(m1*pi, -27, '$(b)$', size =10)
ax2.set_ylim(-4, 4)
ax2.set_yticks([-4, 0, 4])
ax2.set_yticklabels(['$-4$', '$0$', '$4$'])
ax2.set_ylabel('$Currents, $~A')
p1, = ax1.plot(ang[N-3*n:N-n-1], vdiode[N-3*n:N-n-1], 'r',               \
               label='$V_{D0}$')
p2, = ax1.plot(ang[N-3*n:N-n-1], vout[N-3*n:N-n-1], 'r--',               \
               label='$V_{0}$')
p3, = ax2.plot(ang[N-3*n:N-n-1], cindls[N-3*n:N-n-1], 'b',               \
               label = '$I_{ls}$')
```

```
    p4, = ax2.plot(ang[N-3*n:N-n-1], cdiode[N-3*n:N-n-1], 'b--',           \
                   label='$I_{diode}$')
    p5, = ax2.plot(ang[N-3*n:N-n-1], cindr[N-3*n:N-n-1], 'g',              \
                   label='$I_{lr}$')
    ax1.yaxis.get_label().set_color(p1.get_color())
    ax2.yaxis.get_label().set_color(p3.get_color())
    ax1.legend(bbox_to_anchor=(0., 1.02, 1., .102), loc=3, ncol=5,         \
                                mode="expand", borderaxespad=0.)
    ax1.grid(True)
    #
    plt.show()
    #
    #plt.draw()
    #plt.savefig('Fig821.ps', dpi=400, bbox_inches='tight')
    #
```

8.6.4 Program 8.24

```
#
def fdiode(vd):
    """ Diode characteristic.
        Returns diode current and its conductance"""
    #
    glow = 0.1e-07
    ghigh = 0.4e+02
    Vr = -0.5
    #
    if vd >= Vr:
        id = glow*(vd - Vr)
        gd = glow
    else:
        id = ghigh*(vd - Vr)
        gd = ghigh
    #
    return id, gd
#
#
def fswitch(vsw, gsw):
    """Bi-directional switch characteristic.
       Returns switch current and conductance."""
    #
    glow = 0.1e-07
    ghigh = 0.4e+02
    Vr = 0.5
    #
    if vsw >= Vr:
        vd = (gsw*vsw + ghigh*Vr)/(ghigh + gsw)
        iw = ghigh*(vd -Vr)
        gw = ghigh*gsw/(ghigh + gsw)
    else:
        vd = (gsw*vsw + glow*Vr)/(glow + gsw)
        iw = glow*(vd - Vr)
```

```
        gw = glow*gsw/(glow + gsw)
    #
    return iw, gw
#
#
def ch8_mrbuck_fig24_solver(es, grs, gls, grr, glr,                    \
            gcswa, gcswb, grc, gcp, gload, nc1, nc2, ncycle, n):
    """Assumes fdiode is a function.
       Returns the lists vout, cinduct"""
    #
    import numpy as np
    #
    ghigh = 40.0
    glow = 0.1e-07
    erlim = 0.1e-04
    #
    gsser = grs*gls/(grs + gls)
    grser = grr*glr/(grr + glr)
    gpser = grc*gcp/(grc + gcp)
    #
    vsa = 19.9
    vsb = 0.1
    v0 = 12.0
    #
    pils = 0.41628
    pilr = -0.29153
    pvsa = 19.0
    pvsb = 0.1
    pvc = 12.0
    #
    vout = []
    vswitcha = []
    vswitchb = []
    cindr = []
    cindls = []
    cswitcha = []
    cswitchb = []
    #
    for k in range(ncycle):
        #
        for j in range(n):
            #
            if j <= nc1:
                gsa = ghigh
            else:
                gsa = glow
            #
            if j <= nc2:
                gsb = ghigh
            else:
                gsb = glow
            #
            for its in range(1000):
```

```
        #
        imsa, gswa = fswitch(vsa, gsa)
        idwa, gda = fdiode(vsa)
        icswa = gcswa*(vsa - pvsa)
        iswatot = imsa + idwa + icswa
        gswatot = gswa + gda + gcswa
        #
        imsb, gswb = fswitch(vsb, gsb)
        idwb, gdb = fdiode(vsb)
        icswb = gcswb*(vsb - pvsb)
        iswbtot = imsb + idwb + icswb
        gswbtot = gswb + gdb + gcswb
        #
        vr = (grr*es + glr*(vsa + vsb) - pilr)/(grr + glr)
        vs = (grs*vsb + gls*v0 - pils)/(grs + gls)
        vc = (grc*v0 + gcp*pvc)/(grc + gcp)
        #
        ilr = grr*(es - vr)
        ils = grs*(vsb - vs)
        icp = grc*(v0 - vc)
        iload = gload*v0
        #
        irrsa = ilr - iswatot
        irrsb = ilr - ils - iswbtot
        irr0 = ils - icp - iload
        #
        err = np.sqrt(irrsa**2 + irrsb**2 + irr0**2)
        #
        if err < erlim:
            break
        else:
            #
            g12 = grser + gsser
            g11 = gswatot + g12
            g22 = gswbtot + g12
            g33 = gload + gsser + gpser
            gmat = np.array([ [g11, g12, -gsser],
                              [g12, g22, -gsser],
                              [-gsser, -gsser, g33] ])
            #
            bcol = np.array([irrsa, irrsb, irr0])
            xdv = np.linalg.solve(gmat, bcol)
            #
            dvsa = xdv[0]
            dvsb = xdv[1]
            dv0 = xdv[2]
            vsa += dvsa
            vsb += dvsb
            v0 += dv0
    #
    if its >= 998:
        print('time-step= ', (k*n + j), 'j = ', j, 'k = ',        \
                             k, 'solver does not converge')
```

```
                break
            #
            pilr = ilr
            pils = ils
            pvsa = vsa
            pvsb = vsb
            pvc = vc
            #
            cindr.append(ilr)
            cindls.append(ils)
            cswitcha.append(imsa + idwa)
            cswitchb.append(imsb + idwb)
            vout.append(v0)
            vswitcha.append(vsa)
            vswitchb.append(vsb)
            #
        #
    #
    return vout, vswitcha, vswitchb, cindr, cindls, cswitcha, cswitchb
#
def ch8_mrbuck_fig24(es, delt, rr, lr, rs, ls, cswa, cswb,              \
                     rc, cp, rload, nc1, nc2, ncycle, n):
    """ Main program, plots results
        quasi-resonant buck converter.
        Maksimovic & Cuk, IEEE Trans. on Pow.El., vol.6, no.1, Jan. 1991."""
    #
    #calling command:
    #ch8_mrbuck_fig24(20.0, 0.1e-08, 0.005, 0.55e-06, 0.04,              \
    #0.1e-04, 0.1e-07, 0.1e-07, 0.02, 0.47e-05, 7.5, 200, 250, 400, 500)
    #
    from math import pi
    import numpy as np
    import matplotlib.pyplot as plt
    from mpl_toolkits.axes_grid1 import host_subplot
    #
    N = ncycle*n
    #
    dteta = 2.0*pi/n
    grs = 1.0/rs
    grr = 1.0/rr
    grc = 1.0/rc
    gload = 1.0/rload
    gls = delt/ls
    glr = delt/lr
    gcswa = cswa/delt
    gcswb = cswb/delt
    gcp = cp/delt
    #
    vout, vswitcha, vswitchb, cindr, cindls, cswitcha, cswitchb =        \
            ch8_mrbuck_fig24_solver(es, grs, gls, grr, glr, gcswa,       \
                    gcswb, grc, gcp, gload, nc1, nc2, ncycle, n)
    #
    m0 = 2*(ncycle - 3)
```

```
m1 = 2*(ncycle - 2)
m2 = 2*(ncycle - 1)
ang = dteta*np.arange(N)
d1 = 2.0*nc1/n
d2 = 2.0*nc2/n
#
v0 = np.mean(vout[N-2*n:N-1])
print('mean value of output voltage = ', v0)
#
ax1 = host_subplot(2, 1, 1)
ax2 = ax1.twinx()
ax1.set_xlim([m0*pi, m2*pi])
ax1.set_xticks([m0*pi, (m0+d1)*pi, (m0+d2)*pi, m1*pi, m2*pi])
ax1.set_xticklabels(['$0$', '$\\theta_{1}$', '$\ $', '$2\pi$', '$4\pi$'])
ax1.set_ylim([-2, 60])
ax1.set_yticks([0, 30, 60])
ax1.set_yticklabels(['$0$', '$30$', '$60$'])
ax1.set_ylabel('$Voltage, V$')
ax1.text(m1*pi, -24.6, '$(a)$', size =10)
ax2.set_ylim(-10, 10)
ax2.set_yticks([-10, 10])
ax2.set_yticklabels([ '$-10$', '$10$'])
ax2.set_ylabel('$Currents, $~A')
p1, = ax1.plot(ang[N-3*n:N-n-1], vswitcha[N-3*n:N-n-1], 'r',           \
               label = '$V_{swa}$')
p2, = ax1.plot(ang[N-3*n:N-n-1], vswitchb[N-3*n:N-n-1], 'r--',         \
               label = '$V_{swb}$')
p3, = ax2.plot(ang[N-3*n:N-n-1], cswitcha[N-3*n:N-n-1], 'b',           \
               label='$I_{swa}$')
p4, = ax2.plot(ang[N-3*n:N-n-1], cswitchb[N-3*n:N-n-1], 'b--',         \
               label='$I_{swb}$')
ax1.yaxis.get_label().set_color(p1.get_color())
ax2.yaxis.get_label().set_color(p3.get_color())
ax1.legend(bbox_to_anchor=(0., 1.02, 1., .102), loc=3, ncol=4,         \
                              mode="expand", borderaxespad=0.)
ax1.grid(True)
plt.subplots_adjust(hspace = 0.9)
#
ax1 = host_subplot(2, 1, 2)
ax2 = ax1.twinx()
ax1.set_xlim([m0*pi, m2*pi])
ax1.set_xticks([m0*pi, (m0+d1)*pi, (m0+d2)*pi, m1*pi, m2*pi])
ax1.set_xticklabels(['$0$', '$\ $', '$\\theta_{2}$', '$2\pi$', '$4\pi$'])
ax1.set_xlabel('$\omega t, radians$')
ax1.set_ylim([0, 12])
ax1.set_yticks([0, 6, 12])
ax1.set_yticklabels(['$0$', '$\ $', '$12$'])
ax1.set_ylabel('$Voltage, V$')
ax1.text(m1*pi, -7.44, '$(b)$', size =10)
ax2.set_ylim(-10, 10)
ax2.set_yticks([-10, 0, 10])
ax2.set_yticklabels(['$-10$', '$0$', '$10$'])
ax2.set_ylabel('$Currents, $~A')
```

```
p1, = ax1.plot(ang[N-3*n:N-n-1], vout[N-3*n:N-n-1], 'r',                    \
               label='$V_{0}$')
p2, = ax2.plot(ang[N-3*n:N-n-1], cindr[N-3*n:N-n-1], 'b',                   \
               label='$I_{lr}$')
p3, = ax2.plot(ang[N-3*n:N-n-1], cindls[N-3*n:N-n-1], 'b--',                \
               label = '$I_{ls}$')
ax1.yaxis.get_label().set_color(p1.get_color())
ax2.yaxis.get_label().set_color(p2.get_color())
ax1.legend(bbox_to_anchor=(0., 1.02, 1., .102), loc=3, ncol=3,              \
                               mode="expand", borderaxespad=0.)
ax1.grid(True)
#
plt.show()
#
#plt.draw()
#plt.savefig('Fig824.ps', dpi=400, bbox_inches='tight')
#
```

References

[1] R. W. Erickson, D. Maksimovic, *Fundamentals of Power Electronics,* Kluwer Academic Publishers, 2004

[2] R. L. Steigerwald, *A comparison of half-bridge resonant converter topologies,* IEEE Trans. on Power Electronics, vol.3, no.2, April 1988, pp.174-182

[3] Bob Mammano, *Resonant converter topologies,* Texas Instruments Literature SLUP085. 1985

[4] M. M. Jovanovic, *Resonant, quasi-resonant, multi-resonant and soft-switching techniques - merits and limitations,* Int. J. of Electronics, vol.77, no.5, 1994, pp.537-554

[5] S. Y. Yu, R. Chen, A. Viswanathan, *Survey of resonant converter topologies,* Texas Instruments Literature SLUP376, 2018

[6] ST Microelectronics, *An introduction to LLC resonant, half-bridge converter,* Application Note AN2644, Sept.2008

[7] C. Basso, *Understanding the LLC structure in resonant applications,* ON Semiconductor, AND8311/D, Jan.2008

[8] H. Huang. *Designing an LLC resonant, half-bridge, power converter,* Texas Instruments Literature SLUP263, 2010

[9] ON Semiconductor, *Basic principles of LLC resonant, half-bridge converter and DC/Dynamic circuit simulation examples,* Application Note AND9408/D, April 2016

[10] Toshiba Electronic Devices and Storage Corporation, *Resonant circuits and soft-switching (LLC converter and resonant inverter,* Application Note, 2019

[11] A. K. S. Bhat, *Analysis and design of a high-frequency, resonant converter using LCC-type commutation,* IEEE Trans. on Power Electronics, vol.PE2, no.4, Oct.1987, pp.291-301

[12] N. Mohan, T. Undeland, W. Robbins, *Power Electronics Converters, Applications and Design, 2nd Edition,* John Wiley and Sons, Inc., 1995

[13] D. Maksimovic, S. Cuk, *Constant frequency control of quasi-resonant converters,* IEEE Trans. on Power Electronics, vol.6, no.1, Jan.1991, pp.141-150

[14] M. K. Kazimierczuk, *Design-oriented analysis of boost, ZVS, resonant, DC/DC converter,* IEEE Trans.on Power Electronics, vol.3, no.2, Apeil 1988, pp.126-136

9

Inverters

Inverters transfer power from a DC source to an AC load. The thyristor–based, single–phase full-bridge converter and the three-phase, AC–DC converter can transfer power to the AC side. But a source should already exist on the AC side for this kind of power transfer. The inverters examined in this chapter generate an AC voltage when the available source is a direct voltage source. (For this reason, they are referred to as voltage source inverters.) Power electronic inverters are extensively applied in variable-speed induction motor drives, uninterruptible power supplies and grid-connected, renewable energy systems.

9.1 Single-phase Inverters

9.1.1 Square-wave inverter

The full-bridge, square-wave inverter is shown in Figure (9.1). The voltage v_O applied to the load depends on the switch positions as shown in Table 9.1.

Table 9.1 Switching scheme of the square-wave inverter.

Switch positions				Output
SW1	SW2	SW3	SW4	
On	Off	On	Off	E_S
On	On	Off	Off	0
Off	On	OFF	On	$-E_S$
Off	Off	On	On	0

The switch pairs $(SW1, SW4)$ and $(SW2, SW3)$ are complementary. When one switch is turned on, the other switch is turned off. A dead time of about 1–4 μs is inserted between the turn-off of one switch and turn-on of the other switch. The duty ratio of the switches is approximately 0.5 to generate an alternating square-wave output voltage.

Let the fundamental frequency of the periodic, square-wave output voltage be f_1 Hz, so that its time period T_S is $(1/f_1)$ seconds. Thus, switches $SW1, SW3$ are turned on for approximately $T_S/2$ seconds. During this interval, switches $SW2, SW4$ are turned off. In the next half-period, switches $SW1, SW3$ are turned off and switches $SW2, SW4$ are turned on. The square-wave output voltage may be represented by the Fourier series

$$v_0 = \sum_{m=1,2,3,...} \frac{4E_S}{m\pi} \sin(m\omega_1 t), \tag{9.1}$$

where $\omega_1 = 2\pi f_1$ rads/sec. If the load is an RL load (as shown in Figure 9.1), the harmonic components of the load current may be calculated and added together:

$$i_{ls} = \sum_m \frac{4E_S}{m\pi} \frac{1}{R^2 + (m\omega_1 L)^2} \{R \sin(m\omega_1 t) - m\omega_1 L \cos(m\omega_1 t)\}. \tag{9.2}$$

From Figure 9.1, we obtain

$$\begin{aligned} i_{SW1}, i_{SW3} &= \text{ils} \quad \text{for } 0 \leq \omega t \leq \pi \\ &= 0 \quad \text{for } \pi \leq \omega t \leq 2\pi \\ i_{SW2}, i_{SW4} &= 0 \quad \text{for } 0 \leq \omega t \leq \pi \\ &= \text{-ils} \quad \text{for } \pi \leq \omega t \leq 2\pi. \end{aligned}$$

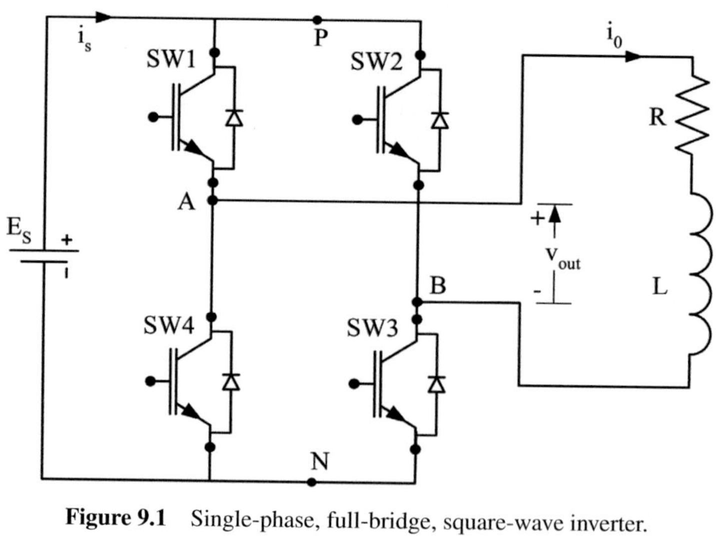

Figure 9.1 Single-phase, full-bridge, square-wave inverter.

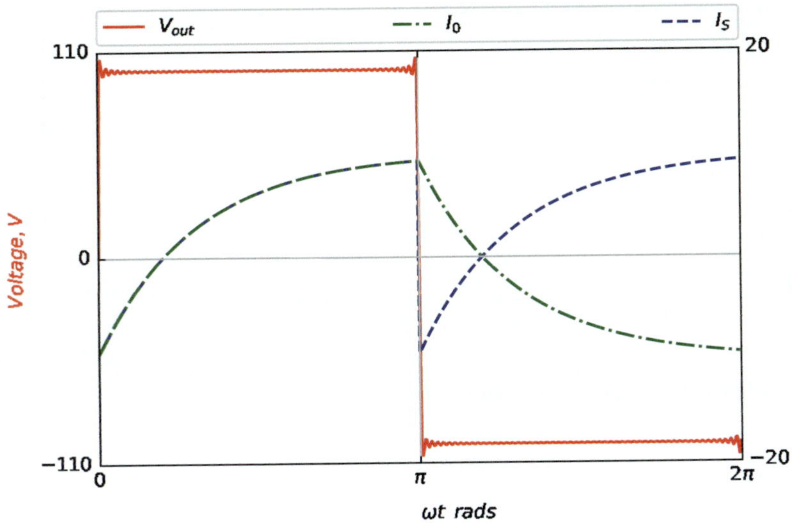

Figure 9.2 Square–wave converter waveforms.

Figure 9.2 (plotted by Program 9.02) shows the output voltage, the load current, and the source current. Although the square-wave inverter may be adequate for some applications, the output voltage is significantly distorted. The THD of the output voltage is 48.3%. It would be worthwhile to eliminate at least a few of the harmonic components.

9.1.2 Amplitude control and harmonic elimination

Specific harmonics in the square-wave output voltage may be eliminated and the fundamental frequency amplitude adjusted by a switching scheme that introduces intervals of zero voltage in the waveform. The scheme is referred to as the voltage cancellation technique for harmonic elimination [1], [4].

Consider, for instance, a switching scheme that introduces zero voltage intervals of α radians in the square-wave output, as shown in Figure 9.3. The Fourier series representation of the output voltage waveform of Figure 9.3 is

$$v_0(t) = \frac{4E_S}{\pi} \sum_{m=1,3,5,\ldots} \frac{\cos(m\alpha)}{m} \sin(m\omega_1 t). \tag{9.3}$$

The mth order harmonic in $v_0(t)$ may be eliminated by choosing $\alpha = (\frac{1}{m}\frac{\pi}{2})$. That is, to eliminate the third harmonic, choose $\alpha = \pi/6$ radians.

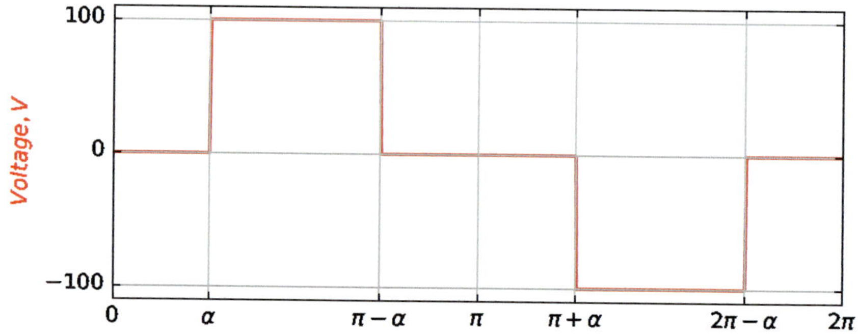

Figure 9.3 Square wave with voltage cancellation of α radians.

However, the amplitude of the fundamental, which is $V_1 = \frac{4E_S}{\pi}\cos(\alpha)$, cannot be controlled.

As the next step, consider the elimination of the third and fifth harmonics in $v_O(t)$. Also, suppose that the amplitude of the fundamental frequency component is required to be $0.8E_S$. The design specifications are obtained by considering the output voltage to be

$$v_O(t) = \frac{4E_S}{\pi}\sum_m \frac{\cos m\alpha_1 - \cos m\alpha_3 + \cos m\alpha_5}{m}\sin(m\omega_1 t). \tag{9.4}$$

Eqn (9.4) is the summation over m of the product of two terms. The first term is the sum of three cosine terms with alternating signs. (To eliminate k harmonics and obtain the specified fundamental frequency amplitude, $(k+1)$ terms are necessary in the summation.) The fundamental frequency amplitude is

$$V_1 = \frac{4E_S}{\pi}\big(\cos\alpha_1 - \cos\alpha_3 + \cos\alpha_5\big) = 0.8E_S. \tag{9.5}$$

That is,

$$\big(\cos\alpha_1 - \cos\alpha_3 + \cos\alpha_5\big) = 0.8 * \left(\frac{\pi}{4}\right). \tag{9.6}$$

To eliminate the third and fifth harmonics, we require that

$$\big(\cos 3\alpha_1 - \cos 3\alpha_3 + \cos 3\alpha_5\big) = 0 \tag{9.7}$$

$$\big(\cos 5\alpha_1 - \cos 5\alpha_3 + \cos 5\alpha_5\big) = 0. \tag{9.8}$$

The nonlinear equations (9.6), (9.7), and (9.8) are solved for the delay angles α_1, α_2, andα_3, using the Newton–Raphson procedure.

The iterative procedure is as follows:

1. Choose a trial set $(\alpha_1, \alpha_3, \alpha_5)$. Calculate the vectors

$$\begin{aligned} irr_1 &= 0.8 * \frac{\pi}{4} - (\cos\alpha_1 - \cos\alpha_3 + \cos\alpha_5) \\ irr_3 &= -(\cos 3\alpha_1 - 3\cos\alpha_3 + 3\cos\alpha_5) \\ irr_5 &= -(\cos 5\alpha_1 - 5\cos\alpha_3 + 5\cos\alpha_5). \end{aligned}$$

2. Solve the equations

$$\begin{bmatrix} -\sin\alpha_1 & \sin\alpha_3 & -\sin\alpha_5 \\ -\sin 3\alpha_1 & \sin 3\alpha_3 & -\sin 3\alpha_5 \\ -\sin 5\alpha_1 & \sin 5\alpha_3 & -\sin 5\alpha_5 \end{bmatrix} \begin{bmatrix} \delta\alpha_1 \\ \delta\alpha_3 \\ \delta\alpha_5 \end{bmatrix} = \begin{bmatrix} irr_1 \\ irr_3 \\ irr_5. \end{bmatrix} \quad (9.9)$$

The corrected set is $\alpha_1 + \delta\alpha_1$, $\alpha_3 + \delta\alpha_3$, $\alpha_5 + \delta\alpha_5$. Return to (1).

After a few iterations, the procedure converges to $\alpha_1 = 31.42°$, $\alpha_3 = 54.57°$ and $\alpha_5 = 69.22°$. The output voltage, plotted by Program 9.04, is shown in Figure 9.4 (a). The load current is also shown in the same figure when the output voltage is applied to an RL load ($R = 10\ \Omega, L = 25$ mH). Figure 9.4 (b) shows the waveforms when the 3rd to 11th harmonics are

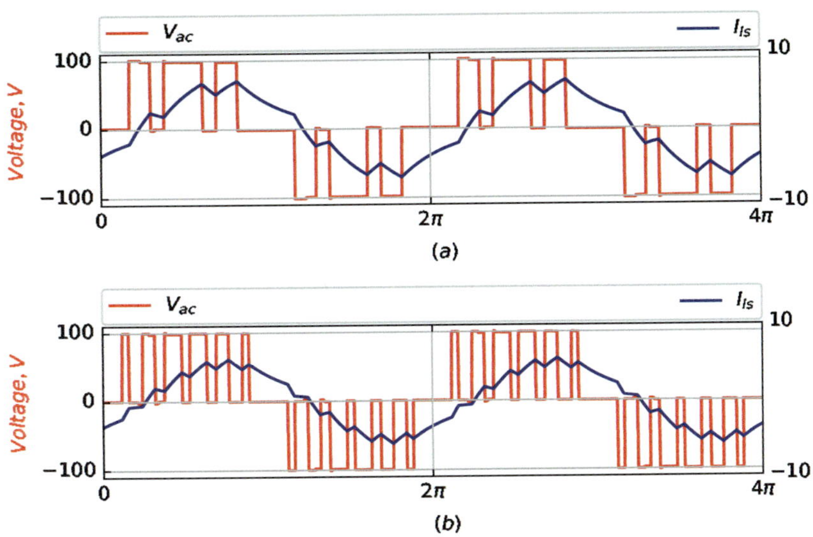

Figure 9.4 Harmonic elimination by voltage cancellation; elimination of (a) 3rd and 5th harmonics; (b) 3rd to 11th harmonics.

eliminated. Elimination of more harmonics is obtained by inserting more intervals of voltage cancellation. The disadvantages are that the RMS value of the output voltage decreases and the switching losses increase.

9.2 Pulse-width Modulation

Pulse-width modulation (PWM) is a technique that generates a periodically repeating sequence of voltage pulses [1, 2, 3, 4]. The pulses are generated at a high frequency. However, the pulse train comprises a low fundamental frequency component and harmonics at much higher frequencies than those of the square-wave voltage. It is relatively simple to filter the high–frequency harmonics. Control of the amplitude of the fundamental frequency voltage is possible, though the control circuits are complex. The switching frequency employed in PWM is high and it leads to higher switching losses.

The single-phase PWM inverter may be based on the full-bridge topology of Figure 9.1. The train of output voltage pulses is obtained by an appropriate switching strategy. The usual switching schemes are (a) bipolar switching and (b) unipolar switching. Both schemes require a high–frequency, alternating, triangular voltage v_{tri} that is referred to as the carrier signal. The peak value of the triangular wave is V_{tri} and its frequency is the switching frequency. The carrier is compared to a reference control signal $v_c = V_c \sin 2\pi f_1 t$. The frequency of the control signal is the fundamental frequency of the alternating output voltage. The frequency of the carrier is $m_c f_1$ where m_c is the frequency modulation ratio. It is usually an integer greater than 9. The amplitude modulation ratio is $m_a = V_c/V_{\text{tri}}$. The range $0 \leq m_a \leq 1$ is the linear range.

9.2.1 Bipolar switching

With reference to Figure 9.1, switches $SW1$ and $SW3$ are turned on when $v_c \geq v_{\text{tri}}$ and a positive pulse of height E_S is generated at the output. Similarly, switches $SW2$ and $SW4$ are turned on when $v_c < v_{\text{tri}}$ and a negative pulse of height $-E_S$ is generated at the output. Figure 9.5 shows the waveforms v_{tri}, v_c and the output v_0.

The output voltage is a train of alternating pulses with their durations modulated by v_c. Fourier analysis of the output voltage (included in Program 9.05) shows that the amplitude spectrum does not depend on the frequency modulation ratio, provided that $m_f > 9$. On the other hand, the amplitude spectrum does depend on the amplitude modulation ratio m_a as

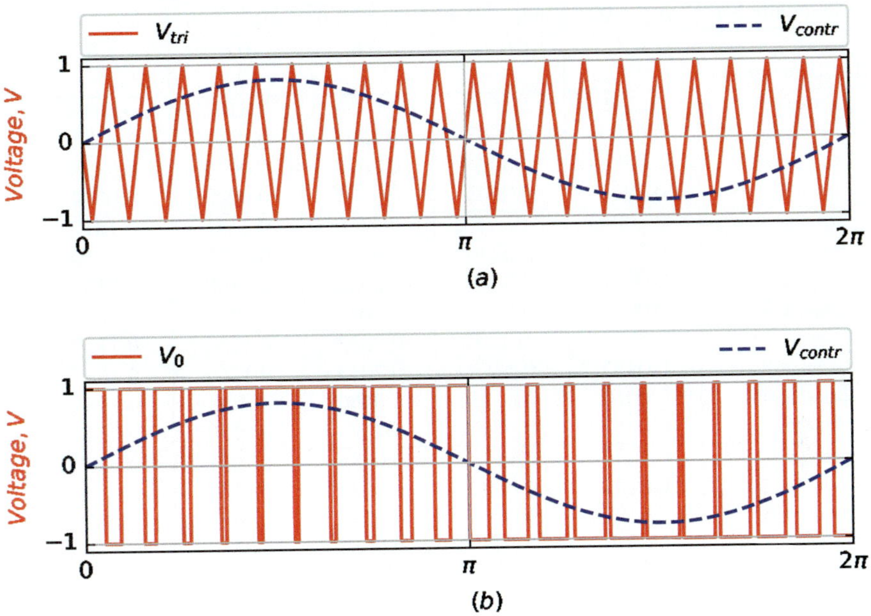

Figure 9.5 Bipolar switching, PWM waveforms.

shown in Table 9.2. By choosing m_f to be an odd integer, we obtain an output voltage that has odd and half-wave symmetries (i.e., $v_0(-t) = -v_0(t)$ and $v_0(t) = -v_0(t + T_1/2)$). Thus, only odd harmonics are present in $v_0(t)$ and the PWM is said to be synchronized.

Inspection of Table 9.2 shows that the frequencies present in the output voltage are [2]:

- Fundamental frequency f_1.
- Harmonics that are odd multiples of m_f: $m_f f_1$, $3m_f f_1$, $5m_f f_1$, ... Even multiples of m_f are absent.
- Sidebands $(m_f \pm 2)f_1$, $(m_f \pm 4)f_1, \ldots,$ on either sides of $m_f f_1$. Similar sidebands are present on either sides of the odd multiples of m_f.
- Sidebands $(2m_f \pm 1)f_1$, $(2m_f \pm 3)f_1, \ldots,$ on either sides of $2m_f f_1$. Similar sidebands are present on either sides of the even multiples of m_f.

The amplitude of the fundamental frequency voltage is $m_a E_S$ in the linear range $0 \leq m_a \leq 1$. The amplitude modulation ratio could be increased beyond unity to obtain higher fundamental frequency output voltage. This

Table 9.2 Normalized amplitudes V_f/E_S as a function of m_a.

Harmonic number	$m_a = 0.2$	$m_a = 0.4$	$m_a = 0.6$	$m_a = 0.8$	$m_a = 1.0$
1 (fundamental)	0.2	0.4	0.6	0.8	1.0
m_f	1.242	1.1502	1.0056	0.8183	0.6012
$m_f \pm 2$	0.0155	0.0607	0.1308	0.2201	0.3175
$m_f \pm 4$	0.0	0.0004	0.0025	0.0076	0.0178
$m_f \pm 6$	0.0013	0.0002	0.0001	0.0002	0.0004
$2m_f \pm 1$	0.1904	0.326	0.3706	0.3141	0.1817
$2m_f \pm 3$	0.0038	0.0236	0.0709	0.1392	0.2119
$2m_f \pm 5$	0.0001	0.0006	0.0035	0.0128	0.0332
$2m_f \pm 7$	0.0004	0.0007	0.0010	0.0004	0.0074
$3m_f$	0.3352	0.123	0.0837	0.1704	0.1129
$3m_f \pm 2$	0.0435	0.1388	0.2033	0.1765	0.0627
$3m_f \pm 4$	0.0010	0.0115	0.0466	0.104	0.1564
$3m_f \pm 6$	0.0012	0.0002	0.003	0.0139	0.0340
$4m_f \pm 1$	0.1632	0.1572	0.0085	0.1051	0.0677
$4m_f \pm 3$	0.0124	0.0699	0.1320	0.1153	0.0116
$4m_f \pm 5$	0.0	0.0065	0.0346	0.0864	0.1327
$4m_f \pm 7$	0.0070	0.0007	0.0064	0.0357	0.1034

is known as over-modulation and it leads to some of the pulses merging together. Over-modulation results in an output voltage that contains more harmonics, including harmonics that are lower order multiples of the fundamental frequency such as the third and fifth harmonics. These are difficult to filter. Moreover, the amplitude of the fundamental frequency voltage increases in a nonlinear manner. The maximum amplitude attainable is $4/\pi$, when the output voltage pulses degenerate into a square wave.

It is interesting to find the relation between the output voltage and the voltages v_{AN}, v_{BN} shown in Figure 9.1. Assuming that the switches are identical, it is easy to show that

$$\begin{aligned} v_{AN} &= \frac{1}{2}(E_S + v_{AB}) \\ v_{BN} &= \frac{1}{2}(E_S - v_{AB}). \end{aligned}$$

Suppose that $E_S = 100$ V, $m_a = 0.8$, and $m_f = 21$ in a full-bridge inverter using bipolar switching. Let the fundamental frequency be 60 Hz. The amplitudes of the fundamental and harmonic voltages are

$$\begin{aligned} v_0(60\,\text{Hz}) &= 0.8 * 100 = 80\ \text{V} \\ v_0(1140\,\text{Hz}) &= 0.22 * 100 = 22\ \text{V} \end{aligned}$$

$$\begin{aligned}
v_O(1260\,\text{Hz}) &= 0.818 * 100 = 81.8\ \text{V}\\
v_O(1380\,\text{Hz}) &= 0.22 * 100 = 22\ \text{V}\\
v_O(2460\,\text{Hz}) &= 0.314 * 100 = 31.4\ \text{V}\\
v_O(2580\,\text{Hz}) &= 0.314 * 100 = 31.4\ \text{V}.
\end{aligned}$$

Division of the amplitudes by $\sqrt{2}$ gives the RMS values. The harmonic amplitudes are quite significant. However, their frequencies are high and they are easily filtered.

9.2.1.1 Simulation of the full-bridge, single-phase inverter with bipolar switching

A brief discussion of the simulation of the full-bridge, single-phase inverter of Figure 9.1 is presented. The parameter values utilized in the simulation are: $E_S = 100$ V, $m_a = 0.8$, $f_1 = 60$ Hz, $V_{\text{tri}} = 1$ V, $m_f = 21$, Load : $R = 1\,\Omega$, and $L = 5$ mH.

Figure 9.6 shows the transient equivalent and the small-signal circuits. The probe voltage is v_a. The calculation of the error current irr_a and the probe voltage correction δv_a are left as exercises. Figure 9.7, plotted by Program 9.07, shows the output voltage, its fundamental frequency component v_1, the load current, and the input current. The output voltage of the inverter is a series of alternating pulses. The pulse height is $+E_S$ when the switches $SW1$ and $SW3$ are turned on and it is $-E_S$ when the switches $SW2$ and $SW4$ are turned on. The load current is approximately a 60–Hz sine wave. The harmonic components in the load current are suppressed by the inductor that presents a high impedance at the switching frequency. The calculated amplitude of the load current is 37.5 A.

The input current flows from the DC source to the load when the switches $SW1$ and $SW3$ are turned on and it is equal to the load current. The polarity of the input current reverses and flows into the DC source when the switches $SW2$ and $SW4$ are turned on, though the absolute value of the input current is still equal to the load current. Thus, the input current is a train of alternating pulses with sinusoidal envelopes. Assuming that are no losses in the inverter, nor energy storage, the volt-ampere balance requires that

$$\begin{aligned}
E_S\, i_{\text{DC}}(t) &= V_{\text{load}} \sin \omega_1 t\, I_{\text{load}} \sin(\omega_1 t - \phi)\\
&= \frac{V_{\text{load}} I_{\text{load}}}{2} * 2 \sin \omega_1 t\, \sin(\omega_1 t - \phi)\\
&= \frac{V_{\text{load}} I_{\text{load}}}{2} \Big(\cos \phi - \cos(2\omega_1 t - \phi) \Big)
\end{aligned}$$

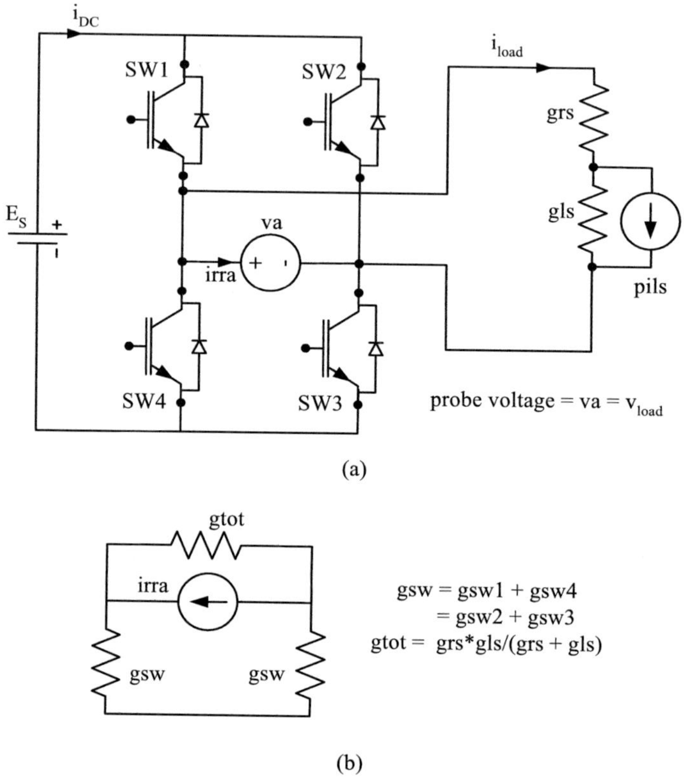

Figure 9.6 Full-bridge, single–phase inverter: Transient and small-signal equivalent circuits.

$$= \frac{V_{\text{load}} I_{\text{load}}}{2} \cos\phi - \frac{V_{\text{load}} I_{\text{load}}}{2} \cos(2\omega_1 t - \phi). \quad (9.10)$$

In eqn (9.10), V_{load}, and I_{load} are the amplitudes of the fundamental frequency output voltage and load current and ϕ is the phase lag of the load current. Thus, the input current is given by

$$i_{\text{DC}}(t) = \frac{V_{\text{load}} I_{\text{load}}}{2E_S} \cos\phi - \frac{V_{\text{load}} I_{\text{load}}}{2E_S} \cos(2\omega_1 t - \phi). \quad (9.11)$$

The input current has a second harmonic current.

9.2.2 Full-bridge, single-phase inverter with unipolar switching

In this scheme, the switches in the two legs of the inverter are not operated simultaneously. The switches $SW1$ and $SW4$ are controlled by comparing

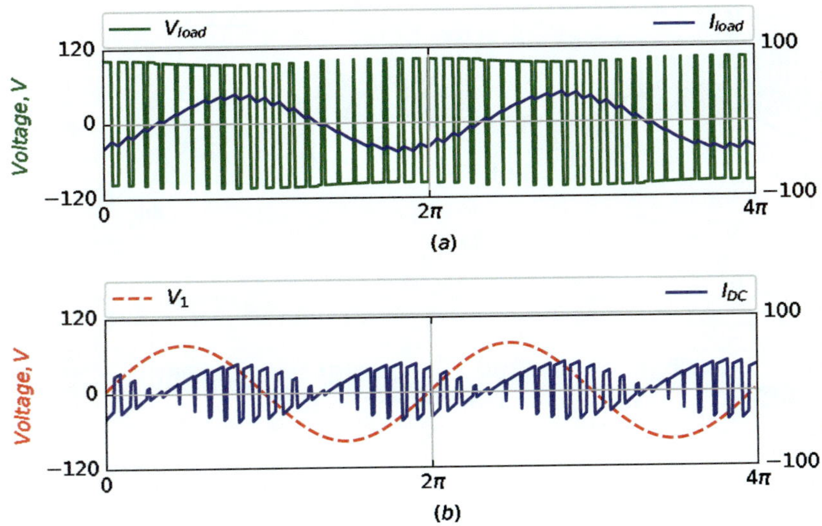

Figure 9.7 Full-bridge, single-phase inverter with bipolar switching: response waveforms.

the sinusoidal reference signal v_c with the triangular carrier v_{tri}. The switches in the other leg $SW2$ and $SW3$ are controlled by comparing $-v_c$ with v_{tri}:

$$\begin{aligned}
&\text{if} \quad v_c > v_{\mathrm{tri}} \text{ then } SW1 \text{ is turned on;} && v_{AN} = E_S \\
&\qquad\qquad \text{else } SW4 \text{ is turned on;} && v_{AN} = 0 \\
&\text{if} \quad -v_c > v_{\mathrm{tri}} \text{ then } SW2 \text{ is turned on;} && v_{BN} = E_S \\
&\qquad\qquad \text{else } SW3 \text{ is turned on;} && v_{BN} = 0.
\end{aligned}$$

The voltage waveforms of v_{AN} and v_{BN} are identical to each other, with v_{BN} lagging v_{AN} by π radians.

The output voltage is $v_0 = v_{AN} - v_{BN}$. Therefore, the harmonic of order m_f in the output voltage is

$$\begin{aligned}
v_0(m_f\omega_1) &= v_{AN}(m_f\omega_1) - v_{BN}(m_f\omega_1) \\
&= v_{AN}\angle\phi_{AN}(m_f) - v_{BN}\angle(\phi_{AN}(m_f) - m_f\pi).
\end{aligned}$$

If m_f is an even integer, the harmonic voltage of order m_f and its multiples are eliminated. The sidebands at odd multiples of m_f are also eliminated. The harmonics that remain in the output voltage are the sidebands at even multiples of the switching frequency, i.e., at $(2mf) \pm 1), (2mf) \pm 3), \ldots, (4mf) \pm 1), (4mf) \pm 3), \ldots$.

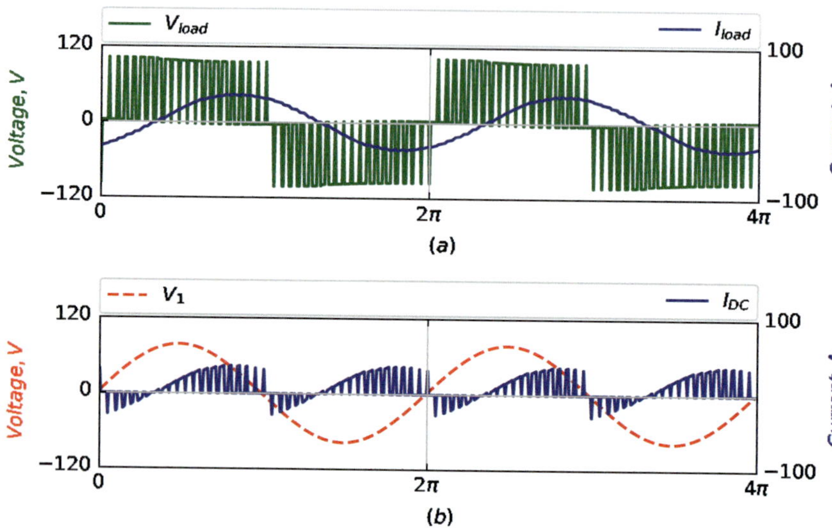

Figure 9.8 Full-bridge, single-phase inverter, unipolar switching: waveforms.

Figure 9.8, plotted by Program 9.08, shows the output voltage, its fundamental frequency component v_1, load current, and input current waveforms when the switching scheme is unipolar switching.

9.3 Three-phase Inverters

The basic Three-phase inverters described in this section are:

1. the square-wave inverter;
2. the triangle-comparison PWM inverter;
3. the space-vector PWM converter;
4. the hybrid PWM inverter.

All of the three-phase inverters work with the same electric circuit shown in Figure 9.9. The differences are in the switching schemes adopted in the inverters. In the circuit of Figure 9.9, the parameter values used in the subsequent simulations are as follows. The input voltage $E_S = 320$ VDC. The inverter supplies a generalized load comprising $R = 2\,\Omega$, $L = 10$ mH per phase and a sinusoidal, balanced emf of 106 V per phase. The emf has a phase lag of 12.4° with respect to the phase voltage generated by the inverter. The inverter consists of three legs (one per phase), and there is a complementary

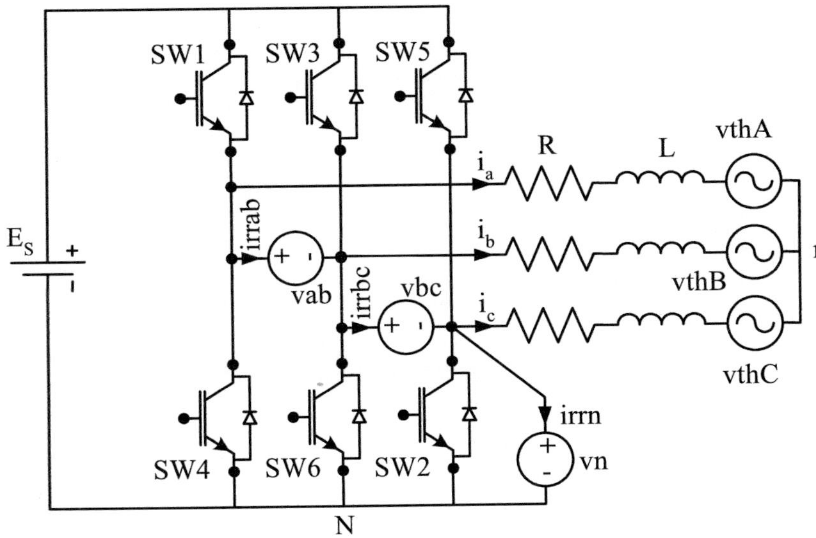

Figure 9.9 Three-phase inverter with generalized load.

switch pair in each leg. Each switch is turned on only after the complementary switch has been turned off and a reasonable dead-time (1 – 3 μs) has elapsed.

Attention is drawn to the numbering of the switches. The upper switches are $SW1$ (phase A), $SW3$ (phase B), and $SW5$ (phase C) for all inverters. The complementary bottom switches are numbered $SW4$, $SW6$, and $SW2$ for the square-wave inverter. This numbering is in accordance with the sequence in which the switches are turned on. For all other inverters, it is convenient to number the bottom switches as $SW2$, $SW4$, and $SW6$. The switching scheme for the square-wave inverter is shown in Figure 9.10.

Figure 9.11 shows the transient equivalent circuit and the probe voltages are v_{AB}, v_{BC}, and v_N. The small-signal equivalent circuit is shown in Figure 9.12. Applying Kirchhoff's laws, we obtain

$$i_a - i_b = gser * \{v_{AB} - (vthA - vthB) + (pila - pilb)/gls\} = I_A \tag{9.12}$$

$$i_a + 2\,i_b = gser * \{v_{BC} - (vthB - vthC) + (pila + 2pilb)/gls\} = I_B, \tag{9.13}$$

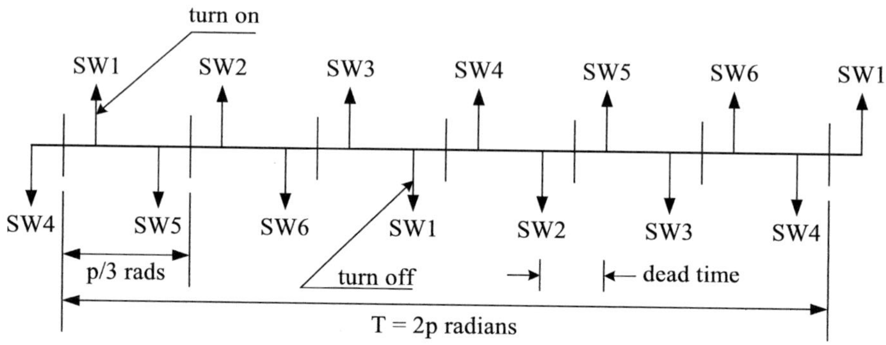

Figure 9.10 Switching scheme for the square-wave generator.

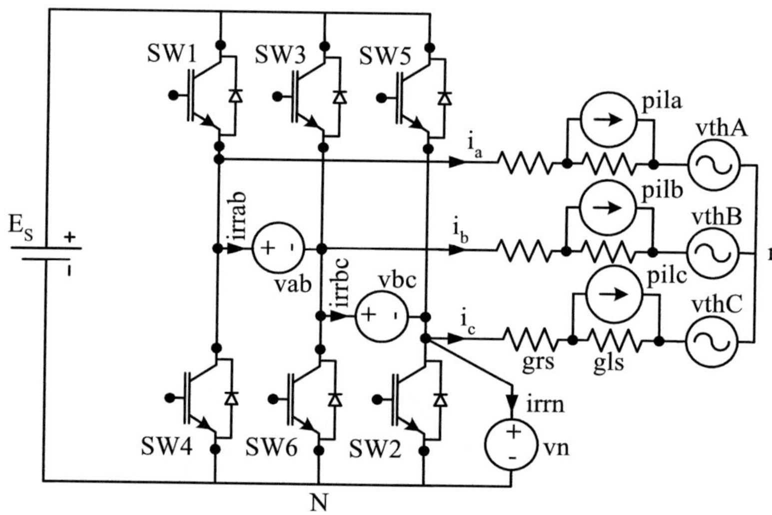

Figure 9.11 Three-phase inverter: transient equivalent circuit.

where i_a, i_b are the load phase currents and $gser = grs * gls/(grs + gls)$. The phase currents are

$$i_b = \frac{1}{3}(I_B - I_A) \tag{9.14}$$

$$i_a = I_A + i_b \tag{9.15}$$

$$i_c = -(i_a + i_b). \tag{9.16}$$

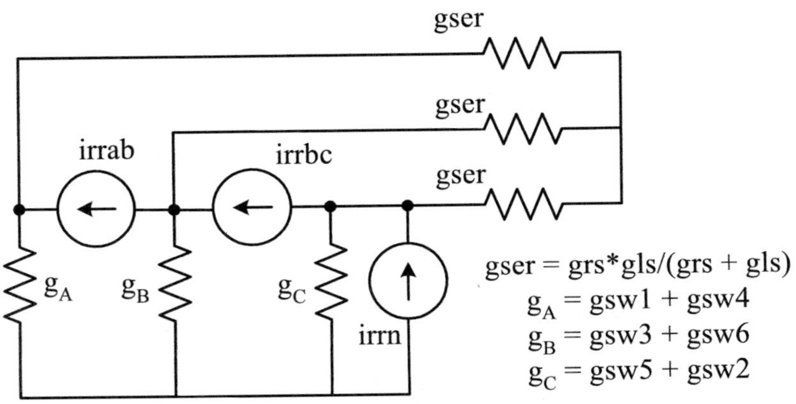

Figure 9.12 Three-phase inverter: small-signal equivalent circuit.

The error currents are

$$i_{SW1} - i_{SW4} - i_a = irr_{ab} = irr_A \quad (9.17)$$

$$i_{SW3} - i_{SW6} - i_b = irr_{bc} - irr_{ab} = irr_B \quad (9.18)$$

$$i_{SW5} - i_{SW2} - i_c = irr_n - irr_{bc} = irr_C. \quad (9.19)$$

Finally, analysis of the small-signal equivalent circuit gives

$$\begin{bmatrix} g_A + \frac{2}{3}gser & g_A + \frac{1}{3}gser & g_A \\ -\frac{1}{3}gser & g_B + \frac{1}{3}gser & g_B \\ -\frac{1}{3}gser & -\frac{2}{3}gser & g_C \end{bmatrix} \begin{bmatrix} \delta v_{AB} \\ \delta v_{BC} \\ \delta v_N \end{bmatrix} = \begin{bmatrix} irr_A \\ irr_B \\ irr_C, \end{bmatrix} \quad (9.20)$$

where $g_A = gsw1+gsw4$, $g_B = gsw3+gsw6$, and $g_C = gsw5+gsw2$.

9.3.1 Three-phase, square-wave inverter

Each switch has a duty ratio of approximately 0.5 and a switching action takes place at intervals of $\pi/3$ radians. Figure 9.13, plotted by Program 9.13, shows the response waveforms. The output voltage applied to each phase of the load is a stepped wave, with each step being $\pm\frac{1}{3}E_S$. The amplitude of the output voltage can only be controlled by varying the input DC voltage. The THD of the inverter's output voltage waveform is 30.4% and that of the load phase currents is 10.4%.

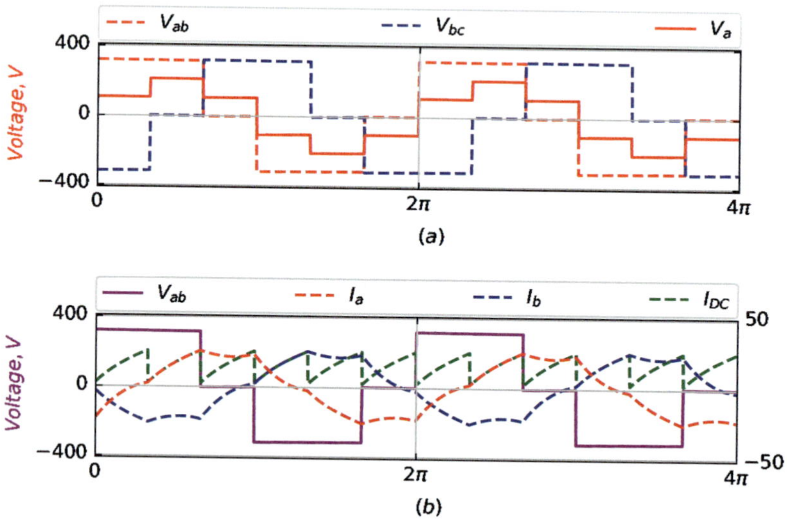

Figure 9.13 Three-phase, square-wave inverter: response waveforms.

9.3.2 Three-phase, triangle-comparison PWM

Three sinusoidal, reference, control voltages are compared with a single triangular carrier waveform to generate the three-phase, balanced output voltages [2]. Suppose the peak value of the carrier is 1 V and its frequency is $f_c = m_f f_1$, where m_f is the modulation frequency ratio and f_1 is the frequency of the reference control signals. The reference control signals are

$$\begin{aligned} v_{\text{contr-A}} &= m_a \sin(2\pi f_1 t) \\ v_{\text{contr-B}} &= m_a \sin(2\pi f_1 t - 2\pi/3) \\ v_{\text{contr-C}} &= m_a \sin(2\pi f_1 t + 2\pi/3). \end{aligned}$$

The switches $SW1$, $SW3$, and $SW5$ are turned on, based on the comparison of the respective reference control signal with the triangular carrier:

if $v_{\text{contr-A}} > v_{\text{tri}}$, then $SW1$ is turned on; else $SW2$ is turned on
if $v_{\text{contr-B}} > v_{\text{tri}}$, then $SW3$ is turned on; else $SW4$ is turned on
if $v_{\text{contr-C}} > v_{\text{tri}}$, then $SW5$ is turned on; else $SW6$ is turned on.

The triangular carrier wave oscillates between V_{tri} and $-V_{\text{tri}}$ at the angular frequency $m_f * \omega_1$ rads/sec, where ω_1 is the fundamental frequency. The

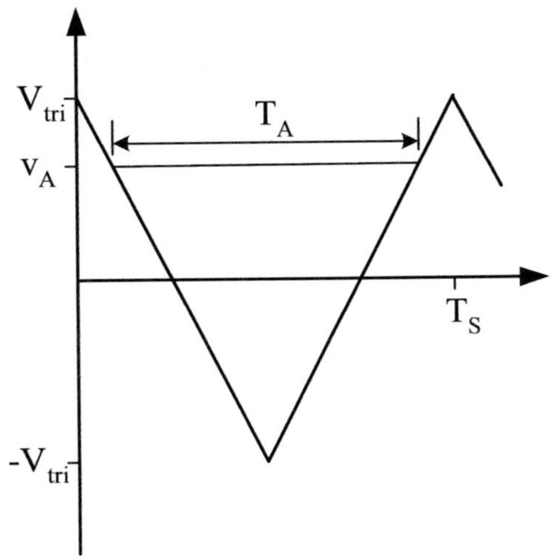

Figure 9.14 Switching interval for $SW1$.

control signals are fundamental frequency sine waves and their amplitudes are $m_a V_{\text{tri}}$, where $m_a \leqslant 1$. The control signals are approximately constant over a period T_S of the carrier frequency. Consider one period of the carrier waveform shown in Figure 9.14. Let the control signal for the switch $SW1$ be v_A. The time interval during which $SW1$ is switched on is T_A. From Figure 9.14, we obtain

$$\frac{T_A}{T_S} = \frac{v_A + V_{\text{tri}}}{2 * V_{\text{tri}}}. \tag{9.21}$$

That is,

$$\begin{aligned} v_A^* &= \frac{v_A}{V_{\text{tri}}} \\ &= 2\frac{T_A}{T_S} - 1. \end{aligned}$$

Similarly

$$\begin{aligned} v_B^* &= 2\frac{T_B}{T_S} - 1 \\ v_C^* &= 2\frac{T_C}{T_S} - 1. \end{aligned}$$

The normalized control signals for triggering $SW1$, $SW3$, and $SW5$ are v_A^*, v_B^*, and v_C^*.

Consider the voltages of the output terminals with reference to the negative input terminal. The harmonics in these voltages are identical to the harmonics in the output voltage of the single-phase, bipolar PWM inverter. The output phase to negative terminal voltages may be written as

$$v_{AN} = \frac{E_S}{2}\left(1 + m_a \sin(\omega_1 t) + k_{mf} \sin(m_f \omega_1 t) + \cdots\right)$$

$$\begin{aligned} v_{BN} &= \frac{E_S}{2}\left(1 + m_a \sin\left(\omega_1 t - \frac{2\pi}{3}\right) + k_{mf}\right. \\ &\qquad \left.\sin\left(m_f\left(\omega_1 t - \frac{2\pi}{3}\right)\right) + \cdots\right) \\ v_{CN} &= \frac{E_S}{2}\left(1 + m_a \sin\left(\omega_1 t + \frac{2\pi}{3}\right) + k_{mf}\right. \\ &\qquad \left.\sin\left(m_f\left(\omega_1 t + \frac{2\pi}{3}\right)\right) + \cdots\right). \end{aligned}$$

The phase to negative terminal voltages have a direct component. However, there is no direct component in the load's phase to neutral voltages or in the line to line voltages.

Only odd harmonics are present in the output voltages if m_f is an odd integer. Further, by choosing m_f to be odd and also a multiple of 3, harmonics of order m_f and odd multiples of m_f are eliminated in the line to line voltages. Table 9.3 provides the harmonic amplitudes as a function of the amplitude modulation ratio.

Figure 9.15, plotted by Program 9.14, shows the response waveforms of the three-phase inverter using the triangle-comparison PWM scheme.

9.3.3 Space-vector PWM

Space-vector PWM (SVPWM) is based on the (a, b, c) to (α, β, γ) transformation applied to three-phase voltages and current in the phase domain [5]. Suppose the three-phase voltages of a node are (v_a, v_b, v_c). The linear transformation to the (α, β, γ) domain is

$$\begin{bmatrix} v_\alpha \\ v_\beta \\ v_\gamma \end{bmatrix} = \sqrt{\frac{2}{3}} \begin{bmatrix} 1 & -1/2 & -1/2 \\ 0 & \sqrt{3}/2 & -\sqrt{3}/2 \\ 1/\sqrt{2} & 1/\sqrt{2} & 1/\sqrt{2} \end{bmatrix} \begin{bmatrix} v_a \\ v_b \\ v_c. \end{bmatrix}. \tag{9.22}$$

The transformation matrix is orthonormal and, therefore, its inverse is equal to its transpose.

Table 9.3 Normalized amplitudes V_{LL}/E_S as a function of m_a.

Harmonic number	$m_a = 0.2$	$m_a = 0.4$	$m_a = 0.6$	$m_a = 0.8$	$m_a = 1.0$
1 (fundamental)	0.1732	0.3462	0.5197	0.6928	8659
m_f	0.0	0.0	0.0	0.0	0.0
$m_f \pm 2$	0.0134	0.0527	0.1137	0.1903	0.2752
$m_f \pm 4$	0.0	0.0005	0.0021	0.0066	0.0155
$m_f \pm 6$	0.0	0.0	0.0	0.0	0.0
$2m_f \pm 1$	0.1648	0.2822	0.3205	0.2723	0.157
$2m_f \pm 3$	0.0	0.0	0.0	0.0	0.0
$2m_f \pm 5$	0.0001	0.0006	0.0031	0.0108	0.0291
$3m_f$	0.0	0.0	0.0	0.0	0.0
$3m_f \pm 2$	0.0377	0.1199	0.1763	0.1527	0.0538
$3m_f \pm 4$	0.0007	0.0101	0.0403	0.0907	0.1358
$3m_f \pm 6$	0.0	0.0	0.0	0.0	0.0
$4m_f \pm 1$	0.1413	0.1361	0.0068	0.0911	0.0584
$4m_f \pm 3$	0.0	0.0	0.0	0.0	0.0
$4m_f \pm 5$	0.0003	0.0056	0.0295	0.0731	0.1027
$4m_f \pm 7$	0.0002	0.0004	0.0029	0.0151	0.0435

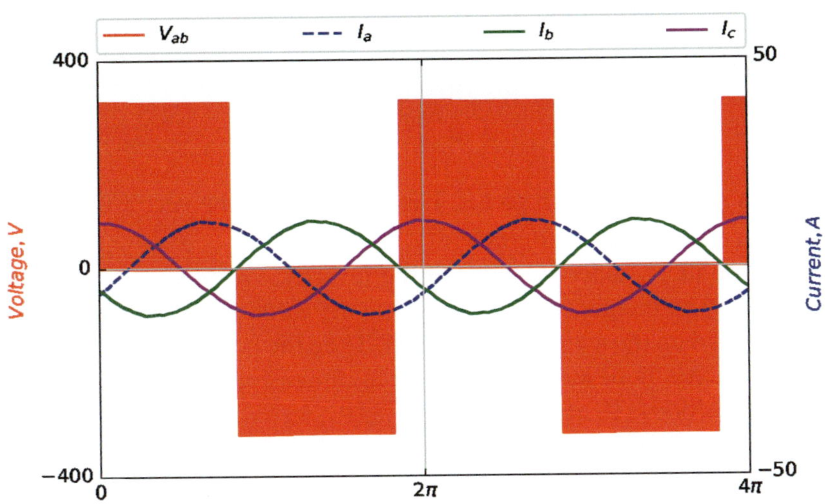

Figure 9.15 Three-phase inverter: triangle-comparison switching scheme.

Most loads supplied by three-phase inverters are balanced loads and for such loads, it is reasonable to assume that $v_a + v_b + v_c = 0$. Then, the transformation of eqn (9.22) reduces to (v_a, v_b, v_c) to (α, β) transformation. That is,

$$\begin{bmatrix} v_\alpha \\ v_\beta \end{bmatrix} = \sqrt{\frac{2}{3}} \begin{bmatrix} 1 & -1/2 & -1/2 \\ 0 & \sqrt{3}/2 & -\sqrt{3}/2 \end{bmatrix} \begin{bmatrix} v_a \\ v_b \\ v_c. \end{bmatrix}. \tag{9.23}$$

The inverse transformation is given by

$$\begin{bmatrix} v_a \\ v_b \\ v_c \end{bmatrix} = \sqrt{\frac{2}{3}} \begin{bmatrix} 1 & 0 \\ -1/2 & \sqrt{3}/2 \\ -1/2 & -\sqrt{3}/2 \end{bmatrix} \begin{bmatrix} v_\alpha \\ v_\beta \end{bmatrix}. \tag{9.24}$$

As an example, the transformation given by eqn (9.22) may be applied to the balanced, phase-domain voltages

$$\begin{aligned} e_a &= E_p \sin(\omega t) \\ e_b &= E_p \sin(\omega t - 2\pi/3) \\ e_c &= E_p \sin(\omega t + 2\pi/3). \end{aligned}$$

The transformed voltages in the (α, β) domain are

$$\begin{aligned} e_\alpha &= \sqrt{\frac{3}{2}} E_p \sin \omega t \\ e_\beta &= \sqrt{\frac{3}{2}} E_p \cos \omega t. \end{aligned}$$

The vector $e_{\alpha\beta}$ is a rotating vector. The amplitude of the vector is $\sqrt{\frac{3}{2}} E_p$ and its angular velocity is ω radians/sec.

Consider the three upper switches of the inverter. We represent the state of each switch by an integer variable that takes the value 1 when the switch is turned on and 0 when the switch is turned off. Then the composite state of the three upper switches is described by the three-dimensional vector that takes the values $V_0 = (000)$, $V_1 = (100)$, $V_2 = (110)$, $V_3 = (010)$, $V_4 = (011)$, $V_5 = (001)$, $V_6 = (101)$, and $V_7 = (111)$. The states V_0 and V_7 generate zero voltages at the inverter's output.

Suppose the composite state is (100). In this state, the upper switch in leg A is turned on and the upper switches in legs B and C are turned off. Then, from Figure 9.9,

$$\begin{aligned} v_{AN} &= E_S &= v_{An} + v_{nN} \\ v_{BN} &= 0 &= v_{Bn} + v_{nN} \\ v_{CN} &= 0 &= v_{Cn} + v_{nN}. \end{aligned}$$

The sum of the phase voltages of the load is $v_{An} + v_{Bn} + v_{Cn} = 0$. Therefore, $v_{nN} = \frac{1}{3}E_S$. Finally, $v_{An} = \frac{2}{3}E_S$, $v_{Bn} = -\frac{1}{3}E_S$, and $v_{Cn} = -\frac{1}{3}E_S$. Transforming the phase voltages of the load to the α, β domain, we get $V_\alpha = \sqrt{\frac{2}{3}}E_S$, $V_\beta = 0$. We define the vector $\sqrt{\frac{2}{3}}E_S\angle 0$ in the α, β domain as the state vector representation of the composite state (100). Table 9.4 provides the phase voltages applied to the load and the state vector equivalents of the eight composite switching states.

Figure 9.16 shows the geometrical representation of the state vectors in the (α, β) plane. The state vectors divide the plane into six sectors. In this same plane, three-phase balanced voltages will appear as a vector rotating in the anti-clockwise direction with angular velocity of ω radians/sec. The space vector PWM scheme uses the state vectors to generate the three-phase voltages at the inverter's output that are equivalent to a reference rotating vector V in the (α, β) plane.

Consider the six sectors in the α, β plane as shown in Figure 9.16 (a). The reference rotating vector takes $T = 2\pi/\omega_1$ sec to make a single rotation. Let each sector be divided into m equal sub-sectors. The rotating vector takes $T_S = T/(6m)$ sec to move from the center of one sub-sector to the center of the next sub-sector, where T_S is the switching period. Suppose the vector is in the first sub-sector, making an angle γ with the α-axis (Figure 9.16 (a)).

Table 9.4 State vectors in the α, β domain.

Switching state	v_{AN}	v_{BN}	v_{CN}	v_Z	v_{An}	v_{Bn}	v_{Cn}	State vector
$V(000)$	0	0	0	0	0	0	0	$0\angle 0$
$V(100)$	E_S	0	0	$\frac{1}{3}E_S$	$\frac{2}{3}E_S$	$-\frac{1}{3}E_S$	$-\frac{1}{3}E_S$	$\sqrt{2/3}\,E_S\angle 0$
$V(110)$	E_S	E_S	0	$\frac{2}{3}E_S$	$\frac{1}{3}E_S$	$\frac{1}{3}E_S$	$-\frac{2}{3}E_S$	$\sqrt{2/3}\,E_S\angle\pi/3$
$V(010)$	0	E_S	0	$\frac{1}{3}E_S$	$-\frac{1}{3}E_S$	$\frac{2}{3}E_S$	$-\frac{1}{3}E_S$	$\sqrt{2/3}\,E_S\angle 2\pi/3$
$V(011)$	0	E_S	E_S	$\frac{2}{3}E_S$	$-\frac{2}{3}E_S$	$\frac{1}{3}E_S$	$\frac{1}{3}E_S$	$\sqrt{2/3}\,E_S\angle\pi$
$V(001)$	0	0	E_S	$\frac{1}{3}E_S$	$-\frac{1}{3}E_S$	$-\frac{1}{3}E_S$	$\frac{2}{3}E_S$	$\sqrt{2/3}\,E_S\angle -2\pi/3$
$V(101)$	E_S	0	E_S	$\frac{2}{3}E_S$	$\frac{1}{3}E_S$	$-\frac{2}{3}E_S$	$\frac{1}{3}E_S$	$\sqrt{2/3}\,E_S\angle -\pi/3$
$V(111)$	E_S	E_S	E_S	E_S	0	0	0	$0\angle 0$

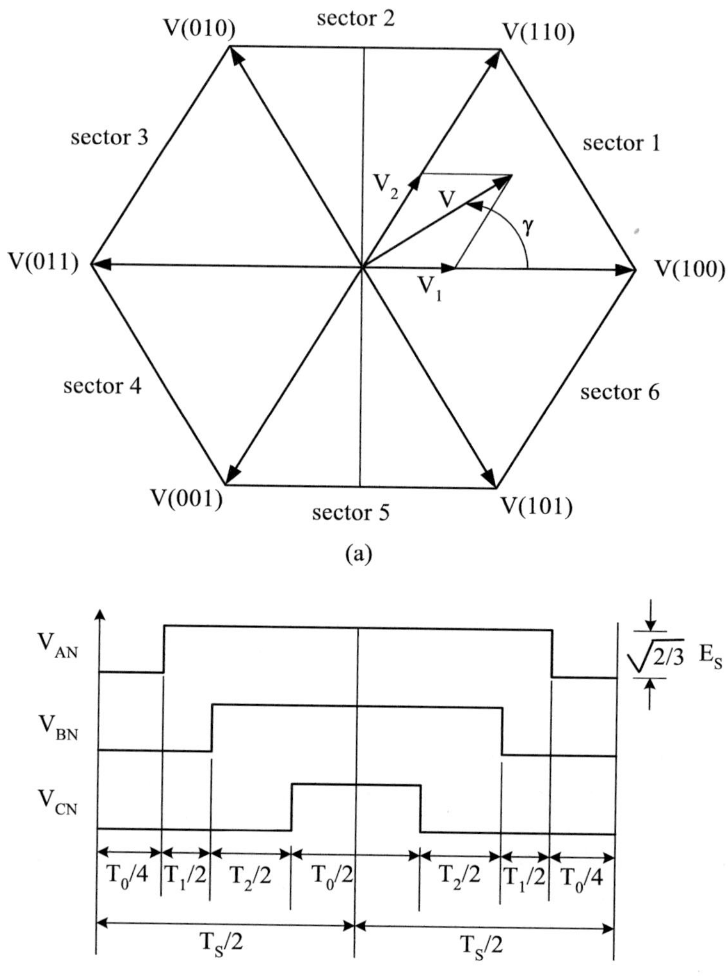

Figure 9.16 Space vector representation in the α, β plane.

The SVPWM technique suggests that the reference vector is, on the average over T_S sec, equivalent to applying the state vector $V(100)$ for T_1 sec and the state vector $V(110)$ for T_2 sec, $T_1, T_2 < T_S$. That is,

$$T_1\sqrt{\frac{2}{3}}E_S\angle 0 + T_2\sqrt{\frac{2}{3}}E_S\angle\pi/3 = T_S V\angle\gamma. \tag{9.25}$$

Converting the equation to the rectangular form, we obtain

$$T_1\sqrt{\frac{2}{3}}E_S(1+j0) + T_2\sqrt{\frac{2}{3}}E_S(\frac{1}{2}+\frac{\sqrt{3}}{2}) = T_S V(\cos\gamma + j\sin\gamma). \quad (9.26)$$

Equating the real and imaginary parts, we obtain

$$T_1 = \frac{a\,T_S \sin(\pi/3-\gamma)}{\sin(\pi/3)} \quad (9.27)$$

$$T_2 = \frac{a\,T_S \sin\gamma}{\sin(\pi/3)}, \quad (9.28)$$

where $a = V/(\sqrt{\frac{2}{3}}\,E_S)$. During the interval $T_0 = T_S - T_1 - T_2$, the zero states $V(000)$ and $V(111)$ are applied to the output terminals.

Figure 9.16 (b) shows the sequence of application of the two adjacent states $V(100)$ and $V(110)$ in the first sector. The switching sequence for each sector is shown in Table 9.5 and Table 9.6. In sectors 2, 4, and 6, the order of the intervals T_1 and T_2 calculated by eqn (9.27) and (9.28) are interchanged. During the first half of a switching period, the sequence starts with all the upper switches turned off. Then, the upper switches are turned on one by one. In the second half of the switching period, the upper switches are turned off in the reverse order.

The maximum amplitude of the rotating vector that can be generated by space-vector PWM is equal to the radius of the inscribed circle in the hexagon

Table 9.5 Switching sequences in sectors 1, 3, and 5.

Sector no.	$T_0/4$	T_1	T_2	$T_0/2$	T_2	T_1	$T_0/4$
1	$V(000)$	$V(100)$	$V(110)$	$V(111)$	$V(110)$	$V(100)$	$V(000)$
3	$V(000)$	$V(010)$	$V(011)$	$V(111)$	$V(011)$	$V(010)$	$V(000)$
5	$V(000)$	$V(001)$	$V(101)$	$V(111)$	$V(101)$	$V(001)$	$V(000)$

Table 9.6 Switching sequences in sectors 2, 4, and 6.

Sector no.	$T_0/4$	T_2	T_1	$T_0/2$	T_1	T_2	$T_0/4$
2	$V(000)$	$V(010)$	$V(110)$	$V(111)$	$V(110)$	$V(010)$	$V(000)$
4	$V(000)$	$V(001)$	$V(011)$	$V(111)$	$V(011)$	$V(001)$	$V(000)$
6	$V(000)$	$V(100)$	$V(101)$	$V(111)$	$V(101)$	$V(100)$	$V(000)$

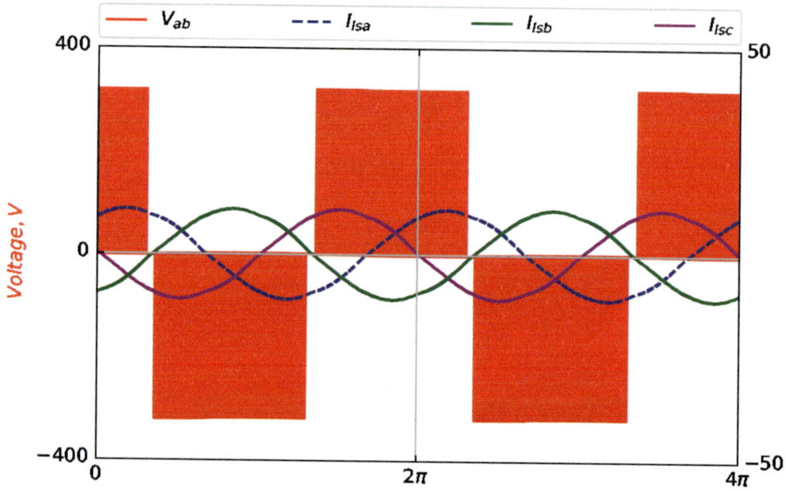

Figure 9.17 Three-phase inverter, space-vector PWM: response waveforms.

of Figure 9.16 (a). Thus,

$$\begin{aligned} V_{\alpha\beta,\max} &= \sqrt{2/3}E_S\frac{\sqrt{3}}{2} \\ &= \sqrt{2}(0.5E_S) \\ \text{But, } V_{\alpha\beta,\max} &= \sqrt{3/2}\,V_{abc,\max} \\ \text{Thus, } \frac{V_{abc,\max}}{0.5E_S} &= \frac{2}{\sqrt{3}} = 1.1547. \end{aligned}$$

This is about 15% greater than the maximum amplitude that can be generated by the triangle-comparison PWM.

Figure 9.17, plotted by Program 9.16, shows the response waveforms of the three-phase inverter using the space-vector PWM. In this simulation, a reference vector is rotated in the $\alpha\beta$ domain and the space-vector PWM has been used to generate the three-phase, balanced voltages at the inverter's output. The three-phase, balanced inductive load with isolated neutral is connected to the inverter.

9.3.4 Hybrid PWM

In the triangle-comparison PWM, reference control signals are compared with a triangular carrier signal to generate the PWM voltage pulses. The hybrid PWM technique derives the set of modified reference control signals that are compared with the same triangular carrier [6, 7]. The resulting PWM pulses are identical to the space-vector PWM pulses.

The space-vector PWM scheme calculates the switch timings sector by sector. In the first sector, the time intervals T_1 and T_2 are obtained using eqn (9.27) and (9.28). Given these intervals, the turn-on time intervals of the upper switches are calculated as

$$\begin{aligned} T_A &= T_0/2 + T_1 + T_2 \\ T_B &= T_0/2 + T_2 \\ T_C &= T_0/2. \end{aligned}$$

The reference control signals in the triangle-comparison PWM that will give the same turn-on time intervals for the upper switches are the modified reference signals being sought by the hybrid PWM scheme. The modified reference signals are:

$$v_A^{**} = \frac{2}{T_S}\left(T_0/2 + T_1 + T_2\right) - 1 \tag{9.29}$$

$$v_B^{**} = \frac{2}{T_S}\left(T_0/2 + T_2\right) - 1 \tag{9.30}$$

$$v_C^{**} = \frac{2}{T_S}\left(T_0/2\right) - 1. \tag{9.31}$$

Consider the time interval T_1 defined by eqn (9.27). In the first sector,

$$\begin{aligned} T_1 &= \frac{V}{\sqrt{2/3}E_S} T_S \left(\cos\gamma - \frac{1}{\sqrt{3}}\sin\gamma\right) \\ &= \frac{1}{\sqrt{2/3}E_S} T_S \left(V_\alpha - \frac{1}{\sqrt{3}}V_\beta\right). \end{aligned}$$

Let the equivalents of V_α, V_β in the three-phase domain be V_a, V_b, V_c. Using eqn (9.23) to substitute for V_α, V_β in terms of V_a, V_b, V_c, we obtain

$$\begin{aligned} T_1 &= \frac{T_S}{2}\left(\frac{V_a}{0.5E_S} - \frac{V_b}{0.5E_S}\right) \\ &= \frac{T_S}{2}\left(v_A^* - v_B^*\right), \end{aligned} \tag{9.32}$$

where v_A^*, v_B^*, v_C^* are the reference control signals in the triangle-comparison PWM scheme. Similarly, it may be shown that

$$T_2 = \frac{T_S}{2}\left(v_B^* - v_C^*\right). \tag{9.33}$$

Now, consider the modified reference signal v_A^{**} given by eqn (9.29). Substituting for T_0, we obtain

$$\begin{aligned} v_A^{**} &= \frac{T_1}{T_S} + \frac{T_2}{T_S} \\ &= \frac{1}{2}\left(v_A^* - v_B^*\right) + \frac{1}{2}\left(v_B^* - v_C^*\right) \\ &= v_A^* + \left(-\frac{(v_A^* + v_C^*)}{2}\right). \end{aligned} \tag{9.34}$$

Similarly, it may be shown that, in the first sector,

$$v_B^{**} = v_B^* + \left(-\frac{(v_A^* + v_C^*)}{2}\right) \tag{9.35}$$

$$v_C^{**} = v_C^* + \left(-\frac{(v_A^* + v_C^*)}{2}\right). \tag{9.36}$$

A zero sequence voltage has to be added to the reference control signals of the triangle-comparison technique to obtain the modified reference control signals.

In the first sector, the zero sequence voltage is given by $-(v_A^* + v_C^*)/2$. In this sector

$$\begin{aligned} v_A^* &= \max\left(v_A^*, v_B^*, v_C^*\right) \\ v_C^* &= \min\left(v_A^*, v_B^*, v_C^*\right). \\ \text{Therefore, } v_Z^* &= -\left(\frac{\max\left(v_A^*, v_B^*, v_C^*\right) + \min\left(v_A^*, v_B^*, v_C^*\right)}{2}\right). \end{aligned} \tag{9.37}$$

Eqn (9.37) is valid in all sectors. The amplitude of the sinusoidal, reference control signals (v_A^*, v_B^*, v_C^*) vary between 0 and $2/\sqrt{3}$. The modified reference control signals are non-sinusoidal and their peak values vary between 0 and 1.

The elegant hybrid PWM scheme that emerges is as follows:

- Let the desired three-phase, sinusoidal output voltages of the inverter be v_A, v_B, v_C. Divide the amplitudes by $0.5E_S$ to obtain the reference control signals (v_A^*, v_B^*, v_C^*).
- The triangular carrier is the same as for the triangle-comparison scheme. The carrier has a peak value of 1 and its frequency is m_f times the

frequency of the reference control signals. One period of the carrier is identified with the time interval T_S, which is the switching period.

- Divide the switching period into n_S sub-intervals and sample the reference control signals at the center of each sub-interval. Assume that the samples remain constant during the sub-interval. Let $v_{\max} = \max(v_A^*, v_B^*, v_C^*)$, $v_{\min} = \min(v_A^*, v_B^*, v_C^*)$ and $v_Z^* = -(v_{\max} + v_{\min})/2$. Form the modified reference control signals $v_A^{**} = v_A^* + v_Z^*$, $v_B^{**} = v_B^* + v_Z^*$, and $v_C^{**} = v_C^* + v_Z^*$.
- Compare the modified reference control signals to the triangular carrier. If $v_A^{**} > v_{\text{tri}}$, turn on the upper switch of phase A; else turn on the lower switch of phase A. Repeat for phases B and C.

The code presented in Program 9.17 simulates the three-phase inverter with the balanced, inductive load connected at its output. The response waveforms are shown in Figure 9.18. The reference control signal and the modified reference control signal for phase A are shown in Figure 9.18 (b).

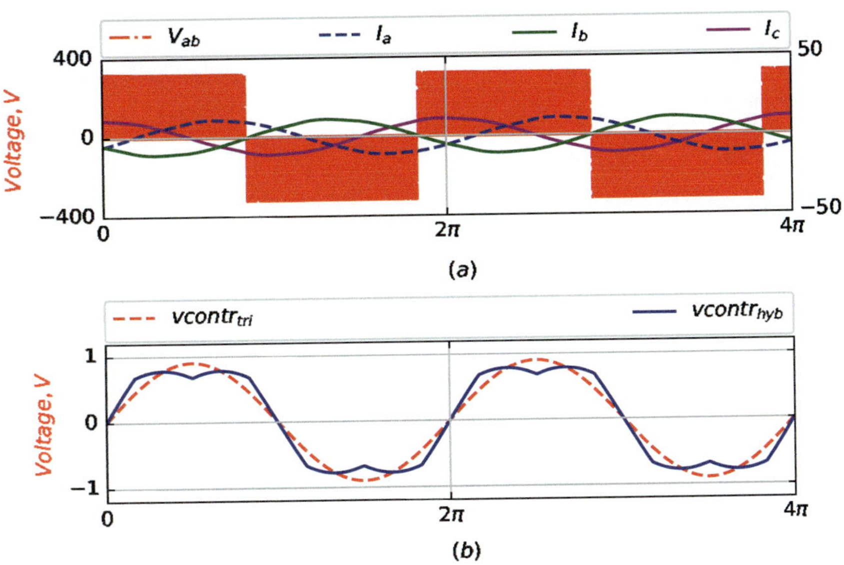

Figure 9.18 Hybrid PWM scheme. (a) Load currents; (b) reference and modified reference control signals.

9.4 Computational Programs

9.4.1 Program 9.02

```
#
def ch9_fig2(es, freq, rs, ls, nterms, n):
    """ Main program, single-phase, full-bridge inverter
        nalpha can be varied."""
    #
    #calling command:
    #ch9_fig2(100.0, 60.0, 10.0, 0.025, 200, 512)
    #
    from math import pi
    import numpy as np
    import matplotlib.pyplot as plt
    from mpl_toolkits.axes_grid1 import host_subplot
    #
    dteta = 2.0*pi/n
    xls = 2.0*pi*freq*ls
    Em = 4.0*es/pi
    #
    ang = []
    vout = []
    cindls = []
    csource = []
    #
    for k in range(n):
        #
        teta = dteta*k
        vsum = 0.0
        csum = 0.0
        #
        for j in range(nterms):
            #
            m = 2*j + 1
            zs = rs**2 + (m*xls)**2
            ilsm = (4.0*es/(m*pi*zs))*(rs*np.sin(m*teta) -        \
                                       m*xls*np.cos(m*teta))
            csum = csum + ilsm
            vsum = vsum + (Em/m)*np.sin(m*teta)
            #
        vout.append(vsum)
        cindls.append(csum)
        if k < n/2:
            csource.append(csum)
        else:
            csource.append(-csum)
        ang.append(teta)
        #
    #
    ax1 = host_subplot(1, 1, 1)
    ax2 = ax1.twinx()
    ax1.set_xlim([0, 2*pi])
```

```
ax1.set_xticks([0, pi, 2*pi])
ax1.set_xticklabels(['$0$', '$\pi$', '$2\pi$'])
ax1.set_ylim([-110, 110])
ax1.set_yticks([-110, 0, 110])
ax1.set_yticklabels(['$-110$', '$0$', '$110$'])
ax1.set_ylabel('$Voltage, V$')
ax2.set_ylim(-20, 20)
ax2.set_yticks([-20, 20])
ax2.set_yticklabels([ '$-20$', '$20$'])
ax2.set_ylabel('$Current, A$')
ax1.text(pi, -140, '$\omega t\,\,\,rads$', size=10)
p1, = ax1.plot(ang, vout, 'r', label = '$V_{out}$')
p2, = ax2.plot(ang, cindls, 'g-.', label='$I_{0}$')
p3, = ax2.plot(ang, csource, 'b--', label = '$I_{S}$')
ax1.yaxis.get_label().set_color(p1.get_color())
ax2.yaxis.get_label().set_color(p2.get_color())
ax1.legend(bbox_to_anchor=(0., 1.02, 1., .102), loc=3, ncol=4, \
                              mode="expand", borderaxespad=0.)
ax1.grid(True)
#
plt.show()
#plt.savefig('Fig902.ps', dpi=400, bbox_inches='tight')
#
```

9.4.2 Program 9.04

```
#
def fdiode(vd):
    """ Diode characteristic.
        Returns diode current and its conductance"""
    #
    glow = 0.1e-07
    ghigh = 0.4e+02
    Vr = -0.5
    #
    if vd >= Vr:
        id = glow*(vd - Vr)
        gd = glow
    else:
        id = ghigh*(vd - Vr)
        gd = ghigh
    #
    return id, gd
#
#
def fswitch(vsw, gsw):
    """Bi-directional switch characteristic.
       Returns switch current and conductance."""
    #
    glow = 0.1e-07
    ghigh = 0.4e+02
    Vr = 0.5
```

```
    #
    ida, gda = fdiode(vsw)
    #
    if vsw >= Vr:
        vd = (gsw*vsw + ghigh*Vr)/(ghigh + gsw)
        iw = ghigh*(vd -Vr)
        gw = gda + ghigh*gsw/(ghigh + gsw)
    else:
        vd = (gsw*vsw + glow*Vr)/(glow + gsw)
        iw = glow*(vd - Vr)
        gw = gda + glow*gsw/(glow + gsw)
    #
    isw = iw + ida
    #
    return isw, gw
#
#
def ch9_fig4_solver(es, grs, gls, nalpha, ncycle, n):
    """Assumes fdiode is a function.
       Returns the lists vout, cinduct"""
    #
    import numpy as np
    #
    ghigh = 40.0
    glow = 0.1e-07
    erlim = 0.1e-04
    nh = n/2
    #
    gstot = grs*gls/(grs + gls)
    pils = -6.48
    #
    vsa = 100.58
    vsb = 99.176
    #
    vout = []
    cindls = []
    #
    for k in range(ncycle):
        #
        for j in range(n):
            #
            if j >= nalpha and j < nh + nalpha:
                gs1 = ghigh
            else:
                gs1 = glow
            #
            if j > (nh - nalpha) and j <= (n - nalpha):
                gs3 = glow
            else:
                gs3 = ghigh
            #
            if j > nalpha and j <= (nh + nalpha):
                gs4 = glow
```

```
else:
    gs4 = ghigh
#
if j >= (nh - nalpha) and j < (n - nalpha):
    gs2 = ghigh
else:
    gs2 = glow
#
for its in range(1000):
    #
    vsw1 = es - vsa
    vsw2 = es - vsb
    isw1, gsw1 = fswitch(vsw1, gs1)
    isw2, gsw2 = fswitch(vsw2, gs2)
    isw3, gsw3 = fswitch(vsb, gs3)
    isw4, gsw4 = fswitch(vsa, gs4)
    id1, gd1 = fdiode(vsw1)
    id2, gd2 = fdiode(vsw2)
    id3, gd3 = fdiode(vsb)
    id4, gd4 = fdiode(vsa)
    is1 = isw1 + id1
    is2 = isw2 + id2
    is3 = isw3 + id3
    is4 = isw4 + id4
    g1 = gsw1 + gd1
    g2 = gsw2 + gd2
    g3 = gsw3 + gd3
    g4 = gsw4 + gd4
    #
    vm = (grs*vsa + gls*vsb - pils)/(grs + gls)
    ils = grs*(vsa - vm)
    irrsa = is1 - is4 - ils
    irrsb = is2 - is3 + ils
    #
    err = np.sqrt(irrsa**2 + irrsb**2)
    #
    if err < erlim:
        break
    else:
        #
        g11 = g1 + g4 + gstot
        g22 = g2 + g3 + gstot
        gmat = np.array([ [g11, -gstot ],
                          [-gstot, g22 ], ])
        #
        bcol = np.array([irrsa, irrsb])
        xdv = np.linalg.solve(gmat, bcol)
        #
        dvsa = xdv[0]
        dvsb = xdv[1]
        vsa += dvsa
        vsb += dvsb
#
```

```
            if its >= 998:
                print('time-step= ', (k*n + j), 'j = ', j, 'k = ',      \
                                                k, 'solver does not converge')
                break
            #
            pils = ils
            #
            cindls.append(ils)
            vout.append(vsa - vsb)
            #
        #
    #
#    print(' pils = ', ils, ' vsa = ', vsa, ' vsb = ', vsb)
    #
    return vout, cindls
#
#
def ch9_fig4(es, freq, rs, ls, ncycle, n):
    """ Main program, single-phase, full-bridge inverter;
        harmonic elimination by voltage cancellation;
        nalpha can be varied."""
    #
    #calling command:
    #ch9_fig4(100.0, 60.0, 10.0, 0.025, 200, 1024)
    #
    from math import pi
    import numpy as np
    import matplotlib.pyplot as plt
    from mpl_toolkits.axes_grid1 import host_subplot
    #
    N = ncycle*n
    delt = 1.0/(freq*n)
    #
    nalpha1 = 89
    nalpha2 = 155
    nalpha3 = 197
    #
    nbeta1 = 60
    nbeta2 = 81
    nbeta3 = 123
    nbeta4 = 162
    nbeta5 = 191
    nbeta6 = 244
    #
    dteta = 2.0*pi/n
    gls = delt/ls
    grs = 1.0/rs
    #
    vout_a = []
    cindls_a = []
    vout1, cindls1 = ch9_fig4_solver(es, grs, gls, nalpha1, ncycle, n)
    vout2, cindls2 = ch9_fig4_solver(es, grs, gls, nalpha2, ncycle, n)
    vout3, cindls3 = ch9_fig4_solver(es, grs, gls, nalpha3, ncycle, n)
```

```
#
for j in range(N):
    vsum = vout1[j] - vout2[j] + vout3[j]
    csum = cindls1[j] - cindls2[j] + cindls3[j]
    vout_a.append(vsum)
    cindls_a.append(csum)
#
vout_b = []
cindls_b = []
vout1, cindls1 = ch9_fig4_solver(es, grs, gls, nbeta1, ncycle, n)
vout2, cindls2 = ch9_fig4_solver(es, grs, gls, nbeta2, ncycle, n)
vout3, cindls3 = ch9_fig4_solver(es, grs, gls, nbeta3, ncycle, n)
vout4, cindls4 = ch9_fig4_solver(es, grs, gls, nbeta4, ncycle, n)
vout5, cindls5 = ch9_fig4_solver(es, grs, gls, nbeta5, ncycle, n)
vout6, cindls6 = ch9_fig4_solver(es, grs, gls, nbeta6, ncycle, n)
#
for j in range(N):
    vsum = vout1[j] - vout2[j] + vout3[j] - vout4[j] +          \
                                              vout5[j] - vout6[j]
    csum = cindls1[j] - cindls2[j] + cindls3[j] - cindls4[j] +   \
                                          cindls5[j] - cindls6[j]
    vout_b.append(vsum)
    cindls_b.append(csum)
#
m0 = 2*(ncycle - 2)
m1 = 2*(ncycle - 1)
m2 = 2*ncycle
ang = dteta*np.arange(N)
#
ax1 = host_subplot(2, 1, 1)
ax2 = ax1.twinx()
ax1.set_xlim([m0*pi, m2*pi])
ax1.set_xticks([m0*pi, m1*pi, m2*pi])
ax1.set_xticklabels(['$0$', '$2\pi$', '$4\pi$'])
ax1.set_ylim([-110, 110])
ax1.set_yticks([-100, 0, 100])
ax1.set_yticklabels(['$-100$', '$0$', '$100$'])
ax1.set_ylabel('$Voltage, V$')
ax1.text(m1*pi, -190, '$(a)$', size=10)
ax2.set_ylim(-10, 10)
ax2.set_yticks([-10, 10])
ax2.set_yticklabels([ '$-10$', '$10$'])
ax2.set_ylabel('$Current, A$')
p1, = ax1.plot(ang[N-2*n:N-1], vout_a[N-2*n:N-1], 'r',            \
                                             label = '$V_{ac}$')
p2, = ax2.plot(ang[N-2*n:N-1], cindls_a[N-2*n:N-1], 'b',          \
                                               label='$I_{ls}$')
ax1.yaxis.get_label().set_color(p1.get_color())
ax2.yaxis.get_label().set_color(p2.get_color())
ax1.legend(bbox_to_anchor=(0., 1.02, 1., .102), loc=3, ncol=2,    \
                             mode="expand", borderaxespad=0.)
ax1.grid(True)
plt.subplots_adjust(hspace=0.8)
```

```
#
ax1 = host_subplot(2, 1, 2)
ax2 = ax1.twinx()
ax1.set_xlim([m0*pi, m2*pi])
ax1.set_xticks([m0*pi, m1*pi, m2*pi])
ax1.set_xticklabels(['$0$', '$2\pi$', '$4\pi$'])
ax1.set_ylim([-110, 110])
ax1.set_yticks([-100, 0, 100])
ax1.set_yticklabels(['$-100$', '$0$', '$100$'])
ax1.set_ylabel('$Voltage, V$')
ax1.text(m1*pi, -190, '$(b)$', size=10)
ax2.set_ylim(-10, 10)
ax2.set_yticks([-10, 10])
ax2.set_yticklabels([ '$-10$', '$10$'])
ax2.set_ylabel('$Current, A$')
p1, = ax1.plot(ang[N-2*n:N-1], vout_b[N-2*n:N-1], 'r',          \
                                          label = '$V_{ac}$')
p2, = ax2.plot(ang[N-2*n:N-1], cindls_b[N-2*n:N-1], 'b',        \
                                          label = '$I_{ls}$')
ax1.yaxis.get_label().set_color(p1.get_color())
ax2.yaxis.get_label().set_color(p2.get_color())
ax1.legend(bbox_to_anchor=(0., 1.02, 1., .102), loc=3, ncol=2,  \
                              mode="expand", borderaxespad=0.)
ax1.grid(True)
#
plt.show()
#plt.savefig('Fig904.ps', dpi=400, bbox_inches='tight')
#
```

9.4.3 Program 9.05

```
#
def ch9_sphase_pwm_table1(ma, ns, mf):
    """single-phase, pwm, bipolar switching;
       table of amplitudes;
       ma=amplitude modulation ratio (0<ma<=1);
       ns=time-steps per period,
       mf=no. of carrier periods per period of control signal"""
    #
    #calling command
    #ch9_sphase_pwm_table1(0.8, 1024, 33)
    #
    from math import pi
    from scipy.fftpack import rfft
    import numpy as np
    import matplotlib.pyplot as plt
    from mpl_toolkits.axes_grid1 import host_subplot
    #
    N = ns*mf
    dteta = 2.0*pi/N
    #
    vtri = []
```

```
ns4 = int(ns/4)
dvtr = 1.0/ns4
for j in range(ns4):
    vm = j*dvtr
    vtri.append(-vm)
for j in range(ns4):
    vm = j*dvtr
    vtri.append(vm - 1.0)
for j in range(ns4):
    vm = j*dvtr
    vtri.append(vm)
for j in range(ns4):
    vm = j*dvtr
    vtri.append(1.0 - vm)
#
vout = []
vcarrier = []
vcontr = []
ang = []
for k in range(mf):
    nk = k*ns
    #
    for j in range(ns):
        teta = dteta*(nk + j)
        vj = vtri[j]
        ang.append(teta)
        vcarrier.append(vj)
        vctr = ma*np.sin(teta)
        vcontr.append(vctr)
        if vctr >= vj:
            vout.append(1.0)
        else:
            vout.append(-1.0)
        #
    #
#
Fv = (2/N)*np.abs(rfft(vout))
#
fvmod =[]
for j in range(ns):
    ks = 2*j
    kc = ks - 1
    fvmod.append(np.sqrt(Fv[kc]**2 + Fv[ks]**2))
#
print('fvmod[1]          ', round(fvmod[1], 4))
#
print('')
#
print('fvmod[mf]         ', round(fvmod[mf], 4))
print('fvmod[mf + 2]     ', round(fvmod[mf + 2], 4))
print('fvmod[mf + 4]     ', round(fvmod[mf + 4], 4))
print('fvmod[mf + 6]     ', round(fvmod[mf + 6], 4))
#
```

```
print('')
#
print('fvmod[2*mf + 1]  ', round(fvmod[2*mf + 1], 4))
print('fvmod[2*mf + 3]  ', round(fvmod[2*mf + 3], 4))
print('fvmod[2*mf + 5]  ', round(fvmod[2*mf + 5], 4))
print('fvmod[2*mf + 7]  ', round(fvmod[2*mf + 7], 4))
#
print('')
#
print('fvmod[3*mf]      ', round(fvmod[3*mf], 4))
print('fvmod[3*mf + 2]  ', round(fvmod[3*mf + 2], 4))
print('fvmod[3*mf + 4]  ', round(fvmod[3*mf + 4], 4))
print('fvmod[3*mf + 6]  ', round(fvmod[3*mf + 6], 4))
#
print('')
#
print('fvmod[4*mf + 1]  ', round(fvmod[4*mf + 1], 4))
print('fvmod[4*mf + 3]  ', round(fvmod[4*mf + 3], 4))
print('fvmod[4*mf + 5]  ', round(fvmod[4*mf + 5], 4))
print('fvmod[4*mf + 7]  ', round(fvmod[4*mf + 7], 4))
#
ax1 = host_subplot(2, 1, 1)
ax1.set_xlim([0, 2*pi])
ax1.set_xticks([0, pi, 2*pi])
ax1.set_xticklabels(['$0$', '$\pi$', '$2\pi$'])
ax1.set_ylim([-1.1, 1.1])
ax1.set_yticks([-1.0, 0, 1.0])
ax1.set_yticklabels(['$-1$', '$0$', '$1$'])
ax1.set_ylabel('$Voltage, V$')
ax1.text(pi, -1.9, '$(a)$', size = 10)
p1, = ax1.plot(ang, vcarrier, 'r', label = '$V_{\rm tri}$')
p2, = ax1.plot(ang, vcontr, 'b--', label='$V_{contr}$')
ax1.yaxis.get_label().set_color(p1.get_color())
ax1.legend(bbox_to_anchor=(0., 1.02, 1., .102), loc=3, ncol=2,     \
                                    mode="expand", borderaxespad=0.)
ax1.grid(True)
plt.subplots_adjust(hspace = 0.9)
#
ax1 = host_subplot(2, 1, 2)
ax1.set_xlim([0, 2*pi])
ax1.set_xticks([0, pi, 2*pi])
ax1.set_xticklabels(['$0$', '$\pi$', '$2\pi$'])
ax1.set_ylim([-1.1, 1.1])
ax1.set_yticks([-1.0, 0, 1.0])
ax1.set_yticklabels(['$-1$', '$0$', '$1$'])
ax1.set_ylabel('$Voltage, V$')
ax1.text(pi, -1.9, '$(b)$', size = 10)
p1, = ax1.plot(ang, vout, 'r', label = '$V_{0}$')
p2, = ax1.plot(ang, vcontr, 'b--', label='$V_{contr}$')
ax1.yaxis.get_label().set_color(p1.get_color())
ax1.legend(bbox_to_anchor=(0., 1.02, 1., .102), loc=3, ncol=2,     \
                                    mode="expand", borderaxespad=0.)
ax1.grid(True)
```

```
    #
    plt.show()
    #plt.savefig('Fig905.ps', dpi=400, bbox_inches='tight')
    #
```

9.4.4 Program 9.07

```
#
def fdiode(vd):
    """ Diode characteristic.
        Returns diode current and its conductance"""
    #
    glow = 0.1e-07
    ghigh = 0.4e+02
    Vr = -0.5
    #
    if vd >= Vr:
        id = glow*(vd - Vr)
        gd = glow
    else:
        id = ghigh*(vd - Vr)
        gd = ghigh
    #
    return id, gd
#
#
def fswitch(vsw, gsw):
    """Bi-directional switch characteristic.
       Returns switch current and conductance."""
    #
    glow = 0.1e-07
    ghigh = 0.4e+02
    Vr = 0.5
    #
    ida, gda = fdiode(vsw)
    #
    if vsw >= Vr:
        vd = (gsw*vsw + ghigh*Vr)/(ghigh + gsw)
        iw = ghigh*(vd -Vr)
        gw = gda + ghigh*gsw/(ghigh + gsw)
    else:
        vd = (gsw*vsw + glow*Vr)/(glow + gsw)
        iw = glow*(vd - Vr)
        gw = gda + glow*gsw/(glow + gsw)
    #
    isw = iw + ida
    #
    return isw, gw
#
#
def ch9_fig7_bi_solver(es, dteta, grs, gls, ma, ns, nm):
    """Assumes fdiode is a function.
```

```
    Returns the lists vout, cinduct"""
#
import numpy as np
#
ghigh = 40.0
glow = 0.1e-07
irlim = 0.1e-06
#
gtot = grs*gls/(grs + gls)
pils = -5.48
#
va = -18.4
#
vtri = []
ns4 = int(ns/4)
dvtr = 1.0/ns4
for j in range(ns4):
    vm = j*dvtr
    vtri.append(-vm)
for j in range(ns4):
    vm = j*dvtr
    vtri.append(vm - 1.0)
for j in range(ns4):
    vm = j*dvtr
    vtri.append(vm)
for j in range(ns4):
    vm = j*dvtr
    vtri.append(1.0 - vm)
#
vload = []
cload = []
cdc = []
#
for k in range(nm):
    #
    nk = ns*k
    #
    for j in range(ns):
        #
        teta = dteta*(nk + j)
        vcontr = ma*np.sin(teta)
        #
        vtj = vtri[j]
        if vcontr >= vtj:
            gsa = ghigh
            gsb = glow
        else:
            gsa = glow
            gsb = ghigh
        #
        for its in range(1000):
            #
            vsw1 = 0.5*(es - va)
```

```
                vsw2 = 0.5*(es + va)
                isw1, gsw1 = fswitch(vsw1, gsa)
                isw2, gsw2 = fswitch(vsw2, gsb)
                #
                vm = (grs*va - pils)/(grs + gls)
                ils = grs*(va - vm)
                irra = isw1 - isw2 - ils
                #
                err = np.abs(irra)
                #
                if err < irlim:
                    break
                else:
                    #
                    dva = irra/(0.5*(gsw1 + gsw2) + gtot)
                    #
                    va += dva
                    #
                #
            #
            if its >= 998:
                print('time-step= ', (k*ns + j), 'j = ', j, 'k = ',      \
                                          k, 'solver does not converge')
                break
            #
            pils = ils
            #
            vload.append(va)
            cload.append(ils)
            cdc.append(isw1 + isw2)
            #
        #
    #
    return vload, cload, cdc
#
#
def ch9_fig7_bi(es, freq, rs, ls, ma, ns, mf, ncycle):
    """ Main program, single-phase, full-bridge inverter;
        bipolar switching."""
    #
    #calling command:
    #ch9_fig7_bi(100.0, 60.0, 1.0, 0.005, 0.8, 16, 21, 40)
    #
    from math import pi
    import numpy as np
    from scipy.fftpack import rfft
    import matplotlib.pyplot as plt
    from mpl_toolkits.axes_grid1 import host_subplot
    #
    n = ns*mf
    nm = mf*ncycle
    N = ncycle*n
    delt = 1.0/(freq*n)
```

```
dteta = 2.0*pi/n
ang = dteta*np.arange(N)
#
grs = 1.0/rs
gls = delt/ls
#
vload, cload, cdc = ch9_fig7_bi_solver(es, dteta,                 \
                                        grs, gls, ma, ns, nm)
#
vout = vload[N-n:N]
Fv = (2/n)*rfft(vout)
vc = Fv[1]
vs = -Fv[2]
V01 = np.sqrt(vc**2 + vs**2)
vf1 = []
for j in range(N):
    teta = ang[j]
    vf1.append(vc*np.cos(teta) + vs*np.sin(teta))
#
Fcs = (2/n)*rfft(cdc[N-n:N])
Is = Fcs[0]/2.0
Is2 = np.sqrt(Fcs[3]**2 + Fcs[4]**2)
#
Fld = (2/n)*rfft(cload[N-n:N])
ics = Fld[1]
isn = -Fld[2]
Ild = np.sqrt(ics**2 + isn**2)
pf = isn/Ild
pwr = V01*Ild*pf/2.0
#
print('amplitude of V01 = ', V01)
print('I_{01} = ', Ild)
print('I_{sd} = ', Is, 'Is2 = ', Is2)
print('power factor = ', pf, 'output power = ', pwr)
#
m0 = 2*(ncycle - 2)
m1 = 2*(ncycle - 1)
m2 = 2*ncycle
#
ax1 = host_subplot(2, 1, 1)
ax2 = ax1.twinx()
ax1.set_xlim([m0*pi, m2*pi])
ax1.set_xticks([m0*pi, m1*pi, m2*pi])
ax1.set_xticklabels(['$0$', '$2\pi$', '$4\pi$'])
ax1.set_ylim([-120, 120])
ax1.set_yticks([-120, 0, 120])
ax1.set_yticklabels(['$-120$', '$0$', '$120$'])
ax1.set_ylabel('$Voltage, V$')
ax1.text(m1*pi, -205, '$(a)$', size = 10)
ax2.set_ylim(-100, 100)
ax2.set_yticks([-100, 100])
ax2.set_yticklabels([ '$-100$', '$100$'])
ax2.set_ylabel('$Current, A$')
```

```
p1, = ax1.plot(ang[N-2*n:N-1], vload[N-2*n:N-1], 'g',              \
                                          label = '$V_{load}$')
p2, = ax2.plot(ang[N-2*n:N-1], cload[N-2*n:N-1], 'b',              \
                                          label = '$I_{load}$')
ax1.yaxis.get_label().set_color(p1.get_color())
ax2.yaxis.get_label().set_color(p2.get_color())
ax1.legend(bbox_to_anchor=(0., 1.02, 1., .102), loc=3, ncol=2,     \
                            mode="expand", borderaxespad=0.)
ax1.grid(True)
plt.subplots_adjust(hspace = 0.8)
#
ax1 = host_subplot(2, 1, 2)
ax2 = ax1.twinx()
ax1.set_xlim([m0*pi, m2*pi])
ax1.set_xticks([m0*pi, m1*pi, m2*pi])
ax1.set_xticklabels(['$0$', '$2\pi$', '$4\pi$'])
ax1.set_ylim([-120, 120])
ax1.set_yticks([-120, 0, 120])
ax1.set_yticklabels(['$-120$', '$0$', '$120$'])
ax1.set_ylabel('$Voltage, V$')
ax1.text(m1*pi, -205, '$(b)$', size = 10)
ax2.set_ylim(-100, 100)
ax2.set_yticks([-100, 100])
ax2.set_yticklabels([ '$-100$', '$100$'])
ax2.set_ylabel('$Current, A$')
p1, = ax1.plot(ang[N-2*n:N-1], vf1[N-2*n:N-1], 'r--',              \
                                              label = '$V_{1}$')
p2, = ax2.plot(ang[N-2*n:N-1], cdc[N-2*n:N-1], 'b',                \
                                             label = '$I_{DC}$')
ax1.yaxis.get_label().set_color(p1.get_color())
ax2.yaxis.get_label().set_color(p2.get_color())
ax1.legend(bbox_to_anchor=(0., 1.02, 1., .102), loc=3, ncol=2,     \
                            mode="expand", borderaxespad=0.)
ax1.grid(True)
#
plt.show()
#plt.savefig('Fig907.ps', dpi=400, bbox_inches='tight')
#
```

9.4.5 Program 9.08

```
#
def fdiode(vd):
    """ Diode characteristic.
        Returns diode current and its conductance"""
    #
    glow = 0.1e-07
    ghigh = 0.4e+02
    Vr = -0.5
    #
    if vd >= Vr:
        id = glow*(vd - Vr)
```

```
            gd = glow
        else:
            id = ghigh*(vd - Vr)
            gd = ghigh
        #
        return id, gd
#
#
def fswitch(vsw, gsw):
    """Bi-directional switch characteristic.
       Returns switch current and conductance."""
    #
    glow = 0.1e-07
    ghigh = 40.0
    Vr = 0.5
    #
    ida, gda = fdiode(vsw)
    #
    if vsw >= Vr:
        vd = (gsw*vsw + ghigh*Vr)/(ghigh + gsw)
        iw = ghigh*(vd -Vr)
        gw = gda + ghigh*gsw/(ghigh + gsw)
    else:
        vd = (gsw*vsw + glow*Vr)/(glow + gsw)
        iw = glow*(vd - Vr)
        gw = gda + glow*gsw/(glow + gsw)
    #
    isw = iw + ida
    #
    return isw, gw
#
#
def ch9_fig8_uni_solver(es, dteta, grs, gls, ma, ns, nm):
    """Assumes fdiode is a function.
       Returns the lists vout, cinduct"""
    #
    import numpy as np
    #
    ghigh = 40.0
    glow = 0.1e-07
    irlim = 0.1e-06
    #
    gtot = grs*gls/(grs + gls)
    pils = -32.253
    #
    va = 3.42
    vb = -1.306
    #
    vtri = []
    ns4 = int(ns/4)
    dvtr = 1.0/ns4
    for j in range(ns4):
        vm = j*dvtr
```

```
    vtri.append(-vm)
for j in range(ns4):
    vm = j*dvtr
    vtri.append(vm - 1.0)
for j in range(ns4):
    vm = j*dvtr
    vtri.append(vm)
for j in range(ns4):
    vm = j*dvtr
    vtri.append(1.0 - vm)
#
vload = []
cload = []
cdc = []
#
for k in range(nm):
    #
    nk = ns*k
    #
    for j in range(ns):
        #
        teta = dteta*(nk + j)
        vcontr = ma*np.sin(teta)
        #
        vtj = vtri[j]
        if vcontr >= vtj:
            gs1 = ghigh
            gs4 = glow
        else:
            gs1 = glow
            gs4 = ghigh
        #
        if -vcontr >= vtj:
            gs2 = ghigh
            gs3 = glow
        else:
            gs2 = glow
            gs3 = ghigh
        #
        for its in range(1000):
            #
            vsw4 = va + vb
            vsw1 = es - vsw4
            vsw2 = es - vb
            isw1, gsw1 = fswitch(vsw1, gs1)
            isw2, gsw2 = fswitch(vsw2, gs2)
            isw3, gsw3 = fswitch(vb, gs3)
            isw4, gsw4 = fswitch(vsw4, gs4)
            #
            vm = (grs*va - pils)/(grs + gls)
            ils = grs*(va - vm)
            irra = isw1 - isw4 - ils
            irrb = (isw1 + isw2) - (isw3 + isw4)
```

```
                #
                err = np.sqrt(irra**2 + irrb**2)
                #
                if err < irlim:
                    break
                else:
                    #
                    gs14 = gsw1 + gsw4
                    gs23 = gsw2 + gsw3
                    gmat = np.array([ [(gs14 + gtot), gs14],
                                      [gs14, (gs14 + gs23)], ])
                    #
                    bcol = np.array([irra, irrb])
                    xdv = np.linalg.solve(gmat, bcol)
                    #
                    dva = xdv[0]
                    dvb = xdv[1]
                    va += dva
                    vb += dvb
                    #
                #
            #
            if its >= 998:
                print('time-step= ', (k*ns + j), 'j = ', j, 'k = ',     \
                                   k, 'solver does not converge')
                break
            #
            pils = ils
            #
            vload.append(va)
            cload.append(ils)
            cdc.append(isw1 + isw2)
            #
        #
    #
    return vload, cload, cdc
#
#
def ch9_fig8_uni(es, freq, rs, ls, ma, ns, mf, ncycle):
    """ Main program, single-phase, full-bridge inverter;
        unipolar switching."""
    #
    #calling command:
    #ch9_fig8_uni(100.0, 60.0, 1.0, 0.005, 0.8, 16, 24, 40)
    #
    from math import pi
    import numpy as np
    from scipy.fftpack import rfft
    import matplotlib.pyplot as plt
    from mpl_toolkits.axes_grid1 import host_subplot
    #
    n = ns*mf
    nm = mf*ncycle
```

```
N = ncycle*n
delt = 1.0/(freq*n)
dteta = 2.0*pi/n
ang = dteta*np.arange(N)
#
grs = 1.0/rs
gls = delt/ls
#
vload, cload, cdc = ch9_fig8_uni_solver(es, dteta, grs,      \
                                        gls, ma, ns, nm)
#
vout = vload[N-n:N]
Fv = (2/n)*rfft(vout)
vc = Fv[1]
vs = -Fv[2]
vf1 = []
for j in range(N):
    teta = ang[j]
    vf1.append(vc*np.cos(teta) + vs*np.sin(teta))
#
m0 = 2*(ncycle - 2)
m1 = 2*(ncycle - 1)
m2 = 2*ncycle
#
ax1 = host_subplot(2, 1, 1)
ax2 = ax1.twinx()
ax1.set_xlim([m0*pi, m2*pi])
ax1.set_xticks([m0*pi, m1*pi, m2*pi])
ax1.set_xticklabels(['$0$', '$2\pi$', '$4\pi$'])
ax1.set_ylim([-120, 120])
ax1.set_yticks([-120, 0, 120])
ax1.set_yticklabels(['$-120$', '$0$', '$120$'])
ax1.set_ylabel('$Voltage, V$')
ax1.text(m1*pi, -205, '$(a)$', size = 10)
ax2.set_ylim(-100, 100)
ax2.set_yticks([-100, 100])
ax2.set_yticklabels([ '$-100$', '$100$'])
ax2.set_ylabel('$Current, A$')
p1, = ax1.plot(ang[N-2*n:N-1], vload[N-2*n:N-1], 'g',        \
                                   label = '$V_{load}$')
p2, = ax2.plot(ang[N-2*n:N-1], cload[N-2*n:N-1], 'b',        \
                                   label = '$I_{load}$')
ax1.yaxis.get_label().set_color(p1.get_color())
ax2.yaxis.get_label().set_color(p2.get_color())
ax1.legend(bbox_to_anchor=(0., 1.02, 1., .102), loc=3, ncol=2,  \
                         mode="expand", borderaxespad=0.)
ax1.grid(True)
plt.subplots_adjust(hspace = 0.8)
#
ax1 = host_subplot(2, 1, 2)
ax2 = ax1.twinx()
ax1.set_xlim([m0*pi, m2*pi])
ax1.set_xticks([m0*pi, m1*pi, m2*pi])
```

```
ax1.set_xticklabels(['$0$', '$2\pi$', '$4\pi$'])
ax1.set_ylim([-120, 120])
ax1.set_yticks([-120, 0, 120])
ax1.set_yticklabels(['$-120$', '$0$', '$120$'])
ax1.set_ylabel('$Voltage, V$')
ax1.text(m1*pi, -205, '$(b)$', size = 10)
ax2.set_ylim(-100, 100)
ax2.set_yticks([-100, 100])
ax2.set_yticklabels([ '$-100$', '$100$'])
ax2.set_ylabel('$Current, A$')
p1, = ax1.plot(ang[N-2*n:N-1], vf1[N-2*n:N-1], 'r--',          \
                                            label = '$V_{1}$')
p2, = ax2.plot(ang[N-2*n:N-1], cdc[N-2*n:N-1], 'b',             \
                                           label = '$I_{DC}$')
ax1.yaxis.get_label().set_color(p1.get_color())
ax2.yaxis.get_label().set_color(p2.get_color())
ax1.legend(bbox_to_anchor=(0., 1.02, 1., .102), loc=3, ncol=2,    \
                             mode="expand", borderaxespad=0.)
ax1.grid(True)
#
plt.show()
#plt.savefig('Fig908.ps', dpi=400, bbox_inches='tight')
#
```

9.4.6 Program 9.13

```
#
def fdiode(vd):
    """ Diode characteristic.
        Returns diode current and its conductance"""
    #
    glow = 0.1e-07
    ghigh = 0.4e+02
    Vr = -0.5
    #
    if vd >= Vr:
        id = glow*(vd - Vr)
        gd = glow
    else:
        id = ghigh*(vd - Vr)
        gd = ghigh
    #
    return id, gd
#
#
def fswitch(vsw, gsw):
    """Bi-directional switch characteristic.
       Returns switch current and conductance."""
    #
    glow = 0.1e-07
    ghigh = 0.4e+02
    Vr = 0.5
```

```
    #
    ida, gda = fdiode(vsw)
    #
    if vsw >= Vr:
        vd = (gsw*vsw + ghigh*Vr)/(ghigh + gsw)
        iw = ghigh*(vd -Vr)
        gw = gda + ghigh*gsw/(ghigh + gsw)
    else:
        vd = (gsw*vsw + glow*Vr)/(glow + gsw)
        iw = glow*(vd - Vr)
        gw = gda + glow*gsw/(glow + gsw)
    #
    isw = iw + ida
    #
    return isw, gw
#
#
def ch9_fig13_sqwave_solver(es, Vtmax, dteta, grs, gls, ncycle, n):
    """Assumes fdiode is a function.
       Returns the lists vout, cinduct"""
    #
    from math import pi
    import numpy as np
    #
    ghigh = 40.0
    glow = 0.1e-07
    irlim = 0.1e-06
    phi = 2.0*pi/3.0
    phse = pi*12.4/180.0
    #
    gser = grs*gls/(grs + gls)
    pila = -22.5
    pilb = -2.65
    #
    vab = 320.43
    vbc = -317.61
    vn = 318.24
    #
    vpha = []
    vlab = []
    vlbc = []
    cpha = []
    cphb = []
    cdc = []
    #
    for k in range(ncycle):
        #
        nk = n*k
        #
        for j in range(n):
            #
            if j >= 1 and j < 359:
                gs1 = ghigh
```

```
    else:
        gs1 = glow
    #
    if j >= 241 and j < 599:
        gs3 = ghigh
    else:
        gs3 = glow
    #
    if j >= 119 and j < 481:
        gs5 = glow
    else:
        gs5 = ghigh
    #
    if j >= 361 and j < 719:
        gs4 = ghigh
    else:
        gs4 = glow
    #
    if j >= 239 and j < 601:
        gs6 = glow
    else:
        gs6 = ghigh
    #
    if j >= 121 and j < 479:
        gs2 = ghigh
    else:
        gs2 = glow
    #
    tetA = dteta*(nk + j)
    tetB = tetA - phi
    tetC = tetA + phi
    vthA = Vtmax*np.sin(tetA - phse)
    vthB = Vtmax*np.sin(tetB - phse)
    vthC = Vtmax*np.sin(tetC - phse)
    #
    for its in range(1000):
        #
        vsw4 = vab + vbc + vn
        vsw6 = vbc + vn
        vsw2 = vn
        vsw1 = es - vsw4
        vsw3 = es - vsw6
        vsw5 = es - vsw2
        isw1, gsw1 = fswitch(vsw1, gs1)
        isw2, gsw2 = fswitch(vsw2, gs2)
        isw3, gsw3 = fswitch(vsw3, gs3)
        isw4, gsw4 = fswitch(vsw4, gs4)
        isw5, gsw5 = fswitch(vsw5, gs5)
        isw6, gsw6 = fswitch(vsw6, gs6)
        #
        Ia = gser*(vab - vthA + vthB + (pila - pilb)/gls)
        Ib = gser*(vbc - vthB + vthC + (pila + 2.0*pilb)/gls)
        #
```

```
            ilsb = (Ib - Ia)/3.0
            ilsa = Ia + ilsb
            ilsc = -(ilsa + ilsb)
            van = vthA + ilsa/gser - pila/gls
            irrA = isw1 - isw4 - ilsa
            irrB = isw3 - isw6 - ilsb
            irrC = isw5 - isw2 - ilsc
            #
            err = np.sqrt(irrA**2 + irrB**2 + irrC**2)
            #
            if err < irlim:
                break
            else:
                #
                g11 = gsw1 + gsw4 + 2.0*gser/3.0
                g12 = gsw1 + gsw4 + gser/3.0
                g13 = gsw1 + gsw4
                g22 = gsw3 + gsw6 + gser/3.0
                g23 = gsw3 + gsw6
                g33 = gsw5 + gsw2
                gmat = np.array([ [g11, g12, g13],
                                  [-gser/3.0, g22, g23],
                                  [-gser/3.0, -2.0*gser/3.0, g33] ])
                #
                bcol = np.array([irrA, irrB, irrC])
                xdv = np.linalg.solve(gmat, bcol)
                #
                dvab = xdv[0]
                dvbc = xdv[1]
                dvn = xdv[2]
                vab += dvab
                vbc += dvbc
                vn += dvn
                #
            #
        #
        if its >= 998:
            print('time-step= ', (nk + j), 'j = ', j, 'k = ',      \
                                         k, 'solver does not converge')
            break
        #
        pila = ilsa
        pilb = ilsb
        #
        vpha.append(van)
        vlab.append(vab)
        vlbc.append(vbc)
        cpha.append(ilsa)
        cphb.append(ilsb)
        cdc.append(isw1 + isw3 + isw5)
        #
    #
#
```

```
    return vpha, vlab, vlbc, cpha, cphb, cdc
#
#
def ch9_fig13_sqwave(es, Vtmax, freq, rs, ls, ncycle, n):
    """ Main program;
        3 phase, square wave inverter."""
    #
    #calling command:
    #ch9_fig13_sqwave(320.0, 106.0, 60.0, 2.0, 0.01, 50, 720)
    #
    from math import pi
    import numpy as np
    from scipy.fftpack import rfft
    import matplotlib.pyplot as plt
    from mpl_toolkits.axes_grid1 import host_subplot
    #
    N = ncycle*n
    delt = 1.0/(freq*n)
    dteta = 2.0*pi/n
    #
    grs = 1.0/rs
    gls = delt/ls
    #
    vpha, vlab, vlbc, cpha, cphb, cdc = ch9_fig13_sqwave_solver(es,    \
                                        Vtmax, dteta, grs, gls, ncycle, n)
    #
    Fc = (2.0/n)*rfft(cpha[N-n:N])
    fcmod = []
    isumsq = 0.0
    Fv = (2.0/n)*rfft(vpha[N-n:N])
    fvmod = []
    vsumsq = 0.0
    for j in range(61):
        jp = j + 1
        ks = 2*jp
        kc = ks - 1
        hcurr = round(np.sqrt(Fc[kc]**2 + Fc[ks]**2), 4)
        isumsq = isumsq + hcurr**2
        hvolt = round(np.sqrt(Fv[kc]**2 + Fv[ks]**2), 4)
        vsumsq = vsumsq + hvolt**2
        if jp == 1:
            hc1 = hcurr
            hv1 = hvolt
        fcmod.append(hcurr)
        fvmod.append(hvolt)
    #
    xthd = round((np.sqrt((isumsq - hc1**2)/hc1**2)), 4)
    thd = round((100*xthd), 4)
    print('THD of phase current = ', thd, '%')
    #
    ythd = round((np.sqrt((vsumsq - hv1**2)/hv1**2)), 4)
    vthd = round((100*ythd), 4)
    print('THD of phase voltage = ', vthd, '%')
```

```
#
m0 = 2*(ncycle - 2)
m1 = 2*(ncycle - 1)
m2 = 2*ncycle
ang = dteta*np.arange(N)
#
ax1 = host_subplot(2, 1, 1)
ax1.set_xlim([m0*pi, m2*pi])
ax1.set_xticks([m0*pi, m1*pi, m2*pi])
ax1.set_xticklabels(['$0$', '$2\pi$', '$4\pi$'])
ax1.set_ylim([-410, 410])
ax1.set_yticks([-400, 0, 400])
ax1.set_yticklabels(['$-400$', '$0$', '$400$'])
ax1.set_ylabel('$Voltage, V$')
ax1.text(m1*pi, -720,'$(a)$', size=10)
p1, = ax1.plot(ang[N-2*n:N-1], vlab[N-2*n:N-1], 'r--',          \
                                            label = '$V_{ab}$')
p2, = ax1.plot(ang[N-2*n:N-1], vlbc[N-2*n:N-1], 'b--',          \
                                            label = '$V_{bc}$')
p3, = ax1.plot(ang[N-2*n:N-1], vpha[N-2*n:N-1], 'r',            \
                                             label = '$V_{a}$')
ax1.yaxis.get_label().set_color(p1.get_color())
ax1.legend(bbox_to_anchor=(0., 1.02, 1., .102), loc=3, ncol=3, \
                            mode="expand", borderaxespad=0.)
ax1.grid(True)
plt.subplots_adjust(hspace=0.9)
#
ax1 = host_subplot(2, 1, 2)
ax2 = ax1.twinx()
ax1.set_xlim([m0*pi, m2*pi])
ax1.set_xticks([m0*pi, m1*pi, m2*pi])
ax1.set_xticklabels(['$0$', '$2\pi$', '$4\pi$'])
ax1.set_ylim([-410, 410])
ax1.set_yticks([-400, 0, 400])
ax1.set_yticklabels(['$-400$', '$0$', '$400$'])
ax1.set_ylabel('$Voltage, V$')
ax1.text(m1*pi, -720,'$(b)$', size=10)
ax2.set_ylim(-50, 50)
ax2.set_yticks([-50, 50])
ax2.set_yticklabels([ '$-50$', '$50$'])
ax2.set_ylabel('$Current, A$')
p1, = ax1.plot(ang[N-2*n:N-1], vlab[N-2*n:N-1], 'm',            \
                                            label = '$V_{ab}$')
p2, = ax2.plot(ang[N-2*n:N-1], cpha[N-2*n:N-1], 'r--',          \
                                              label = '$I_{a}$')
p3, = ax2.plot(ang[N-2*n:N-1], cphb[N-2*n:N-1], 'b--',          \
                                              label = '$I_{b}$')
p4, = ax2.plot(ang[N-2*n:N-1], cdc[N-2*n:N-1], 'g--',           \
                                             label = '$I_{DC}$')
ax1.yaxis.get_label().set_color(p1.get_color())
ax2.yaxis.get_label().set_color(p2.get_color())
ax1.legend(bbox_to_anchor=(0., 1.02, 1., .102), loc=3, ncol=4,  \
                            mode="expand", borderaxespad=0.)
```

```
    ax1.grid(True)
    #
    plt.show()
    #plt.savefig('Fig913.ps', dpi=400, bbox_inches='tight')
    #
```

9.4.7 Program 9.14

```
#
def fdiode(vd):
    """ Diode characteristic.
        Returns diode current and its conductance"""
    #
    glow = 0.1e-07
    ghigh = 0.4e+02
    Vr = -0.5
    #
    if vd >= Vr:
        id = glow*(vd - Vr)
        gd = glow
    else:
        id = ghigh*(vd - Vr)
        gd = ghigh
    #
    return id, gd
#
#
def fswitch(vsw, gsw):
    """Bi-directional switch characteristic.
       Returns switch current and conductance."""
    #
    glow = 0.1e-07
    ghigh = 0.4e+02
    Vr = 0.5
    #
    ida, gda = fdiode(vsw)
    #
    if vsw >= Vr:
        vd = (gsw*vsw + ghigh*Vr)/(ghigh + gsw)
        iw = ghigh*(vd -Vr)
        gw = gda + ghigh*gsw/(ghigh + gsw)
    else:
        vd = (gsw*vsw + glow*Vr)/(glow + gsw)
        iw = glow*(vd - Vr)
        gw = gda + glow*gsw/(glow + gsw)
    #
    isw = iw + ida
    #
    return isw, gw
#
#
def ch9_fig14_tri_solver(Es, Vtmax, dteta, dtm, grs, gls, ma, ns, nm):
```

```
"""Assumes fdiode is a function.
   Returns the lists vout, cinduct"""
#
from math import pi
import numpy as np
#
ghigh = 40.0
glow = 0.1e-07
irlim = 0.1e-06
phi = 2.0*pi/3.0
phse = pi*12.4/180.0
#
gtot = grs*gls/(grs + gls)
pila = -5.48
pilb = 4.0
pilc = -(pila + pilb)
#
vab = -18.4
vbc = 5.0
vn = 0.0
#
vtri = []
ns4 = int(ns/4)
dvtr = 1.0/ns4
for j in range(ns4):
    vm = j*dvtr
    vtri.append(-vm)
for j in range(ns4):
    vm = j*dvtr
    vtri.append(vm - 1.0)
for j in range(ns4):
    vm = j*dvtr
    vtri.append(vm)
for j in range(ns4):
    vm = j*dvtr
    vtri.append(1.0 - vm)
#
esab = []
vlab = []
vlbc = []
cpha = []
cphb = []
cphc = []
angnk = []
#
for k in range(nm):
    #
    nk = ns*k
    #
    sumab = []
    ank = dteta*(ns*(k + 0.5))
    angnk.append(ank)
    #
```

```
for j in range(ns):
    #
    tetA = dteta*(nk + j)
    tetB = tetA - phi
    tetC = tetA + phi
    vctrA = ma*np.sin(tetA)
    vctrB = ma*np.sin(tetB)
    vctrC = ma*np.sin(tetC)
    vthA = Vtmax*np.sin(tetA - phse)
    vthB = Vtmax*np.sin(tetB - phse)
    vthC = Vtmax*np.sin(tetC - phse)
    #
    vtj = vtri[j]
    if vctrA >= vtj:
        gs1 = ghigh
        gs2 = glow
    else:
        gs1 = glow
        gs2 = ghigh
    #
    if vctrB >= vtj:
        gs3 = ghigh
        gs4 = glow
    else:
        gs3 = glow
        gs4 = ghigh
    #
    if vctrC >= vtj:
        gs5 = ghigh
        gs6 = glow
    else:
        gs5 = glow
        gs6 = ghigh
    #
    for its in range(1000):
        #
        vsw2 = vab + vbc + vn
        vsw1 = Es - vsw2
        vsw4 = vbc + vn
        vsw3 = Es - vsw4
        vsw6 = vn
        vsw5 = Es - vsw6
        isw1, gsw1 = fswitch(vsw1, gs1)
        isw2, gsw2 = fswitch(vsw2, gs2)
        isw3, gsw3 = fswitch(vsw3, gs3)
        isw4, gsw4 = fswitch(vsw4, gs4)
        isw5, gsw5 = fswitch(vsw5, gs5)
        isw6, gsw6 = fswitch(vsw6, gs6)
        #
        Ia = gtot*(vab - vthA + vthB + (pila - pilb)/gls)
        Ib = gtot*(vbc - vthB + vthC + (pila + 2.0*pilb)/gls)
        #
        ilsb = (Ib - Ia)/3.0
```

```
            ilsa = Ia + ilsb
            ilsc = -(ilsa + ilsb)
            #
            irrA = isw1 - isw2 - ilsa
            irrB = isw3 - isw4 - ilsb
            irrC = isw5 - isw6 - ilsc
            #
            err = np.sqrt(irrA**2 + irrB**2 + irrC**2)
            #
            if err < irlim:
                break
            else:
                #
                g11 = gsw1 + gsw2 + 2.0*gtot/3.0
                g12 = gsw1 + gsw2 + gtot/3.0
                g13 = gsw1 + gsw2
                g22 = gsw3 + gsw4 + gtot/3.0
                g23 = gsw3 + gsw4
                g33 = gsw5 + gsw6
                gmat = np.array([ [g11, g12, g13 ],
                                  [-gtot/3.0, g22, g23 ],
                                  [-gtot/3.0, -2.0*gtot/3.0, g33] ])
                #
                bcol = np.array([irrA, irrB, irrC])
                xdv = np.linalg.solve(gmat, bcol)
                #
                dvab = xdv[0]
                dvbc = xdv[1]
                dvn = xdv[2]
                vab += dvab
                vbc += dvbc
                vn += dvn
                #
            #
        #
        if its >= 998:
            print('time-step= ', (k*ns + j), 'j = ', j, 'k = ',      \
                                        k, 'solver does not converge')
            break
        #
        pila = ilsa
        pilb = ilsb
        pilc = ilsc
        #
        vlab.append(vab)
        vlbc.append(vbc)
        cpha.append(ilsa)
        cphb.append(ilsb)
        cphc.append(ilsc)
        #
        sumab.append(vab)
        #
    #
```

```
        esab.append(np.mean(sumab))
        #
    #
    return esab, vlab, vlbc, cpha, cphb, cphc, angnk
#
#
def ch9_fig14_tri(Es, Vtmax, freq, rs, ls, ma, ns, mf, ncycle):
    """ Main program, plots results
        3-phase inverter;
        triangle comparison switching."""
    #
    #calling command:
    #ch9_fig14_tri(320.0, 106.0, 60.0, 2.0, 0.01, 0.9, 24, 321, 50)
    #
    from math import pi
    import numpy as np
    import matplotlib.pyplot as plt
    from mpl_toolkits.axes_grid1 import host_subplot
    #
    n = ns*mf
    nm = mf*ncycle
    N = ncycle*n
    delt = 1.0/(freq*n)
    dteta = 2.0*pi/n
    dtm = 2.0*pi/mf
    #
    grs = 1.0/rs
    gls = delt/ls
    #
    esab, vlab, vlbc, cpha, cphb, cphc, angnk = ch9_fig14_tri_solver(Es,  \
                                    Vtmax, dteta, dtm, grs, gls, ma, ns, nm)
    #
    print('peak value of cpha = ', np.max(cpha[N-2*n:N]))
    print('peak value of cphb = ', np.max(cphb[N-2*n:N]))
    print('peak value of cphc = ', np.max(cphc[N-2*n:N]))
    print('peak value of esab = ', np.max(esab[nm-2*mf:nm]))
    print('mean value of esab = ', np.mean(esab[nm-2*mf:nm]))
    print('minimum value of esab = ', np.min(esab[nm-2*mf:nm]))
    #
    m0 = 2*(ncycle - 2)
    m1 = 2*(ncycle - 1)
    m2 = 2*ncycle
    ang = dteta*np.arange(N)
    #
    ax1 = host_subplot(1, 1, 1)
    ax2 = ax1.twinx()
    ax1.set_xlim([m0*pi, m2*pi])
    ax1.set_xticks([m0*pi, m1*pi, m2*pi])
    ax1.set_xticklabels(['$0$', '$2\pi$', '$4\pi$'])
    ax1.set_ylim([-400, 400])
    ax1.set_yticks([-400, 0, 400])
    ax1.set_yticklabels(['$-400$', '$0$', '$400$'])
    ax1.set_ylabel('$Voltage, V$')
```

```
    ax2.set_ylim(-50, 50)
    ax2.set_yticks([-50, 50])
    ax2.set_yticklabels([ '$-50$', '$50$'])
    ax2.set_ylabel('$Current, A$')
    p1, = ax1.plot(ang[N-2*n:N], vlab[N-2*n:N], 'r',                  \
                                                  label = '$V_{ab}$')
    p2, = ax2.plot(ang[N-2*n:N], cpha[N-2*n:N], 'b--',                \
                                                  label = '$I_{a}$')
    p3, = ax2.plot(ang[N-2*n:N], cphb[N-2*n:N], 'g',                  \
                                                  label = '$I_{b}$')
    p4, = ax2.plot(ang[N-2*n:N], cphc[N-2*n:N], 'm',                  \
                                                  label = '$I_{c}$')
    ax1.yaxis.get_label().set_color(p1.get_color())
    ax2.yaxis.get_label().set_color(p2.get_color())
    ax1.legend(bbox_to_anchor=(0., 1.02, 1., .102), loc=3, ncol=4,    \
                                    mode="expand", borderaxespad=0.)
    ax1.grid(True)
    #
    plt.show()
    #plt.savefig('Fig915.ps', dpi=400, bbox_inches='tight')
    #
```

9.4.8 Program 9.16

```
#
def fdiode(vd):
    """ Diode characteristic.
        Returns diode current and its conductance"""
    #
    glow = 0.1e-07
    ghigh = 0.4e+02
    Vr = -0.5
    #
    if vd >= Vr:
        id = glow*(vd - Vr)
        gd = glow
    else:
        id = ghigh*(vd - Vr)
        gd = ghigh
    #
    return id, gd
#
#
def fswitch(vsw, gsw):
    """Bi-directional switch characteristic.
       Returns switch current and conductance."""
    #
    glow = 0.1e-07
    ghigh = 0.4e+02
    Vr = 0.5
    #
    ida, gda = fdiode(vsw)
```

```
    #
    if vsw >= Vr:
        vd = (gsw*vsw + ghigh*Vr)/(ghigh + gsw)
        iw = ghigh*(vd -Vr)
        gw = gda + ghigh*gsw/(ghigh + gsw)
    else:
        vd = (gsw*vsw + glow*Vr)/(glow + gsw)
        iw = glow*(vd - Vr)
        gw = gda + glow*gsw/(glow + gsw)
    #
    isw = iw + ida
    #
    return isw, gw
#
#
def ch9_fig16_svpwm_solver(es, Vtmax, dteta_m, dteta_n, grs, gls, ma,   \
                                              m, ns, ncycle):
    """Assumes fdiode is a function.
       Returns the lists vout, cinduct"""
    #
    from math import pi
    import numpy as np
    #
    ghigh = 40.0
    glow = 0.1e-07
    irlim = 0.1e-06
    nsect = 6
    pi3 = pi/3
    phse = pi*12.4/180.0 - 0.5*pi
    #
    gtot = grs*gls/(grs + gls)
    pila = -5.48
    pilb = 4.0
    pilc = -(pila + pilb)
    #
    vab = -18.4
    vbc = 5.0
    vn = 0.0
    #
    esab = []
    etha = []
    vlab = []
    vlbc = []
    cpha = []
    cphb = []
    cphc = []
    #
    for kcy in range(ncycle):
        #
        ang_kcy = 2*pi*kcy
        #
        for kst in range(nsect):
            #
```

```
ang_kst =ang_kcy + kst*pi3
#
for km in range(m):
    #
    ang_km = ang_kst + dteta_m*(km + 0.5)
    #
    gma = dteta_m*(km + 0.5)
    nx = int(ma*ns*np.sin(pi3 - gma))
    ny = int(ma*ns*np.sin(gma))
    #
    if (kst % 2) == 0:
        n1 = nx
        n2 = ny
    else:
        n1 = ny
        n2 = nx
    #
    if n1 >= ns:
        n1 = ns - 1
    if n1 <= 0:
        n1 = 1
    if n2 >= ns:
        n2 = ns - 1
    if n2 <= 0:
        n2 = 1
    #
    nq0 = int((ns - (n1 + n2))/4)
    nh1 = int(n1/2)
    nh2 = int(n2/2)
    nmax = np.zeros(ns, int)
    nmid = np.zeros(ns, int)
    nmin = np.zeros(ns, int)
    nmax[nq0:ns-nq0] = int(1)
    nmid[nq0+nh1:ns-nq0-nh1] = int(1)
    nmin[nq0+nh1+nh2:ns-nq0-nh1-nh2] = int(1)
    #
    if kst == 0:
        na = nmax
        nb = nmid
        nc = nmin
    elif kst == 1:
        na = nmid
        nb = nmax
        nc = nmin
    elif kst == 2:
        na = nmin
        nb = nmax
        nc = nmid
    elif kst == 3:
        na = nmin
        nb = nmid
        nc = nmax
    elif kst == 4:
```

```
        na = nmid
        nb = nmin
        nc = nmax
    else:
        na = nmax
        nb = nmin
        nc = nmid
    #
    sumab = []
    #
    for j in range(ns):
        #
        ang_ns = ang_km + j*dteta_n
        #
        vthA = Vtmax*np.sin(ang_ns - phse)
        vthB = Vtmax*np.sin(ang_ns - 2.0*pi/3.0 - phse)
        vthC = Vtmax*np.sin(ang_ns + 2.0*pi/3.0 - phse)
        #
        if na[j] == 1:
            gs1 = ghigh
            gs2 = glow
        else:
            gs1 = glow
            gs2 = ghigh
        #
        if nb[j] == 1:
            gs3 = ghigh
            gs4 = glow
        else:
            gs3 = glow
            gs4 = ghigh
        #
        if nc[j] == 1:
            gs5 = ghigh
            gs6 = glow
        else:
            gs5 = glow
            gs6 = ghigh
        #
        for its in range(1000):
            #
            vsw2 = vab + vbc + vn
            vsw4 = vbc + vn
            vsw6 = vn
            vsw1 = es - vsw2
            vsw3 = es - vsw4
            vsw5 = es - vsw6
            isw1, gsw1 = fswitch(vsw1, gs1)
            isw2, gsw2 = fswitch(vsw2, gs2)
            isw3, gsw3 = fswitch(vsw3, gs3)
            isw4, gsw4 = fswitch(vsw4, gs4)
            isw5, gsw5 = fswitch(vsw5, gs5)
            isw6, gsw6 = fswitch(vsw6, gs6)
```

```
        #
        Ia = gtot*(vab - vthA + vthB + (pila - pilb)/gls)
        Ib = gtot*(vbc - vthB + vthC + (pila +           \
                                        2.0*pilb)/gls)
        #
        ilsb = (Ib - Ia)/3.0
        ilsa = Ia + ilsb
        ilsc = -(ilsa + ilsb)
        #
        irrA = isw1 - isw2 - ilsa
        irrB = isw3 - isw4 - ilsb
        irrC = isw5 - isw6 - ilsc
        #
        err = np.sqrt(irrA**2 + irrB**2 + irrC**2)
        #
        if err < irlim:
            break
        else:
            #
            g11 = gsw1 + gsw2 + 2.0*gtot/3.0
            g12 = gsw1 + gsw2 + gtot/3.0
            g13 = gsw1 + gsw2
            g22 = gsw3 + gsw4 + gtot/3.0
            g23 = gsw3 + gsw4
            g33 = gsw5 + gsw6
            gmat = np.array([ [g11, g12, g13 ],
                              [-gtot/3.0, g22, g23 ],
                              [-gtot/3.0, -2.0*gtot/3.0, g33] ])
            #
            bcol = np.array([irrA, irrB, irrC])
            xdv = np.linalg.solve(gmat, bcol)
            #
            dvab = xdv[0]
            dvbc = xdv[1]
            dvn = xdv[2]
            vab += dvab
            vbc += dvbc
            vn += dvn
            #
        #
    #
    if its >= 998:
        print('time-step= ', (kst*m + j), 'j = ', j, 'm = ',      \
              m, 'kst = ', kst,  'solver does not converge')
        break
    #
    pila = ilsa
    pilb = ilsb
    pilc = ilsc
    #
    etha.append(vthA)
    vlab.append(vab)
    vlbc.append(vbc)
```

```
                    cpha.append(ilsa)
                    cphb.append(ilsb)
                    cphc.append(ilsc)
                    #
                    sumab.append(vab)
                #
                esab.append(np.mean(sumab))
                #
            #
        #
    #
    return etha, vlab, vlbc, cpha, cphb, cphc
#
#
def ch9_fig16_svpwm(es, Vtmax, freq, rs, ls, ma, m, ns, ncycle):
    """ Main program;
        3-phase inverter;
        space-vector switching."""
    #
    #calling command:
    #ch9_fig16_svpwm(320.0, 106.0, 60.0, 2.0, 0.01, 0.866, 36, 24, 50)
    #
    from math import pi
    import numpy as np
    import matplotlib.pyplot as plt
    from mpl_toolkits.axes_grid1 import host_subplot
    #
    nsect = 6
    nm = m*nsect
    n = ns*nm
    N = ncycle*n
    dteta_m = 2.0*pi/(m*nsect)
    dteta_n = 2.0*pi/n
    #
    delt = 1.0/(freq*n)
    grs = 1.0/rs
    gls = delt/ls
    #
    etha, vlab, vlbc, cpha, cphb, cphc=ch9_fig16_svpwm_solver(es, Vtmax, \
                dteta_m, dteta_n, grs, gls, ma, m, ns, ncycle)
    #
    print('peak value of cpha = ', np.max(cpha[N-2*n:N]))
    print('peak value of cphb = ', np.max(cphb[N-2*n:N]))
    print('peak value of cphc = ', np.max(cphc[N-2*n:N]))
    #
    m0 = 2*(ncycle - 2)
    m1 = 2*(ncycle - 1)
    m2 = 2*ncycle
    ang = dteta_n*np.arange(N)
    #
    ax1 = host_subplot(1, 1, 1)
    ax2 = ax1.twinx()
    ax1.set_xlim([m0*pi, m2*pi])
```

```
    ax1.set_xticks([m0*pi, m1*pi, m2*pi])
    ax1.set_xticklabels(['$0$', '$2\pi$', '$4\pi$'])
    ax1.set_ylim([-400, 400])
    ax1.set_yticks([-400, 0, 400])
    ax1.set_yticklabels(['$-400$', '$0$', '$400$'])
    ax1.set_ylabel('$Voltage, V$')
    ax2.set_ylim(-50, 50)
    ax2.set_yticks([-50, 50])
    ax2.set_yticklabels([ '$-50$', '$50$'])
    ax2.set_ylabel('$Current, A$')
    p1, = ax1.plot(ang[N-2*n:N], vlab[N-2*n:N], 'r', label = '$V_{ab}$')
    p2, = ax2.plot(ang[N-2*n:N], cpha[N-2*n:N], 'b--', label='$I_{lsa}$')
    p3, = ax2.plot(ang[N-2*n:N], cphb[N-2*n:N], 'g', label='$I_{lsb}$')
    p4, = ax2.plot(ang[N-2*n:N], cphc[N-2*n:N], 'm', label='$I_{lsc}$')
    ax1.yaxis.get_label().set_color(p1.get_color())
    ax2.yaxis.get_label().set_color(p2.get_color())
    ax1.legend(bbox_to_anchor=(0., 1.02, 1., .102), loc=3, ncol=4,    \
                                      mode="expand", borderaxespad=0.)
    ax1.grid(True)
    #
    plt.show()
    #plt.savefig('Fig917.ps', dpi=400, bbox_inches='tight')
    #
```

9.4.9 Program 9.17

```
#
def fdiode(vd):
    """ Diode characteristic.
        Returns diode current and its conductance"""
    #
    glow = 0.1e-07
    ghigh = 0.4e+02
    Vr = -0.5
    #
    if vd >= Vr:
        id = glow*(vd - Vr)
        gd = glow
    else:
        id = ghigh*(vd - Vr)
        gd = ghigh
    #
    return id, gd
#
#
def fswitch(vsw, gsw):
    """Bi-directional switch characteristic.
       Returns switch current and conductance."""
    #
    glow = 0.1e-07
    ghigh = 0.4e+02
    Vr = 0.5
```

```
    #
    ida, gda = fdiode(vsw)
    #
    if vsw >= Vr:
        vd = (gsw*vsw + ghigh*Vr)/(ghigh + gsw)
        iw = ghigh*(vd -Vr)
        gw = gda + ghigh*gsw/(ghigh + gsw)
    else:
        vd = (gsw*vsw + glow*Vr)/(glow + gsw)
        iw = glow*(vd - Vr)
        gw = gda + glow*gsw/(glow + gsw)
    #
    isw = iw + ida
    #
    return isw, gw
#
#
def ch9_fig17_hpwm_solver(es, Vtmax, dteta, dtm, grs, gls, ma, ns, nm):
    """Assumes fdiode is a function.
       Returns the lists vout, cinduct"""
    #
    from math import pi
    import numpy as np
    #
    ghigh = 40.0
    glow = 0.1e-07
    irlim = 0.1e-06
    phi = 2.0*pi/3.0
    phse = pi*12.4/180.0
    #
    gtot = grs*gls/(grs + gls)
    pila = -5.48
    pilb = 4.0
    pilc = -(pila + pilb)
    #
    vab = -18.4
    vbc = 5.0
    vn = 0.0
    #
    vtrmax = 1.0
    vtri = []
    ns4 = int(ns/4)
    dvtr = vtrmax/ns4
    for j in range(ns4):
        vm = j*dvtr
        vtri.append(-vm)
    for j in range(ns4):
        vm = j*dvtr
        vtri.append(vm - vtrmax)
    for j in range(ns4):
        vm = j*dvtr
        vtri.append(vm)
    for j in range(ns4):
```

```
    vm = j*dvtr
    vtri.append(vtrmax - vm)
#
vlab = []
vlbc = []
cpha = []
cphb = []
cphc = []
angnk = []
#
ectrA = []
ecstrA = []
#
for k in range(nm):
    #
    nk = ns*k
    #
    ank = dteta*(ns*(k + 0.5))
    angnk.append(ank)
    #
    for j in range(ns):
        #
        tetA = dteta*(nk + j)
        tetB = tetA - phi
        tetC = tetA + phi
        vctrA = ma*np.sin(tetA)
        vctrB = ma*np.sin(tetB)
        vctrC = ma*np.sin(tetC)
        #
        vmax = max(vctrA, vctrB, vctrC)
        vmin = min(vctrA, vctrB, vctrC)
        vz = -(vmax + vmin)/2.0
        vcstrA = vctrA + vz
        vcstrB = vctrB + vz
        vcstrC = vctrC + vz
        #
        vthA = Vtmax*np.sin(tetA - phse)
        vthB = Vtmax*np.sin(tetB - phse)
        vthC = Vtmax*np.sin(tetC - phse)
        #
        vtj = vtri[j]
        if vcstrA >= vtj:
            gs1 = ghigh
            gs2 = glow
        else:
            gs1 = glow
            gs2 = ghigh
        #
        if vcstrB >= vtj:
            gs3 = ghigh
            gs4 = glow
        else:
            gs3 = glow
```

```
        gs4 = ghigh
    #
    if vcstrC >= vtj:
        gs5 = ghigh
        gs6 = glow
    else:
        gs5 = glow
        gs6 = ghigh
    #
    for its in range(1000):
        #
        vsw2 = vab + vbc + vn
        vsw1 = es - vsw2
        vsw4 = vbc + vn
        vsw3 = es - vsw4
        vsw6 = vn
        vsw5 = es - vsw6
        isw1, gsw1 = fswitch(vsw1, gs1)
        isw2, gsw2 = fswitch(vsw2, gs2)
        isw3, gsw3 = fswitch(vsw3, gs3)
        isw4, gsw4 = fswitch(vsw4, gs4)
        isw5, gsw5 = fswitch(vsw5, gs5)
        isw6, gsw6 = fswitch(vsw6, gs6)
        #
        Ia = gtot*(vab - vthA + vthB + (pila - pilb)/gls)
        Ib = gtot*(vbc - vthB + vthC + (pila + 2.0*pilb)/gls)
        #
        ilsb = (Ib - Ia)/3.0
        ilsa = Ia + ilsb
        ilsc = -(ilsa + ilsb)
        #
        irrA = isw1 - isw2 - ilsa
        irrB = isw3 - isw4 - ilsb
        irrC = isw5 - isw6 - ilsc
        #
        err = np.sqrt(irrA**2 + irrB**2 + irrC**2)
        #
        if err < irlim:
            break
        else:
            #
            g11 = gsw1 + gsw2 + 2.0*gtot/3.0
            g12 = gsw1 + gsw2 + gtot/3.0
            g13 = gsw1 + gsw2
            g22 = gsw3 + gsw4 + gtot/3.0
            g23 = gsw3 + gsw4
            g33 = gsw5 + gsw6
            gmat = np.array([[g11, g12, g13 ],
                             [-gtot/3, g22, g23 ],
                             [-gtot/3, -2*gtot/3, g33]])
            #
            bcol = np.array([irrA, irrB, irrC])
            xdv = np.linalg.solve(gmat, bcol)
```

```
                #
                dvab = xdv[0]
                dvbc = xdv[1]
                dvn = xdv[2]
                vab += dvab
                vbc += dvbc
                vn += dvn
                #
            #
        #
        if its >= 998:
            print('time-step= ', (k*ns + j), 'j = ', j, 'k = ',    \
                                        k, 'solver does not converge')
            break
        #
        pila = ilsa
        pilb = ilsb
        pilc = ilsc
        #
        vlab.append(vab)
        vlbc.append(vbc)
        cpha.append(ilsa)
        cphb.append(ilsb)
        cphc.append(ilsc)
        #
        ectrA.append(vctrA)
        ecstrA.append(vcstrA)
        #
    #
  #
  return vlab, vlbc, cpha, cphb, cphc, angnk, ectrA, ecstrA
#
#
def ch9_fig17_hpwm(es, Vtmax, freq, rs, ls, ma, ns, mf, ncycle):
    """ Main program, plots results
        3-phase inverter
        Triangle comparison with zero sequence added
        to the control signals."""
    #
    #calling command:
    #ch9_fig17_hpwm(320.0, 106.0, 60.0, 2.0, 0.01, 0.9, 24, 321, 50)
    #
    from math import pi
    import numpy as np
    import matplotlib.pyplot as plt
    from mpl_toolkits.axes_grid1 import host_subplot
    #
    n = ns*mf
    nm = mf*ncycle
    N = ncycle*n
    delt = 1.0/(freq*n)
    dteta = 2.0*pi/n
    dtm = 2.0*pi/mf
```

```
#
grs = 1.0/rs
gls = delt/ls
#
vlab, vlbc, cpha, cphb, cphc, angnk, ectrA, ecstrA =                \
   ch9_fig17_hpwm_solver(es, Vtmax, dteta, dtm, grs, gls, ma, ns, nm)
#
print('peak value of cpha = ', np.max(cpha[N-2*n:N]))
print('peak value of cphb = ', np.max(cphb[N-2*n:N]))
print('peak value of cphc = ', np.max(cphc[N-2*n:N]))
#
m0 = 2*(ncycle - 2)
m1 = 2*(ncycle - 1)
m2 = 2*ncycle
ang = dteta*np.arange(N)
#
ax1 = host_subplot(2, 1, 1)
ax2 = ax1.twinx()
ax1.set_xlim([m0*pi, m2*pi])
ax1.set_xticks([m0*pi, m1*pi, m2*pi])
ax1.set_xticklabels(['$0$', '$2\pi$', '$4\pi$'])
ax1.set_ylim([-400, 400])
ax1.set_yticks([-400, 0, 400])
ax1.set_yticklabels(['$-400$', '$0$', '$400$'])
ax1.set_ylabel('$Voltage, V$')

ax1.text(m1*pi, -720, '$(a)$', size =10)
ax2.set_ylim(-50, 50)
ax2.set_yticks([-50, 50])
ax2.set_yticklabels([ '$-50$', '$50$'])
ax2.set_ylabel('$Current, A$')
p1, = ax1.plot(ang[N-2*n:N], vlab[N-2*n:N], 'r-.',              \
                                        label = '$V_{ab}$')
p2, = ax2.plot(ang[N-2*n:N], cpha[N-2*n:N], 'b--',              \
                                         label = '$I_{a}$')
p3, = ax2.plot(ang[N-2*n:N], cphb[N-2*n:N], 'g',                \
                                         label = '$I_{b}$')
p4, = ax2.plot(ang[N-2*n:N], cphc[N-2*n:N], 'm',                \
                                         label = '$I_{c}$')
ax1.yaxis.get_label().set_color(p1.get_color())
ax2.yaxis.get_label().set_color(p2.get_color())
ax1.legend(bbox_to_anchor=(0., 1.02, 1., .102), loc=3, ncol=5, \
                             mode="expand", borderaxespad=0.)
ax1.grid(True)
plt.subplots_adjust(hspace = 0.8)
#
ax1 = host_subplot(2, 1, 2)
ax1.set_xlim([m0*pi, m2*pi])
ax1.set_xticks([m0*pi, m1*pi, m2*pi])
ax1.set_xticklabels(['$0$', '$2\pi$', '$4\pi$'])
ax1.set_ylim([-1.2, 1.2])
ax1.set_yticks([-1, 0, 1])
ax1.set_yticklabels(['$-1$', '$0$', '$1$'])
```

```
ax1.set_ylabel('$Voltage, V$')
ax1.text(m1*pi, -2.1, '$(b)$', size =10)
p1, = ax1.plot(ang[N-2*n:N], ectrA[N-2*n:N], 'r--',              \
                                    label = '$vcontr_{tri}$')
p2, = ax1.plot(ang[N-2*n:N], ecstrA[N-2*n:N], 'b',               \
                                    label = '$vcontr_{hyb}$')
ax1.yaxis.get_label().set_color(p1.get_color())
ax1.legend(bbox_to_anchor=(0., 1.02, 1., .102), loc=3, ncol=5,     \
                             mode="expand", borderaxespad=0.)
ax1.grid(True)
#

plt.show()
#plt.savefig('Fig918.ps', dpi=400, bbox_inches='tight')
#
```

References

[1] J. G. Kassakian, M. F. Schlecht, G. C. Verghese, *Principles of Power Electronics,* 1st Ed., Pearson India, 2009.

[2] N. Mohan, T. M. Undeland, W. P. Robbins, *Power Electronics: Converters, Applications and Design,* 3rd Ed., John Wiley and Sons Inc., U. S. A., 2003.

[3] P. T. Krein, *Elements of Power Electronics,* International 2nd Ed., Oxford University Press, 1997.

[4] D. W. Hart, *Power Electronics,* McGraw-Hill, New York, U. S. A., 2011.

[5] H. W. van der Broeck, H. Skudelny, G. K. Stanke, *Analysis and realization of a pulsewidth modulator based on voltage space vectors,* IEEE Trans. on Indust. Appls., vol.24, no.1, Jan/Feb 1988, pp.142-150.

[6] Vladimir Blasko, *Analysis of a hybrid pwm based on modified space-vector and triangle-comparison methods,* IEEE Trans. on Indust. Appls., vol.33, no.3, May/June, pp.756-764, 1997.

[7] C. B. Jacobina, A. M. N. Lima, E. R. C. da Silva, R. N. C. Alves, P. F. Seixas, *Digital scalar pulse-width modulation: A simple approach to introduce non-sinusoidal modulating waveforms,* IEEE Trans on Power Electronics, vol.16, no.3, pp.351-359, May 2001.

Index

A

air gap, 38
  distributed, 48
American Wire Gauge, AWG, 47
Ampere's law, 36
Ampere-turns, 45
area product
  inductor, 51
  transformer, 54

B

bare-wire area, 47, 52, 54

C

capacitance
  filter, 138
capacitor
  equivalent conductance, 84
  transient equivalent, 84
circuit
  half-wave AC/DC
    capacitive filter, 93
  small-signal equivalent, 87, 91, 121, 125, 136
  thyristor, 157
  transient equivalent, 84, 86, 90, 123, 134
commutation, 81, 117, 131, 175
  angle, 82
  interval, 82, 118, 131
compensating network, 243
  pole-zero placement, 246
  Type II, 245
  Type IIIA, 245
  Type IIIB, 245
conductance
  equivalent, 86
converter
  double-ended full-bridge forward, 357
    simulation, 360
  flyback, 367
    simulation, 368
  full-wave AC/DC, 113
    capacitive load, 119
    general single-phase, 123
    response waveforms, 122
    simplified form, 114
    simulation, 120
  half-wave AC/DC, 73
    RL load, 74, 75
    freewheeling diode, 79, 90
    generalized load, 77, 78
    simulation, 83, 85
  isolated single-ended forward, 340
    leakage inductances, 346
    resonant reset, 351
    simulation, 343

isolated single-ended hybrid-bridge, 355
non-isolated boost, 257
current mode control, 273
feedback control in CCM, 269
in DCM, 279
model in CCM, 265
simulation, 263
voltage mode control, 269
non-isolated buck, 229
feedback control in CCM, 243
in DCM, 251
model in CCM, 239
simulation, 236
non-isolated buck–boost
feedback control in CCM, 294
in DCM, 299
model in CCM, 292
simulation, 290
non-isolated buck–boost, 283
push–pull, 362
high magnetizing current, 365
low magnetizing current, 363
six-pulse, 129
three-phase
thyristor-based, 168
three-phase thyristor-based control, 179
thyristor-based, 157
inverter mode, 167
single-phase, full-wave, 162
converters
isolated single-ended forward
auxiliary reset, 340
resonant, 421
three-phase AC/DC
diode-based, 126
simplified, 127
simulation, 134
core
cross-sectional area, 53
ferromagnetic, 38
gapped, 41
loss equation, 46
magnetic, 42
MPP, 41
tape-wound, 41
window area, 51
core geometry factor, 53
Curie temperature, 40
current
alternating, 45
DC bias, 46
eddy current, 46
inductor, 46
non-sinusoidal, 130
past-history, 87
phase-shifted, square wave, 163
square wave, 114
current density, 36

D

de-magnetize, 40, 48
dead time, 422
Design examples, 56
Filter inductor, 60
Flyback transformer, 67
isolation transformer, 56
three-winding transformer, 64
diode, *see* Power diodes
bridge, 113
distortion, 117
total harmonic (THD), 117

E
electrical conductivity, 52
energy loss, 44
estimator
    recursive least square, 182

F
Faraday's law, 36, 55
flux density
    saturation, 54
flux linkage, 39
Fourier analysis, 130
Fourier series, 114
frequency
    fundamental, 115
    resonant, 426
    switching, 53, 426
fringing flux, 48
fundamental frequency model, 423

G
GTO thyristors, 11
    turn-off, 12

H
harmonic components, 115
Henry, 40
hysteresis
    major loop, 42, 43
    minor loop, 44
    upper limiting curve, 42

I
inductance, 40
    internal, 131
inductor, 35
    core geometry factor, 52
    electrical constraints constant, 52
    equivalent conductance, 84
    high frequency, 41
    regulation, 51
    transient equivalent, 84
Insulated gate bipolar transistor, 22
    NPT structure, 22
    PT structure, 22
    turn on, 25
    turn-off, 25
inverters, 465
    amplitude modulation ratio, 470
    frequency modulation ratio, 470
    over-modulation, 472
    PWM, 470
        carrier signal, 470
        reference control signal, 470
        space-vector, 482
        switching sequence, 487
        triangle comparison, 480
    single phase
        bipolar PWM, 470
        harmonic amplitudes, 472
        harmonic elimination, 467
        simulation, 473
        square wave, 465
        unipolar pwm, 474
    Three-phase, 476
        hybrid PWM, 489
        square-wave, 479

K
Kirchhoff, 117, 124

L
laminations, 46

LCC converter, 436
  fundamental frequency equivalent, 436
  voltage gain, 436
Litz wire, 47
LLC converter, 425
  AC equivalent load, 432
  half-bridge, 425
  linear model, 428
  modes, 428
  simulation, 432
loss, 45
  eddy current, 46
  hysteresis, 45

M

magnetic
  circuits, 38
  field intensity, 36, 42
    coercive force, 42
  flux, 36
  flux density, 35, 42, 46
    peak, 46
    residual, 42
    saturation, 42, 45
  materials, 41
magnetic materials
  amorphous metallic alloys, 41
  ferrites, 41
    MnZn ferrites, 41
    NiZn ferrites, 41
  permalloy, 41
  permendur, 41
  silicin steels, 41
magnetization
  characteristics, 42
  initial, 42
  lower limiting curve, 43
Maxwell's equations, 35
  electric field, 35
  Gauss's law for electric fields, 36
  Gauss's law for magnetic fields, 36
mean length of turn, MLT, 52
mode
  continuous conduction CCM, 229
  discontinuous conduction, 229
MOSFET, 14
  body diode, 14
  circuit model, 18
  n-channel, enhancement type MOSFET, 14
  the field effect, 15
    accumulation layer, 15
  turn on, 19
  turn-off, 21
multi-resonant converters, 443
multi-resonant conveters
  ZVS buck, 443
mylar, 47

N

natural conduction, instants, 162, 168
Newton–Raphson, 76, 85
nodal equations, 126

P

parallel strands, 47
permeability, 40
  incremental, 41
  relative
    effective, 48

permittivity, 36
point of common coupling, pcc, 123
power
    active, 164
    input, 54, 55
    output, 54
    reactive, 164
    total, 54
Power diode, 1
Power diodes
    I–V characteristic, 1
    CAL diode, 6
    diffused, 1
    epitaxial, 1
    Schottky diode, 5
    switching characteristics, 3
        reverse recovery charge, 5
        turn on, 3
        turn-off, 5
power factor, 130, 172
    displacement, 130, 172
probe voltage, 85, 120, 123, 134
pulses
    positive sine wave, 116

Q

quasi-resonant converters, 438
    ZCS, 438
    ZVS, 438
    ZVS buck, 438
        simulation, 441

R

reluctance, 40, 48
resistivity, 46
resonant tank, 421
    parallel Two-element, 424
    series Two-element, 421
        gain curves, 424
    three-element, 421
    Two-element, 421
response waveforms, 126

S

skin effect, 46
    penetration depth, 46

T

Tesla, 40
Thyristors, 6
    simulation model, 10
    turn-off, 9
Thyristors*I–V* characteristic, 7
transform
    (a, b, c) to (α, β, γ), 482
    Fourier, 80
    inverse LaPlace, 77
    LaPlace, 74, 77
transformer, 35
    area product, 53
    core geometry factor, 56
    electrical constraints constant, 56
    high frequency, 41
    primary winding
        resistance, 55
        turns, 53
    regulation, 55

V

voltage
    output, 138
voltage source
    alternating, 114

non-ideal, 119
single-phase
non-ideal, 164
three-phase AC
non-ideal, 131

W

waveform symmetry
half-wave, 114, 130
odd, 114, 130
Webers, 39
Wide bandgap devices, 26
GaN devices, 27
HEMT, 27
Homo-epitaxial devices, 29
Silicon carbide devices, 30
SiC MOSFETs, 31
SiC rectifiers, 30
window utilization factor, 51, 54

About the Author

S. Raghuram Naidu received the Bachelor's and Master's diplomas from IIT Madras, India in 1966, and IISc Bangalore, India in 1970. He earned the Ph.D. in Electrical Engineering from the University of Liverpool, UK in 1975. He was a Visiting Post-doctoral Fellow with the University of British Columbia, Canada during 1986-1988 and a Visiting Scientist with the University of Wisconsin, USA during 2000-2001.

Following graduation, he practiced briefly as an engineer in Madras, India. He became a Professor of Electrical Engineering with the Federal University of Campina Grande, Brazil in 1975 and remained there until retirement in 2015. During his tenure as professor, he lectured and researched in several areas of Electrical Power Engineering, such as: High Voltage Measurements, Electrical Insulation, Magnetism, Power System Transients, Power Electronics and Power Quality.